Key Challenges in Geography

EUROGEO Book Series

This book series addresses relevant topics in the wide field of geography, which connects the physical, human and technological sciences to enhance teaching, research, and decision making. Geography provides answers to how aspects of these sciences are interconnected and are forming spatial patterns and processes that have impact on global, regional and local issues and thus affect present and future generations. Moreover, by dealing with places, people and cultures, Geography explores international issues ranging from physical, urban and rural environments and their evolution, to climate, pollution, development and political economy.

Key Challenges in Geography is an initiative of the European Association of Geographers (EUROGEO), an organization dealing with examining geographical issues from a European perspective, representing European Geographers working in different professional activities and at all levels of education. EUROGEO's goal and the core part of its statutory activities is to make European Geography a worldwide reference and standard. The book series serves as a platform for members of EUROGEO as well as affiliated National Geographical Associations in Europe, but is equally open to contributions from non-members.

The book series addresses topics of contemporary relevance in the wide field of geography. It has a global scope and includes contributions from a wide range of theoretical and applied geographical disciplines.

Key Challenges in Geography aims to:

- present collections of chapters on topics that reflect the significance of Geography as a discipline;
- provide disciplinary and interdisciplinary titles related to geographical, environmental, cultural, economic, political, urban and technological research with a European dimension, but not exclusive;
- deliver thought-provoking contributions related to cross-disciplinary approaches and interconnected works that explore the complex interactions among geography, technology, politics, environment and human conditions;
- publish volumes tackling urgent topics to geographers and policy makers alike;
- publish comprehensive monographs, edited volumes and textbooks refereed by European and worldwide experts specialized in the subjects and themes of the books;
- provide a forum for geographers worldwide to communicate on all aspects of research and applications of geography, with a European dimension, but not exclusive.

All books/chapters will undergo a blind review process with a minimum of two reviewers.

An author/editor questionnaire, instructions for authors and a book proposal form can be obtained by contacting the Publisher.

More information about this series at http://www.springer.com/series/15694

Marko Krevs
Editor

Hidden Geographies

euro
geo

Springer

Editor
Marko Krevs
Department of Geography
Faculty of Arts
University of Ljubljana
Ljubljana, Slovenia

ISSN 2522-8420 ISSN 2522-8439 (electronic)
Key Challenges in Geography
ISBN 978-3-030-74589-9 ISBN 978-3-030-74590-5 (eBook)
https://doi.org/10.1007/978-3-030-74590-5

This Springer imprint is published by the registered company Springer Nature Switzerland AG
The registered company address is: Gewerbestrasse 11, 6330 Cham, Switzerland

Acknowledgements

A book is always a collective endeavour, but with 51 authors, 23 reviewers, and 24 chapters, it becomes a real challenge. Corona times added about half a year delay—not so much directly, but due to resulting adjustments in our professional and private lives, which have taken up so much of the time needed to focus on preparing the book. The greatest thanks are due to the authors, their original contributions and their patient and constructive cooperation in improving the materials. A big—although anonymous—thank you goes to all the reviewers who have made an important contribution to the quality of the texts. In addition, I would like to thank at least a few who played crucial roles, by highlighting their contributions. First, to my colleagues on the organising team for the Hidden Geographies conference in Ljubljana, in August 2019, especially Tatjana Resnik Planinc and Mojca Ilc Klun—for their support of the idea of focusing the conference on hidden geographies. Thanks also for the financial support of the team for the technical preparation of the texts before they could be delivered to the publisher. Thanks to the Department of Geography at the Faculty of Arts, University of Ljubljana, for including the conference in the programme of activities related to the centenary of the Department. Without this occasion, a conference with such international participation would probably not take place. Thanks to EUROGEO, the co-organiser of the above-mentioned conference, for the proposal to publish the book in the Springer series they are editing. Especially to Rafael de Miguel González, the President of EUROGEO, who also helped in the first steps of the implementation of this idea. Special thanks are due to Kostis C. Koutsopoulos, one of the editors of the series, who meticulously read all the chapters and provided critical and honest comments—which contributed to several improvements in the focus and argumentation of the chapters. Once the materials for the chapters were collected, Višnja Jerman patiently helped with the final technical and linguistic improvements to the texts and with the preparation of the package that was sent to the publisher. Finally, I thank the Springer team, especially Doris Bleier, Carmen Spelbos, Sanjiev Kumar Mathiyazhagan and Madanagopal Deenadayalan, for their patient explanations and support of procedures that were new and uncomfortable for me, including extreme patience due to the aforementioned delay caused by the Corona situation.

Editor
Marko Krevs

Contents

Introducing Hidden Geographies

Conceptualising Hidden Geographies

Marko Krevs

Abstract After several decades of rather sporadic use in the scientific literature, the concept of hidden geographies is still usually based on provisional definitions that support the specific geographical hiddenness of the topic presented in a publication. This chapter focuses on hidden geographies, with the aim of providing a usable, not necessarily definitive understanding and definition of the concept. After a conceptual-semantic view at hidden geographies, the meanings of the concept and the term are presented, based on the analysis of literature, which provides a colourful variety of connotations and names of the concept in practise. In the discussion, some of the contexts underlying the concept under study are highlighted, as well as questions regarding its understanding and use, such as understanding the blurred line between hidden and revealed geography, and the roles of geography and geoinformatics in revealing or hiding geographies. Finally, a general definition and some specific definitions are proposed, linked to four layers of understanding of the concept: undiscovered, uncognised, unpublished and deliberately hidden geographies.

Keywords Hidden geography · Senses · Perception · Cognition · Data · Knowledge · Geoinformatics · Definition

1 Introduction

We could hardly find stronger foundations to illustrate the importance of geography than Aristotle's assertion that place gives order to the world (Bonnet 2014: xii) and Immanuel Kant's understanding that it (geography) is one of the two basic forms of human knowledge (Bonnett 2008: 2). Although the statements may be outdated, simplistic and, in the latter case, also biased (Kant introduced the study of geography at the University of Königsberg, in 1756, and lectured on this subject for forty years;

M. Krevs (✉)
Faculty of Arts, University of Ljubljana, Ljubljana, Slovenia
e-mail: marko.krevs@ff.uni-lj.si

M. Krevs (ed.), *Hidden Geographies*, Key Challenges in Geography,
https://doi.org/10.1007/978-3-030-74590-5_1

May 1970), they leave little doubt: when geographical information is missing, or hidden, human knowledge, abilities and activities are considerably curtailed.

From a very broad perspective of our *collective* (civilisational) knowledge, we may like to think that humans have become quite familiar with the Earth, given that its entire surface has been surveyed, remotely sensed and mapped. Yet a steady and increasing stream of new (geographical) information and knowledge reminds us of the extent of what remains unknown about our planet and its changing nature.

The successful functioning of people and societies requires a successful perception and cognition of the circumstances in the physical and social environment in which we live. To be successful in everyday life, though, as *individuals*, we need not know everything. As Tuan (1977: 85) put it, the geographical "knowledge we have as individuals and as members of a particular society remains very limited, selective, and biased". Despite imperfect geographical information and knowledge and limited abilities to think geographically we—as individuals and societies—have to cope with everyday or long-term challenges. To do so, we need both geographical knowledge and the abilities to act effectively in new situations. This knowledge is accumulated and strengthened through formal and informal education as well as through personal or collectively shared experiences. This justifies the importance of *learning just-in-case* about previously unknown geographies, like through geography lessons in schools, reading literature or following the media news.

In a new, unknown situation, the knowledge we have may be insufficient to act effectively by following a learned routine. In new situations, learning may also involve getting to know and evaluating previously unknown geographies, if we need to consider them, and if circumstances allow. Such *situational and selective learning* usually only takes place when the environment is so unfamiliar or extraordinary that we consciously update our existing *mental maps* of our living environment (Holloway and Hubbard 2001: 48). This way of learning and acting in new situations requires special knowledge and abilities, the results of our lifelong learning and training, e.g. *geographic awareness*, *spatial contextual awareness* (Freksa et al. 2007) and *spatial abilities* (Tuan 1977).

From a personal point of view, many geographies, even those at the local level, are unknown to us—to our senses and cognition. Consequently, we as individuals live on a planet where unknown geographies prevail. Such geographies are, therefore, not an occasional or minor information-related problem. But living with them is not necessarily a problem, and our everyday encounters with them need not be very dramatic. We function despite them. Sometimes we try to reveal them, while at the same time we also contribute to them throughout our lives.

In this chapter, we focus on unknown geographies, which we call *hidden geographies*. Our aim is to contribute to their conceptualisation, to the awareness of the relevance of the concept and to the debate on its possible meanings, uses and implications. The concept is only sporadically used in literature and other media. It may seem neglected from a scientific point of view, and by textbooks, perhaps because it seems such a straightforward subject. Pointing out the missing geographical information, which is related to our limited knowledge and the weaknesses of our spatial abilities, can be consciously or subconsciously avoided. However, hidden geographies can have many meanings that relate to a number of different types of hiddenness, ranging from the physically or visually unperceivable to a lack of geographic knowledge and abilities. Understanding hidden geographies, then, is not straightforward. We can see them as one of the very central reasons or motivations for geographical exploration and discovery, including scientific research in geography and related spatial disciplines. When we imagine the possible or known impacts such missing information and knowledge have on our understanding of the world, our spatial decisions and our behaviour, we believe that they deserve closer consideration.

In this chapter, we do the following:

- take a conceptual-semantic look at hidden geographies;
- present the meanings of the concept and the term hidden geographies on the basis of the analysis of literature;
- provide a general definition and some specific definitions of hidden geographies;
- comment on the roles of geography and geoinformatics in revealing and hiding geographies.

First, we focus on semantics, relating the concept of hidden geographies to various meanings of the words constituting it. As demonstrated in the following section, such elementary meanings cover quite a rich and extensive domain, and will probably—and successfully enough—serve as its basic understanding. After clarifying the idea behind hidden geographies, an overview of its use and naming in practise is given, based mainly on the examination of the scientific literature. A summary of the characteristics of the concept, extracted from the outcomes of the semantic and literature analyses, exposes selected aspects that we consider to be particularly important for our debate, such as a blurred line between geographies that are hidden and revealed, and the life cycle of hidden geography. Finally, a general definition of the concept of hidden geography and some definitions of its specific meanings are given, based on what we have learned about the concept so far. In the discussion that follows, the roles of geography and geoinformatics in revealing or hiding geographies, as well as in solving the increasing problem of informational overburdening, are discussed.

2 Concept and Term

As we are approaching such a loosely defined concept as hidden geographies, we distinguish between the *concept as the idea*, in terms of both a generalisation and a deeper meaning, and the term used as the *name of the idea*, in how it signifies the concept. It is the concept, its interpretation and definition, that is the focus of this chapter. The term is considered to be a rather practical matter.

Various terms with different connotations are used in the literature to address the concept. It seems quite unlikely that a common understanding of the concept will emerge naturally, without involving a theoretical debate. As "there is no such thing as intuitive understanding of ordinary language—at any rate if we are speaking about the more subtle kind of communication" (Hansen 2006: 6), we start with a basic semantic analysis of the words used in the phrase hidden geographies.

2.1 Geography

Geography as a discipline is defined in many ways, from Eratosthenes's "writing the Earth" to a contemporary definition such as "the study of the ways in which space is involved in the operation and outcome of social and biophysical processes" (Gregory 2009: 288). In the search for a comprehensive understanding of what is geography as a discipline and what it does, Bonnett (2008) emphasises that it is about studying the world, "an attempt to find and impose order on a seemingly chaotic world" (ibid.: 6). As a holistic discipline, geography plays an important integrative role and resists the current intellectual specialisation/fragmentation processes. This conceptualisation, therefore, emphasises *geography as a discipline*, its complexity, its ways of studying and explaining the Earth, its holistic and integrative character and the spatiality of its focus.

Another denotation of the word geography brings an important perspective to our discussion: *geography as a particular object of enquiry*; "something has a geography like it has a history" (Gregory 2009: 287). As used in previous literature, the word geography, within the phrase, hidden geographies, usually refers to this version of the meaning of the concept. The most obvious or at least the most common analogy to the latter denotation of geography is a map. This analogy may work well for many geographers, looking at the world through the lens of spatial science, like Haggett (1995). When Haggett writes about his conception of geography, he uses characterisations that include "patterns of human occupation of the earth", "science of distributions", "emphasis on space and on geometry". If maps are representations of spatial structures, then geography is "the art of the mappable" (Johnston and Sidaway 2016: 102). But such views are problematized by many geographers as oversimplifying and narrow—"to root geography in mapping is a suffocating fantasy" (Bonnett 2008: 91). Doubts have been expressed about how well a map can represent actual places or landscapes, and how well it can serve human perception, cognition of and behaviour in these places (e.g. Tuan 1977). The constraints on the design and use of maps relate to the representation and cognition of *known, revealed geographies.* These constraints may actually lead in the opposite direction to the goal of mapping, and as a result the geographies represented remain hidden, at least at the cognitive level.

In setting the scene for further debate, we retain a broad definition of geography, somewhat broader than suggested by the mapping analogy. As it relates to absolute and contextual locations/spatial distributions, the terms absolute and contextual geography are used. *Absolute geography* is defined by the position of a phenomenon, e.g. using GPS coordinates, position on a map, street address or similar. Usually, this position is static, in 2D space, but can also be dynamic and/or in 3D space. Absolute geography refers to individual places or distribution of several places. *Contextual geography*[1] is defined by the position (or spatial distribution) of a phenomenon in relation to another phenomenon or phenomena. The relationship can be spatial, such as any combination of distance, direction, adjacency, vicinity, spatial inclusion/overlay, and the like, as in the example the fuel station is 400 meters north of the house. We may think of such cases in terms of *relative location*. But the contextual relationship can also be *temporal* and/or based on the *characteristics* of the phenomena or places involved, as in, for example, the fuel is more expensive at the fuel stations on the highways. The contextual information can also be entirely *subjective*, personal, based on memories or emotions, as in, for example, I have my beloved places all around the city. Many absolute and contextual geographies can potentially be mapped. Absolute geographies must be based on, or transformable to a coordinate system we are able to apply. Contextual geographies must be transformed into absolute geographies before mapping, or we can map them without absolute georeference, as in the case of a hand-drawn treasure map or a sketch map showing a picnic spot. In the latter cases, geography can be revealed to the reader if he knows the geographical context in order to make the necessary georeferencing required for the practical use of such a map.

There is another level of the detachment of a geography from a mapping analogy. As mentioned later in the literature overview, some places—such as those that are legendary, mystical, imaginary—exist, often only in imagination, beliefs, narratives, or propaganda, but from time to time also in reality. Sometimes such *imagined geographies* can even be mapped, but it seems more likely that their absolute location is not known. This is perhaps not even necessary for them to function on an imaginary, symbolic level. Such geographies occasionally become the focus of geographical research, and they are also an interesting topic in the context of hidden geographies.

2.2 Hidden Geographies

The word hidden has several synonyms and antonyms (Table 1). Let us take *hidden* and *revealed* as representative examples. Hidden may be a good candidate for the role of constituting a *term* representing the concept of hidden geographies because it has so many possible connotations, it has already been in use, but not often or consistently enough to settle down and narrow down its meaning.

Hidden geographies could refer to hidden disciplinary (geographical) reasoning, methods or activities. But, as already mentioned, they have so far more often referred

[1] Freksa et al. (2007) defines spatial contextual awareness as the information constructed on the basis of inter-relations between the spatial environment, a cognitive agent, and a cartographic map. The spatial environment is the physical space of performing a task; the cognitive agent is the person or entity completing a task; and the map is the representation of the environment which is used as a tool to complete the task. Although this understanding of spatial context is narrower (limited to the map analogy of geography) than used in our definition of contextual geography, it involves the cognitive relation between the "agent" and the space, which is a basis of our wider, cognition-based understanding of the concept of hidden geographies.

Table 1 Examples of synonyms and antonyms of hidden (Cambridge Dictionary 2018; Thesaurus.com 2018)

Synonyms	Antonyms
Unseen, secret, not easy to find, buried, clandestine, concealed, covered, covert, dark, invisible, latent, mysterious, obscure, private, secluded, underground, undisclosed, unknown, cloaked, close, clouded, disguised, eclipsed, masked, occult, screened, sequestered, shrouded, veiled, withheld, abstruse, cryptic, esoteric, hermetical, unperceivable, in the dark, indiscernible, mystical, out of view, recondite, shadowy, surreptitious, ulterior, undercover, undetected, unexposed, unrevealed	Above board, apparent, bare, bright, clear, disclosed, exposed, known, manifest, obvious, open, plain, public, revealed, sociable, unconcealed, visible, unmasked, exhibited, out, seen, showing, uncovered

to hidden locations and distributions of certain phenomena, objects or places. From the list of synonyms of hidden and the variety of meanings of geography, we can extract a very wide range of existing and also potential nuances of hidden geographies' meanings. Some of them do not seem usable, at least not in a scientific text; a few others not included in this list are touched upon in the following sections.

As a way of beginning to understand hidden geographies, we argue that it is important to consider whether what is hidden results from a deliberate act of hiding, or whether the hiddenness arises from an inability to perceive. This consideration serves as the basis for a model with 4 layers of hiddenness of geographies (Fig. 1), which is used in our further discussion.

Although expressed in terms of geography as a spatial aspect of phenomena, these layers can also be transformed to support the discussion of geography as a discipline, for example, as layers of hidden geographical disciplinary knowledge, models, interpretations.

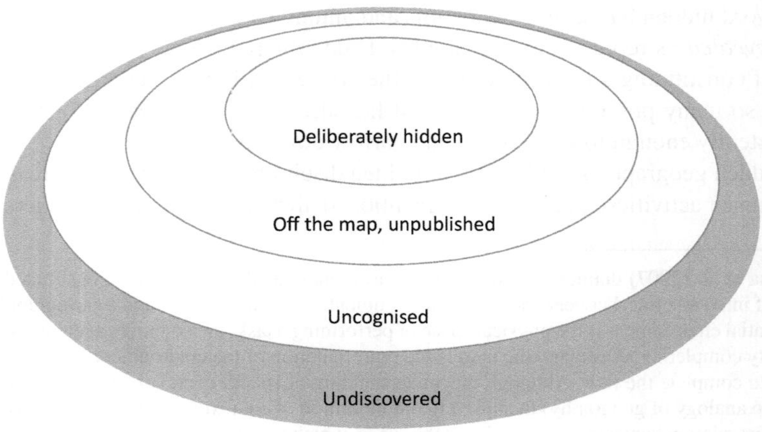

Fig. 1 Elementary layers of hiddenness of geographies

Undiscovered geographies exist and, despite no one knowing about them,[2] they can have a minor or significant impact on other phenomena and processes in the landscape. *Uncognised geographies* include all hidden geographies that are discovered (by someone) but are not known, especially not at higher cognitive levels, to others, individually or collectively. *Unpublished* geographies are not accessible to cognition because geographic information about phenomena has not been made available to the public, either in visual (such as on a map), acoustic or other forms accessible to human senses and perception. Unpublished geographies are often also *deliberately hidden*. Such a schematic representation helps to organise the general understanding and discussion of the concept. However, it should be used flexibly enough to allow for some exceptions to the presented hierarchy of the layers, which are discussed later in this chapter.

3 Connotations of the Concept

Having examined the words in the phrase hidden geographies, we now look at how it occurs in literature. A systematic analysis of all the specific connotations is, of course, beyond the scope of this chapter. By reviewing examples of the occasional use and naming of the concept,[3] we hope to demonstrate its breadth and perhaps stimulate deeper analysis and theorising in the future.

The examples are grouped unevenly, according to the varying presence in the literature, and pragmatically, to support the presentation of the four layers of the concept (Fig. 1). The grouping is not intended to suggest particular types of hidden geographies. As these layers sometimes overlap, some of the examples could be used to represent more than one layer.

3.1 Undiscovered Geographies

It seems quite reasonable to expect little scientific literature on geographies of material or non-material phenomena that exist, but have not yet been discovered. However, this is only partly true. Apart from the (not necessarily scientific) search for legendary places, even hidden continents, as in the case of Atlantis (since Plato) and more recently Zealandia (Sutherland 2017), or the search for the real Ithaca (Bittlestone et al. 2005), a considerable part of the literature as well as technological and methodological developments focuses on searching for *assumed geographies*. Assumptions

[2] In fact, this is not necessarily the case, e.g. when locals are familiar with the place(s) that are undiscovered for others.

[3] We may say we choose the Serresian approach (from Steven Connor's introduction to Serres 2008: 5), immersing into a cloud of words and phrases found in the literature, so that the empirical essence of the concept is learned from diversity of existing connotations.

about the existence of phenomena and their geographies are based on deduction from what is already known, on induction from experiential information, or on less rational ways of knowing.

The geographical information on which such publications focus can remain undiscovered for many reasons, like so far unmeasurable, historically distant, highly subjective phenomena, or from a lack of interest in geographies of these phenomena. Most of the examples from the literature mentioned in our chapter actually present the results and the act of revealing such assumed geographies.

As long as the focus is on assumed geographies or on methods and technologies designed to reveal these geographies, the latter remain undiscovered. Once a geography is revealed and published, this technically means that it is no longer hidden. However, even focusing on such cases indirectly reminds us of the existence of undiscovered geographies, at least in two ways: a geography that is now being revealed was previously undiscovered; the revealed geographies often represent intensively *changeable geographies* and will therefore soon—perhaps already at the time of publication—be outdated and undiscovered/hidden again.

Remote Sensing, Re-interpreting Throughout the history of geographic discoveries, new ways of geographic data collection, analyses, visualisations, (re-)interpretations of available knowledge and similar advances have been close to the core of activities aimed at revealing so far undiscovered geographies. Today rapid development of remote sensing and geoinformatics, in general, plays important role in bringing perpetual geographic discoveries.

Remote sensing essentially aims to reveal hidden geographies (including the renewal of existing data) detectable on the Earth's (or another planet's) surface, with some techniques able to penetrate to specific depths below the surface (e.g. soils, water bodies) or above the surface (e.g. atmospheric phenomena). Numerous publications and online services demonstrate the usability of such information in scientific, professional and everyday situations. A very typical example of publication that announces the revealing of still (at the time of publication) undiscovered geographies with remotely sensed imagery is presented by Posada-Swafford (2005) reporting on a joint NASA-Costa Rica project. Using multiple sensors, from high-definition infrared cameras, subterranean sensors, a radar laser to detect the thickness of the rainforest and spectrometers to analyse the air—to reveal so far hidden geographies and support policies in the fields of geology, environment, urban planning, disaster prevention and archaeology.

Lu et al. (2017) emphasise the importance of a seemingly old-fashioned remote sensing technology—a historical *archive of aerial photographs*—in a contemporary discovery and interpretation of ancient hidden linear cultural relics in the alluvial plain of eastern Henan, China.

The availability of *LiDAR* data has made it possible to reveal many previously hidden geographies, especially those hidden under forests, including natural (e.g. geomorphological) and man-made features (like archaeological sites, e.g. in Štular 2011). Johnson and Ouimet (2013) report such effects of the public availability of LiDAR data in the northeastern United States. Indeed, many discoveries of historical geographies are *rediscoveries* of once mapped phenomena. For example, the once *known geography* of the Isonzo Front became at least partly forgotten, unrecognisable, even physically hidden due to physical decay and reforestation. Many remnants of the *forgotten geographies* are now being rediscovered using LiDAR data (Petrovič et al. 2018).

Undiscovered geographies may be revealed *unintentionally*, like discovering "hidden forests" representing about 10% of the dryland biomes, increasing global forest area estimates by 9% (Bastin et al. 2017), only by using different forest detection techniques and very high temporal and spatial resolution of satellite imagery.

Another way to reveal previously undiscovered geographies can be through *re-interpretation of the existing explanations* in a particular area or situation. Herrick (2017) deals with hidden geographies of global health, and specifically of suffering, that could be revealed if less-archetypal and non-medical spaces were included in critical medical-anthropological explanations of the production of global health. *Meta-analyses* of genetic risk factors of male infertility revealed "the hidden epigenetic geography of the Y chromosome" and provided evidence for its association with ethnicity and geographical region of the tested population (Navarro-Costa and Plancha 2011).

Internet Geographies Pickren (2020) argues that the internet actually produces geographic space (*internet geographies*) and has a particular impact on people and places. This means that it not only plays its role in revealing hidden geographies, but produces new ones that may remain hidden e.g. due to the inaccessibility of the internet (to some people) and the increasing inability to cope with over-informatisation. Fraser (2019) discusses *digital geographies,* data colonialism and their relationship to critical data studies. Better understanding of the impact of digital on existing geographies, including reshaping many of them, mediating the production of geographic knowledge and reconfiguring research relationships, can be linked to a better understanding of rapidly growing and usually hidden digital geographies.

Some additional examples of revealing undiscovered internet geographies are presented in Sect. 5.

3.2 Uncognised Geographies

Undiscovered geographies obviously cannot be cognised, and could, therefore, be part of uncognised geographies. Our decision to limit our discussion of uncognised geographies to those that have been discovered, emphasises the division between undiscovered geographies that cannot be cognised and discovered geographies that could be cognised, but are not.

Cognition of a geography may be constrained due to the lack of relevant information and the personal ability to bring it to a higher level of cognition,[4] from basic data to knowledge or even higher, to its creative application. In a narrower sense, geography is revealed to someone when he has reached the *knowledge* level, which is the lowest cognitive level in Bloom's taxonomy (Anderson et al. 2001). This enables him to learn the facts, remember them, without necessarily understanding them. However, only the *comprehension* level (or higher) involves understanding, summarising, generalising the facts learned; and *application* level allows the acquired knowledge, facts, rules to be applied in new situations and in solving problems. Which level is better suited to separate hidden and revealed geographies remains a question of further debate. In our opinion, this decision should be situation-specific: de facto *revealed geography*, at the lower level of knowledge, may function as revealed if we only need to repeat it, e.g. when passing on information to someone else. In case we need to relate this geography to other phenomena in the landscape or use it in spatial decision-making, a higher level of cognition will be required to make it *functionally revealed*, otherwise this geography will remain *functionally hidden*.

Unpublished and deliberately hidden geographies (discussed in the following two sections) may be considered as subsets of uncognised geographies, the latter representing the broadest understanding of the concept in relation to discovered but hidden geographies. Such a *wide, cognition-based understanding of the concept* considerably expands the pool of examples from which we can learn in our attempt to uncover the horizons of its possible meanings and uses. Topics related to *uncognised geographies* in the reviewed literature range from legendary places, placeless places, places hidden due to lack of knowledge or attention, personal geographies, sense of place, geographies of deprivation, immigration, gender, historical and forgotten geographies, social and economic geographies, hidden geographies revealed due to methodological or technological developments, hidden geographical assumptions, or hidden geography used as a figure of speech.

Legendary, Imagined, Mythical, Religious, Sacred Some *legendary, mythical, imaginary places* may not exist in reality because their main function is symbolic or spiritual. But some of them have a physical existence and location. For Said (1979), *imagined* (or imaginative) geographies relate to the perception (of the real world) and power of those who construct descriptions that influence perception. Both conceptions of imagined geographies, as imaginary and as perceived, are embedded in oral traditions, rituals, allegories; they are more interwoven with real landscapes and lives than we may realise.

Religious, mythical, sacred places or worlds/realms often remain unseen, hidden, even secret—either associated with religious experience or organisation, hidden

[4] We may combine DIKW (data-information-knowledge-wisdom) model (Zeleny 1987) with Bloom's original or revised taxonomy of cognitive domain (Anderson et al. 2001) to position the level of cognition. Level 0 is represented by basic data, with no cognition assumed. Level 1 would be information, descriptive knowledge, that includes basic answer to a question "where?", without necessarily understanding what it means. At this level the two models overlap, combining information from the first model with knowledge from the second. Next levels follow Bloom's hierarchy, from comprehension to application, analysis and evaluation/synthesis.

geographies within religious buildings or within people's homes (Flood 1993). Imaginary (as well as more materialised legendary or mythical) landscapes are also very common in literature, art, and popular culture.

A geography of paradise, a mythical place Shangri-La, for "spiritual accomplishment and renewal" of a Buddhist, turned into a revealed geography as late as in 1998, as Beyul Pemako, "Hidden-Land Arrayed like Lotuses", in Tibetan Tsangpo Gorge (Baker and Dalai Lama 2004; Baker 2016). The place was famous, but its geographical location was obviously a *secret*, deliberately hidden from the general public.

Every culture, including contemporary ones, produces new places that are given special meanings, some of which remain hidden to the cognition of many people. The search for sacred in the landscape, the recognition of spiritual, sacred places and their use as ritual sites has also been practised by all civilisations since prehistoric times. Revealing such hidden geographies, "sacred geography … embedded in the landscape" (Devereux 2010), by learning to look through the eyes of our ancestors, can bring interesting insights into the historical relationships between man and nature, as well as contribute to a better understanding of the historical spatial behaviour and human-driven processes in the landscape.

Inaccessible to Senses and Cognition, Placeless, Restricted, Lack of Attention, Out of Sight In contrast to deliberately ignored geographies, there are places or geographies ignored simply because they are *inaccessible to the senses and cognition*. Many texts on place-related peculiarities, which are often exposed as tourist attractions or discoveries, reveal such geographies, as curious travel destinations, "World's hidden wonders" (Foer et al. 2016), improbable places (Elborough and Horsfield 2016), cursed places (Le Carrer 2015), "enacting Northern European peripheries" that have so far been almost untouched by globalisation and tourism (Bærenholdt and Granås 2008), camel wrestlers in Western Anatolia as carriers of the "geography of a hidden cultural heritage" (Çalışkan 2009) and the exploration of the Arctic and the dissemination of geographical knowledge from indigenous' narratives of the Beaufort Sea (Martin 2020).

A place is perhaps simply unperceived because it lacks—in the eyes, reason or in the heart of the observer—its own place-defining character that would distinguish it from other places. *Placeless* (Relph 1976) places can particularly enrich the concept of hidden geographies because they refer to places of which we may know the location, or even know what they look like. But it can easily happen that we are not sure if this is the right place or if it is another one that looks and feels just the same. Hiddenness in this case has a special connotation, which has to do with disconnecting from our feelings and also from reason, consequently also from a deeper memorisation. Such places are not off the published maps, but they may be *off our mental maps*, or only slightly connected to them. The concept found its echo especially in the geographical and architectural literature and brought many illustrative terms and aspects such as geography of nowhere (Kunstler 1993), non-places (Augé 1995), indifferent sameness (Casey 1998), "traveling from nowhere to nowhere" (Bonnett 2014: xi).

Our cognition of places may be hindered by political, military or other restrictions. Rajaram and Grundy-Warr (2007) explore hidden geographies of borderscapes, connecting empirical, everyday procedures, with political, state sovereignty context. Border is shown "as a moral construct rich with panic, danger, and patriotism", with daily practises with migrants calling for questioning justice and its limits.

The phenomena and their geographical distributions may remain inaccessible to senses and cognition due to *lack of attention* or personal interest, or external distractions. Hidden Journeys Project (2012) motivated the participants to use any of the presentation techniques (from photos and paintings to written descriptions) to report on their geographical exploration during their flights. Avoiding missed opportunities to discover so far hidden geographies (at least hidden to certain individuals) does, of course, not only refer to situations during the flight. But rarely does local observation from the ground provide such an overview of the spatial distribution of a phenomenon as from an airplane.

A particular variant of geographic hiddenness refers to phenomena that are physically *out of sight*. Examples can be found in relation to each of the four sub-categories of hidden geographies. Even many published geographies remain uncognised due to this circumstance. *Subsurface* phenomena, with their *subsurface* (*underground/subterranean* or *underwater*) *geographies*, represent a highly illustrative variant of such hiddenness. Subsurface geographies refer, for example, to soils, underground escape routes, lead mines, karstic caves, underground waters, underwater infrastructure; some of these are presented elsewhere in this review.

Lack of Knowledge Among several reasons why even a revealed geography may remain partially or even completely hidden from our cognition are deficient geographic knowledge and skills. Although such cases are not explicitly mentioned in the reviewed literature, examples are easy to find in our everyday practises. A common example is when we receive information relating to a particular place whose position we do not know exactly. Subconsciously, we may only remember its macro-location, e.g. the continent or a country. If we are interested enough, we could search for the missing contextual and geographic information. Otherwise, the memorised geographic information in such an inappropriate spatial scale remains partially or almost completely hidden from our cognition, and remains rather unusable.

A similarly common example is the national border line, which is drawn on maps and as such is not deliberately hidden from people. However, it is often not clearly visible in the landscape and its exact position may be unknown to a person crossing it. Existing information, even a potential punishment for crossing such a line, does not solve the problem of missing, incomplete or erroneous cognition. Such a geography (of a borderline) may indeed be hidden to a person and have consequences for his spatial behaviour and quality of life, regardless of why this spatial information was hidden for his cognition.

Geography is often understood as revealed when the information about it is published. Even if it seems to be a rather simplistic and unnatural principle, it can be found in all possible situations of our lives. Take the borderline discussed above: you are punished because the information about this line is published somewhere and it is your responsibility to find out about it. Similarly, when driving, you are expected to be familiar with national/local traffic regulations. Geography students are often expected to know everything that is shown on maps or in textbooks they use in a course. The fact that a particular geography is revealed to them through publication is sufficient to expect them to be familiar with it. In fact, several social, cultural, economic, political rules function this way, depending on the geographical specificities. The argument "did not know because I am not from here" is usually not enough to avoid conflict or even punishment.

Hidden to Majority, Local, Personal, Experiential, Emotional Geographies of phenomena present in our environment may be known to those who are in direct contact with them or live in the area where they occur, but they are unknown, hidden from most others. These are actually very common examples of the concept. In terms of spatial scale, they usually represent phenomena that are known locally or regionally, but are hidden from the rest of the world. Examples include hidden geographies of de facto stateless enclaves in India and Bangladesh (Shewly 2017), people living on "no man's land" between national borders (Bonnett 2014), religious creativity in West London faith communities (Gilbert et al. 2019) and spatial dispersion of the immigrants in Louisville and across the USA (Klayko 2015).

Individual geographies, including the (spatial) behaviour, movement, emotions of the individuals, or their sense of place, are hidden from most other people. Several authors focused on personal geographies (Harmon 2003), places of belonging (Hooks 2008), hidden geographies of spatial belonging, inclusion and exclusion (Morgan 2000), emotional geographies (Davidson et al. 2007), senses of place (Eyles 1985), topophilia, topophobia (Tuan 1977, 1990). But only a few, explicitly, discussed the hiddenness of such geographies.

Lucherini (2019) discusses autobiographies as a way to revealing (personal) hidden and emotional geographies, focusing especially on the lifeworlds of people with type I diabetes. Although the author recognises autobiographies as crucial sources for studying everyday life experiences, including their spatial dimension, many everyday problems and concerns related to illness remain hidden. A hidden geography of electoral behaviour is explored by Arrington and Grofman (1999). The electoral preferences of individuals and local communities may differ from real electoral decisions because "voters have an incentive to vote strategically". Such "hidden partisanship" is reflected in the geographical pattern of election results, and in its more difficult predictability.

A hidden geography can be something we experience, even participate in its functioning, *without seeing the bigger picture*, the geographical spread or the spatial density of the phenomenon. Once again, this is one of the frequent, and at the same time disturbing examples of hidden geographies that reveal our geographical ignorance of our directly experienced lives. Labour migrations from the new EU member states to the UK have been statistically observed and perceived by people in contact with the actual process. But it is only the evidence of discrepancies between Worker Registration Scheme and the allocation of the National Insurance Number that reveals the hidden geography of self-employment and entrepreneurial activity of these migrants, with their considerable concentrations also in urban, and not in (expected) rural and peripheral areas (Harris et al. 2015). Another surprisingly intensive process, a dynamic hidden geography of booming alternative food networks and the establishment of farmers' markets in and around Prague, is revealed by Fendrychová and Jehlička (2018).

Deprivation, Marginalisation, Exclusion, Immigration *Geographies of deprived, disabled, marginalised* individuals or social groups are usually hidden and coincide with problems of their inclusion or exclusion. Matthews et al. (2000) point to hidden, alternative geography of exclusion and disenfranchisement of children growing up in rural areas, emphasising "the sharp disjunction between the symbolism and expectation of the Good Life … and the realities and experiences of growing-up in small, remote, poorly serviced and fractured communities". Hall (2004) explores hidden social geographies of people with learning disabilities, "one of the most marginalized groups in society", their exclusion from and integration into mainstream socio-spaces, and their creation of safe spaces formed by themselves. Larreche and Ercolani (2019) show the nocturnal geographies of the LGBT scene in Bahía Blanca, Argentina, "complex cartographies that involve aspects of spatial prestige, internal tensions and microcultural fixations". Buzar (2007) reveals hidden geographies of deprivation, based on energy poverty in the post-socialist Eastern Europe, and Betto et al. (2020) on measuring energy poverty in Italy. Hidden geographies of the changing lifeworlds of women with multiple sclerosis (Dyck 1995) refer not only to concrete places and their distribution (as we have basically defined one of the possible meanings of geography), but also to abstract and generalised representation of their spatial behaviour, specificities of and obstacles to their spatial movements and similar.

Lawrence (1995) unmasks the hidden geography of rural homelessness, which exceeds the intensity of the phenomenon in urban areas (the study was conducted in the USA), and questions representations of rural space as "especially emphatic in their valorization of privacy, property and independence". In a study on the hidden geography of immigrants at risk of homelessness in Vancouver, Fiedler et al. (2006) use hidden to refer to the visual concealment of immigrants and their hiddenness to the statistics (because they have housing, but the cost of which exceeds their income), hidden spatial distribution of their homes, as well as ignorance of public policies to meet the housing needs. Wu (2008) examines the hidden geographies of everyday practises and the creation of social identities of Taiwanese migrant women in Toronto, looking at their "everyday language practice, the reconstruction of food culture and the exploration of culinary practice, the negotiation of home practice, and the creation of new spaces for new identity claims".

Gender *Hidden gender geographies* are very persistent, often a part of traditional social relationships "as they always used to be". Townsend (2002) discusses the hidden geographies of women's lives on Iberian farms in the 1990s (in relation to the publication of the results of a grounded research by Garcia-Ramon et al. 1994). These women's multiple functions are poorly paid and often not visible in any statistics. Some of their activities are defined by local cultures (at home, as mothers), activities on the farm are understood as help, not work, and additional obligations include their work outside the farm, often at home or seasonally. As a result, their hidden lives are "cyclical and discontinuous" in comparison to the lives of men, which are "linear and accumulative" and visible.

Historical, Forgotten Hidden geography can be a result of historical processes that are lost to today's observers as *forgotten geographies*, partly due to physical decay, re-vegetation or other types of landscape changes, as well as the lack of information, collective memory, or awareness of these processes. In his reflection on the Royal Geographical Society's exhibition, Driver (2012) reminds us that when we look at the history of exploration, referring to times when many hidden geographies were revealed, we must expect to be confronted also with hidden histories, which rise questions such as "what is made visible and what is obscured in standard narratives of exploration". Known facts about cooperation of individual Czech modernists with their colleagues in Vienna and Berlin from the late nineteenth to the mid-twentieth century are transformed into revealed geography of the cooperation of Czech modernists (David-Fox 2000) only with a cultural-geographical interpretation of them collectively. The hidden geography reveals itself through generalisation by taking a broader view of the subject. Walker (2001) points to the "geographic magic done by the hand of nature and humankind" in and around Golden Gate, which can be seen in remnants like "dots of brick and mortar tucked into the coastline" that can only be reconstructed by calling them up "from hidden places of collective memory". Another example of hidden historical geographies deals with a complex system of underground drainage channels in the lead mines of Derbyshire, created between the seventeenth and nineteenth centuries to enable mineral extraction. It not only had a major impact on the need for original legal solutions related to vertical conceptualisations of space, but also contributed to a "vertical turn", and volumetric turn, in geography to deal with subterranean landscapes appropriately (Endfield and van Lieshout 2020). The concept of hidden geographies has been used here in at least two ways: historically, by being revealed and renewed a few centuries ago; and rediscovered, reawakened from oblivion in contemporary literature. Stuart (2010) reveals hidden geographies behind the "remaking of the river in accordance with the requirements of capital", including a number of overlapping administrations, regulatory organisations, the establishment of system of locks and other engineered resolutions to the conflicting uses of the river Thames since the seventeenth century.

Economic and Social Many examples can be categorised as hidden economic or social geographies. Community economies in South-East Asia, which include "pre-capitalist", "relict practices" such as mutual aid, reciprocity and sharing as predominant features of everyday community life, are "seen as standing in the way of modern economic growth", but in reality "contribute to local social safety nets and act to support households" in times of difficulty (Gibson et al. 2017). Murphy (2007) reveals the "materiality of computer mediated retailing", hidden geography of e-commerce, how a "seemingly simple act of doorstep food delivery is explicated in urban form, and in transportation and communication infrastructures". Looking at a city through the eyes of burglars brings quite a peculiar socio-economic-geographical perspective to hidden geographies (Manaugh 2016).

Problem of Definition, Reinterpretation, Methods Hidden geography can be the result of methodological problems in defining the spatial extent or position of a phenomenon, such as the borders/margins of settlements (Krevs 2005; Benchetrit and Stadnicki 2014), the shape of buildings (Ai et al. 2013), or even more fundamentally, the phenomenon itself, such as the definition of urban settlements and resulting geographies. Van Duijne (2019) reveals hidden urbanisation in India, which is due to the politicised process of rural-to-urban settlement reclassification, influenced by village leaders who resist being absorbed by the expanding cities, resulting in consequent effects on urban policy. Specificity of this use of the concept of hidden geography is based on the controversial imposition of an external (e.g. by experts, politicians) explanation of the *real geography*, which is different from the geography people know and live with. *Imposed geography* is not merely theoretical, but has concrete consequences for people's lives that are not necessarily positive for them. The external imposer actually initiates the transformation of a *known geography* into a new one, which was previously at least partially hidden from the people's cognition. At the same time, the real geographical process is *obscured* by an administrative model of reality (e.g. on a map) and, at least from this point of view, becomes partially hidden.

Revealing previously hidden geography can be based on complex (geographical) analyses, such as in a study on the transformation of invisible spatial patterns into maps of wind dissipation and wind energy potential (Kılıç 2019). Liu et al. (2018) used complex machine learning techniques of spectroscopy and chemometric data to efficiently discriminate the geographical origin of extra-virgin olive oils, Deng et al. (2020) similarly analysed chemometric data to discriminate the designated geographical origin of tea and Maione et al. (2019) the geographical origin of honey. These examples aim to confirm a known or assumed geography. It is only when a fraud about the geographical origin of products has been uncovered that a previously hidden (false or fraudulent) geography is revealed.

Hidden Geographic Assumptions, Models One of the rare examples of hidden geographies that refers to geography as a discipline was found in a critical discussion by Agnew (1995) on the geographic terminology and models in the social sciences. Agnew argued about the hidden use of inevitable geographical assumptions, "powerful *unconscious geographies*" (italicised by the author of this chapter), that undermine the characterisations of the social sciences as "spaceless", the dangers of over-commitment to fixed geographical models and fixed spaces as containers of society.

Figure of Speech Authors can use hidden geographies, or paraphrased, e.g. as hidden worlds, to illustrate something known to certain people (e.g. scientists in a certain field, local population), but is supposedly unknown to the audience they are addressing. For example, as in "explore nature's mysterious hidden worlds" (2019), the authors emphasise nature's hiddenness to the knowledge of some people—used as a figure of speech to obtain a particular narrative effect, and not necessarily related to geography.

3.3 Unpublished Geographies

Unpublished geographies are considered a subset of uncognised geographies. When geographical information about a discovered phenomenon has not been made available to the public, either in visual (e.g. on a map), acoustic or other forms accessible to the human senses and perception, it remains also inaccessible to individual and collective cognition. The public does not have to include everyone, but only those who are interested, which potentially contributes to the relativity and vagueness of the conceptualisation.

A subset of unpublished geography, the *off-the-map-based* understanding of a hidden geography, refers to phenomena that have been omitted from maps. In this manner, Bonnett defines hidden geographies in relation to places that don't appear on maps, and as the inverse of lost places (2014: 37), or non-existent places (ibid., 3), which appear on maps, but can't be found in reality.

Several topics relating to *unpublished geographies* in the reviewed literature are presented in the adjacent sections, under uncognised and deliberately hidden geographies. Of course, examples of unpublished geographies are hard to find, so we refer to circumstantial evidence—previously hidden and eventually revealed/published geographies. Among such examples are geographies unpublished due to stigmatisation or ignorance, and geographies of phenomena that are not permanently located.

Stigmatisation, Ignorance *Ignored geography* can refer to stigmatised phenomena such as the suicide rate, and occasional publications can appear as the discovery of a hidden geography (Florida 2013), which they are for those who are not aware of it. A similarly avoided issue, "underpolicing" of rape in parts of US cities with the majority of the African American population, is highlighted by Brownlow (2017) in a case study in St. Louis.

Van Schendel (2002) contrasts the production of the "geographies of knowing" with "geographies of ignorance" by emphasising the focus of analyses on certain areas, while (consciously) ignoring the others. South Asia belonged to the latter, being dispensed after the Second World War, became peripheral and consequently the growth of knowledge about it slowed down.

No Permanent Location Places can be hidden and their geography unpublished just because they are *not permanently located* on the Earth's surface, such as "floating places", including floating islands and luxury cruise ships (Bonnett 2014). Or they are constructed and planned to be used only for a limited time (but can remain for a long time), like "ephemeral places", from improvised townships for refugees, temporary workers, occasional festival towns, to "dogging places" (ibid.).

3.4 Deliberately Hidden Geographies

The deliberate hiding of geographies is something that human societies and individuals have practised throughout history. Today the reasons for this may not be so different from historical ones; they range from gaining strategic advantages, as in military or economic situations, to other kinds of selfish prestige. Maintaining security and protecting places defined as special in a given society or by an individual are also among the reasons for actively hiding geographies. Even the reasons that we might regard as rather modern, such as the preservation of vulnerable natural or cultural phenomena, have been used in history and increasingly so today.

With the development of methods for the collection and dissemination of geographical information, the methods of its active hiding have also changed. However, the basic principles of hiding are still similar: either by hiding the geographical information from others by keeping it unpublished, by misleading publication, by physically or legally restricting access to the hidden places, by physically hiding or camouflaging objects or by urging not to go/to search further (e.g. by scaring away).

In our model (Fig. 1), deliberately hidden geographies are a subset of unpublished geographies and the most specific case of discovered but hidden geographies. In the above list of basic principles of hiding geographic information, an exception to this hierarchy is mentioned: deliberate hiding may involve publishing when misleading or incomplete geographic information is published.

Examples of deliberately hidden geographies are as difficult to find as unpublished geographies, even in the cases just mentioned, when they are published. The following examples mainly show situations where such geographies become revealed, or where access restrictions contribute to hindered cognition of these geographies. The topics relate to issues such as violence and fear, deliberate ignorance, restricted information/access, environmental issues, protection of personal data and deceptive geographies.

Violence and Fear A historical example of hiding to survive is presented by Schmid (2020), revealing hidden places and metro underground tunnels in Brooklyn on the way of Missouri slaves towards freedom. A specific part of the literature deals with hidden geographies related to violence, wars, especially civil wars, genocides, mostly coupled with geopolitical interests of the economic and military world powers. Such a deliberately hidden and (internationally) *ignored human geography* of the Congo is illustrated by Wanzola (2013), which highlights some of the most important immanent injustices of the modern world. The hidden geographies of informal power in the Greater Middle East (Anceschi et al. 2014), a perpetual playground of the world's superpowers throughout the history, are another example. Allouche (2019) discusses visible and hidden forms of violence related to state-building, nation making and post-colonial hydropolitics in India and Israel. War-induced hidden geographies can refer to long-lasting historical remnants of war, such as unexploded bombs (e.g. Mombauer 2017). Radio station Slovenia 1 (https://

radioprvi.rtvslo.si/) reported such news about discoveries of the bombs from the 1st or 2nd World War almost daily, as recently as until 2019.

Hidden geographies of violence and crime do not always have political origins. Percy-Smith and Matthews (2001) reveal the hidden geographies of fear, "tyrannical spaces", "no-go areas" in urban neighbourhoods due to bullying of young people in a selected British town. The problem of urban violence, which also involves other age and social groups, is a rather global phenomenon and is often swept under the carpet by local authorities, keeping this geography hidden, which is contrary to the benefits of the (local) population. Geographies of violence and fear do not have to be related to people alone. Bloch and Martínez (2020) report on geographical analysis of dog killings by the police in Los Angeles. The problem "remains largely unknown and all but completely absent from the academic literature on state violence", which the authors attribute in their study to the concentration of the canicides in low-income areas and African American communities.

Restricted, Military Geographies Military sites and activities are usually hidden from the public around the world. Even in cases of their very obvious physical presence, e.g. in settlements inhabited by non-military residents, there are restrictions on access, photographing or sometimes even mentioning them in public communications. When confronted with deliberately hidden geography, such as trespassing a military zone or a place of illegal activities, this can have unpleasant, even dangerous consequences.

The military uses a very illustrative term to describe the areas that remain hidden to an observer due to intervening obstructions, including natural and man-made obstacles: dead ground (Salovaara-Moring 2009). An example of hidden military geography is revealed by Paglen (2006) which refers to the "black world" consisting of secret military bases north of Las Vegas. Some activities there are "so highly classified their very existence is a state secret. For public purposes, they do not exist".

Secrets are not necessarily limited to military installations and activities, but also to their environmental and social impacts, as in the study of nuclear landscapes and indigenous inhabitants in the south-western United States (Kuletz 1998). Environmental issues, such as climate change, can be related to military secrets also in other rather peculiar ways, as exemplified by the Cold War military base Camp Century (Colgan et al. 2016). The intention to remain hidden forever deep in the ice sheet in Greenland has failed as the melting ice will bring it to the surface in the next 75 years. The revelation refers not only to the visual aspect, but also to the "remobilization of physical, chemical, biological, and radiological wastes abandoned at the site".

Environmental Issues, Exceptional Natural and Cultural Goods A large area of intentionally hidden geographies is related to environmental issues, such as pollution and its impact on public health and the economy. Studies of air pollution with fine particulate matter (PM2.5) have shown its strong link to fossil fuel combustion, but little is known about the "socioeconomic factors driving the growth of PM2.5 emissions, which has been hidden from the public due to the geographical separation of production and consumption activities", revealing physical PM2.5 emissions along supply chains to the consumers (Yang and Chen 2017).

Natural and cultural assets of exceptional value and vulnerable social groups can sometimes become protected by restricting access to them or even hiding their location. This may often be the only way to protect them effectively. An example of the hiding of exceptional, vulnerable natural assets for their protection are the so-called "hidden caves", which are described by the Underground Caves Protection Act (Zakon… 2004) as "caves in which the natural cave environment is so vulnerable that it could be damaged or endangered by any entry of persons into the cave". To enter these caves, you need a special permit from the Ministry of the Environment. Their locations may be known to speleologists and the local population, but on the public portal of the world's first online Cave Cadastre (https://www.katasterjam.si/) you will not have access to the data about these 5 (out of 13,659 registered) caves.

Hide our Place: Social Network Discussion About Living on One of the Islands in the Adriatic Sea, Originally in Croatian

Thread by local 1: it is so beautiful here

Non-local 1: you do not know that you live in paradise!

Local 2: we know, don't tell anyone

Local 1: yes, it is our secret!

Local 2: we have known this since forever

(Facebook, hidden link, accessed 30 October 2020)

Personal Data Protection An obvious case for hiding geographies is the protection of privacy of people's movements across space. After decades of increasing misuse of personal data, people are becoming increasingly sensitive to this problem. During the COVID-19 pandemic in 2020, countless online debates criticised the use of mobile applications that use spatial information about people's movements, even though they could help to detect and warn of occurrences of encounters with infected people.

Lies and Deceptive Geographies Many cartography courses include books like How to lie with maps (Monmonier and De Blij 1996) into the essential reading lists. Hiding geographies by *omissions*, *deceptive representations*, or the *inclusion of false information* on maps, or in other media, can be *intentional* or *accidental*. For example, a geography may be *deceptive* due to poor ability to read maps or due to the use of the received geographical information at a higher cognitive level, not just by an intentional act of hiding.

In a time of the COVID-19 pandemic, the chaotic flow of information from official institutions brought frequent expressions of doubt about the reliability of the information. Fears, perhaps justified, that this flow of information, which changed our lives so drastically in just a few weeks, also contained many fake messages—"deliberately false", "misleading by design", "manipulating the audience's cognitive processes" (Gelfert 2018). Even if this was only partially true and the mistakes were the result of lack of knowledge and methodological clumsiness, the data were heavily mapped and contributed to questionable geographical representations and cognitions of a real spatial process, which probably remained hidden.

4 Defining Hidden Geographies

Learning about the concept examined in the chapter so far has revealed a wide range of its possible meanings and uses. A summary of the characteristics of the concept is presented and briefly discussed in order to support a critical, better informed argument about hidden geographies, followed by an attempt to propose a definition of the concept.

4.1 Summarising the Characteristics of Hidden Geographies

After meandering between different instances of the concept, a simple Aristotelian approach based on elementary questions seems appropriate to organise what we know.

What is hidden The most general view of the object of hiddenness represents two fundamental aspects of what geography means, geography as a discipline and as the spatial location/distribution of a phenomenon.

Geography as discipline, geographical modes of representation, argumentation and interpretation, including geographical models and assumptions used in scientific and everyday discourse, are often discussed in the literature. However, there are rare attempts to view them in the context of hidden geographies.

Hidden *geographies as a spatial aspect of phenomena* refer to individual places and their absolute or contextual location or to multiple places and their absolute or contextual spatial distribution (as defined at the beginning of the chapter in Sect. 2.1). Geographies can refer to material/physical and non-material phenomena. Phenomena and their geographies can be the result of imagination, beliefs, reasoning (e.g. in extrapolations of known facts or speculations). Some such imaginary geographies may be mapped (and have been mapped in the course of history), and therefore, their hidden opposites may be considered a special kind of hidden geographies.

Hidden by whom and to whom Who can refer to the provider, the recipient and/or the carrier/user of information about a hidden or revealed geography. These roles can be performed by individuals or collectively, by institutions, social groups, societies.

When we think about a collective, shared group-specific/social/civilisational knowledge (about the geography of a phenomenon), we use a rather abstract way of saying who actually knows. We allude more to anticipated than to actual knowledge, by pointing out that a particular group/society has been informed, e.g. by making certain information publicly available in a way that is expected to be received and cognised by all. In the reviewed literature, as well as in occasional debates among geographers (e.g. at a conference on this topic, Hidden Geographies 2019), such a collective, abstract understanding of the concept seems to have prevailed in practise so far.

When a geography is intentionally hidden, the concept of hidden geography gets a particular socio-relational emphasis. The hidden place exists, and someone (or several people) knows about it, but prevents access to this information or simply does not communicate it.

Why is it hidden Among the many reasons for the hiddenness of geographies presented in this chapter, the following cover the majority of situations. Geographies are hidden before they are discovered. When discovered, they may remain hidden due to deliberate hiding or insufficient cognition. The reasons for deliberate hiding range from selfish, e.g. acquiring a particular personal, social, economic, political, military advantage, to altruistic, e.g. to protect fragile natural or man-made goods. Insufficient cognition (insufficient to reveal a geography in a given situation) may be the result of ignorance, lack of attention or interest, sensory/perceptual deficiencies and other obstacles/disruptions in accessing the geographic information. Even when the geographic information is received, cognition may be insufficient if there is a lack of prior geographic knowledge and/or the ability to think geographically. Among the main sources of hiddenness of geographies are also the following two processes: forgetting (as individuals or collectively) and the difference between real geography and known geography as its representation, both processes usually increasing over time (as discussed in more detail below).

How is it hidden What must be missing to get a hidden geography? The most basic reason for the existence of a hidden geography is that it has not yet been discovered. The next basic step, the lack of access to raw data or information (such as a map) about the geography of a phenomenon obviously prevents its revealing. Geographies can be hidden intentionally or by providing false, misleading geographical information. Finally, published, even online ubiquitous geographies can also remain hidden due to insufficient geoinformation abilities or cognition.

When we obtain spatial data about a phenomenon, various outcomes are possible. If our knowledge allows us to effectively use the data, cognitively transform them into a revealed geography, then the geography is no longer hidden to us. In all other situations, it remains at least partially hidden, even completely hidden to our cognition in case we have no spatial knowledge to help us use these data.

As discussed, an individual has to reach at least the basic *knowledge* level of Bloom's cognitive taxonomy in order to memorise and transfer a learned geographic information. If we are not able to use this knowledge in our reasoning and activities, the situation is practically comparable to that when geography is hidden from us. We only possess the isolated knowledge in a rather unusable form.

In order to be able to use the acquired information more competently in solving problems, spatial decision-making and similar, at least the *comprehension* level has to be acquired, or higher. Effectively linked knowledge (about a revealed geography) enables us to actively use it, whenever needed, even in new situations. Therefore, at this or higher cognitive level, we can also link the revealing of a previously hidden geography with personal spatial or geographical abilities that enable a person to function spatially in an efficient way.

When is it hidden The apparent dualism in the use of the phrases hidden geography and revealed geography suggests that a hidden geography ceases to exist when it is revealed. However, from a *cognitive-based perspective*, both phrases denote very blurred and transient phenomena. Transience is characteristic of both the phenomenon/landscape and our knowledge of revealed geographies, for different reasons, as illustrated in a life cycle of hidden geographies (Fig. 2). The knowledge is only temporary and may gradually fade (through forgetting) and/or become increasingly inaccurate due to the changes in the real landscape. Over time, the hiddenness of a geography tends to increase and even predominate. Within the inner life circle (marked 2B in Fig. 2), hidden geographies are not only a consequence of hiding information from potential recipients, but also of their willingness and ability to perceive this information and bring it to a higher level of cognition.

The hiddenness of a geography can be quite *persistent*. When it is revealed (to someone, to certain people, even to the general public through publication), it remains hidden to some people. When it is revealed to someone, it may further remain hidden to his perception (e.g. in case he does not pay attention to this information) and his cognition. This could be due to that person's lack of spatial, cartographic and geographical knowledge and abilities, but also to situational circumstances such as stress or external distractions. Information is revealed, loses its hiddenness, but does not necessarily reach higher cognitive levels to become usable. As a result, it may be poorly understood, stored and transmitted.

Spatial knowledge is often *unreliable* and *transient*. When we learn about a geography that is not related to our personal experience, like in case when it is related to some distant landscapes, the mental images and cognition of such geography are rather abstract and may only seemingly correlate with *real geography*. Even when crossing experienced places, which are often local and closer to human perception, people can get lost due to incomplete or otherwise unreliable mental maps, perhaps combined with a lack of attention. When a real geography changes, people may retain a mental image of the earlier, less and less correlated landscape. Besides, they may also be unaware of this change. *Known geography* tends to discord with the *real* geography of a phenomenon, and therefore, the former becomes false (or at least partially right) and the latter becomes hidden.

In summary, the changeability of real geographies, the fading of knowledge of a known geography due to forgetting, the loss of bits of the original information, the addition of new bits that can change the original information, and other factors that determine the adequate *correlation between real and known geographies* contribute to the following conclusion: hidden geographies obviously have a tendency to overshadow the knowledge of real geographies.

Where is it hidden Even a hidden geography has a geography of its own. Where can relate to the place/space referred by the hidden geography, or to the distribution of those who know (or do not know) about this geography.

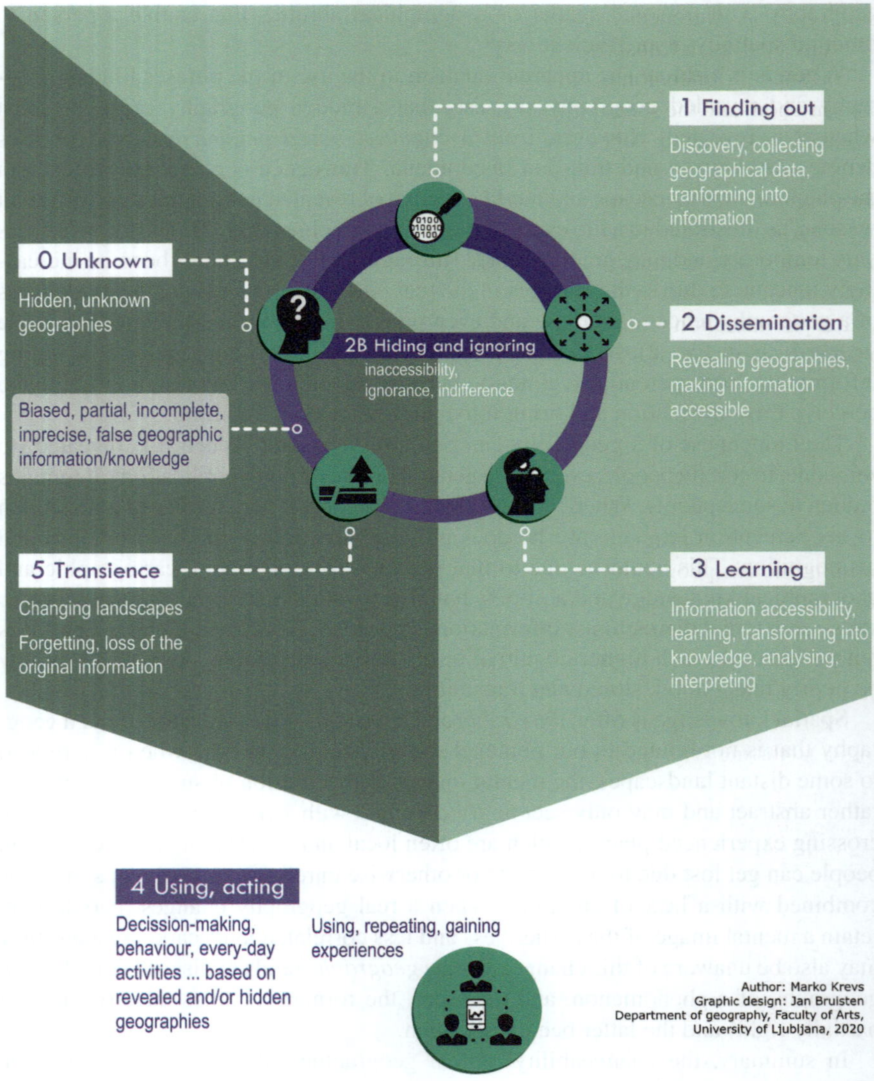

Fig. 2 Life cycle of hidden geographies based on cognition-based definition

When studying deliberate hiding of geographical information, then the geography of hiddenness strongly relates to the spatial distribution of those without access to the information.

Exploring the hiddenness of geography as a discipline may focus on the geographical way of thinking, and geographic/geoinformatics abilities, which may correlate with (spatial) differences in the existence of geography education and training, or its different success in transferring geographic knowledge and skills.

The spatial dimension of the concept is, therefore, a result of the geography of a phenomenon, and the spread of the information and knowledge about it. These are not necessarily correlated. A phenomenon may be located at a single point on the Earth's surface and well known to the majority of (e.g. informed, educated) population. A phenomenon may be spread globally, but remains hidden from everyone, or from the majority. Information about a phenomenon and its spatial distribution may be diffused in relation to the distance of the people from where the phenomenon is located, or with almost no relation to it. The *geography of hiddenness of a geography* (of a certain phenomenon) can, therefore, often be quite complex and unpredictable, even in times of ubiquity of geographical information.

4.2 Definitions

Looking at the continuous debates about what geography is, one cannot expect defining hidden geographies to be a one-time effort. Therefore, our attempt should be seen as a motivation for further debate rather than as something final. Several paths to understanding the concept of hidden geographies have been explored in this chapter, and some of them could already serve as working definitions in practise. In this section, we only briefly synthesise some of them in a more formal form. First, we focus on four of them, following the model of the layers of hidden geographies (Fig. 1), starting with the *undiscovered geographies*.

> Undiscovered geography refers to an existing or imagined geography of a material or non-material phenomenon that has not yet been discovered. It can refer to the hiddenness from a particular individual or society, or to the geography of a phenomenon that has not been discovered due to the inaccessibility of the information, usually caused by lack of knowledge or appropriate technology/methodology.

Uncognised geographies are the most general of the hidden geographies that have already been discovered.

> Uncognised geography is a discovered geography which is hidden from cognition.

The definition of *uncognised geography* is defined in more detail as follows:

> Uncognised geography is an absolute or contextual geography of a material or non-material phenomenon which is hidden from cognition. The necessary level of cognition to reveal a hidden geography is situation specific and requires at least knowledge level, but normally the comprehension level or higher. When such a situation-specific level of cognition is reached, geography is functionally revealed, otherwise it remains functionally hidden.

A specific case of uncognised geography is *unpublished geography*.

> Unpublished geography is geography uncognised due to hiding geographic information from the public.

A more detailed version of the definition of *unpublished geography* is as follows:

> Unpublished geography is an absolute or contextual geography of a material or non-material phenomenon which has not been made available to the public in visual (e.g. on a map), acoustic or other form accessible to the human senses and perception.

The narrowest understanding of the concept in the model is represented by *deliberately hidden geographies.*

> Deliberately hidden geography is intentionally unpublished geography, or intentionally published in such a way that its cognition is hindered.

A more detailed version of the definition of *deliberately hidden geography* is as follows:

> Deliberately/intentionally hidden geography refers to the absolute or contextual geography of a material or non-material phenomenon that is deliberately hidden by certain people from other people. Usually it refers to the hiding of geographical information by omitting it from maps or by preventing access to maps containing this information. However, it may also refer to forms of communication other than visual.

Finally, a more general definition of the concept is as follows:

> Hidden geography refers to an existing or imagined, absolute or contextual geography of a material or non-material phenomenon which is hidden for one or more reasons, e.g. because it is undiscovered, deliberately hidden, unpublished or otherwise hidden from cognition. Such hiddenness may for example be the result of ignorance or lack of knowledge, forgetting information, memorising and using outdated information while the real geography changed, as well as deliberately hidden geographies.

The above definitions refer to geography as spatial location/distribution of a phenomenon. They can also be adequately adapted to refer to geography as a discipline.

5 Hidden Geographies, Geography and Geoinformatics

One of the essential missions of geography as a discipline is to reveal, explain, interpret geographies of phenomena, and complex interrelations between them. Hidden geographies can therefore be seen as a measure of the lack of efficiency of the geographic discipline. However, this measure would not be fair or appropriate to the discipline. Among the reasons, at least the following three should be highlighted: geography and geoinformatics (including cartography) play an obviously controversial role in deliberate hiding of certain (known) geographies; the number of hidden geographies is increasing rapidly, already due to increase in the world population and the diversity of its activities; it is not the amount of revealed geographies that really matters, but rather the proportion of revealed and cognised (geographical) *relevant information* that we need to function in our lives and our ability to use it.

Geography, together with geoinformatics (in its broadest definition including cartography and remote sensing), has never been a mere *provider of geographic information*. But this area of their activities is still important and is developing rapidly, updating already known and revealing fascinating previously hidden geographies. The development includes the discovery and presentation (of geographies) of new phenomena, the use of new visualisation methods, the application of new measuring or analytical techniques and the mapping of barely measurable qualitative information. The global *geographical information coverage of physical environments*, mostly based on various remotely sensed data, is quite well established and is used in environmental modelling on different temporal and spatial scales. The global *geographical information coverage of human activities* is much less accessible to researchers or individual users and is collected and used in often ethically questionable ways (e.g. based on tracking online behaviour and spatial movements of people without their consent or awareness). However, there are attempts to retrieve useful geographical information from publicly available online sources and to provide efficient demonstrations of certain global phenomena related to human activities.

An illustration of the extreme reaches of the geoinformation technologies, which reveal previously hidden human geographies, is based on the activities within the GDELT project (Leetaru 2020a). This online platform focuses on automated, machine-learning-driven retrieval, analysis and presentation of global online news coverage in 65 languages (Leetaru 2015a). It focuses on sensing the social/media reactions to selected social or natural phenomena, revealing geographies that tend to be very unpredictable and transient and consequently hidden most of the time, uncertain, even quite soon after their revelation. The database is extensive in terms of media and the languages, although still selective. In the GDELT analysis of the languages used in the news from 2015 to 2018 (Fig. 3), more than 850 million articles referred to 6.6 billion locations (Leetaru 2018a, b). The resulting map "reminds us of just how much of the world we miss when we look only at English" (ibid.); most of the planet is covered by news media using other languages. "To penetrate into local news and perspectives", to reveal a certain hidden geography, "one must utilize local news in local languages, making use of mass machine translation to process it all" (Leetaru 2015b). Some other examples of revealing the hidden geographies of planetary media news in the GDELT project include media responses to the Covid-19 pandemic (Fig. 4), wildlife crime activities (Leetaru 2015c), global climate change (Leetaru 2015d) and an animated map of global protest activities at the country level over the period 1979–2019 (Leetaru 2019).

Under conditions of the ubiquity of geographical information, we face the problem of its efficient use, given its quantity and the lack of knowledge required for efficient handling. Knowledge relates to our personal skills and cognition and, increasingly, to procedures, many of which involve artificial intelligence built into the digital tools we use. In addition to the more traditional geographic, cartographic and spatial abilities, we are, therefore, increasingly developing *geoinformatics abilities*.

The problem of the efficient use of imperfect geographical information in terms of spatial ability was discussed long before the current situation of information overload (e.g. in Tuan 1977). Conditional, personalised, adaptive selectivity of geographic

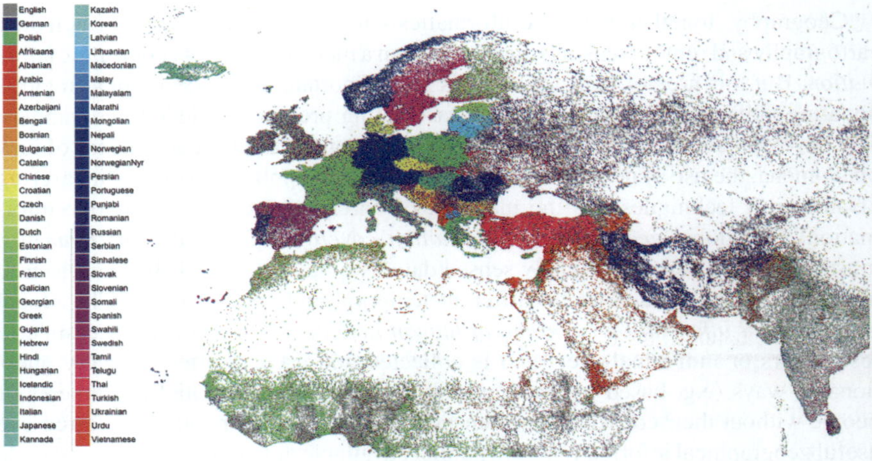

Fig. 3 A linguistic diversity of world news: a map of the languages used in media news for the period 2015–2018 (reproduced with permission, Leetaru 2018a)

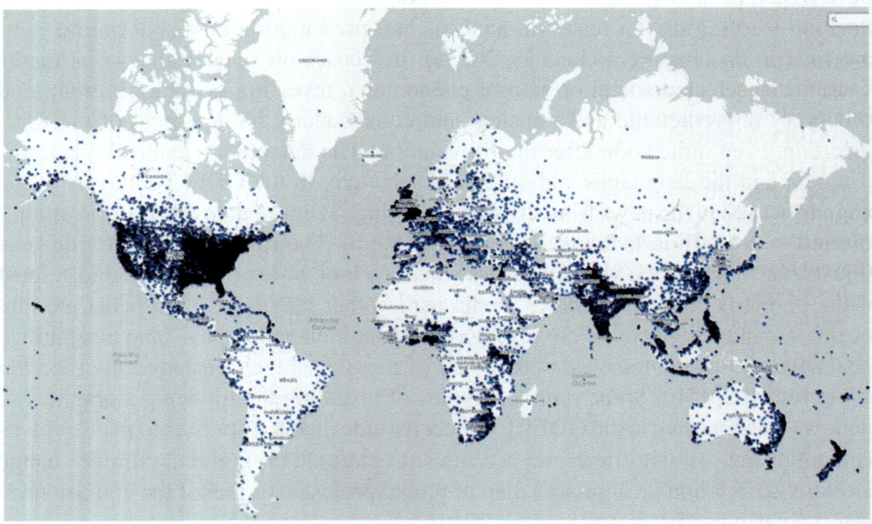

Fig. 4 Covid-19-related news from 1 to 5 April 2020: a world map of a transient geography of media response to the pandemics (reproduced with permission, Leetaru 2020b)

information, when we need it and at what level of precision we need it (Holloway and Hubbard 2001), will probably remain the most important way of adapting—people and tools—to such circumstances. The growing volume of geographical information increasingly requires tools that support intelligent filtering, pondering, interpreting the information to support informed decisions based on the specificity of the geographical question, while taking personal or professional preferences into account. The tools and embedded artificial intelligence should actually mimic the human reasoning: they should search for "underlying missing pieces of information", which are to be found or completed to solve a problem (Tschopp 2018). Such reasoning will increasingly need to address also hidden geographies, either by searching for previously missing geographical information (hidden geographies) or by an appropriate weighting and circumstantial interpretation of known geographies.

Contemporary development of an *efficient digital support for our cognition* seems to lag behind the explosion of available geographic data. This leads to an increasing amount of available geographic information, which in reality—for the reasons mentioned above—is inaccessible for our cognition, as geographies remain functionally hidden to us. Perhaps this situation is not as temporary as it seems. Under the circumstances discussed, we come up against the limits of human cognition and increasing creativity of artificial intelligence (Gobet and Sala 2019). Which inevitably leads in the direction of increasing human dependence on artificial intelligence, "humanity has been driven into a state of dependency where no way back is foreseen" (Tschopp 2018). Along this process, a *dynamic adaptation of the geographical and geoinformatics abilities* of individuals and societies will be needed, so they could function efficiently. The roles of geography and geoinformatics in achieving such abilities appear to be quite challenging.

6 Conclusions

The presented analyses show that the concept of hidden geographies, despite its sporadic occurrence, has a broad range of uses in the existing literature. With this rather detailed overview and resulting interpretations of the situation, we believe that the assumption from the introduction of the chapter—about the importance of the missing (hidden) geographical information for human knowledge, abilities and activities—is well illustrated. Unknown geographic information seems to be a continuing motivator for geographic and geoinformatics exploration.

The relevance of the concept is evident as a reflection of the many ongoing debates on a more general level—on the importance of geography as a discipline, on the importance of geographical information in our lives, and on the rapid development of technologies for intensive exploration of our planet. All this complements the strong views of Aristotle and Kant set out in the introduction, and thus underlines the potential weight of the consequences that a situation in which geographical information is lacking can have.

The conceptual-semantic view on the concept of hidden geography provided a comprehensive and solid basis for our discussion. In fact, it opened the floor for further debate on an aspect that had hardly been touched on in the literature so far, or in this chapter, namely hidden geography as a hidden discipline or a way of geographical thinking. On the other hand, the literature review provided an exhaustive set of examples of connotations of the concept, mainly based on geography as a spatial

Table 2 Examples of the terms used to connote hidden or revealed geography, or to indicate a particular aspect of the meaning of the concept

Connoting hidden geography	Connoting revealed geography
Undiscovered geography	de facto revealed geography
Uncognised geography	Functionally revealed geography
Unpublished geography	Rediscovered geography
Deliberately/intentionally hidden geography	Learned geography
Imagined geography	Known geography
Assumed geography	Real geography
Changeable geography	True geography
Functionally hidden geography	
Imposed geography	
False geography	
Deceptive geography	
Fraudulent geography	
Unconscious geography	
Cognition-based hidden geography	
Missing, absent geographical information	
Ignored geography	
Deceptive geography	
Omitted geography	
Obscured geography	
Absolute and contextual geography	
Personal, private geography	

distribution of phenomena, suggesting its relevance and usability in various contexts. Table 2 gives only a small insight into the wealth of connotations and the variety of situations in which we encounter hidden geographies in this chapter.

The model of the layers of hidden geographies, hierarchically ordered from undiscovered to uncognised, unpublished and deliberately hidden geographies, has provided a firm and practical structure for conducting the discussion. The examples used to illustrate these four layers of the concept have been grouped rather unevenly and sometimes arbitrarily and are not intended to suggest a typology of how the concept is used. Also, the hierarchy of the layers should not be taken too rigidly, to allow exceptions, as already explained in the chapter. But the four layers as different understandings of the concept seem convincing and may be combined in the future with other relevant perspectives of the concept, such as distinctions based on individual/collective perception, or whether geographies are real or imagined.

The definitions of the specific meanings of the concept, corresponding to the four layers of the model mentioned above and the general definition of the overarching understanding of hidden geographies are condensed syntheses of the discussions in this chapter. They can be used as they are in further research or stimulate further debate—in either case, the main aim of the chapter will prove to be justified.

Many questions for further debate and research on hidden geographies arise from the discussion in this chapter, from those related to conceptualisation alone to the term we might use for this concept in the future, and questions about the roles of geography, geoinformatics and related disciplines in better supporting geographical abilities in general and the efficient dealing with hidden geographies in particular.

To conclude the chapter, one of the relevant topics was selected to illustrate—using a very practical and current example—the specificity of the concept, especially in the concrete situation in which it escalates. What are the challenges in cases where we are thinking of hiding a geography? And what are the consequences of our decisions—if we decide for or against hiding?

Hiding geography does not always have a negative connotation or selfish origin. However, keeping a geography of something hidden can be an interesting ethical challenge. A typical example: by telling everyone how great a solitary natural destination you visited was, you may be helping to increase the number of visitors and degrade what you so highly valued. There are many other examples, from protection of very fragile landscapes/ecosystems to places where someone has found gold nuggets or mushrooms, or suppliers of very high quality natural food in limited quantities. You may choose to keep the information about such places to yourself/from others because you want to, you need to, or because you believe it is the right thing to do.

One of my students, a mountain runner, has completed a project on the best areas for "solitary running" in and around my city Ljubljana (Čufer 2020). The multi-criteria evaluation was based on the tracking data of several runners and selected other data on the environmental factors of running. Should she in principle publish the results, and threaten the very central feature of the areas she likes? And should she publish the results in times of COVID-19 pandemic, when people who lack such information have to search for such areas themselves, and—to avoid encounters with others as much as possible—have walked and ran all over the forested and other open spaces in the municipality? Will awareness (of the existence of such areas) form the basis for their rational use and protection, or will it lead to negligent use and destroy the core of their value (as has been the case in so many places to date)? Does democracy work and does it trigger ethical behaviour in these kind of situations? These are only a small selection of questions that we hope will be addressed in future studies.

Those readers who expected a simple explanation of what hidden geography is are hopefully a little less confused than before they set out to read this chapter. But the explanation of the concept is not so simple, and the possible uses are far more diverse than probably anticipated. The main aim was not a "boundary-drawing exercise" (Gregory 2009: 287), to bring definition(s) that could serve to establish a generally accepted concept. Rather, it was to show the existing uses and connotations, to stimulate further discussion and to raise awareness of the existence of a potentially usable concept. We can only guess whether this idea will continue to grow and what place the concept/term will assume in current geographical discourse.

Acknowledgements Thanks to Mark Nuttall for his suggestions, especially on hidden geographies in relation to religions, popular culture and military secrets, and for his help in tightening up parts of the text—which I fear is still far from what he had in mind.

References

Agnew, J. 1995. The hidden geographies of social science and the myth of the 'geographical turn'. *Environment & Planning D: Society & Space* 13 (4): 379–380.

Ai, T., X. Cheng, P. Liu, and M. Yang. 2013. A shape analysis and template matching of building features by the Fourier transform method. *Computers, Environment and Urban Systems* 41: 219–233. https://doi.org/10.1016/j.compenvurbsys.2013.07.002.

Anderson, L.W., D.R. Krathwohl, P.W. Airasian, K.A. Cruikshank, R.E. Mayer, P.R. Pintrich, J. Raths, and M.C. Wittrock. 2001. *A taxonomy for learning, teaching, and assessing: A revision of Bloom's taxonomy of educational objectives*. New York: Pearson, Allyn & Bacon.

Allouche, J. 2019. State building, nation making and post-colonial hydropolitics in India and Israel: Visible and hidden forms of violence at multiple scales. *Political Geography* 75. https://doi.org/10.1016/j.polgeo.2019.102051.

Anceschi, L., G. Gervasio, and A. Teti (eds.). 2014. *Informal power in the Greater Middle East: Hidden geographies*. London and New York: Routledge.

Arrington, T.S., and B. Grofman. 1999. Party registration choices as a function of the geographic distribution of partisanship: A model of 'hidden partisanship' and an illustrative test. *Political Geography* 18 (2): 173–185.

Augé, M. 1995. *Non-places. Introduction to an anthropology of supermodernity*. London: Verso.

Bærenholdt, J.O., and B. Granås (eds.). 2008. *Mobility and place: Enacting Northern European peripheries*. Aldershot: Ashgate.

Baker, I. 2016. The geography of paradise: Hidden-lands in Himalayan myth, history, and Buddhist practice. An illustrated talk. https://www.facebook.com/events/lekdanling-unit-b-14-collent-street-hackney-london-e9-6sg/an-illustrated-talk-the-geography-of-paradise-hidden-lands-in-himalayan-myth-his/1537695659884404/. Video: https://meridian-trust.org/video/the-geography-of-paradise-ian-baker/. Accessed 10 Apr 2020.

Baker, I., and X.I.V. Dalai Lama. 2004. *The heart of the world: A journey to the last secret place*. London: Penguin Press.

Bastin, J.-F., et al. 2017. The extent of forest in dryland biomes. *Science* 356 (6338): 635–638. https://doi.org/10.1126/science.aam6527.

Benchetrit, M., and R. Stadnicki. 2014. Visualizing the margins of Gulf cities. *City* 18 (6): 701–707. https://doi.org/10.1080/13604813.2014.962888.

Betto, F., P. Garengo, and A. Lorenzoni. 2020. A new measure of Italian hidden energy poverty. *Energy Policy* 138. https://doi.org/10.1016/j.enpol.2019.111237.

Bittlestone, R., J. Diggle, and J. Underhill. 2005. *Odysseus unbound. The search for Hommer's Ithaca*. Cambridge: Cambridge University Press.

Bloch, S., and D.E. Martínez. 2020. Canicide by Cop: A geographical analysis of canine killings by police in Los Angeles. *Geoforum (In press)*. https://doi.org/10.1016/j.geoforum.2020.02.009.

Bonnett, A. 2008. *What is geography?*. London: Sage.

Bonnett, A. 2014. *Unruly places. Lost places, secret cities, and other inscrutable geographies*. New York: Houghton Mifflin Harcourt Publishing Company.

Brownlow, A. 2017. The uneven geographies of America's hidden rape crisis: A district-level analysis of underpolicing in St. Louis. *Annals of the American Association of Geographers* 108 (2): 411–423. https://doi.org/10.1080/24694452.2017.1365588.

Buzar, S. 2007. The 'hidden' geographies of energy poverty in post-socialism: Between institutions and households. *Geoforum* 38 (2): 224–240. https://doi.org/10.1016/j.geoforum.2006.02.007.

Çalışkan, V. 2009. Geography of a hidden cultural heritage: Camel Wrestlers in Western Anatolia. *Journal of International Social Research* 2 (8): 123–137.
Cambridge Dictionary. 2018. Hidden. https://dictionary.cambridge.org/dictionary/english/hidden. Accessed 1 Nov 2018.
Le Carrer, O. 2015. *Atlas of cursed places: A travel guide to dangerous and frightful destinations*. New York: Black Dog & Leventhal.
Casey, E. 1998. *The fate of place. A philosophical history*. Berkeley: University of California Press.
Colgan, W., H. Machguth, M. MacFerrin, J. Colgan, D. van As, and J. MacGregor. 2016. The abandoned ice sheet base at Camp Century, Greenland, in a warming climate: Reconsidering Camp Century. *Geophysical Research Letters* 43. https://doi.org/10.1002/2016GL069688.
Čufer, A. 2020. Geoinformacijska podpora odločanju pri iskanju najmanj obiskanih območij za rekreacijo v MOL (Geoinformatic decision-support in the search for the least visited recreational areas in the Urban Municipality of Ljubljana). Research report. Department of Geography, Faculty of Arts, University of Ljubljana.
David-Fox, K. 2000. Prague-Vienna, Prague-Berlin: The hidden geography of Czech modernism. *Slavic Review* 59 (4): 735–760. https://doi.org/10.2307/2697417.
Davidson, J., L. Bondi, and M. Smith (eds.). 2007. *Emotional geographies*. Aldershot and Burlington: Ashgate.
Deng, X., Z. Liu, Y. Zhan, K. Ni, and K.M. Rogers. 2020. Predictive geographical authentication of green tea with protected designation of origin using a random forest model. *Food Control* 107. https://doi.org/10.1016/j.foodcont.2019.106807.
Devereux, P. 2010. *Sacred geography deciphering hidden codes in the landscape*. London: Gaia.
Driver, F. 2012. Hidden histories made visible? Reflections on a geographical exhibition. *Transactions of the Institute of British Geographers* 38 (3): 420–435. https://doi.org/10.1111/j.1475-5661.2012.00529.x.
van Duijne, R.J. 2019. Why India's urbanization is hidden: Observations from "rural" Bihar. *World Development* 123. https://doi.org/10.1016/j.worlddev.2019.104610.
Dyck, I. 1995. Hidden geographies: The changing lifeworlds of women with multiple sclerosis. *Social Science and Medicine* 40 (3): 307–320. https://doi.org/10.1016/0277-9536(94)E0091-6.
Elborough, T., and A. Horsfield. 2016. *Atlas of improbable places: A journey to the world's most unusual corners*. London: Aurum Press Ltd.
Endfield, G.H., and Van Lieshout, C. 2020. Water and vertical territory: The volatile and hidden historical geographies of Derbyshire's lead mining soughs, 1650s–1830s. *Geopolitics* 25 (1): 65–87. https://doi.org/10.1080/14650045.2018.1486299.
Explore nature's mysterious hidden worlds. 2019. News, Geography, College of Life and Environmental Sciences. http://geography.exeter.ac.uk/news/articles/explorenaturesmysterioush.html. Accessed 10 Apr 2020.
Eyles, J. 1985. *Senses of place*. Warrington: Silverbrook Press.
Fendrychová, L., and P. Jehlička. 2018. Revealing the hidden geography of alternative food networks: The travelling concept of farmers' markets. *Geoforum* 95: 1–10.
Fiedler, R., N. Schuurman, and J. Hyndman. 2006. Hidden homelessness: An indicator-based approach for examining the geographies of recent immigrants at-risk of homelessness in Greater Vancouver. *Cities* 23 (3): 205–216.
Flood, G. (ed.). 1993. *Mapping invisible worlds*. Edinburgh: Edinburgh University Press.
Florida, R. 2013. The hidden geography of America's surging suicide rate. CITYLAB. https://www.citylab.com/equity/2013/05/hidden-geography-americas-surging-suicide-rate/5489/. Accessed 10 Apr 2020.
Foer, J., D. Thuras, and E. Morton. 2016. *Atlas Obscura*. New York: Workman Publishing.
Fraser, A. 2019. Curating digital geographies in an era of data colonialism. *Geoforum* 104: 193–200.
Freksa, C., A. Klippel, and S. Winter. 2007. A cognitive perspective on spatial context. In *Spatial cognition: Specialization and integration*, ed. A.G. Cohn, C. Freksa, and B. Nebel. Dagstuhl Seminar Proceedings 05491. https://drops.dagstuhl.de/opus/volltexte/2007/980/pdf/05491.FreksaChristian.Paper.980.pdf. Accessed 10 Apr 2020.

Garcia-Ramon, M.-D., J. Cruz Villalon, I. Salamaña Serra, and M. Villarino Perez. 1994. La mujer y agricultura en España: Género, trabajo y contexto regional [*Women and agriculture in Spain: Gender, work and regional context*]. Barcelo-na: Oikos Tau.
Gelfert, A. 2018. Fake news: A definition. *Informal Logic* 38 (1). https://doi.org/10.22329/il.v38i1.5068.
Gibson, K., A. Hill, and L. Law. 2017. Community economies in Southeast Asia. A hidden economic geography. In *Routledge handbook of Southeast Asian development*, ed. A. McGregor, L. Law, and F. Miller, 131–141. London and New York: Routledge.
Gilbert, D., C. Dwyer, N. Ahmed, L. Cuch Graces, and N. Hyacinth. 2019. The hidden geographies of religious creativity: Place-making and material culture in West London faith communities. *Cultural Geographies* 26 (1): 23–41. https://doi.org/10.1177/1474474018787278.
Gobet, F., and G. Sala. 2019. How artificial intelligence can help us understand human creativity. *Frontiers in Psychology* 10: 1401. https://doi.org/10.3389/fpsyg.2019.01401.
Gregory, D. 2009. Geography. In *The dictionary of human geography*, 5th ed., ed. Gregory et al., 287–295. Chichester: Wiley-Blackwell.
Haggett, P. 1995. *The geographer's art*. Oxford: Blackwell.
Hall, E. 2004. Social geographies of learning disability: Narratives of exclusion and inclusion. *Area* 36 (3): 298–306. https://doi.org/10.1111/j.0004-0894.2004.00227.x.
Hansen, F.E. 2006. *Layers of musical meaning*. Copenhagen: Museum Tusculanum Press.
Harmon, K. 2003. *You are here: Personal geographies and other maps of the imagination*. Princeton: Princeton Architectural Press.
Harris, C., D. Moran, and J.R. Bryson. 2015. Polish labour migration to the UK: Data discrepancies, migrant distributions, and indicators of entrepreneurial activity. *Growth & Change* 46 (2): 196–217. https://doi.org/10.1111/grow.12087.
Herrick, C. 2017. When places come first: Suffering, archetypal space and the problematic production of global health. *Transactions of the Institute of British Geographers* 42 (4): 530–543. https://doi.org/10.1111/tran.12186.
Hidden Geographies. 2019. Collection of abstracts. In *Conference "hidden geographies"*. Ljubljana, 29–31 August 2019. https://t.ly/HiddenGeographies. Accessed 21 Apr 2020.
Hidden Journeys Project promotes geography. 2012. Geospatial world. https://www.geospatialworld.net/news/hidden-journeys-project-promotes-geography/. Accessed 10 Apr 2020.
Holloway, L., and P. Hubbard. 2001. *People and place. The extraordinary geographies of everyday life*. Harlow: Pearson Education Limited.
Hooks, B. 2008. *Belonging: A culture of place*. London and New York: Routledge.
Johnson, K.M., and W.B. Ouimet. 2013. Rediscovering the lost archaeological landscape of southern New England using airborne light detection and ranging (LiDAR). *Journal of Archaeological Science* 43: 9–20. https://doi.org/10.1016/j.jas.2013.12.004.
Johnston, R., and J.D. Sidaway. 2016. *Geography and geographers. Anglo-American human geography since 1945*, 7th ed. London and New York: Routledge.
Kılıç, B. 2019. Determination of wind dissipation maps and wind energy potential in Burdur province of Turkey using geographic information system (GIS). *Sustainable Energy Technologies and Assessments* 36. https://doi.org/10.1016/j.seta.2019.100555.
Klayko, B. 2015. This map reveals the hidden geography of immigration in Louisville and across the United States. Broken Sidewalk. https://brokensidewalk.com/2015/immigration-map/. Accessed 10 Apr 2020.
Krevs, M. 2005. Where is the border of a settlement? In Zbornik radova I kongresa geografa Bosne i Hercegovine, 226–233. Sarajevo: Geografsko društvo Federacije Bosne i Hercegovine.
Kuletz, V.L. 1998. *The Tainted desert: Environmental and Social Ruin in the American West*. London and New York: Routledge.
Kunstler, J. 1993. *The geography of nowhere. The rise and decline of America's man-made landscape*. New York: Simon and Schuster.
Larreche, I.J., and P. Ercolani. 2019. Un paréntesis en Geografía. Cartografías de la noche LGBT en Bahía Blanca (Argentina) (A geographical gap. Maps of LGBT nights in Bahía Blanca

(Argentina)). *Investigaciones Geograficas* 72: 151–166. https://doi.org/10.14198/INGEO2019.72.07.

Lawrence, M. 1995. Rural homelessness: A geography without a geography. *Journal of Rural Studies* 11 (3): 297–307. https://doi.org/10.1016/0743-0167(95)00025-I.

Leetaru, K. 2015a. GDELT 2.0: Our global world in realtime. https://blog.gdeltproject.org/gdelt-2-0-our-global-world-in-realtime/. Accessed 31 Aug 2020.

Leetaru, K. 2015b. Linguistic geography: Mapping the languages of GDELT February-July 2015. https://blog.gdeltproject.org/linguistic-geography-mapping-the-languages-of-gdelt-february-july-2015/. Accessed 31 Aug 2020.

Leetaru, K. 2015c. Mapping poaching: Three months of wildlife crime. https://blog.gdeltproject.org/mapping-poaching-three-months-of-wildlife-crime/. Accessed 31 Aug 2020.

Leetaru, K. 2015d. Map of global climate change news coverage. https://blog.gdeltproject.org/map-of-global-climate-change-news-coverage/. Accessed 31 Aug 2020.

Leetaru, K. 2018a. Mapping the linguistic geography of GDELT: 2015–2018. https://blog.gdeltproject.org/mapping-the-linguistic-geography-of-gdelt-2015-2018/. Accessed 31 Aug 2020.

Leetaru, K. 2018b. Mapping the hidden geography of language through the news. Forbes. https://www.forbes.com/sites/kalevleetaru/2018/05/17/mapping-the-hidden-geography-of-language-through-the-news/#73e923bb69d6. Accessed 10 Apr 2020.

Leetaru, K. 2019. Mapping global protest trends 1979–2019 through one billion news articles. https://blog.gdeltproject.org/mapping-global-protest-trends-1979-2019-through-one-billion-news-articles/. Accessed 31 Aug 2020.

Leetaru, K. 2020a. The GDELT project. https://www.gdeltproject.org/. Accessed 31 May 2020.

Leetaru, K. 2020b. One-click mapping of Covid-19 news coverage with BigQuery + Carto. GDELT project, 6 April 2020. https://blog.gdeltproject.org/one-click-mapping-of-covid-19-news-coverage-with-bigquery-carto/. Accessed 31 Aug 2020.

Liu, W., C. Liu, J. Yu, Y. Zhang, and L. Zheng. 2018. Discrimination of geographical origin of extra virgin olive oils using terahertz spectroscopy combined with chemometrics. *Food Chemistry* 251: 86–92. https://doi.org/10.1016/j.foodchem.2018.01.081.

Lu, P., R. Yang, P. Chen, Y. Guo, and R. Lasaponara. 2017. On the use of historical archive of aerial photographs for the discovery and interpretation of ancient hidden linear cultural relics in the alluvial plain of eastern Henan, China. *Journal of Cultural Heritage* 23 (Supplement): 20–27.

Lucherini, M. 2019. Representing diabetes: 'Brightside' and 'chaos' in autobiography. *Emotion, Space and Society* 31: 10–17. https://doi.org/10.1016/j.emospa.2019.02.004.

Maione, C., F. Barbosa, and R.M. Barbosa. 2019. Predicting the botanical and geographical origin of honey with multivariate data analysis and machine learning techniques: A review. *Computers and Electronics in Agriculture* 157: 436–446.

Manaugh, G. 2016. *A burglar's guide to the city*. New York: Farrar, Straus & Giroux Inc.

Martin, P.R. 2020. Indigenous tales of the Beaufort Sea: Arctic exploration and the circulation of geographical knowledge. *Journal of Historical Geography* 67: 24–35. https://doi.org/10.1016/j.jhg.2019.10.012.

Matthews, H., M. Taylor, K. Sherwood, F. Tucker, and M. Limb. 2000. Growing-up in the countryside: Children and the rural idyll. *Journal of Rural Studies* 16 (2): 141–153. https://doi.org/10.1016/S0743-0167(99)00059-5.

May, J. A. 1970. *Kant's concept of geography and its relation to recent geographical thought*. Toronto: University of Toronto Press. https://www.jstor.org/stable/10.3138/j.ctvcj2shg.5. Accessed 4 July 2020.

Mombauer, M. 2017. Germany's hidden bombs. Open Learn. https://www.open.edu/openlearn/society-politics-law/geography/germanys-hidden-bombs. Accessed 10 Apr 2020.

Monmonier, M.S., and H.J. De Blij. 1996. *How to lie with maps*, 2nd ed. Chicago and London: University of Chicago Press.

Morgan, J. 2000. To which space do I belong. Imagining citizenship in one curriculum subject. *The Curriculum Journal* 11 (1): 55–68.

Murphy, A.J. 2007. Grounding the virtual: The material effects of electronic grocery shopping. *Geoforum* 38 (5): 941–953. https://doi.org/10.1016/j.geoforum.2006.12.012.

Navarro-Costa, P., and C.E. Plancha. 2011. Heterochromatin: The hidden epigenetic geography of the Y chromosome. *Human Reproduction Update, Letter to the Editor* 17 (3): 434. https://doi.org/10.1093/humupd/dmr002.

Paglen, T. 2006. A most peculiar airline, flight tracking. http://vectors.usc.edu/issues/04_issue/trevorpaglen/. Accessed 10 Apr 2020.

Percy-Smith, B., and H. Matthews. 2001. Tyrannical spaces: Young people, bullying and urban neighbourhoods. *Local Environment* 6 (1): 49–63. https://doi.org/10.1080/13549830120024242.

Petrovič, D., K. Kozmus Trajkovski, and Ž. Kokalj. 2018. Different visualizations and used data sources of the mountain battlefield on the Soča/Isonzo front line. *International Journal of Cartography* 4 (3): 336–347. https://doi.org/10.1080/23729333.2018.1512185.

Pickren, G. 2020. Internet geographies. In *International encyclopedia of human geography*, 2nd ed., 389–395.

Posada-Swafford, A. 2005. RADAR; CARTA 2005 is initiative operated by NASA, exploring physical and chemical properties of Costa Rica. Latin Trade. Freedom Magazines. https://www.thefreelibrary.com/Hidden+geography-a0135706061. Accessed 10 Apr 2020.

Rajaram, P.K., and C. Grundy-Warr. 2007. *Borderscapes: Hidden geographies and politics at territory's edge.* Borderlines 29. Minneapolis: University of Minnesota Press.

Relph, E. 1976. *Place and placelessness.* London: Pion.

Said, E. 1979. Imaginative geography and its representations: Orientalizing the oriental. In *Orientalism.* New York: Vintage.

Salovaara-Moring, I. 2009. Dead ground: Media studies and hidden geographies of knowledge. *Television & New Media* 10 (1): 144–146. https://doi.org/10.1177/1527476408325369.

van Schendel, W. 2002. Geographies of knowing, geographies of ignorance: Jumping scale in Southeast Asia. *Environment and Planning D: Society and Space* 20: 647–668.

Schmid, E. 2020. Uncovering the underground railroad's hidden past in the Metro East. St. Louis Public Radio. https://news.stlpublicradio.org/post/uncovering-underground-railroad-s-hidden-past-metro-east#stream/0. Accessed 10 Apr 2020.

Serres, M. 2008. *The five senses. A philosophy of mingled bodies.* London: Continuum.

Shewly, H.J. 2017. Life in de facto statelessness in enclaves in India and Bangladesh. *Singapore Journal of Tropical Geography* 38 (1): 108–122. https://doi.org/10.1111/sjtg.12181.

Stuart, O. 2010. Navigability and the improvement of the river Thames, 1605–1815. *Geographical Journal* 176 (2): 164–177. https://doi.org/10.1111/j.1475-4959.2010.00354.x.

Sutherland, R. 2017. Explorers probe hidden continent of Zealandia. The Conversation. http://theconversation.com/explorers-probe-hidden-continent-of-zealandia-83406. Accessed 10 Apr 2020.

Štular, B. 2011. The use of lidar-derived relief models in archaeological topogra-phy. *Arheološki vestnik* 62 (1): 393–432.

Thesaurus.com. 2018. Hidden. https://www.thesaurus.com/browse/hidden. Accessed 1 Nov 2018.

Tschopp, M. 2018. Human cognition and artificial intelligence—A plea for science. https://www.scip.ch/en/?labs.20180419. Accessed 8 Nov 2020.

Townsend, J.G. 2002. Hidden workers, hidden women, hidden geographies. *Antipode* 34 (5): 1001–1003. https://doi.org/10.1111/1467-8330.00290.

Tuan, Y. 1977. (ninth printing 2002). *Space and place. The perspective of Experience.* Minneapolis: University of Minnesota Press.

Tuan, Y. 1990. *Topophilia: A study of environmental perception, attitudes and values.* New York: Columbia University Press.

Walker, R. 2001. Golden gate: A hidden geography. Shaping San Francisco's Digital Archive. Found SF. http://foundsf.org/index.php. Accessed 10 Apr 2020.

Wanzola, H. 2013. Rediscovering the hidden world: The changing human geography of Kongo, English ed. Amazon Media.

Wu, C.Y. 2008. The spatiality of social identities: Taiwanese migrant women practice everyday spaces in Toronto. Dissertation/Thesis. Department of Geography, University of Toronto. https://tspace.library.utoronto.ca/bitstream/1807/17271/1/Wu_Chin-Yen_200811_PhD_thesis.pdf. Accessed 10 Apr 2020.

Yang, S., and B. Chen. 2017. Driving forces of particulate matter emissions in China. *Energy Procedia* 105: 4601–4606. https://doi.org/10.1016/j.egypro.2017.03.995.

Zakon o varstvu podzemnih jam (ZVPJ) (Underground Caves Protection Act). 2004. Official Gazette of the Republic of Slovenia, No. 2/04, 61/06—ZDru-1, 46/14—ZON-C and 21/18—ZNOrg. http://www.pisrs.si/Pis.web/pregledPredpisa?id=ZAKO2068&d-49682-p=1&d-49682-o=2&d-49682-s=3. Accessed 10 April 2020.

Zeleny, M. 1987. Management support systems: Towards integrated knowledge management. *Human Systems Management* 7 (1): 59–70. https://doi.org/10.3233/HSM-1987-7108.

Cognitive Geography. Space Reflected in the Mind

Adriana Galvani, Margarita Zaleshina, and Alexander Zaleshin

Abstract Modern science has already reached such a high level that technological solutions in practice expand the limits of the capabilities of human intelligence and move forward in the creation of artificial intelligence (AI). The topic of automated decision-making in navigation is constantly being included in global research and is funded by several state and commercial institutions. Investigating the ability of the brain to orient spatially is very important because, in the future, AI will require not only the usage of typical template solutions for orientation, but also the ability to handle fuzzy spatial information. Hidden geographic objects can be defined as those objects that are not explicitly recognized, but at the same time, they are perceived as sensory stimuli and actively processed and integrated into the brain. Based on only part of an incomplete object, the brain itself is able to think of its continuation, and this helps in finding the way and in remembering the way. However, people increasingly cognize and represent the Earth through devices, while humanity is losing the innate ability to perceive the environment, which is present in the animal world. Animals have a multicomponent sensory contact with natural elements, they form their own mental maps based on orientation by the Sun, by stars, by the surrounding landscape, i.e. by the same natural landmarks that people followed in ancient times. Even the ancient Mayans and Aztecs were more able to perceive spatial knowledge on different scales than many modern users of new electronic devices. Presently it is important to understand how the mind perceives, draws and creates a picture of the surrounding space in memory. Using cognitive descriptions in geography, it will be possible to integrate algorithms of innate spatial perception with numerous technological applications, expanding the use of AI for complex spatial tasks.

A. Galvani (✉)
University of Bologna, Bologna, Italy
e-mail: adriana.galvani@unibo.it

M. Zaleshina · A. Zaleshin
Moscow Institute of Physics and Technology, Moscow, Russia
e-mail: terbiosorg@gmail.com

A. Zaleshin
e-mail: alex.ihna@yandex.ru

M. Krevs (ed.), *Hidden Geographies*, Key Challenges in Geography,
https://doi.org/10.1007/978-3-030-74590-5_2

Keywords Cognitive geography · Mental maps · Brain · Spatial perception · Ethology

1 Introduction

In this paper, we propose a holistic consideration of the cognitive problems of geography, considering the importance of both the macrocosm—at the level of animals' spatial behaviour and space exploration by humans, and the microcosm—at the level of mental perception and neural networks.

Modern global communication networks are extremely heaped and reach out to the inhabitants of the entire planet; but sets and methods of communication in a single brain can be more sophisticated, more effective, and more complex. Researchers still don't understand the real potentialities of the human brain, if not in trans-disciplinary teams, although both the EU and the US are funding many kinds of research and development in this field. The world's largest companies—for example, Google—are investing heavily in improving their natural and artificial intelligence. But all face the difficult challenges of an unexplored world, since every human brain is a mystery even to its owner.

Nowadays, the rapid development of medical and scientific technology—such as electroencephalography, functional magnetic resonance imaging, computed tomography and optogenetics—allows researchers to better study the structure of the brain. Therefore, it became possible to examine the basis of human behaviour at the level of brain structures.

In addition, the rapid development of geographic information technologies allows to increase the volume of big data processing on spatial objects and to increase the accuracy of tracking movements using global navigation satellite systems (GPS, GLONASS, and Galileo).

We believe that the study of human activity using geoinformation systems can help in decoding neural network structures, and vice versa, the explanation of the principles of the origin and propagation of neural activity can help in the development of better ways to interact with the Earth's space.

2 Mental Perception of Spatial Data

Alessandro Galvani, a physicist at the University of Trento, has explained in simple terms the complexity of creation: "All the creatures derive from the decay of stars; when they, at the end of their life, are falling off, the elements which are composed of, don't dissolve, but they congregate under other forms, all the forms of the universe" (unpublished quote).

Similar concept is affirmed in philosophy by Gregory Bateson (Fairlamb and Bateson 1979; Turner and Bateson 1980; Bateson 1999), who investigated the unity and relationship of man and nature. In addition, Bateson applied principles of cybernetics to the field of ecological anthropology and the concept of homeostasis. He wrote: "People live in space, and space forms their bodies and minds. Starting from the world around us, we need the help of geography—which the only real explanation of what occurs to us derives from. In conclusion, we can say that we will start from geography to know ourselves" (1999: 171). Perhaps the space that is more worth exploring is not the space outside, but the space inside, or something embedded in our mind and about which we know very little.

Geography of the outside world exists, but is there a geography of the inside? How is geography mentally understood? Does the spatial structure of neural networks affect cognition? Is it possible to combine these three geographies (external, mental and neural)? How can they interact?

Environmental cognition, together with the use or management of the territory, gives meaning to the history of people and models their consciousness and culture.

Modern anthropology no longer takes into consideration man solely in his human condition, but the whole human being in its destiny and vocation.

Geography comprehends not only the physical features on Earth, but even immaterial aspects of mind, and some technologies of psychology and psychiatry could be applied on it. Main topics of geographies of the mind were shown in (James et al. 1976).

The purely physical base of space expands human possibilities of observation, which inflate the reactions that trigger brain connections and consequently intellectual abilities.

The spatial analysis is associated with the mind which organizes the observations and the connections; the analysis is able to modulate new perceptions into different organizations. The connections identified in space are easy to infer: spatial elements are perceived in a global or assembled way, the perception itself selects connections through a process that takes place first in the space, then in the brain. This process was named by Claude Bernard (French physician and physiologist, 1813–1878) as *intérieur-terrain* which replaced the term life forces that are creating self-regulation in the nervous system. These findings are considered an antecedent of the contemporary cognitive science.

The brain is in fact continually connecting the internal and external contexts: there is the external space, the brain observes it, the brain interprets it; describes it; interiorizes it, and finally uses and modifies it.

There are, according to the modern neurological studies, internal mechanisms of interpretation of space which make it different, depending on its perceptions and uses. The connection between space and mind has not yet been entirely explored, but the space of the mind is broader than the outside space, and it is more flexible, more blended—one could say that in the mind there are more landscapes than in reality.

The individual and his environment are a dynamic configuration in its totality; all the parts are closely interrelated and continually interchanging, and under such constant interaction it is very difficult to say where a line of demarcation could be drawn (Linton 1945).

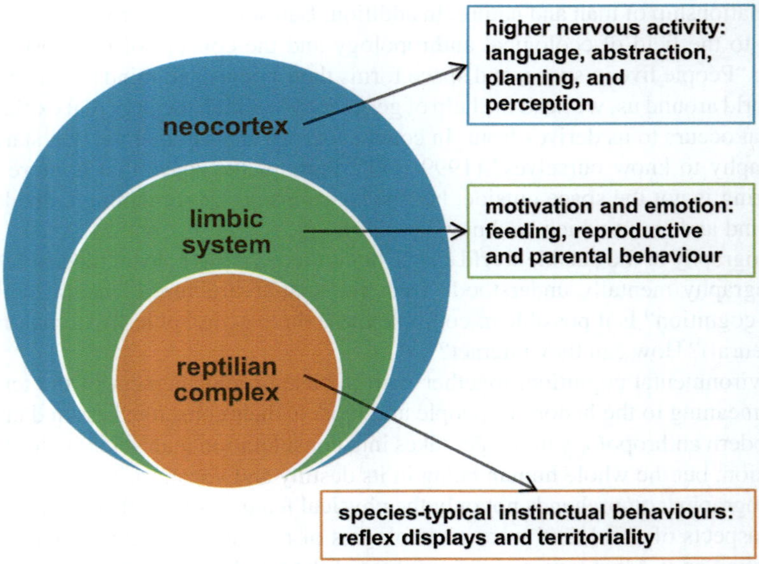

Fig. 1 Paul McLean's Triune Brain model

As demonstrated by scientists, human and animal brains maintain memory of all past knowledge (naturally hidden in the subconscious) of millions of Earth's lives, especially correlated to natural world features. Such example are animal migrations, following always the same paths, guided by natural characteristics, like Sun and stars (Berthold 1991, 2003).

Philosophy calls this "innatism", which can be explained through the "three brains theory", or the "Triune Brain Concept" (McLean 1990) (Fig. 1).

"The division of the brain into three parts constitutes what MacLean calls the Triune Brain of mammals (meaning one evolved brain could entail three brains, from an evolutionary perspective). The three stages of functioning are related in a way that the most recent one evolved into a neo-mammalian neocortex of the cerebral hemispheres, which surrounds the older paleo-mammalian limbic system, which, in turn, surrounds the ancient reptilian upper brainstem" (Gould 2002).

According to this concept, we are what we have learned and what is hidden in the core of our heads. The last goal could be to restore the innate skills and combine them with new technical instruments—an orchestra of three brains, playing in different ways. Special education could merge the knowledge of the natural world with new technological abilities.

This could be confirmed with the example of birds, which are among the oldest living species. Surely, they are the most capable of organizing their scheduled trips along thousands of kilometres, even if other species are more mobile. They use innate capacities related to Earth's movements, or solar and stars' relative positions. Some of these capacities are also present in human ancestral mobility, as demonstrated by Bedouins and by the skilled navigators during the time of the great oceanic discoveries.

Researchers suppose that birds can use mental maps in their flights (Golledge et al. 1996; Montello and Freundschuh 2004). As a result, birds follow the same way year after year in wintering season, searching for a better climate and richness of food. The reason is always food, even for humans, who have been migrating since their first appearance on Earth.

A theory of cognitive mapping has developed that depends only on accepted properties of hippocampal function, namely, long-term potentiation, the place cell phenomenon, and the associative or recurrent connections made among pyramidal cells. It is also possible that the cognitive mapping functions of the hippocampus are carried out by parallel graph searching algorithms implemented as neural processes (Eckardt 1980).

Surely, animals maintain those natural capacities; people, on the other hand, have already translated natural abilities into technical functions, attaining the excellence of GPS and GIS maps elaboration (www.esri.com). Modern behavioural geographers have started working with perception and are already using mental maps to study the visualization of cities and the spatial preferences of travellers. In the era of smartphones and global Internet coverage, applications such as ArcGIS Online (https://www.arcgis.com/index.html) or CartoDB (https://carto.com) allow users to use the technology online (Pánek 2016).

3 Spatial Orientation in Humans and Animals

The mystery of birds is hidden in prehistoric times millions of years ago, in the pre-humanoid era, considering that birds are among the oldest creatures. Their brain structure is sculptured in a way that they do not need to learn migrating habits from elders. It has been discovered that young birds, which never had the opportunity to follow adults on their trips, are able to find the same ancestral path, going away and coming back.

What is also strange is that even caged birds are feeling the urge to migrate at the same time as free birds. Scientists captured animals to release them away from the usual track of the species. Once released, animals retrieved the usual track and resumed the desired trip.

They fly thousands of kilometres, spending lots of energy and fighting against natural elements, risks and predators. In the eighteenth century, scholars assumed that birds do have a special endowment—not gene, since it was not discovered at that time—but a specific hormone of movement. Later, scientists had supposed that birds have specific skills in perceiving sunlight, following stars or magnetic poles, and finding natural direction which satisfy their needs.

The influence of natural landscape on spatial perception and, consequently, on the trajectory of movement was shown in our previous study (Zaleshina et.al. 2019).

Multiscale brain and space interactions are shown in Fig. 2.

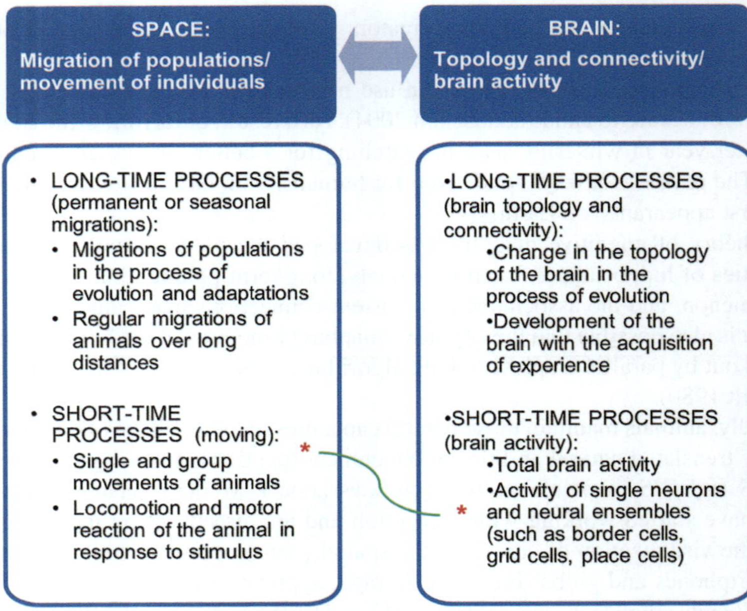

Fig. 2 Multiscale brain and space interactions

Bedouins can also travel through deserts for long distances. Civilization and urban habits such as motorization and economic changes have rendered Bedouins sedentary, so they do not follow any more directions according to the natural elements.

Modern way of life is changing human habits. We can presume that all humans have had the same abilities, remembering the hominids who reached Europe and Asia from Africa. The dispersal from Africa has been a long-lasting event, so long that it continues until now. It has permitted the survival of the best species and the disappearance of frail species. Hominids didn't have any instrumentation, so we can presume they followed their instinct, supported by the natural perception of temperatures, winds' direction, and Sun's and stars' position. History tells about the exceptional abilities of pre-Amerinds such as Incas, Mayan or Aztecs, not only to perceive, but also to measure natural phenomena. The point to which they arrived in mathematical and astronomical measures was so elevated that not only we are unable to repeat them, but even unable to understand them. They predicted natural events for millennia ahead and they understood astronomical phenomena which we can only interpret today with sophisticated instruments.

The principles of topological psychology were built as a set of constructive concepts based on experimental investigations as well as real psychological cases (Lewin 1936). Our attachment to Earth remains in our constant sense of nostalgia for something to be found elsewhere, or for our birthplace. Migrants try to return home. When we feel unsatisfied, we try to change life or, often, to change our place of living. No one is, probably, satisfied with what they have, or the place where they live. Everyone has the feeling of something that exists somewhere else, which could fulfil their lives in the best way possible. Surely, the sense of dissatisfaction with routine life, or the sense of going, like some migrant birds or nomads feel, is inexplicable. Birds are the ancestors of humans; they know where to go, unlike men who do not know where to go; they only feel to go *somewhere*. This sense of otherness can instil an illness of unrest.

The scheme of spatial perception in animal movements is shown in Fig. 3.

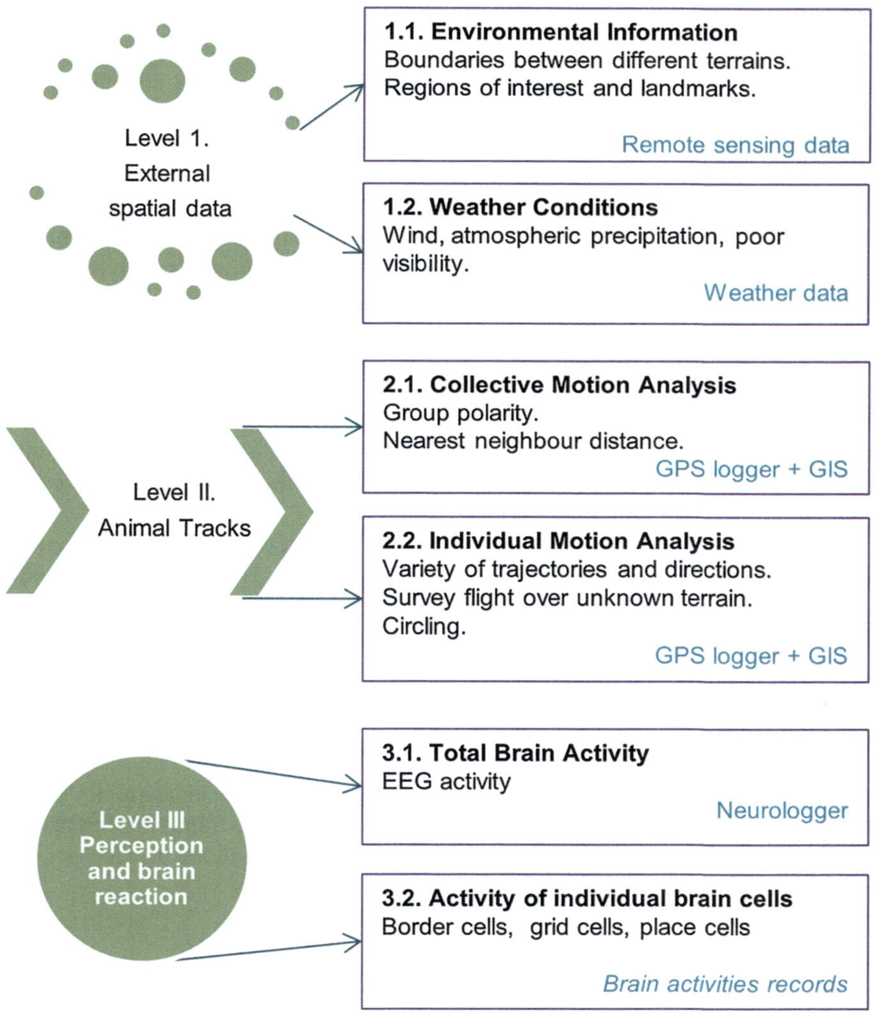

Fig. 3 Spatial perception in animal movements

Animal models bring important solutions for human psychiatric studies, for example, to help to determine the generality of emotion and personality, establishing that they are not only humans. Human interest in animal behaviour deeply extends researches into our common evolutionary roots. There are some kinds of similarities between wolves and humans in the behaviour of social interaction. It is evident that humans are afraid of wolves, but also that the latter are afraid of humans. Wolves illustrate how some actions which may be categorized as more instinctive, and others, which are considered to be more learned—on the continuum that ethologists define as ranging from fixed reflex to flexible intelligence—have been retained in the ancestral part of the genome. Much learning in wolves occurs because of social transmission and remains a question for future inquiry.

At the third level of analysis, of the physical ecosystem, wolf populations are influenced by biotic (e.g. prey, competing species, diseases) and abiotic factors (e.g. winter severity, fire cycles, drought cycles). From a theoretical perspective, each of these systems is viewed as subjective because individuals harmonize within groups, subpopulations and ecosystems. Surely, the effects revealed by experiments on wolves could be reproduced on several other animals, especially migrating animals. Unfortunately, at the time when technological processes facilitate new skills, we are losing some abilities. In humans, there is a sort of balance between learned skills and innate skills which become progressively unknown. Questions of animal awareness and evolutionary continuity of mental experience were shown in (Friedman and Griffin 1977).

Discoveries of Columbus, Vasco da Gama and other great navigators were surely related not only to commercial expansion, but probably unconsciously, also to the sense of understanding where our world starts and finishes, as far as our dreams can bring us.

One thing is to take part in events, the other to see them on television; one thing to be among people, another to speak on a smartphone; one thing is to go on a virtual vacation to the Maldives or Hawaii by opening YouTube, another to travel to China like Marco Polo, or going to Antarctica and feel the freezing like Amundsen. Bedouins and nomads were able to make long trips following the horizon without instruments. This was undoubtedly a capacity which belonged to all hominids in prehistoric times.

Understanding animal behaviour is key to our survival, especially behavioural genetics which is a rapidly expanding area in the study of animal adaptivity (Jeong and Di Rienzo 2014). Answers to questions about how behavioural dissimilarity is related to the interaction of genetic variation and environmental variation are still elusive (Packard 2019).

4 Positioning Systems in the Brain

In 2014, Edward and May-Britt Moser, and O'Keefe won the Nobel prize for medicine and physiology for their studies on the positioning system in the animal brain. Moser's discoveries, made at Trondheim University in Norway, confirm that the animal brain can store a lot of information on spatial organization: during space exploration by mice, pathways in form of grids are permanently stored into mental patterns, so that animals can restore previous experiences and retrace an already made way (Leutgeb et al. 2005).

Such information is stored in space cells and organized into grids cells.

All this has been found out on mice and it is supposed to exist also in humans, on whom the experiments are more challenging. It is certainly obvious to suppose that the same capacities are present even in humans, since many animal species do have a spatial memory inherited from ancestral times, even from the first appearance of those species on Earth.

The results of many years of research are the discovery of place cells, and later, grid cells on brain, or further, border cells.

The discovery derives from studies on mice. Researchers put electrodes on the heads of rats along their way on the search for food. The first time, the animals start searching their way to food, they navigate without a sure direction, but when they repeat the same path, they learn the correct direction, and neural activities at a specific place mark the right path. The fascinating result of research is not only that animals are able to find the shorter way to the final goal, but also that the neuronal activities appear at specific points which correspond to the nodes of triangles whose composition forms hexagons. Place cells become grid cells.

Scientists let us suppose that even the human brain has such potentialities and performing activities. Geometrical precision is the first passage from terrain to maps, assuming the methodology of the first measures of territories for governmental purposes. Same schemes are found and revealed through electrical connections on caged mice, recorded when they are in search of food.

From that, it could be argued that the brain functions like a GPS, and that it can create maps.

One would suppose that it is very easy for us to read maps, but there is a great gap between having a direction capacity in the brain and reading maps created by technicians. This is the gap which exists between instinct and science, from which we must learn at school how to read and create maps.

Without a doubt, animals follow only instinct, whereas people do not maintain any trust on instinct, and only trust science. However, relying on science, we little by little lose our instinctual abilities.

5 Digitalization of the Mental Representation of Spatial Relationships

The process of digitizing maps represents a cultural revolution as decisive as the language. Design principles for presenting spatial information should bear similarities across these domains but also be somewhat specific to each one (Taylor, Brunyé and Taylor 2008). External visuospatial representations have to bear many similarities to those that reside in the mind (Tversky 2005).

It corresponds to the transformation of spoken language into written language that has changed civilization forever. The social evolution of language and human relations have many correspondences in human movements. To find the links between modern styles and ancestral behaviour requires intercultural studies, which implies scientific, technical and humanistic fundaments of knowledge.

New instruments could advance better if scientists coordinated biological research with medical, engineering and geographical research. The connections which constantly occur on Earth are materially reflecting the hidden performances of the nerve cells, which are interlaced through electrical stimulations.

One such opportunity is the research on Artificial Intelligence (AI). This is possible through research on the brain, including the application of deep learning in neural networks (Fig. 4).

Synapses in brain are the highways for mental information's communication, resembling highways, streets and avenues that connect the elements of a place. The nervous net is a model of constantly interchanging communication, which mixes memory, sensations, emotions, desires, dreams, or forgotten past experiences. It is a complex system of virtual grids, like the net of material communication which we are drawing in several ways on Earth, with movements and human experiences. We can add the media nets, but these are considered to be material ones because they are made through electrical connections. They become virtual when they stimulate

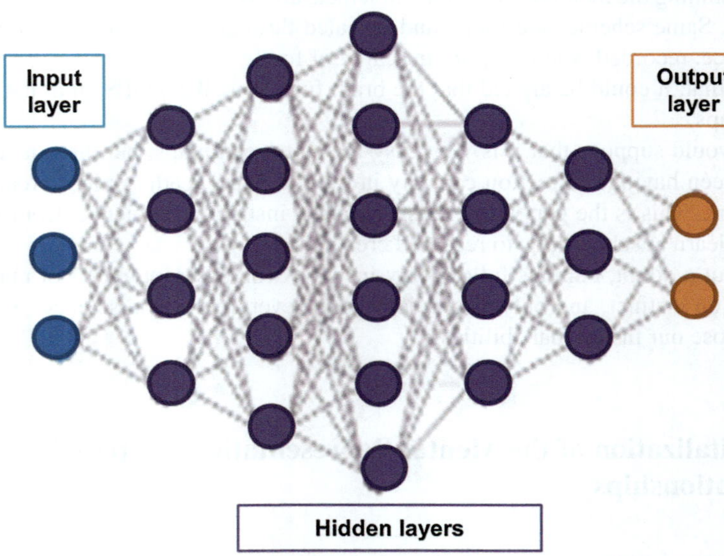

Fig. 4 Deep Neural Network. The network of highly interconnected layers—with input layer on the left, output layer on the right, and customizable hidden layers in the middle (*Source* https://www.esri.com/training)

ideas and reflections in people; only when they stimulate dialogues among people, they become real and material, even if intangible.

We could consider GIS a help in the efforts to augment the set of prospects of both human and artificial intelligence.

6 Conclusions

Body and mind are closely related, just as cognitive geography and brain are connected at another level. In essence, we never will put an end neither to the knowledge of the world nor to our brain. We must recognize that in the mind, there are many multiscale spaces at the same moment.

Exploring the world online/virtually gives us an incredible quantity of images, but they are stored in the brain in different ways, increased by the fourth—time—dimension, and the fifth—memory—dimension. Thus, we are constantly building many worlds, but we will never be satisfied. We are continually searching for other worlds and other existences. With the help of science, we move and do more, we have more goods, we ourselves have become more technically qualified, and we develop our societies. But at the same time, we are losing spontaneity, becoming more intelligent with artificial devices and less intelligent without them.

With the progress of civilization, our contact with the natural world has gradually decreased, as has our sense of involvement with natural biomes. At the same time, animals that are increasingly dependent on global anthropogenic influence continue to be subject to natural elements. Climate change also affects the existence of humans and animals; it is likely to change the usual bodily properties and familiar behaviour, and consequently, psychology.

A goal could be to merge both abilities to become lords of the universe, since people think they are superior to other species, but they are not. They are the only living species among millions of other living species, without supremacy. We are able to learn, as are animals, but we have a duty to enhance this capacity for the best life of all living creatures. And cognitive geography can provide new tools for this.

References

Bateson, G. 1999. *Steps to an ecology of mind*. University of Chicago Press. https://doi.org/10.7208/chicago/9780226924601.001.0001

Berthold, P. 1991. Genetic control of migratory behaviour in birds. *Trends in Ecology & Evolution* 6 (8): 254–257. https://doi.org/10.1016/0169-5347(91)90072-6.

Berthold, P. 2003. Genetic basis and evolutionary aspects of bird migration. *Advances in the Study of Behavior*. https://doi.org/10.1016/S0065-3454(03)33004-9.

Eckardt, M.J. 1980. The hippocampus as a cognitive map. *The Journal of Nervous and Mental Disease* 168 (3): 191–192. https://doi.org/10.1097/00005053-198003000-00018.

Fairlamb, H.L., and G. Bateson. 1979. Mind and nature: A necessary unity. *MLN* 94 (5): 1218. https://doi.org/10.2307/2906578.
Friedman, H., and D.R. Griffin. 1977. The question of animal awareness. Evolutionary continuity of mental experience. *The Journal of Wildlife Management* 41(2): 332. doi: https://doi.org/10.2307/3800619.
Golledge, R.G., R.L. Klatzky, and J.M. Loomis. 1996. *The construction of cognitive maps.* doi: https://doi.org/10.1007/978-0-585-33485-1.
Gould, S.J. 2002. The structure of evolutionary theory. *Harvard University Press.* https://doi.org/10.2307/j.ctvjsf433.
James, P.E., D. Lowenthal, and M.J. Bowden. 1976. Geographies of the mind: essays in historical geosophy in honor of John Kirtland Wright. *Geographical Review* 66 (4): 497. https://doi.org/10.2307/213673.
Jeong, C., and A. Di Rienzo. 2014. Adaptations to local environments in modern human populations. *Current Opinion in Genetics and Development*. https://doi.org/10.1016/j.gde.2014.06.011.
Leutgeb, S., et al. 2005. Place cells, spatial maps and the population code for memory. *Current Opinion in Neurobiology*. https://doi.org/10.1016/j.conb.2005.10.002.
Lewin, K. 1936. *Principles of topological psychology*. New York: McGraw-Hill. https://doi.org/10.1037/10019-000.
Linton, R. 1945. The cultural background of personality. *Appleton-Century* 54.
McLean, P. 1990. The triune brain in evolution. *The Triune Brain in Evolution.*
Montello, D., and S. Freundschuh. 2004. Cognition of geographic information. *A Research Agenda for Geographic Information Science*. https://doi.org/10.1201/9781420038330.ch3.
Packard, J.M. 2019. Wolves. *Encyclopedia of Animal Behavior. Elsevier*. https://doi.org/10.1016/B978-0-12-809633-8.90078-5.
Pánek, J. 2016. From mental maps to geoparticipation. *Cartographic Journal*. https://doi.org/10.1080/00087041.2016.1243862.
Taylor, H.A., T.T. Brunyé, and S.T. Taylor. 2008. Spatial mental representation: implications for navigation system design. *Reviews of Human Factors and Ergonomics* 4 (1): 1–40. https://doi.org/10.1518/155723408X342835.
Turner, B., and G. Bateson. 1980. Mind and nature. *Rain* 36: 11. https://doi.org/10.2307/3032188.
Tversky, B. 2005. Functional significance of visuospatial representations. In *The Cambridge handbook of visuospatial thinking*, 1–34. Cambridge University Press. doi: https://doi.org/10.1017/CBO9780511610448.002.
Zaleshina, M., A. Zaleshin, and A. Galvani. 2019. Navigational strategies in transition from initial route. In *Geospatial Challenges in the 21st Century*, eds. Koutsopoulos, K., de Miguel González, R., and K. Donert, 335–353. Cham: Springer International Publishing. doi: https://doi.org/10.1007/978-3-030-04750-4_17.

Symbolization in (of) Space. Theory and Practices

Jernej Zupančič

Abstract The article discusses the spatial symbols and the process of symbolization. The symbolization process is a process of transformation, where hidden elements become significant, recognizable and usable in some particular ways. Symbolization is a process of establishing a symbolic link between an objective matter (place, object or/and space) and the subjective nature of human perception of material things. The article has two main parts. In the first part, there is a brief presentation of the theoretical premises on the appearance of the symbolic space and about the symbolization process. Symbolic space is known in geography, indirectly even quite often used, but theoretically and empirically fairly rarely seen. Within the national space, there is at least an indicative hierarchy of spatial symbols. The hierarchy is a logical reflection of the importance of the relationship between spatial elements, and at the same time, the reflection of the spatial values of the society. All spatial-relevant identities need frameworks and reasons for identifying or identifying with certain points, zones or objects. At the same time, they are managed: most important spatial symbols are, as a rule, well-visited. The second part deals with spatial symbols in Slovenia and is based on some empirical research. They show their appearance, characteristics of the site and estimate their influence.

Keywords Spatial symbol · Spatial landscape · Symbolization · Slovenia · Cultural geography · (Spatial) identity

1 Introduction

This chapter presents and analyses the symbolic geographical elements in the landscape; their appearance, design, role and use. Symbolization is mostly a planned process in which a certain human community assigns a special (symbolic) value to a selected natural or artificial object, location or even a limited area. This changes its role and meaning and enables different uses of such objects. The most important

J. Zupančič (✉)
Faculty of Arts, Department of Geography, University of Ljubljana, Aškerčeva 2, 1000 Ljubljana, Slovenia
e-mail: jernej.zupancic@ff.uni-lj.si

M. Krevs (ed.), *Hidden Geographies*, Key Challenges in Geography,
https://doi.org/10.1007/978-3-030-74590-5_3

property of symbolic objects is a certain representativeness. Most symbolic objects have a certain symbolic meaning only for a particular community: local, regional or national. Symbolism is thus a special connection between a physical phenomenon (object) placed somewhere in a landscape, and a social community that has a specific relationship to that object. For others (who do not belong to the community), this symbolic object is just interesting, its symbolic value is recognized only indirectly, through the attitude towards the community that has chosen (or has even been created by that community) such a geographical object as the symbolic one.

The topic of symbolic objects in the landscape or society and, of course, symbolization as a process of creating symbolic elements is often underrated and considered sporadically and unsystematically, even though it offers relatively extensive opportunities for application. But the symbolization of space is one of the human needs. It is the characteristic and a quite common reaction of people in the cultural landscape. In fact, we deal with symbolic objects quite often, without specifically thinking about why they have a particular symbolic role at all. It is simply accepted, for example, that certain natural phenomena such as famous karst caves, lakes, mountains or, on the other hand, town squares with famous buildings from a certain historical period or perhaps the locations of historical events spontaneously have special significance. In this chapter, we explain that the vast majority of such objects (spots or even areas) have acquired a symbolic meaning in a completely planned way, as a result of the desired establishment of the value of selected objects. It is true that, for example, squares in cities were created during the construction and growth of the city several centuries ago, but the symbolic meaning was given to them by society only later and undoubtedly with the aim of presenting the specific city, location and building as a symbol of its community (such as a state or region) to ensure, promote and represent it. In Ljubljana, there is such a particular object, known as Cukrarna that has been renovated for at least a decade. Decades ago, the building was completely inconspicuous and unattractive to a foreign observer. However, as several Slovene writers and poets found their place there some hundred years ago and Slovene cultural history is proud of them, the building is being restored into a representative cultural building of the city of Ljubljana, because the location was recognized as symbolically important. Symbolization of this location (i.e. the establishment of the symbolic value of Cukrarna for Slovenian cultural history) is a planned activity of Slovenian society and the state (which fully finances the expensive project), because it expects a certain cultural benefit. It is important because it will represent the city and Slovenian culture as well. Many national homes amongst the Slovene diaspora in the USA are not particularly remarkable in appearance, but they are of great importance for the Slovene diaspora (and only for it!). These are spatial symbols, a material witness of Slovenian settlements (Klemenčič 2011). Some spatial symbols were even constructed explicitly for representative purposes. Today, the well-known lookout tower Vinarium near Lendava (northeast Slovenia) was built to symbolize the landscape overlooking its imposing position, height and shape.

At first glance, symbolic objects, because of their visibility and representativeness, seem to be the exact opposite of what we would like to highlight in the context of hidden geographies. However, this is not quite the case. Many natural and also built, anthropogenic objects can be completely anonymous, hidden and unrecognized for long periods of time. Only the process of symbolization as a conscious activity of a certain social community recognizes them as valuable and gradually shapes them into spatial symbols.

The chapter is divided into two main parts. In the first part, the theoretical framework of thinking about the geographical aspects of symbolic space and symbolic elements on the one hand and the process of symbolization on the other hand, is explained. The broader theoretical background is mainly in the field of social and cultural geography. Symbolic space (as well as individual symbolic objects or locations) is a part of a wide human functional space; while in some theoretical concepts it stands out as a special space. The second part of the chapter presents some practical aspects of symbolic space and the process of symbolization, in the case of Slovenia. The study was created on the basis of our research and work in a geographical seminar with students of geography in the years 2014 to 2016. Both the theoretical and empirical part is mainly the result of own research.

2 Some Theoretical Aspects of Symbolization and Spatial Symbols

This part discusses the theoretical aspects of symbolization as a process and spatial symbols as objects of observation. The aim is to explain the nature of spatial symbols (which could be, according to their size, form, role and position: the objects, places and landscapes), analyses their phenomenological character and structural features, presents locational, cultural and historical circumstances of their occurrence, and assess the rules of their formation and the possibilities of application. In the context of hidden geography, we discuss hidden objects in the cultural landscape and the process of their transformation into visible, public and important ones. Spatial symbols are those hidden elements that, according to the decisions of organized social communities, have managed to break out of geographical anonymity.

The field of symbolization and spatial symbols is specific, but quite broad. Due to the limited space, in this chapter, we limit ourselves mainly to the aspects of the creation of spatial symbols, which later will be empirically verified. The presented theoretical aspects were formed in a seminar, otherwise intended for the theory of geography. Spatial symbols were the main object of study, and symbolization was the main process. Slovenia is otherwise a small country in the European and global dimensions, so that building theoretical constructions may be risky in some regards. On the other hand, Slovenia has at its disposal a sufficient geographical diversity of cultural landscapes, which allows us to draw some conclusions of a theoretical nature possible.

Within the process of symbolization, we find out how, under the influence of the human community, the selected elements of the cultural landscape change into representative ones for space and community. Unknown or even hidden elements become public and important. Symbolization represents a planned anthropogenic intervention, by which the selected element changes physiognomically and functionally. The concept of symbolic landscape falls primarily within the field of cultural geography. At the same time, this is an example of the transformation of a lesser-known and hidden landscape into a public and representative one.

The spatial symbols are the anchors of spatial identification both for the people and territory they settled. They have a highly emotive value amongst the community members. The community builds various relations to the spatial symbols. This is manifested reflectively (the response to objects perceived), experientially (evokes intense experiences) and generatively (invokes certain reactions and thereby generates new phenomena). Amongst the most important functions of a spatial symbol is its role in territorial identification, which refers both to the space/landscape or territory and the people occupying, managing or merely using the space. The geographical objects, considered as spatial symbols are a scene of intense interaction between community and their environment. The spatial symbols are natural and artificial objects that got particular symbolic importance in the eyes of the surrounding community. This community has particular perceptions and imaginations to the material objects in the surrounding (Horton and Kraftl 2014), which later create various forms of human reaction to that objects (Levy 2008). Once the objects become symbols, they get new functions. A spatial symbol becomes a subject of management, with the measures as protection, preservation, restauration (if necessary), promotion and application (Norton 2002: 158). Many spatial symbols have become, due to their symbolic significance, attractive for tourists. Very important characteristics of spatial symbols are their geographical names (Jordan 2006).

Spatial symbols (they can be in the form of various objects, places or broader landscapes with which certain groups have a special relationship) are primarily a cultural phenomenon. Each symbolic object or even landscape is related to a particular social group or society; it is in almost any case socially-related (De Meo 2008). Social geography also provides a theoretical basis for human (social) behaviour and place-related thinking (Jones 2009). There are different interpretations of the term social space (Gill 2001; compare the explanation in: De Meo and Buleon 2005). But there are various explanations of the term landscape too (Wylie 2007). A symbolic landscape is a special part of social space in which humans realize certain functions. Social geography divides social space into three forms: residential, functional (Freytag et al. 2016) and symbolic (De Meo 2014). A residential (or living) space is defined as the location of a (permanent) residence for which a permanent or regular presence, ownership or lease relationship to the land/house or flat unit is presumed (Del Casino 2009). In practice, this is a residential neighbourhood, village or street with a rounded functional surface. Residential space is relatively small in scope; however, due to its permanent presence, a resident develops very intense and diverse interactions. A residential landscape is a home, with strong practical and emotive attachments (Holloway and Hubbard 2001). There are different perceptions that make larger or closer spatial frame of considering the term home (Hubbard and Kitchin 2010). A functional landscape is the venue for conducting various roles and realizing the needs and interests of people, such as work, education and training, supplying goods, health care, social care, spiritual care, technical and personal services, recreation, sports, entertainment, social, cultural and political life, depending on the interests and capabilities of participants and the available choice in the environment, considering the habits and technical options for reaching the locations where such functions are conducted. The functional space is multi-local and their extent is fluctuating during time (Weichhart 2009). The people are changing their occupations through the particular periods of the year, and so they change their spatial behaviour. The concept of spatial-related functions has been introduced by the Munich school of social geography (Ruppert et al. 1977) and has a long history as well by having been adopted in Central Europe and is still valued as a reasonable geographical theoretical concept (Freytag et al. 2016) adopted in applied geography (Kellner 2006). A functional space, therefore, differs from person to person, varies with seasons, and can be distributed spatially, depending on geographic circumstances, into various spatially related forms: linearly, spherically, in the form of hotspots or dispersed. Under normal circumstances, the starting point is the place of residence or simply home, while functional space fluctuates, depending on chosen functions and the places, where particular functions can be offered. Spatial symbols have a constant location, form and size; they are therefore a spatial constant in the collective perception (Zupančič 2017: 45–51). It is further to suppose that the spatial distribution of spatial symbols is at least partially dispersed.

We need to dedicate a few words to spatial symbols. The cultural landscape can be imagined as a diverse mosaic. It consists of various physical- and socio-geographical elements. Only some of them have the opportunity to become spatial symbols. We assume that these predispositions are mainly in the properties of elements such as visibility, representativeness and reachability. All three properties are a necessary precondition for the creation of spatial symbols, they appear together and are connected.

Visibility is a property that derives from the importance of visual effects on people's perception. We want to see objects. People believe in well-visible elements; the closer they are, the more they like them. The visible effect triggers a series of associations, performances and emotions in the observer. In particular, those elements that evoke feelings of comfort, pleasantness, pride and satisfaction are acceptable to the community. A spatial symbol has to express a certain degree of peculiarity or uniqueness regarding its form, aesthetic virtue, besides the content (for example: cultural and historical significance and character). In some cases, however, elements that trigger negative emotions also appear symbolic. Such are, in particular, some monuments of historical heritage, reminiscent of, for example, tragic events from the past. The visibility criterion requires a certain shape, which distinguishes this element from the others on the one hand, and exposes it due to its characteristic features on the other hand. Regardless of whether the selected element is exceptional or typical, this spatial symbol should first enable the identification and then the identification of the community with it. Sometimes people idealize the objective picture: they see it in a better light than it really is. Imagination of particular spatial elements or places and idealization became an important topic within modern post-structural geographies (Murdoch 2006). Not every object is necessarily well visible a priori. It just needs to have the potential to become visible. In any case, it must be visible after the establishment of this object as a spatial symbol. Objects that the community has established as symbolically important will, as a rule, be made more visible by various measures. Making the spatial symbols (more) visible is a typical and necessary activity during the process of symbolization.

The main expectable characteristic of the spatial symbol is representability. This is an objective ability to represent either a typical or exceptional piece of landscape (then, it is representing a broader expanded typology of a distant landscape, or some natural or/and cultural phenomenon) or the community that is behind the phenomenon (for example, they are owners, or users or holders of some tradition, historical/cultural/political notion), or they are heirs of the community from the past. In order for an object to become representative, the surrounding community must recognize it as appropriate. It is essential that a community recognizes, interprets, and mentally adopts an object as its own thereby giving it a high degree of emotiveness (Chivallon 2008: 23–29). This emotive interaction is relatively permanent and long-term, as well as consensual due to its collective adoption. Representativeness is primarily meant for the content or notion. The chosen element must be what it represents more broadly. For example, the karst landscapes can only be represented by a selected karst object, such as an underground cave or another phenomenon that can be considered as characteristic or exceptional one. Historical events and circumstances can be represented primarily by an original object or place from that time, which is in certain ways connected with (our) ancestors and space. Only in this case will the current community consider such an object as its historical and cultural heritage and will be ready to assign it a special—symbolic—status and meaning. Time is also considered when discussing the content category. Older constructed objects have very often an advantage in human perception. The old vineyard cottage will probably be the undisputed favourite of the representativeness of the home wine-growing area. Furthermore, a certain category of representativeness is also a certain collective character and the collective use of the selected object. In the collective memory of the community, those objects that were the scene of meetings and events with greater participation of community members or their ancestors, are better considered. Even better if such facilities are meeting places even recently. It is, therefore, not surprising that the objects such as sacral objects (churches), cultural institutions (theatre, opera), national homes or town halls acquire symbolic significance. Some are already officially called representative objects, such as the seats of parliament, ministries and public (state, regional) services.

Accessibility is a category that highlights primarily a geographically relative location. In this sense, spatial symbols are those objects or places that can be reached by community members. Reachability needs to be understood dynamically. An object does not have to be well accessible a priori, but it certainly has to be reachable a posteriori, that is, after it has been assigned a certain symbolic role. It is therefore important that such an object can be made accessible in a certain way. Particularly for some monuments of historical significance, they were poorly accessible or inaccessible at the time of anonymity (which helped to keep them hidden), but when they became, due to their content, important, the access to them was also arranged. In some circumstances, such as in post-crisis areas, for example, the spatial symbols compete for the most visible and attractive location: all in order to demonstrate exactly our own spatial symbols (Zupančič 2015a, b).

Additionally, there is the need to discuss the process of symbolization, which is a sequence of processes in which a hidden and unknown object is transformed into a spatial symbol. The spatial symbol is selected by the community. Until the organized community selects the object and declares it important and symbolic, the element may actually be hidden, unknown, or at least disregarded. Recognition by the community gives it a certain public value and thus ensures its transformation. This process is generally planned and deliberate, since spatial symbols have a great significance. A certain time is required for the community to accept them as symbols and identify with them. The steps (phases) shown here are adapted to the situation at the local and regional level. There, due to its proximity, people can come into contact with the objects easily, and therefore, quite often. Community members can gain personal experience with an object that can become later a spatial symbol. At the level of wider territorial units and at the national level, the principles are the same, but they are implemented in an adapted manner. Above all, the role of institutions (especially schools and the media) is much greater, especially when people need to be taught about the role and meaning of individual symbols. Only a sufficiently broad knowledge of even distant objects allows people to accept them as national spatial symbols.

Symbolisation is performed by the following steps or stages:

- experience and selection (of objects with appropriate prerequisites),
- identification of content and significance,
- interpretation of content and determination of relevance,
- representation.

In some cases, these phases are followed possibly by additional phases:

- imagination and
- mystification (Zupančič 2017).

Experience and Selection When selecting objects, it is necessary to assess whether they have the content important enough for symbolic meaning, if it is possible to arrange the appropriate access and also to prepare the form so that the object will be visually arranged, safe and attractive. Content is the key; accessibility and visual appearance and tidiness can be adjusted accordingly later. The community assumes the importance of particular objects by rethinking the experiences they have with the object during a longer period. Experiences are very important in the initial phase. It is often a particular tradition that reinforces positive experiences amongst community members. The selected objects are normally relatively common (especially for local and regional sites), have a representative value, and are recognized by the community as typical or extraordinary. Further, if the location is not accessible well, it might be made accessible later. Even better, if selected objects are in accessible locations. Objects of historical, cultural and political importance are, as a rule, subject to many compromises and long-lasting selection. It is not entirely clear for national spatial symbols (at least in the Slovenian case) when they really became this. During selection, the community tentatively recognizes an object as important, distinctive, widespread and visible and estimates that it has a good chance of being a spatial symbol of the community in a particular field.

Identification of Content and Significance It is the stage in which already selected elements are reasonably found, explored, determined or even created content and defined their role and significance; all this in relation to the community. This places the selected object in a specific area for which the belief in the community is somehow important. It is equally important to disseminate knowledge and information about these objects amongst community members to the extent that they are accepted by the community as subjectively and objectively important and representative objects, so that people can identify with them. At the local level, knowledge is disseminated through different social opportunities, such as events and social networks, and the school and media can also play an important role.

Interpretation of Content and Determination of Relevance This is primarily a phase of creating an in-depth explanation of the selected spatial symbols. Normally, this is a step for professional elaboration, where suggested spatial elements must be analyzed in detail. They have to explain, why exactly that object represents an important issue for the community and for the territory. The interpretation follows the specific needs of the community, so it is tailored and carefully articulated. Sometimes, the perceptions of usual people could have more impacts to the decision makers than the efforts of professionals. Some objects are already known and accepted amongst community members, so the people consider them as spatial symbols. Here, the folktales, myths and similar can play a reasonable role, regardless of the truth about these objects. That is why the professional elaboration and interpretation could be demanding in some regards. If people want to pledge elements chosen by their popular perception as spatial symbols (because they probably have high emotions to the object), it is difficult to stop them. Once the object is institutionally proclaimed as a spatial symbol, the community begins to make closer ties to our symbol. At this stage, objects can also be ranked by their/the importance. National significance is achieved only by those spatial symbols about which the state or nation achieves wide knowledge and is actually accepted amongst people.

(Re)presentation This is the stage of putting the selected and proclaimed spatial symbol object to the practice. In the broadest sense, it is the use of spatial symbols for different (re)presentational purposes. This stage marks a turning point when a previously hidden object becomes a public, known, accepted, and in many ways an important spatial symbol. The community begins to manage and manipulate these spatial symbols, and the objects themselves undergo various interventions. Some are subject to various adaptations in order to be as accessible and attractive as possible and thus their content (meaning) would have a better effect. Others receive conservation: an intervention that protects the authentic form and looks in the long run and prevents the possibility of changes and, above all, their damage. However (especially when it comes to monuments of cultural and historical significance), conservation is a very demanding form of social intervention into space. The spatial symbol becomes a factor. The selected objects get an economic value, become marketed and are also economically ranked. But the dominant function is the representation of the region or the community. Some objects of this type become local, regional and national brands; branding is a typical function of spatial symbols.

Imagination This is a stage reached by only rare symbolic objects, when additional attributes begin to be spontaneously or intentionally attributed to them, which are not experientially affirmed, but the community creates a form of idealized beliefs about them. These are subject to some natural monuments (which are usually the result of flawed research and behaviour or are carried over from earlier periods) and some historical monuments (to which imaginary explanations add more fame and sense). This category also includes objects that are intentionally prepared for a representative purpose. They are used specifically in an urban space when urban spatial identity is sought or proven. In fact, it is a literal production of space by which the community plans in advance to demonstrate its ideas, economic power, or ideology. This is a marking of space, a demonstrative act of proving of ourness of particular object or place, and exposed to the observers.

Mystification It is a phase reserved for rare symbolic elements: to those most recognized and known ones. These objects gain sublime importance. It is usually tied to local and national mythologies and can lead to the mythologization of this symbolic object. The interpretation is adapted and the reality seems to be less important and the excess of the chosen symbolic object is emphasized. However, mystification does not diminish the rational meaning of the spatial symbol. It only adds an additional value and meaning dimension to the object. In national mythologies, selected spots and objects are a kind of material infrastructure, to them the myths can be placed. Myths need a rational space of their position and presentation, make it truly ours. Mythologized spatial symbols usually increase their influence, scope and meaning.

3 Sources, Methods and Working Attempt

Before presenting some empirical aspects of the study on the process of symbolization and spatial symbols, a brief explanation of sources, work steps, and methods follows. We estimate that qualitative approaches are especially suitable for the study of spatial symbols. Quantitative analyses can be helpful too, but cannot have a primary role. The basic material for the presented results on spatial symbols in Slovenia (in the next section) was the material obtained in the seminar of geography students. The topic of this seminar for three consecutive years (2014–2016) were spatial symbols at different levels. The seminar was not prepared in advance for this study. The author of this text systematically elaborated, edited and analyzed the given results (reports, comments and minutes of the seminar). Totally 89 students participated with their contributions, processed 12 statistical regions and 136 municipalities in Slovenia. The contributions at least briefly mentioned 632 different objects with a certain symbolic meaning. In addition, we used the mental map method in one group of these students (N = 23; in 2016) just before the start of the seminar. The students of the 4th year of study already had some knowledge about the spatial situation of Slovenia, so a relatively systematic and objectified approach was expected when calling for up to 20 different spatial symbols in Slovenia, which best represent the Slovenian state and Slovene nation. (These statements are an integral part of the aforementioned 632 objects mentioned before). This allowed us a certain quantitative approach. We expected students to mention spatial symbols according to their acquired geographical knowledge and less according to their regional preferences and experiences. This attempt can largely be applied to further research in similar studies (compare to Jackson 2001). Due to the low number of participants, we can only consider the results obtained as indicative, and cannot make any broader generalizations on our findings.

The task of the participating students was to find relevant interlocutors for certain statistical regions and municipalities, who can be called opinion leaders (such as local representatives, leaders of institutions and politicians). They were presumed to know their environment well because of their job, field of work, experiences and knowledge. For each unit, they had to list at least five objects of a symbolic nature and also give an interpretation of why they considered them to be spatial symbols. In addition to this collected statement, the students had to give their (geographical) opinion and explanation of the identified spatial symbols. Then they ranked them and presented a brief argument about the symbolic significance of each chosen spatial symbol. They had to determine (at least with a rough estimate) the permanence and intensity of experience of the symbolic significance of these symbols. The fourth obligation was to describe any recent and possible use of selected spatial symbols.

In this way, we got a relatively transparent overview for most of Slovenia about how local opinion leaders define spatial symbols in their domestic and well-known environment and which objects are important for the regional and national level. In the latter, the quantitative approach was more appropriate at least for ranking elements at the state level. At the level of local communities, the ranking has—as it turned out—slightly different rules. For spatial symbols of national importance, it was possible indirectly (by analyzing the papers from the seminar) to partially explain the characteristics of the selected objects. Of course, we also used other sources of information about these spatial symbols, too.

4 Symbolization and Symbol Places in Practice. The Case of Slovenia

In the empirical section, we present spatial symbols within Slovenia. The way how we got the list of spatial symbols, their interpretation and ranking is described in the previous section. We, therefore, consider that the results are comparable across Slovenia, even though the subject is very subjective. As mentioned, there is a reasonable difference in principles, why and how the local opinion leaders consider the spatial symbols in their local environment on one hand, and how they explain and rank the spatial symbols on the national level.

Lets see the spatial symbols on the national level first. Four objects stood out prominently at national level: Triglav, Bled, Postojna Cave (Postojnska jama) and Karst Polje (Lake) of Cerknica (Cerkniško jezero). All are natural phenomena that acquired a symbolic significance with very intentional and long-lasting symbolization, when these otherwise recognizable phenomena (due to their aesthetic value as a tourist attraction) were adopted as national symbols, apparently with the intention of positioning Slovenia and Slovenians in a specific macro-regional European context. This is a case of fairly predictable national marketing with symbols. Amongst them, the undisputed champion is Triglav, the highest mountain peak in Slovenia. As a stylized logo, it appears on national symbols, its name is used by schools, societies and companies, and it also appears on the national coat of arms and flag. No other element in the Slovenian landscape has a comparable standing and presence. Triglav is now by far the most emblematic Slovenian toponym. This name is contained in schools, sports, educational, political and cultural clubs and organizations all around the world, wherever people of Slovenian descent have settled (Zupancic 2015a, b: 90–94). Furthermore, these are enterprises and organizations at the national level. Triglav is also the most visited Slovenian mountain; the role of the national symbol is, therefore, a factor in promoting the flow of tourism. Every Slovene should visit it, in order to show a national belonging! The symbolic significance of Triglav is interpreted through the Alpine character of the country and the people. Slovenians want Slovenia to be regarded as an Alpine nation, as this places them in the macro-regional context of Alpine countries and thus Central Europe, which is a constant geopolitical motif in Slovenian politics. The Alpine character represents a certain prestige, aesthetic element, strength, determination and other similar characteristics. Even though only a small part of the national territory is truly mountainous (Alpine) in terms of mountain landscape, the country is establishing itself as an Alpine country. But Triglav is a national spatial symbol (as a figure) and symbolic landscape (as a characteristic image) to the same extent that Slovenes have settled in geographically real alpine landscape (in the north-western part of the country) as well as for those living in the Pannonian lowland (in the north-eastern part of the country) or somewhere amongst the hilly countryside of Dolenjska (to the south-east). Triglav, originally the highest mountain peak within Slovenia became a generalized national symbolic object (could be considered as a spot or even landscape). The second symbolic object, Bled, stands out with its Alpine features. A large glacier lake with an island, church and castle, surrounded by the Julian Alps, which together with Triglav, form an Alpine duo, further reinforcing the position of Slovenia in its desired regional European context. In the case of Bled, its aesthetic features stand out even more, thus placing both symbols in the context of the most representative and exclusive of Alpine landscapes. It is essential that both symbols have a seemingly greater typical Alpine background (Figs. 1, 2, 3 and 4).

The third and fourth national symbolic landscapes are Postojna Cave and Lake Cerknica, two karst phenomena that owe their representational attributes to the uniqueness of their phenomena. Postojna Cave is one of the most visited tourist caves in the world, renowned for the beauty of its speleothem formations, with the large number of tourists placing Postojna Cave—and thus Slovenia—on the global

Fig. 1 Triglav, the Slovenian highest peak (2864 m), counts to the most respectable sites amongst Slovenes. "Every true Slovene should visit it!" This mountain is the most emblematic Slovenian symbolic landscape and a national symbol. Stylized, it is used on the coat of arms and national flag (Photo: J. Zupančič)

Fig. 2 Vineyard cottages are typical of the Slovene vineyard landscape. In Slovenia, they are often transformed into weekend houses. Due to their local symbolic significance, they maintain an idyllic old form and represent a hidden economic potential (Photo: J. Zupančič)

Fig. 3 The artificial Bukovnica Lake acquired symbolic value according to the plan in order to complete the tourist offer of Dobrovnik municipality. The site became known by springs and energy points along the lake and are attributed to mystical miraculous power (Photo: J. Zupančič)

Fig. 4 Old castles with historical significance have a very good chance of becoming spatial symbols. In addition to its excellent location (very close to the world-famous Postojna Cave), Predjama Castle also has a mysterious hidden story about a knight, Erazem of Predjama, and his adventures. On this basis, several mediaeval events take place in front of the castle (Photo: J. Zupančič)

tourist map. The karst character of the country is interpreted only in parallel. Lake Cerknica is a global phenomenon primarily due to the karst hydrographical system. And because the phenomenon is scientifically really well explained!

Among spatial symbols of national importance, the anthropogenic creations, the cities of Piran and Ljubljana, only come in fifth and sixth place. Piran, a well-preserved Mediterranean-styled city, located at the head of a small peninsula in the North Adriatic, is placed on the symbolic identity map as an aesthetic element of the coastal Mediterranean world, which Slovenia wants to be a part of. The city itself otherwise has no historical reason for the symbolism of Slovenian territory or ethnic character (until the mid-1950s, the city was populated mainly by people of an Italian linguistic and cultural descent). Despite old urban traditions and a typical coastal economy, it remained a small city with practically no hinterland. Koper, for example, another Slovenian coastal city, and now one of Slovenia's vital urban cores, and a truly Slovenian maritime window with the exposed port with links to the vast hinterland deep inside Central Europe (Austria, Southern Germany, the Czech Republic, part of Hungary and Slovakia and even Poland, besides Slovenia itself of course) (Klemenčič and Zupančič 2016), plays almost no real symbolic role, despite undoubtedly its great economic and strategic importance! On the other hand, Ljubljana, the capital of Slovenia, without any doubt holds such a role. Ljubljana is demographically the most important centre, with its politically dominant role and its cultural standing. Thanks to its central geographical location at a crossroads, the centrality of the city within Slovenian national space is understood as a logical consequence. In its simplified landscape, it is usually presented as an image with the mediaeval part of the town with the hill and castle, the latter also included in the Ljubljana coat of arms. The city as a whole (symbolic landscape) is reduced (and simplified) to its smaller, most attractive part or even just a simple image of the Castle Hill with its castle (spatial symbol). This reduction of a typical image is a very common phenomenon during the process of symbolization. Furthermore, Slovenia's national spatial symbols (as well as places, landscapes) are largely accepted by Slovene communities abroad (Klemenčič and Zupančič 2016), even though these people have practically hardly any real direct experience with these places.

The regional response to national symbolic landscapes of national character is also interesting. Triglav and Bled are ubiquitously recognized as the most important Slovenian symbolic landscapes in all regions, with the interpretation of the Alpine character and representational characteristics being accepted by them throughout the country. Postojna Cave is recognized as a nationally important symbol in western and central parts of Slovenia, and less so in the eastern parts of Slovenia. Lake Cerknica is recognized in all parts of the country equally, but, it seems, primarily by highly educated respondents. A similar attitude is noticeable with Ljubljana, where it is recognized most in the Central Slovenian region, whereas in western and eastern parts of Slovenia it is recognized as a national landscape to a similar extent only by highly educated respondents. According to respondents (even small numbers), when dealing in terms of mental maps, the spatial symbols appear to be bigger and more visible than they really are. This can be explained through the effect of visual reflection of important spatial symbols, but probably also as a learned reaction:

all important locations on the geographic map are drawn bigger because they are important!

The local and regional levels include spatial symbols with highlighted representativeness. They originate from frequent, widespread, and therefore typical landscapes, with respondents describing them as close, local and useful, in addition to the general conditions (accessibility, acceptability, aesthetic elements and recognisability). It seems that experience and the perspective of utilization of chosen symbols, which often find their place in marketing for local and regional products, played an important role in the symbolization process for these elements. These are actually empirical landscapes, which people recognize as appropriate to present a typical and thus recognizable image; so a phenomenon as such. Local and regional spatial symbols reflect, to a relatively high degree, a specific geographic reality. Stereotypes are therefore a summary of frequent experience, intended for the recognition amongst the competition within the country and, to a lesser extent, internationally. The spatial symbols found during the mentioned seminar at a local/regional level can be categorized in four groups according to their motif:

- historical and architectural objects,
- typical regional animals and plants,
- typical regional landscapes,
- local/regional products.

Historical and architectural characteristics are attractive due to their uniqueness and simple recognisability in the immediate and wider environment. Even better if they can be accompanied by sufficiently imposing buildings or objects. Historical and cultural monuments require appropriate interpretation, which can be a divisive issue, especially when it comes to elements taken from World War II, since various ideologies and memories are associated with it. Spatial symbols of this type are often subject to more diligent interventions, reconstructions and even the revival of specific folklore, traditions or specific events. These landscapes and related events are subject to very creative approaches. A lot of them were hidden in the near past, and then, after some improvement, explained as symbolically important. Among them, however, are also those that were quite favoured and known in the recent past, but are now sinking into oblivion. They are becoming hidden spatial symbols (for example: Baza 20 near Kočevje, a memorial site from Slovene resistance during World War II).

The next group is (as well as the fourth) a particular one. The motifs of typical plants, animals or natural phenomena are a relatively common occurrence, even though they lack emotional intensity. Birch trees in Bela Krajina are a recognizable spatial phenomenon, similar to the black pine of the Karst region, the chamois in Alpine regions or storks in the Pannonian Basin. They give the impression that these elements act more often as regional and not local symbols. These objects were mentioned relatively frequently by respondents, although in reality they cannot really be considered as spatial symbols. However, we mention them because they have appeared very often at the local and regional level. On the other hand, they can be assigned all three key properties of spatial symbols: visuality, representativeness, and location.

The third characteristic group of spatial symbols is represented by characteristic landscapes. Landscape is presented as a typical image, which a wider environment can identify with and use to represent itself. The selection of such landscapes is a sort of compromise. The selected elements do not stand out as special, but as characteristic—and there are many such elements. The selection is made easier by elements of general accessibility, aesthetic elements and availability. Characteristic examples are the cultivated terraces with orchards and vineyards at Brda, olive groves in the Slovenian Istria, stone architecture with vineyards in the Karst region, floating mills on the Mura River, vineyards with wind rattles in the Slovenian Hills, fields in Prekmurje or the wetland fields and meadows on the Ljubljana Marshes. An important imperative is the distinction between neighbouring (different) regions. Inhabitants often have a protective attitude to such regions/objects. Inclinations for various types of protectionism and an intentional search for an old landscape (which is supposedly more authentic) can, in some cases, not only preserve but also reconstruct or recreate such characteristic landscapes.

The fourth group of spatially related symbols includes characteristic local products, including cuisine. Even though they do not directly represent a spatial image (landscape) or the region, they are strongly and fundamentally associated with the landscape—they are its product and surrogate at the same time, simplifying the representative characteristics of a region and its people. The visual effect is accompanied by usability, giving this element a higher representative effectiveness. Local and regional products (such as locally or regionally specific food products—prosciutto and *kraški teran* (regional wine) in the Karst region, *cviček* (regional wine) in Lower Carniola, *gibanica* (typical, very sweet dessert from Prekmurje, north-eastern Slovenia), doughnuts from Trojane and some cheeses from Bohinj or traditional industrial or craft products (Idrija lace, wooden dry goods (wooden accessories for the kitchen) from Ribnica and pottery from Prekmurje). Links to industrial companies are rarer (Krka, pharmaceutical company, Novo Mesto, Alpina sportswear, Žiri). The greater popularity of traditional craft products can be explained as a function of time and general usability. The product must therefore have local significance (exposure, primary), must have been produced over a long period of time and must have general usability outside its region. The latter is essential since the recognisability of a local symbol is reflective: it refers to the possibility that an object (symbol) can be recognized in a wider and not only local environment.

5 Conclusions

In the age of visual communication, the image of any geographic phenomenon can be more easily transferred, and thus represents a simplified access to the target audience. Spatial symbols quickly become a part of regional marketing, aimed at improving recognisability and forming its identity. Spatial symbols and landscapes are carefully chosen, are not abstract, but require an explicit and recognizable form and interpretation of content (meaning), which, together, enable a community to be identified.

Spatial symbols are anchors for identity, but require a certain degree of usability. When putting in practice, these geographical objects began a new life. Formerly hidden stories become consumed by people, who differ them according to their needs.

The symbolization process is a process of transformation, where hidden elements become significant, recognizable and usable in some particular ways, from a tourist attraction to the specific representative role. The process of symbolization is similar for symbols of national and local significance. However, there are greater differences in the selection of objects and their later use. The empirical study has shown the following: (a) the determination of national spatial symbols is conducted differently than when determining symbols at local level; (b) national spatial symbols are mostly adopted and interpreted, whereas local symbols are formed and selected; (c) the intensity of an emotive relationship to national symbols is relatively permanent, whereas it is more variable and adaptable to local symbols; (d) usability is very important for local symbols, but not so for national symbols, becoming important indirectly through visiting such locations.

There are a few more things to note. The symbolic objects shown in the second and fourth category are very well known amongst Slovenes, most of all at the local and regional level. For many visitors, however, they are much less known or are practically hidden. This is a real example of the dichotomy of hidden elements of the cultural landscape for visitors and well-known representative and public symbolic objects for the local population. This dichotomy of secrecy on the one hand and the public on the other hand, of course, has its own rational logic, despite the fact that, for example, local products are very often used as local markers of society and space. In some cases, however, these objects even appear in branding, which proves their symbolic character and significance.

References

Chivallon, C. 2008. L'espace, le reel et l'imaginaire: a-t-on encore besoin de la géographie culturelle? *Annales de Géographie* 660–661: 67–89.

De Meo, G. 2014. *Introduction á la géographie sociale*. Paris: Armand Colin.

De Meo, G., and P. Buléon. 2005. *L'espace social. Lecture géographique des sociétés*. Paris: Armand Collin.

Del Casino, V.J. 2009. *Social geography*. Chicester: Wiley-Blackwell.

De Meo, G. 2008. La géographie culturelle: quelle aproche sociale? Annales de Géographie, No. 660-661, 47–66.

Freytag, T., et al. 2016. *Humangeographie kompakt*. Berlin, Heidelberg: Springer.

Gill, V. 2001. Social geographies: Space and society, New York, Prentice hill.

Holloway, L., and P. Hubbard. 2001. *People and place. The extraordinary geography of everyday life*. Harlow: Pearson Education Limited.

Horton, J., and P. Kraftl. 2014. *Cultural geographies. An introduction*. New York: Routledge.

Hubbard, P., and R. Kitchin (eds.). 2010. *Key thinkers on space and place*. London: Sage Publications.

Jackson, P. 2001. *Maps of meaning: An introduction to cultural geography*. London, New York: Routledge.
Jones, M. 2009. Phase space: Geography, relational thinking, and beyond. *Progress in Human Geography* 31 (1): 487–506.
Jordan, P. 2006. Anmerkungen zum Konzept einer Sprachengeographie. *Dela* 25: 73–86.
Kellner, K. 2006. Management consulting. Ein neuer Weg in der Angewandten Sozial- und Wirtschaftsgeographie – Legoland Deutschland. *Dela* 25: 97–108.
Klemenčič, M. 2011. *Zgodovina skupnosti slovenskih Američanov v Pueblu, Kolorado*. Maribor – Ljubljana: Univerza v Mariboru & Inštitut za narodnostna vprašanja.
Klemenčič, M., and J. Zupančič. 2016. Slovene minority issues. *Journal of Geography, Politics and Society* 6 (3): 5–17.
Levy, J. 2008. La géographie culturellea-t-elle sense? *Annales de Géographie* 660–661: 28–46.
Murdoch, J. 2006. *Post-structuralist geography*. London: Sage Publications.
Norton, W. 2002. Explaining landscape change: Group identity and behavior. *The Behavior Analyst Today* 3 (2): 155–160.
Ruppert, Karl, et al. 1977. *Sozialgeographie*. Braunschweig: Georg Westremann Verlag.
Weichhart, P. 2009. Multilokalität Konzepte. Theoriebezüge und Forschungsfragen. *Informationen zur Raumentwicklung* 1/2: 1–14.
Wylie, J. 2007. *Landscape*. London, New York: Routledge.
Zupančič, J. 2015a. Crisis urbanization and empty spaces: The spatial aspects of ethnic conflicts. In *Ethnic conflicts and global interventions (Terrorism, Hot Spots and Conflict-related Issues)*, ed. Maureen Wells, 21–39. New York: Nova Science Publishers.
Zupančič, J. 2015b. Spatial structure of Slovene ethnic settlements in the USA: The role and sense of spatiality among diaspora members. *Studia Historica Slovenica* 15 (1): 87–104.
Zupančič, J. 2017. *Socialna geografija. Človek, prostor in čas. (Social Geography. The Man, Space and Time.)* Ljubljana: Znanstvena založba Filozofske fakultete.

Geographies of the Sacred: Mystical Literature as a Source of Knowledge

Gianfranco Battisti

Abstract The reconstruction of the landscapes of the past, a classical theme of historical geography, is a typical case of concealed geography, intrinsically hidden from the view of the contemporary scholar. His work meets with increasing difficulties as time passes, considering the inevitable transformations that have occurred in the territory under study. The Holy Land, i.e. the land where the biblical events took place, is a typical example of this. Due to the scarcity of traditional geographical, cartographic and bibliographical sources, it is nowadays to archaeological research that we owe the most significant contributions, although nothing will ever replace coeval testimonials. This work studies the possibility of resorting to mystical literature as a source of knowledge. It is a grey area comprised between fiction and popular piety. After framing the question in the two main structures of religion and agnosticism, we shall summarize the stance of the Catholic Church, concentrating on the New Testament. We shall in particular compare the written works attributed to A. K. Emmerich and M. Valtorta, which provide us with a wealth of landscape descriptions.

Keywords Historical geography · Holy Land · Geography of religion · Mystical visions · Catherine Emmerich · Maria Valtorta

1 Foreword

Talking about hidden geographies means facing realities beyond the ordinary perception sphere, which is veiled with obstacles of various nature. They are real barriers, which may be either physical and/or communicative and can be found either in space or in time—hence their interest to the geographer. A famous book reminds us that "the past is a foreign country" (Lowenthal 1985), and as such yet to be explored. Historical geography is per se concealed not only from our experience but, to a large extent, also from our knowledge.

G. Battisti (✉)
Department of Humanistic Studies, University of Trieste, Trieste, Italy
e-mail: gbattisti@units.it

M. Krevs (ed.), *Hidden Geographies*, Key Challenges in Geography,
https://doi.org/10.1007/978-3-030-74590-5_4

Traces of the past are scarce, being the result of an endless selection which often takes place beyond our will and even our awareness. Direct traces are increasingly scant the further we go back in time. We can generally see the remains of buildings and other changes in the landscape, even of a destructive nature, brought about by man's activities (Brunhes 1910). As for nature, it is subject to geomorphological processes which can often cause radical changes. This makes it difficult for scholars to reconstruct the landscape patterns of the past and determine the location of historical events. See as an example Vinci (2005), who naively tried to identify in Scandinavia the true site of ancient Troy on the basis of linguistic and geomorphological analogies with Homer's descriptions.[1]

When material testimonies are either unavailable or have been lost we can, however, sometimes partially rely on written testimonies, especially of contemporaries, as the main historians of classical Greece already knew. Thucydides, Herodotus and Polybius are the ones who "more than any others identified in the degree of autopsy the degree of historical accuracy" (Neci 1994: 35).[2] As a consequence, the stories they narrate are "essentially about facts contemporary to their authors" (ibid.), of which they are, at least partially, eyewitnesses.

This criterion also applies to the information on territories contained in their works, which have been considered so important in historical geography as well that some count these authors among the scholars of this discipline. The question of sources, of their availability and reliability therefore unites history and geography to such an extent that it endorses the opinion of those who theorize the existence of just one discipline—geohistory.

2 An Unconventional Source

As far as historical geography is concerned, the first scholar who investigated the landscapes of the classical era was Philipp Cluever in the seventeenth century (Cluveri 1619). In this work, which is far from being over, even before archaeological finds and geological and pedological surveys we resort to the archival sources. In the second half of the nineteenth century, Schliemann went in search of Troy on the basis of the Iliad. Although his historical conclusions have been debunked, recent geological prospections seem to support the morphological indications of the site contained in the poem (Kraft et al. 1982, 2002).

[1] This without considering the alterations suffered by the area due to bradyseisms during the approximately 3,300 years separating us from these events. For the recent discovery in Sweden of a Mesolithic settlement submerged by the sea, see Petrone (2016).

[2] It is a typical requirement of the Greek world, which was unknown, e.g. to the Jewish one.

As we can see, the rediscovery of literary works as "mines" of geographical information that started to take place in the 1970s represents the development of seeds that were planted several centuries ago.[3]

To the modern humanistic geographer (Pocock 1981), literature takes on a particular value as it brings his attention to those facts and relations of the human world that he might ignore (Lando 1993: 3). This observation appears very fitting with reference to that niche production known under the name of mystical literature. A peculiar category, little known and even less studied, it is often classified among the works of fantasy, while for some people it is an object of fideistic veneration that defies the arguments of science. It includes the writings left to us by the so-called *mystics*, men and women present in every epoch that somehow surpassed natural limits, coming to contemplate the dimension of the sacred.[4]

According to Tuan (1976) and Seamon (1976), the mystic is a mediator between natural and supernatural reality. Similarly, the geographer comes to be a sort of "intellectual mediator between literature and science; the one who breaks down artistic experiences into thematic issues that can be scientifically treated and investigated" (Lando 1993: 3). In fact, the humanistic approach tends to bring out the particular relationship, always of a spiritual nature, that is established between man and the places where he lives fundamental experiences. Crucial is the personal experience which often assumes the character of a collective phenomenon. Hence, a sort of mysticism of places which, however, even if partly linked to natural religiosity, leads to paths far from the literature we are examining.

3 The Universe of Mystics

Mystical literature faces a double judgement (which is often also a bias) from both religious and scientific authorities. The latter is rightly skeptical against revelations coming through ways not verifiable using the traditional tools, regardless of their origin. And alongside the religious one, there is a vast profane literature that revolves around communication with spirits.

Over the past three centuries attempts have been made several times to develop it as a disciplinary corpus, thanks to authors such as the Swedish scientist Emanuel Swedenborg (1688–1772) and the French pedagogist Denizard Rivail (1804–1869),

[3] Do not forget that on the basis of a Strabo's passage several authors indicated in Homer the origin of Greek geography.

[4] The 32nd Italian Geographical Congress (Rome 2017) hosted a session dedicated to the theme. A working group of the Association of Italian Geographers (AGEI) was subsequently formed (https://www.ageiweb.it/gruppi-di-lavoro/geografie-del-sacro-nuove-prospettive-per-la-ricerca-geografica-proposta-di-istituzione-di-un-gruppo-di-lavoro-agei-coordinamento-prof-gianfranco-battisti/). For Italian studies on geography of religion, see Galliano (2002, 2003).

better known under the pseudonym of Allan Kardec, until the birth of parapsychology,[5] promoted by the American botanist Joseph B. Rhine (1895–1980). The scientific community continues, however, to reject the existence of a paranormal phenomenology in the absence of objectively verifiable results.

Upon careful analysis, the information collected using mediums and psychics generally tends to go beyond the scientific domain to invade the proper sphere of religions, which they try to integrate[6] or even supplant.[7]

F. M. Dermine offers us an in-depth examination of mystical experiences, both in Christian and non-Christian fields, focusing on the methods and purposes. In the first case, the experiences are characterized by spontaneity, fleetingness and immediaty. "Not being the fruit of imagination or research the phrases and visions come abitually in a complete unpredictable way, preventing every thought and every possible cooperation, including prayer" (Dermine 2003: 327). Basically, it is never man who takes the initiative. As for the contents, there is "the primacy of unitive love over knowledge and presence over experience" (ibid.).

Generally, "not even in prophetic phenomena can we find the primacy of the knowledge of certain things that are future or hidden" (ibid.: 458). "It is therefore not a question of intellectual communication but a vital process, in which God approaches man; in this process it is possible for contents affecting the intellect and the understanding the mistery of God to manifest themselves" (ibid.: 457). It is to these contents, often marginal for the theologian, that the present note is addressed. "In non-Christian (or natural) mystical phenomenology," on the contrary, "the primacy of knowledge seems to undergo some flaws in Judaism and Sufism" (ibid.: 467).

4 The Position of the Magisterium

The Roman Catholich Church[8] appears particularly well equipped to investigate the subject: systematic and orderly documentation concerning supernatural events began with the Council of Trent (1545–1563). It recognizes that "the charism of prophecy is given to the Church at all times; it must be examined but it also cannot be despised. In this regard, it should be borne in mind that in the biblical sensory prophecy does not mean predicting the future but explaining God's will for the present and therefore it shows the path towards the future" (Ratzinger 2000).

[5] Despite the opposition of the scientific environment, under the direction of anthropologist Margaret Mead (also disputed in her own disciplinary field) in 1969 the Parapsychological Association (founded in 1957) obtained affiliation with the American Association for the Advancement of Science.

[6] E.g. Swedenborg's theses have been quashed by the Lutheran Church.

[7] This is the case with theosophy, which Elena Petrovna von Hahn Blavatsky (1831–1891), a Russian medium later converted to buddhism and the founder of the Theosophical Society (1875–) defines as a synthesis of science, religion and philosophy.

[8] Of course, Christianity does not have a monopoly on visions and revelations.

The correspondence to the Scriptures of the various revelations is traditionally entrusted to the theologians of the Congregation for the Doctrine of the Faith (former *Holy Office*). On the other hand, the Congregation for the Causes of the Saints takes care of the smallest details of the overall examination of the life of candidates to the honours of the altar. In order to verify the events reported as supernatural, it resorts to the advice of men of science who work with the utmost seriousness.[9]

In the first place, we must distinguish private revelations, aiming at the personal improvement of the receiver, from prophetic revelations. The latter "commit the seer to transmit to the Church and the world the messages received, to admonish and exhort, often in preparation of important events to come" (Rahner 1995: 41). Generally, these writings are compiled under supervision of the priests acting as their spiritual guides. In a certain way, they become part of the patrimony of the Church even when the texts remain at the disposal of private individuals. Traditionally, the ecclesiastical authorities reserved the right to approve or forbid their publication, in line with the judgement on consistency with the Scriptures and the opportunity of the moment. See, for example, the gradualness used in spreading revelations from Fatima.

The theological commentary on this message explains better the position of Catholicism, which substantially distinguishes between public and private revelations. The former term designs the revelation "addressed to all mankind which found its literary expression in the two parts of the Bible, the New and Old Testament. Private revelations instead refer to all visions and revelations occurring after the end of the New Testament" (Ratzinger, quot.).[10]

Over the centuries several of them have been approved by the Church,[11] also obtaining recognition in the liturgy. Caution is necessary in a matter which, in the course of history, has had serious consequences, such as division among believers caused by heresies as well as social unrest caused by popular credulity.

Intended as moments "of actualization of Christ's presence in every age" (ibid.), the works of the mystics are predominantly aimed at promoting faith and familiarity with the divine and are therefore concerned with personal conscience and beliefs. As mentioned before, as geographers we are interested in the informative part instead, the frame indispensable to make the fundamental message understood and accepted by those who receive it and those who will read it later. Borrowing from theatrical language, what we look at is scenography rather than representation. Regarding the

---

[9] Of the approximately 7,500 healings reported in Lourdes in a century and a half, of which some 2,000 are unexplained according to the Paris Medical Bureau, the Catholic Church declared just 70 miracles.

[10] "Their role is not (…) to *complete* the definitive Revelation of Christ but to help to live it more fully in a specific historical epoch" (C.C.C. 1992, n. 67). Consequently, while public revelation demands our faith (…) a certainty which cannot occur in any human form of knowledge", private revelation demands "rather an assent of human faith in conformity with the rules of prudence" (Ratzinger, quot.). This assent is not required even if there has been ecclesial approval for its publication.

[11] The judgement concerning alleged supernatural revelations and related writings is in the first instance subject to the jurisdiction of the diocesan bishop, and in special cases to the Episcopal Conference and the Congregation for the Doctrine of the Faith.

anthropological (psychological) character of private revelations, theological anthropology distinguishes between three forms of perception or vision: sensory vision, that is the external bodily perception, and spiritual vision (*visio sensibilis—imaginativa—intellectualis*)" (ibid.). In sensory visions, which in the vast majority present as interlocutors Jesus Christ and the Virgin Mary,[12] the seer sometimes testifies to have experienced, with their sight and even with all their senses,[13] the physical environment where the holy characters lived their earthly existence.

5 The Difficult Balance Between Science and Faith

The spread of Christianity has made the Holy Land a relevant theme for all kinds of scholars, geographers included. Especially since the Crusades, its representation has marked significant pages in the history of cartography (Lago and Galliano 1995). However, the scale used in the times closest to the events narrated in the Gospels (therefore in the pre-geodesic era) gives us far too little detail to satisfy our thirst for knowledge. Today, our detection capability is incomparably greater, but over 2,000 years separate us from the events recorded by the apostles. One does not need to be a specialist to understand how the territories in question have undergone massive changes, due to both man (repeated cycles of construction and destruction)[14] and nature, which acts through morphological processes and climatic changes. Noteworthy is the attention paid to this area by Carl Ritter (1779–1859), one of the founders of modern geography. Four out of nineteen volumes of his Erdkunde (2nd edition) deal with the area including the Sinai Peninsula, Palestine and Syria (Ritter 1966). Also noteworthy is that these descriptions, printed between 1848 and 1854, were written without the author ever setting foot there.

As we all know, the advent of positivism has destroyed the faith in the historicity of the facts reported in the Scriptures to the point that the very real existence of Christ has been seriously questioned.[15]

In the past century, especially after the birth of the state of Israel, Palestine has become the object of ever more numerous archaeological campaigns. These

---

[12] According to the leading scholar of the subject the reported apparitions of the Virgin amount to about 2,000 (Laurentin and Sbalchiero 2007). The extent of the phenomenon worldwide is reported in Orth (2015). Until today, only 20 of them have obtained official recognition.

[13] In these cases, we speak more properly of bilocation: without moving his body the mystic somehow splits himself in two, achieving at the height of his powers a transfer in space and sometimes also in time.

[14] The well-known processes of territorialisation and deterritorialisation (Raffestin 1984), widely investigated by Angelo Turco, with particular reference to Africa.

[15] The scientific criticism of the historicity of Gospels begins with the publication by G. E. Lessing, under the title Fragments of an anonymous (1774–1777), of the texts written by H. S. Reimarus (Apology, or defense of rational worshippers of Christ) (Reimarus 1972).

Fig. 1 Nature and man, each for his part, contribute to constantly changing the environment: Jerusalem (courtesy: P. Benedetti, Trieste, 2009)

have brought to light countless testimonies of the civilizations that in a continuous succession have established themselves in the region over about four millennia (Fig. 1).

Jewish scholars were soon able to find solid confirmation of quite a large number of facts from the Old Testament (Keller 1955), although skepticism has not entirely disappeared among the local academic environment. The findings in the first centuries of our era were no less significant. In addition to the sacred places of Jerusalem and Betlehem, already known in Roman times, nowadays the remains of Peter's house, the chief of the apostles, can be visited in Capernaum on the lake of Genezareth. The same goes for Mary's house in Nazareth, the ruins of Magdala, the birthplace of Mary Magdalene and the pool of Bethesda, where the healing of the paralytic must have taken place (Gv 19,13).

Although the patient work of many scholars (archaeologists, geologists, ecologists) is producing meticulous and increasingly precise fruits, only a contemporary testimony can give us an animated as well as detailed vision of the geographical background on which the events reported by the religious tradition took place. But on the subject the testimonies, at least in the form that geography can recognize as its own, are lacking. We would need literary documents but these are scarce, with the exception of holy scriptures, which have a completely different purpose and are (rightly or wrongly) no longer considered historically reliable sources.[16] In this information vacuum the testimonies of the visionaries, who claim to have witnessed some scenes of sacred history, can prove of great value. The mystics offer us glimpses of what only contemporaries could grasp. Thus, they enable us to make significant progress in terms of knowledge of the historical geography of these territories, obviously when there is a substantial verification of their reports.

Of course, among them there are some who saw (and transcribed) more and others who did less, focusing more on the spiritual aspects of their experiences. This is a

[16] From a geographical point of view the Gospels give us almost no information as happens, by the way, also in the works of Flavius Josephus (Josephus etc. 1966).

prevailing feature up to the whole of the eighteenth century. Moreover, common to all of them is the centrality of the life of Christ, especially his passion, an event that has the city of Jerusalem as its background.

6 The Geographical Contents

As the events of sacred history took place within the framework of profane history, in the texts of mystics we can often find useful information as to the location of events, sometimes accompanied by descriptions of these places. Both events and places may either have been already described in other sources or be partially or completely new to the scholar.

In these reports, we can therefore identify a series of contributions to geographical knowledge, for example:

- descriptions of landscapes (background environments to events of sacred history);
- description of itineraries (followed by historical figures, especially taken from the biblical history);
- territorial details (naturalistic or humanistic elements, agrarian landscape, artefacts, settlements);
- precise localization of the events narrated (for example, the site of the Ascension of Jesus).

These events are all the more important in view of the transformations undergone by territories in the time elapsed since the events narrated in the visions actually took place. It is therefore no wonder that sometimes information related by a mystic differs from the opinions common among scholars. In any case, all the elements provided must then be verified according to the usual criteria followed by modern science.

Several mystics have experienced, through visions and bilocations, Palestine as it was in the time of Jesus. Let us mention the British gentlewoman Richeldis de Faverches (1061): visions of Nazareth; Anna Katherina Emmerich (1774–1824): the whole of Palestine (Bouflet 2007); Maria Natalia Magdolna (1945): bilocations in Nazareth, Gethsemane and Calvary (Matera 2019); Maria Valtorta (1943–1951): visions and bilocation all over Palestine; Renato Baron (1990): bilocation in Nazareth (Speziale 2018).

Considering only the life of Christ, three works stand out among all the others for their depth, completeness and details of descriptions. They are the main texts attributed to the blessed A. Katharina Emmerich and the servant of God Teresa Neumann (1898–1962)—(Steiner 1967) both of them, as is well known, stigmatized—and Maria Valtorta. The former two lived in Germany, the latter in Italy. They received their visions in the following time intervals, respectively, 1819–1824, 1926–1962, 1943–1945.

Unlike most mystics, especially the stigmatized, who periodically “relive” the scenes of the Passion in their bodies, both for Emmerich and Valtorta the visions mentioned seem to concentrate in time, as if they were but a particular chapter of their personal history. Both of them provide us with an impressive wealth of information on people, events and places. In Valtorta in particular, geographical details concerning geological, hydrographic, meteorological and botanical facts as well as antropology and settlements are given with such abundance as to enable the reader to experience with her, to a certain extent, the episodes related. For this reason, we have concentrated our attention on these two authors.

It was established that neither visited the Holy Land. Being seriously ill, they were both bedridden for a large part of their lives. Moreover, they lacked the cultural background necessary to anyone who wants to write a detailed life of Christ, nor could they have access to the available literature. There is therefore no reasonable explanation for the descriptions they have bequeathed to us. This, of course, does not prove the existence of a supernatural source, but excludes the possibility of rejecting this hypothesis a priori. The accuracy of the details concerning places and circumstances and their reliability, confirmed by our present knowledge, is, however, a solid hint to the credibility of their visions.

Being neither theologians nor apologetes, we shall avoid delving into questions that do not pertain to our discipline. Our contribution as geographers can legitimately have two aims: a) increase our knowledge of the landscapes in Palestina at the onset of the Christian era; b) compare, whenever possible, the descriptions provided by various mystics, in particular referring to our present-day knowledge of the historical geography of the Middle East. That said, we shall here attempt to draw a brief comparison between the geographical descriptions in the writings of these two mystics.

7 Discussion

According to our premise, the assertions of our two mystics are to be analysed through the types of investigation deemed appropriate in every single instance. The first and foremost criterion is, of course, the examination of their coherence with the bibliografical and cartographic sources available at present.

As an example, let us quote (a) the localization of the site of the Ascension of Jesus (Valtorta: Poema, X, 23: 209–215, Lc 24, 50–52); (b) some details on the town of Lébona (Valtorta: Poema, VIII, 31; Emmerick: *quoted by* Lavère 2017: 227–228); some imagines of the city of Antiochia (Valtorta, Poema, V, 10).

A second criterion takes into consideration geological and pedological characteristics. Notable examples are (a) the agricultural landscape in the surroundings of Hebron (Valtorta: see Appendix 2); (b) the morphology of Golgotha and part of Jerusalem at the beginning of the first century (Emmerich: see Appendix 1).

A farther criterion comes from archaelogy. See the aqueduct of Engannim (Valtorta: Poem, III, 53); Mary's house near Ephesus[17] (Emmerick: The Life, etc. 1970: 346–347; Carroll 2002), the urban structure of the town of Tiberias (Valtorta, Poem, 2: 64).

It is worth noting that the description of Mary's house near Ephesus, as well as that of Tiberias, precedes by several years the actual discovery and/or archaelogical documentation of their remains. In the case of Tiberias, the first reconstruction dates back to 1999 (Lavère 2017: 369–370). There are many similar examples, among which the hot springs of Mammat Gader (Valtorta, Poem, V: 46; Lavère 2012: 96–97). Other sites described by the visionaries have yet to be found or recognized as conforming to their descriptions (function of the buildings, names of the owners, etc.).

Although there is a certain uniformity as to their spiritual message as such, each mystic receives visions on different aspects of sacred history. Emmerich's visions concern both the Old and the New Testament, with wide digressions, interesting very ancient times and territories far away from the Holy Land. Valtorta instead concentrates on the period from the birth of the Virgin Mary to her Assumption and in particular on the three years of Jesus' preaching. Contrary to what one may think, with the mystics we have different works, not repetitive versions of the same narration, although of course their narrations often overlap. Therefore, this is not one single film, screened several times with the same scenes, only each time to a new audience.

Emmerich's visions were collected and initially reworked by Clemens Maria Brentano (1778–1842), a romantic writer and a friend of Goethe's. Considering their editorial history and the subsequent interventions, a large part of the writings attributed to the mystic are to be considered at least of third-hand, so that today it would be extremely difficult to ascertain what part of them can be attributed to her with a sufficient degree of certainty. To solve this question, it would be necessary to study in particular the immense wealth of manuscripts left by Brentano. Critics have investigated the role of this author and his rather questionable past. Doubts on the texts published have been raised from the beginning and have not been solved up until now. However, Emmerich's name has become popular since she has inspired Mel Gibson, the director of the film The Passion (2004).

As far as geographers are concerned, in the biblical events related by Brentano, many mistakes can be found, both geological and topographical (Brentano 1875). There are also a great many elements taken from minor religious historiography of the medieval kind, which echo the "wondrous tales" of chroniclers and travellers of a pre-scientific epoch.

In contrast, in the visions surely written with her own hand, Valtorta left an account of the public life of Jesus that meets the needs of today's man, immersed in a scientific culture. Approximations and contradictions are no longer allowed, being correctly

[17] Located on the hill of the nightingales, in the present Selcuk district, it was known to locals as *Panagla Kaulu*: now it is the shrine of *Meryem Ana Evi*.

identified as elements likely to undermine the credibility both of the mystic him-/herself and the "truths" he or she is revealing. First published without the authorization of the Holy Office, at the time mandatory for Catholics, Valtorta's revelations were put on the Index of Prohibited Books. Shortly thereafter the Index was closed but the negative judgement was later to be reaffirmed, as a part of the ecclesiastical environment remains hostile to the dissemination of texts of purported supernatural origin. Nonetheless, the work has been almost prodigiously spread worldwide. Although printed and distributed by a small Italian publisher, it has been translated into about 30 foreign languages and counting, and from the late 1980s a—not only quantitatively—significant bibliography has been building up around it.

Unlike other mystics, hers are not only theological reflections or devotional texts. The wealth of concrete information disseminated in the text has attracted the specialist attention of a patrol of authors moving in a religious perspective but with a scientific background and even academic curriculum (Biagini 2018; Debroise 2016; Hopfen 2003; Laurentin et al. 2012; Lavère 2016, 2017). Independently or in collaboration with each other and moving in different directions they are gradually honing their knowledge of the subject.

Geographical data provide tremendous support for Valtorta's writings (Fig. 2). Analysing more than 4,000 references scattered throughout her work, Aulagner (1991), a French economist, has meticulously reconstructed a perfectly coherent chronology of Jesus's life which allows the dating of the facts reported by the four

Fig. 2 Clemens Maria Brentano, who transcribed the visions of Emmerich, portrayed by Emilie Linder (circa 1837)

Fig. 3 A recent study on Valtorta visions

evangelists. For his part, the Italian physicist Liberato De Caro (2005–) is currently refining this dating work. In turn, Hans J. Hopfen (1995), an agronomic engineer, formerly in force at the FAO, thanks to repeated inspections has reconstructed a detailed map of historical Palestine on which he succeeded in locating most of the approximately 500 toponyms (localities, regions, reliefs, rivers, etc.) reported by Valtorta.

8 Conclusion

In many cases, the two seers report the same events so that a comparison is feasible. We thus discover that topographical details often disagree and the same is true of numerous scenes. This poses a clear problem to the scholar willing to verify the reliability of their visions. While the facts narrated substantially correspond, this is not the case for the more strictly geographical aspect.

It is also necessary to note the extreme precision that characterizes Valtorta and at the same time correctly evaluate the inevitable misinterpretations that occurred in Brentano's reconstruction work. Compared to the latter, Valtorta also seems to have a much higher ability to grasp the significant geographical elements of landscapes and to report them in a clear, at times intensely lyrical language. It can be easily appreciated in the passage reported in the Appendix to enable the readers to form their own opinion.

Not intending to discuss here the value of private revelations, a theme that the Catholic Church itself treats with great care, we must, however, acknowledge that there is a vast literature concerning visions of particular moments in the sacred history. In it, we can sometimes find detailed geographical information, which partly corresponds to our knowledge on the subject, partly goes beyond it. What is amazing is that in some cases the progress of field research on the ground confirms its credibility. After remaining for a long time confined to a restricted circuit, nowadays these writings are becoming increasingly widespread and thus available to various categories of scholars. In my opinion, they deserve the attention of geographers. Close analysis may prove them to be more or less truthful. In any case, as we all know, even negative judgements contribute to the progress of science.

Appendix—Excerpts from the Visions[18]

(1) *Anna Katharina Emmerich* (La Dolorosa etc. 1933: 546–547)

Some localities of ancient Jerusalem

The first gate to the east of Jerusalem, at the Southeast corner of the Temple, leads to the suburb of Ophel. The Sheep's door is the one that, to the north, is the closest to the northeast corner of the Temple. Between these two doors another one was opened quite recently, which leads to some streets east of the Temple, inhabited, for the most part, by stone-breakers and other workers. Their houses lean on the foundations of the temple and almost all of them belong to Nicodemus who built them. The workers pay him a rent either in cash or by working for him, because they are constantly in contact with him and his friend Joseph of Arimatea, who owns in his native country large stone quarries which he exploits. Nicodemus recently had a beautiful door built which leads to these ways and is called Mary's gate. It had just been finished and Jesus was the first to go through it as he entered the city on Palm Sunday. So he passed through Nicodemus's new gate and was buried in the new tomb of Joseph of Arimatea, where no one had yet been laid. This gate was bricked up later, but according to an ancient tradition Christians would one day go through that door once again. Today there still exists near there a door which the Turks call the Golden Gate.

(2) *Maria Valtorta* (Valtorta 1986, vol. 1: 132)

I, 32. Arrival at Zechariah's house

I am in a mountainous place. They are not high mountains but neither can they be called hills. They already have yokes and inlets like real mountains, similar to those we can see in our Tuscan-Umbrian Appennines.

[18] Author's translations.

The vegetation is thick and beautiful and there is an abundance of cool waters, that maintain green the pastures and fertile the orchards, which are almost all cultivated with apple trees, figs and grapes: around the houses, a small vineyard. As for the season it must be spring, because the grapes are already quite big, like grains of vetch, and the flowers on the apple trees now appear as many green balls, and on top of the branches of the figs there are the first fruits still embryonic, but already well formed. As to the lawns, they are a soft carpet of a thousand colours.

On them, the sheep graze, or rest, white spots on the emerald of the grass.

Maria climbs, on her little donkey, up a good enough path road, probably the main road. She climbs, because the village, quite neat-looking, lies higher. My internal warning tells me: "This place is Hebron." You told me about Montana. But I do not know what to do with it. It is mentioned to me by this name. I don't know if Hebron is the whole area or just the village. I feel this way and so I say.

References

Aulagner, J. 1991. *Avec Jésus au jour le jour*. Montsurs: Résiac.

Biagini, E. 2018. *Maria Valtorta, la testimone della vita di Cristo.* Isola del Liri (Fr): CEV.

Bouflet, J. 2007. *Anna Katarina Emmerick che condivise la passione di Cristo.* Milano: San Paolo.

Brentano, C. 1875. *Das bittere Leiden unseres Herrn Jesu Christi: nach den Betrachtungen der gottseligen Anna Katharina Emmerich, Augustinerin des Klosters Agnetenberg zu Dülmen; nebst dem Lebens umriß dieser Begnadigten.* Regensburg: Cotta.

Brunhes, J. 1910. *La géographie humaine. Essai de classification positive. Principes et exemples.* Paris: Alcan.

Carroll, D. 2002. *Mary's house: The extraordinary story behind the discovery of the house where the Virgin Mary lived and died.* Christian Classics.

Cluveri, Ph. 1619. *Sicilia antiqua cunminoribus insulis ei adjacentibus.* Lugduni Batavorum: Exofficina Elseviriana.

Congregazione per la Dottrina della Fede. 1996. *Acta Apostolicae Sedis – AAA* LXXXVIII (12): 96–957.

Debroise, F.-M. 2016. *Maria Valtorta visionnaire et mystique pour notretemps.* Plouisy: Rassemblement à Son Image.

Dermine, F.-H. 2003. *Mistici e veggenti. Esperienze dell'aldilà a confronto.* Città del Vaticano: L.E.V.

Galliano, G. (ed.). 2002. *Geografia e religione. Una lettura alternativa del territorio*, Geotema. n.18.

Galliano, G. (ed.). 2003. *Orizzonti spirituali e itinerari terrestri.* Geotema 21.

Hopfen, H.J. 2003. *Indice e Carta della Palestina per "L'Evangelo come mi è stato rivelato" di Maria.* Isola del Liri (Fr): CEV.

Josephus with an English translation by St J. Thackeray, M.A. 1966. London: Heinemann, Cambridge: Harvard University Press (9 vols.).

Keller, W. 1955. *The bible as history.* London: Hodder & Stoughton.

Kraft, J., I. Kayan, and O. Erol. 1982. *Troy. The archaeological geology*, ed. G. Rapp, and J. Gifford. Princeton University.

Kraft, J., I. Kayan, H. Bruckner, and G. Rapp. 2002. Sedimentary facies patterns and the interpretation of paleogeographies of Ancient Troia. In *Troia and the Troad*, ed. G. Wagner, E. Pernicka, and H.P. Werpmann, 361–377. Heidelberg.

La Dolorosa Passione di N. S. Gesù Cristo. Dalle visioni di Anna Caterina Emmerich. 1933. Torino: R. Berruti & C.
Lago, L., and G. Galliano. 1995. *La terra santa e la sua immagine nella cartografia antica.* Firenze: Alinari.
Lando, F. (a cura di). 1993. *Fatto e finzione. Geografia e letteratura.* Milano: ETAS Libri.
Laurentin, R., and P. Sbalchiero (sous la direction). 2007. *Dictionnaire des "apparitions" de la Vierge Marie.* Paris: Fayard.
Laurentin, R., F.-M. Debroise, and J.-F. Lavere. 2012. *Dictionnaire des personnages del'E'vangile selon Maria Valtorta.* Paris: Salvator.
Lavère, J.F. 2002, 2016. *L'enigma Valtorta.* Isola del Liri (Fr): CEV (2 vols.).
Lavère, J.F. 2017. *Dictionnaire géographique de l'Évangile d'après Maria Valtorta. Lieux, Monuments, Peuples.* Plouisy: Éditions Maria Valtorta.
Lowenthal, D. 1985. *The past is a foreign country.* Cambridge University Press.
Matera, M. 2019. *Rivelazioni profetiche di suor Maria Natalia Magdolna. Mistica del XX secolo.* Milano: Sugarco.
Neci, G. 1994. Il motivo dell'autopsia nella storiografia greca. In Silvana Fasce (a cura di), *Temi e discussioni di geografia antica*, 7–51. Genova: ECIG.
Orth, M. 2015. How the Virgin Mary became the world's most powerful woman. *National Geographic* 12. https://www.nationalgeographic.com/magazine/2015/12/virgin-mary-worlds-most-powerful-woman/. Accessed 10 April 2020.
Our Lady of Walsingham. https://santiebeati.it/Detailed/95420.html.
Paolo II, Giovanni. 1992 October 11. *Catechismo della Chiesa Cattolica, promulgated.* Città del Vaticano.
Petrone, A. 2016. https://www.scienzenotizie.it/2016/11/15/svezia-scoperta-una-citta-sommersa-nel-mar-baltico-0818308. Accessed 10 April 2020.
Pocock, D.C.D. (ed.). 1981. *Humanistic geography and literature. Essays on the experience of place.* London: Croom Helm.
Raffestin, C. 1984. Territorializzazione, deterritorializzazione, riterritorializzazione e informazione. In *Regione e regionalizzazione: colloquio internazionale*, ed. Angelo Turco, 69–82. Verona, Milano: Franco Angeli.
Rahner, K. 1995. *Visionen und Prophezieungen, Freiburg, Herder, 1958* Italian edition: *Visioni e profezie: mistica ed esperienza della trascendenza.* Milano: Vita e Pensiero.
Ratzinger, Card J. 2000. Commento teologico. In Congregazione per la dottrina della fede, *Il messaggio di Fatima.* http://www.vatican.va/roman_curia/congregations/cfaith/documents/rc_con_cfaith_doc_20000626_message-fatima_it.html. Accessed 29 March 2020.
Reimarus, H.S. 1972. *Apologie oder Schutzschrift für die vernünftigen Verehrer Gottes.* Insel Verlag. Italian edition: *I frammenti dell'Anonimo di Wolfenbuttel pubblicati da G. E. Lessing-* a cura di Fausto Parente. Napoli: Bibliopolis.
Ritter, C. 1966. *The comparative geography of Palestine and the Sinaitic Peninsula.* Edinburgh: T&T Clark (4 vols.).
Seamon, D. 1976. The phenomenological investigation of imaginative literature: a commentary. In *Environmental knowing. Theories, research and methods*, ed. G.T. Moore, and R.G. Golledge, 286–290. Strousdsburg, Pennsylvania: Hutchinson and Ross.
Speziale, V. 2018. *Renato Baron. Vita, missione e croce con la regina dell'amore.* Udine: Segno.
Steiner, J. 1967. *Therese Neumann (A portrait based on Authentic Accounts).* Regensburg: Schnell & Steiner.
Tuan, Y. 1976. Humanistic geography. *Annals, A.A.G.* 66: 266–276.
Valtorta, M. 1986–2001. *L'Evangelo come mi è stato rivelato.* Isola del Liri (Fr): CEV (10 vols.).
Vinci, F. 2005. *The Baltic origins of Homer's epic tales. The "Illiad", the "Odyssey" and the migration of myth.* Rochester (Vermont): Inner Traditions.

Hidden Environmental Geographies

Soils—A Hidden World: A Case Study of Soil Sequences on Carbonate Parent Materials in Slovenia

Blaž Repe

Abstract Of all natural or physical elements of a landscape, soils are the most hidden ones. Topography, waters and living world are well visible, obvious and considered as the most important and attractive landscape features. While soils beneath our feet on the other hand are hidden from our sight and furthermore are even considered boring. All soil properties, processes and links to other elements are too often neglected, do not get a proper attention and due to ignorance, are treated poorly. Soil sequence or catena is a concept that deals with these hidden interlinks and reveals new information about landscape and soils. The study area comprises the eastern foot slopes of the Polhograjsko hills, on the western edge of the Ljubljana basin. Both entities are the part of the Pre-Alpine region, which extends south of the Alps, across the entire country. The lower parts of the Polhograjsko hills slopes consist of the Carboniferous and Permian, hard, siliceous clay and sandstones. On top of this, thick layers of hard and compact Triassic limestone and dolomites represent the upper part of slopes and summits. Soil genesis in the foot of the Polhograjsko hills and its slopes show a modified soil sequence (pedosequences), already described by Stritar in 1991. He states this is the most important soil sequence in Slovenia because it appears in all-natural regions (high and low mountains and hills and especially in karstic areas). The developing stages of soil on hard carbonate rocks show considerable regularity in appearance (limestones and dolomites of different periods) throughout the regions of Slovenia. The variety of soil groups on limestones and dolomites is closely connected or conditioned by the specific geomorphology of these surfaces, which are modulated by irregularly distributed sinkholes, smaller valleys and precipitous, larger plateaus, terraces and undulating valleys. This variegated relief by itself would not be a cause for such sudden changes and variations in soil forms where it is not for the associated micro-relief of the rock. Due to the high micro-variation of geological and topographical features, a term mosaic soil spatial distribution is used.

B. Repe (✉)
Department of Geography, Faculty of Arts, University of Ljubljana, Aškerčeva 2, 1000 Ljubljana, EU, Slovenia
e-mail: blaz.repe@ff.uni-lj.si

M. Krevs (ed.), *Hidden Geographies*, Key Challenges in Geography,
https://doi.org/10.1007/978-3-030-74590-5_5

Keywords Soil-forming factors · Soil sequences · Toposequence · Carbonate parent material · Mosaic soil distribution · Slovenia

1 Introduction to the Hidden World of Soil Connections

Soils never get a proper attention, they deserve. On one hand, the reason is somehow obvious. Of all natural or physical elements of a landscape, soils are the most hidden ones. Topography, waters and living world are well visible, obvious and considered as the most important and attractive landscape features (Smrekar et al. 2006, 2014; Erhartič 2010). While soils beneath our feet on the other hand are hidden from our sight and furthermore are even considered boring (Repe 2008c, 2009). Strange enough, when evaluating, it is very difficult to put a price on topography, air or water. But the value of land and therefore soils as a property is very high and precious to us. Thus, it is difficult to understand, why all soil properties, processes and links to other elements are too often neglected, do not get a proper attention and due to ignorance are poorly investigated (Repe 2008b, 2010). Soil sequence or catena is a concept that deals with these hidden interlinks and reveals new information about landscape and soils.

The concept of soil properties being connected to other physical landscape elements can be traced back to the nineteenth century. Dokuchaiev in Russia and independently Hilgard in America discovered that a given parent material may form different soils depending on the environmental conditions, particularly climate and vegetation. This was redefined later by adding additional independent factors to the following soil formation equation (Jenny 1941):

$$S = f(cl, o, r, p, t, \ldots)$$

This formula was adapted to accommodate the variety of soil-forming factors found in Slovenia (Stritar 1991; Prus 2000):

$$S = f(p, w, r, cl, o, m)\, t$$

where soil (S) is a function (f) of the parent material (p), water (w), relief/topography (r), climate (cl), organisms (o), and man (m), all working in time frame (t). One factor can predominate locally, while others can be barely noticeable in the short term. Due to the enormous natural (and social) variability, Slovenia's soil-forming factors, soil genesis, and soil processes are very specific.

In 1935, with catena, Milne (Milne 1935) introduced a similar concept by interlocking of soils on a landscape. He used this term to describe a regular repetition of soils in a given landscape. He described two types of catenas. In the first one, the parent rock was uniform and therefore the differences in soils within the catena were the result of differences in drainage and differences due to lateral movement of materials both at the surface and in the subsurface. The second type of catena involved more than one type of parent material that influenced soil genesis. The concept of catena has been further modified (Bushnell 1942) and is now used almost interchangeably with toposequence (Hall 1983). In this chapter, we consider the soil-landscape system mainly according to the changes in topography, especially position. Apart from this, there are minor changes in parent material, but all types of rock belong to carbonate sedimentary group.

Slovenia's moderate and humid climate causes relatively slow processes of soil formation. Minerals therefore enter the pedosphere chemically only partially changed or even completely unaltered. Since carbonate rocks prevail, the most commonly soil types in Slovenia are Eutric Leptosols, Chromic Cambisols and Eutric Cambisols. Together they cover more than 56% of territory. On the other hand, topography is certainly Slovenia's most prominent natural feature. It provides the general image of the county and is at the same time woven into all other landscape elements. Topography as a soil-forming factor influences our soil cover indirectly. It has a direct and crucial impact on soil water conditions and micro- and mezzo-climatic conditions. The basic interaction between slope processes, landforms in fluvial geomorphic systems and soil formation or other soil parameters is more or less known. The chapter deals with the question what impact the relief and its elements on soils in karst geomorphic system has where slope processes are different.

2 Soil-Forming Factors in the Foot Slopes of the Polhograjsko Hills

The study area comprises the eastern foot slopes of the Polhograjsko hills, on the western edge of the Ljubljana basin (Fig. 1). Both entities are the part of the Pre-Alpine region, which extends south of the Alps, across the entire country. The Pre-Alpine area is characterized by middle-range hills (up to 1000 m.a.s.l.) and swift changes of parent material, especially older siliceous and younger carbonate rocks (Ogrin and Plut 2009). The studied soil toposequence is located in the transition between two mesoregions: the Ljubljana Basin and the Polhograjsko hills, west of Ljubljana capital city. The sequence is typical for this region and it represents a soil genesis on different topographic and slope positions on the carbonate parent material (Gabrovec 1989) (Fig. 2).

2.1 Lithology and Topography

Parent material is considered the most important of Slovenia's soil-forming factors. The moderate and humid climate results in relatively slow soil genesis, especially as regards the weathering of parent material and minerals. Chemically unaltered or only partially changed, the mechanically disintegrated smaller pieces (sandy, silty) of

Fig. 1 Side valley entering the Polhograjsko hills (*photo* Blaž Repe)

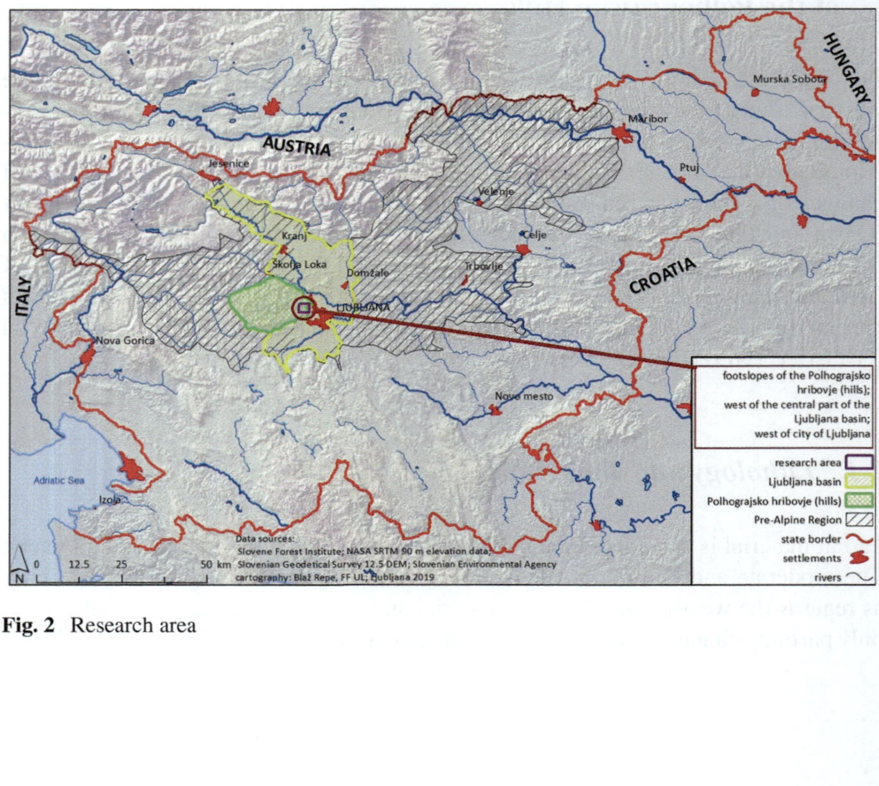

Fig. 2 Research area

Fig. 3 View of the foot slopes landscape (*photo* Blaž Repe)

parent material enter the pedosphere, where they heavily influence the soil properties with their own. Because the research area is karstic and composed of permeable and soluble limestone and dolomite, there is also a direct chemical influence on the soils and water. The parent material also serves as the most important factor for distinguishing among different soil types. In addition to this, the parent material also significantly affects other soil-forming factors (Stritar 1991; Lovrenčak 1994; Kočevar and Vidic 1998; Prus 2000; Repe 2006a).

The foot slopes of the Polhograjsko hills is set of 2–3 km wide range of slopes, rising from the sediments of the Ljubljana basin to the first summits (600–700 m.a.s.l.) (Fig. 3). The hills were formed during the Alpine orogenesis and were lifted above the basin, which in addition, started to descend in Pliocene. During the Pleistocene ice ages, the basin was filled with glacial and alluvial gravel and sand, mostly of the carbonate origin. The hard rock bottom of the basin consists of the Carboniferous and Permian, hard, siliceous clay and sandstones. The same parent material is also typical for the lower parts of the Polhograjsko hills slopes. On top of this, thick layers of hard and compact Triassic limestone and dolomites represent the upper part of slopes and summits. On these rocks, karstic features and hydrology developed (Grad and Ferjančič 1976). In some parts, carbonate material reaches the basin and in some, siliceous rocks extend nearly to the summits, what makes according to the soil cover, the area very diverse and complicated (Repe 2006a) (Fig. 4).

Topography as a factor of soil formation influences our soil cover indirectly. It has a direct and crucial impact on soil water conditions (quantity, movement direction, the effect on soil processes, etc.) and micro- and mezzo-climatic conditions (solar

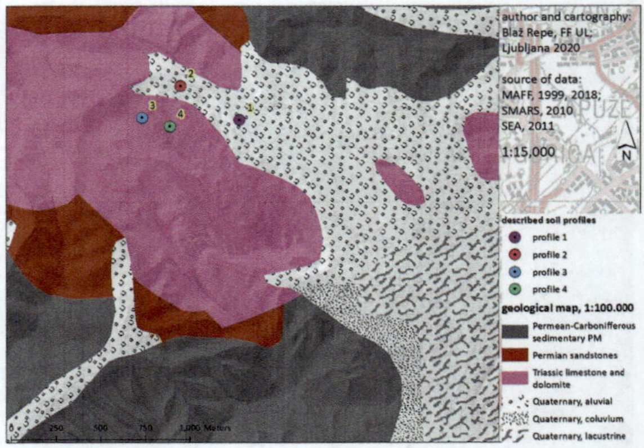

Fig. 4 Geologic/lithologic map of the research area with the locations of the soil profiles

radiation, temperatures, the quantity and distribution of precipitation, wind, etc.) (Vrščaj et al. 2017). The area represents the lowest part of Slovenia's Pre-Alpine hills, which in the highest part, hardly reaches 1000 m.a.s.l. (Tošč, Pasja Ravan), which puts most of the area from 500 to 700 m above local erosion base (the Ljubljana Basin). The average height is only 540 m.a.s.l., but the average slope inclination is more than 20° (Repe 2006a). The area is dissected by ravines and small-scale valleys of occasionally appearing streams, which are all of the extremely torrential nature. They are the main reason for floods in the settlements on the edge of the Ljubljana basin and the capital itself (Radinja 1996). The transition between the ground of the basin and the slopes of the hills is covered with thick layer of colluvial and delluvial material, consisting mainly of the carbonate gravel and siliceous silt and sand (Figs. 5, 6, 7, 8).

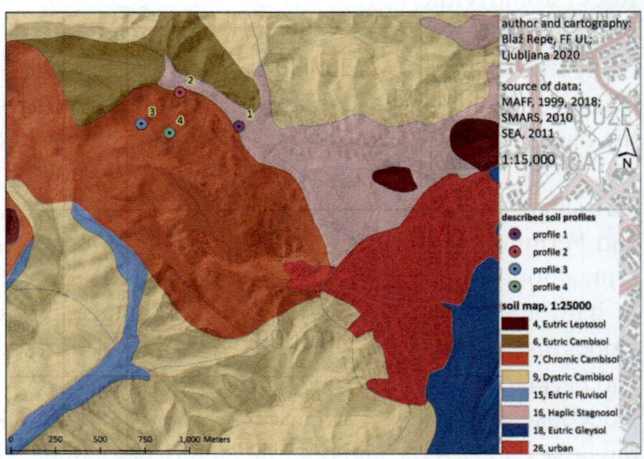

Fig. 5 Soil map of the research area with the locations of the soil profiles

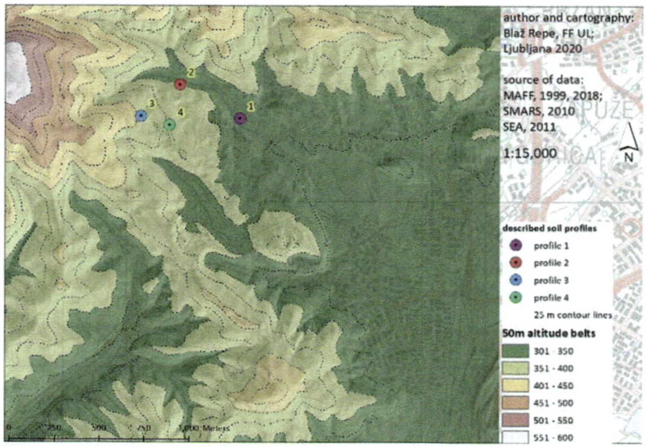

Fig. 6 50 m altitude belts of the research area with the locations of the soil profiles

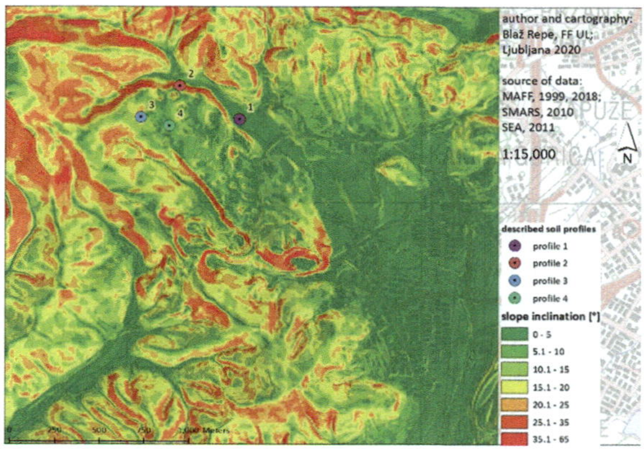

Fig. 7 Slope inclination of the research area with the locations of the soil profiles

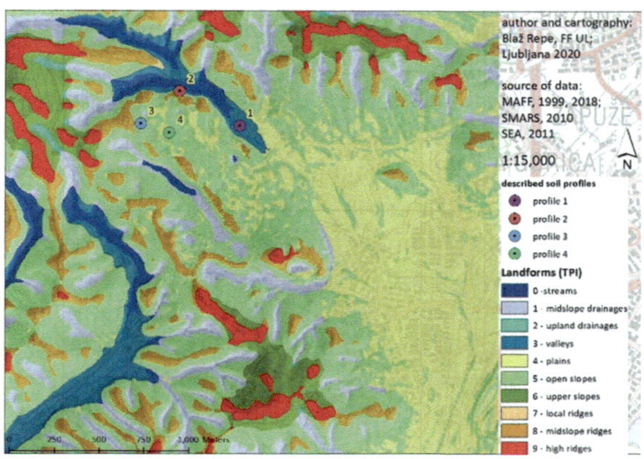

Fig. 8 Mezzo scale landforms of the research area with the locations of the soil profiles

2.2 Other Soil-Forming Factors

The research area belongs to the temperate, humid climate of the central Slovenia. It could be described as moderately continental or moderately warm and humid climate with warm summers (Ogrin 1996). Slopes are well drained with udic regime, but the valleys and basins are often aquic (Vrščaj et al. 2017). The average temperature of the coldest month (January) is −2 °C, while the warmest month is July (18 °C). The average annual precipitation is above average for the central Slovenia and reaches up to 1700 mm. Distribution of the precipitation does not differ much between seasons, but the wettest months are autumn (November, 200 mm) and winter months are the driest (February, 90 mm) (ARSO 2005).

Due to its importance as a soil-forming factor, water plays an important role while discussing soils in Slovenia. As stated in the previous paragraph, the climate is humid. Nevertheless, most of the soils in Slovenia and in the research area belong to the automorphic soil order, where precipitation water drains freely from the profile and processes of reduction are not present (Prus 2000; Repe 2004). We can name two main reasons for this. The first one is related to the parent material. There is the karstic surface of permeable carbonate rocks, with neither surface, stagnant, nor running water (with rare exceptions). On its way into the corroded and cavernous rocks, precipitated water only traverses the pedosphere (Vrščaj et al. 2017). The source reason is topography. Water is attracted from hilly or mountainous towards lower parts and the local erosion basis. On its way, water carries weathered material and detached soil particles. Soils are shallower and automorphic. Waters with denuded and eroded material runs into the local valleys and near Ljubljana basin are considerably deeper, and again, hydromorphic conditions can frequently occur. Since Slovenia is by far more elevated, hilly, and diverse, there is little hydromorphic soil, only 13.4% (Ministry of Agriculture and the Environment 1999; Repe 2006b).

The vegetation and land use are typical for the Pre-Alpine hills, which is a dense forest cover. The dominant species is beech (*Fagus sylvatica*), present in more than ¾ of the wood reserves. In the research area, we can find two main types of beech forest. Over carbonate parent material and on shallow, slightly acid or neutral soils, *Hacquetio-Fagetum* and *Lamio orvalae-Fagetum* associations are dominant (Fig. 9). On southern, towards sun-exposed slopes, thermophyllus *Ostryo carpinifoliae-Fagetum* is also present (Zupančič et al. 1972, 1973). These forests frequently have a protective function, due to the steep slopes and heavy erosion processes (Ogrin et al. 2017).

Due to the steep slope inclinations, shallow and sometimes acid soils, lack of surface waters (karst) and remoteness from the centres, former fields, orchards, meadows and pastures have been abandoned long ago. They exist only in the very close vicinity of the very few settlements and isolated farms (Repe 2006a) (Fig. 10).

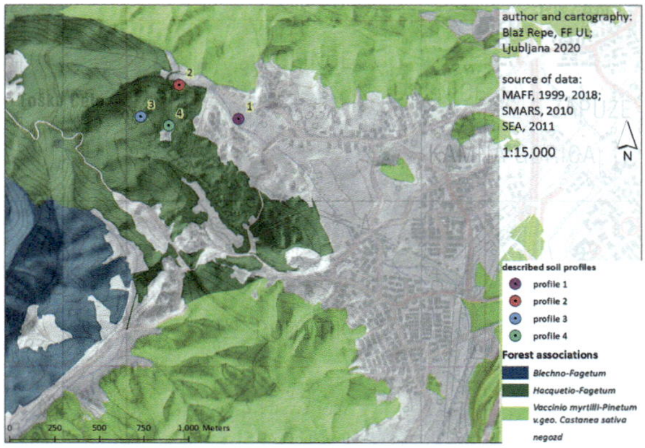

Fig. 9 Forest associations of the research area with the locations of the soil profiles

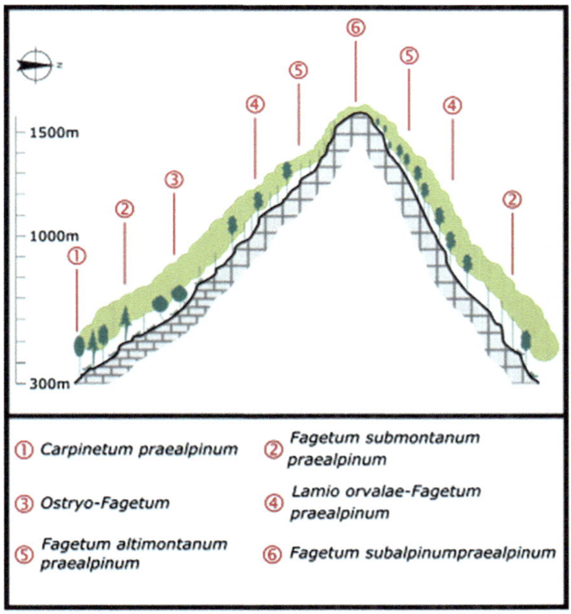

Fig. 10 Prevailing forest associations on the carbonate parent material in the pre Alpine areas (Marinček 1987)

3 Soils of the Research Area

Soils in the foot of the Polhograjsko hills have been explored and studied mainly through the field research beginning in 2005 (Repe 2006a) and constantly continuing through many visits with students every year until present. Four different soil types and four different soil profiles have been chosen. They follow the general sequence of topographic and geologic/lithologic features:

Slope inclination	Flat	Very steep	Steep	Flat, undulating
Altitude	Lowland	Lower mid-slope	Higher mid-slope	Higher plateau
Parent material	Fine and coarse Unconsolidated carbonate	Hard carbonate	Hard carbonate	Hard carbonate

3.1 Soil Profile 1—Regosol

The first soil profile is located (N 46°04′47.0” E 14°26′07.9”) in the transition between ravine bottom and slope (slope inclination 4°, the altitude of 335 m.a.s.l.), on a mixed colluvial material, next to the dry riverbed. The land use is hygrophyllus forest, mixed beech (*Fagus sylvatica*) , alder (*Alnus glutionsa*) and ash (*Fraxinus excelsior*) forest (Fig. 11, Table 1).

Morphology of the profile 1 is as follows: A_h horizon is 25 cm thick, ochric horizon (boundary is gradual); texture class is loam, dusky red (2.5YR 3/2, moist; colour comes mainly from purplish Permian sandstone) colour; slightly moist water

Fig. 11 Location of the soil profile 1

Table 1 Selected soil properties and sequence of horizons of profile 1

Horizon	Depth [cm]	Textural class	$CaCO_3$ [%]	pH [KCl]	Organic matter [%]	Base saturation [%]
A_h	0–25	loam	0.51	6.3	3.1	84.6
AC	25–35	loam	0.66	6.5	1.4	83.6
C_1	35–50	silt loam	0.70	6.5	–	–
C_2	50–(60)	loamy silt	1.4	6.6	–	–

Fig. 12 Soil profile 1

content; moderate subangular structure; soft consistency; roots are common; many rounded coarse fragments. AC horizon is 10 cm thick (boundary is gradual); texture class is loam; yellowish red (5YR 4/6, moist) colour; slightly moist water content; moderate subangular structure; soft consistency; roots are common; many rounded coarse fragments. C_1 horizon is 15 cm thick (boundary is gradual); texture class is silty loam; yellowish brown (10YR 5/6) colour; slightly moist water content; weak subangular structure; many rounded coarse fragments. C_2 horizon is 10 cm thick; rounded coarse fragments of mainly siliceous Permian sandstone (75%) and Triassic limestone (25%) are dominant.

According to the WRB soil classification (IUSS Working Group WRB 2015), the name of the soil is Eutric Anocolluvic Skeletic REGOSOL (Nechic, Ochric, Siltic) (Fig. 12).

3.2 *Soil Profile 2—Leptosol*

The second profile is located (N 46°04′37.9" E 14°26′07.1") on a very steep lower slope (inclination 45°, the altitude of 350 m.a.s.l.), on very hard but fractured mixed Triassic dolomite and limestone, with heavy erosion processes present. The land use

Fig. 13 Location of the soil profile 2

Table 2 Selected soil properties and sequence of horizons of profile 2

Horizon	Depth [cm]	Textural class	$CaCO_3$ [%]	pH [KCl]	Organic matter [%]	Base saturation [%]
OC	0–7	Silt loam	23,	6.8	28	82.7
R	7–(50)	–	–	–	–	–

is very sparse and light forest, mixed beech (*Fagus sylvatica*) and spruce (*Abies alba*) forest with winter heat (*Erica carnea*) and black hellebore (*Helleborus niger*) forest (Fig. 13, Table 2).

Morphology of the profile 2 is as follows: OC horizon is extremely shallow 7 cm thick (boundary is abrupt and irregular); 10YR 2.5/1 (very dark brown, moist) colour; texture class is sandy loam; granular and weakly developed structure; fine roots are common; content of coarse fragments (fresh angular gravel, >75%) is very high and they are present as a surface cover (heavy sheet erosion); very strongly calcareous (dolomite); organic matter is raw and poorly decomposed. R horizon is 43 cm thick, compact Triassic dolomite, cracked at the contact.

According to the WRB soil classification (IUSS Working Group WRB 2015), the name of the soil is Eutric Dolomitic Skeletic Lithic LEPTOSOL (Nechic, Protic) (Fig. 14).

3.3 Soil Profile 3—Phaeozem

The third soil profile is located (N 46°04′39.6” E 14°25′59.3”) on steep higher middle slope (inclination 18°, the altitude of 390 m.a.s.l.), on hard Triassic dolomitic limestone. The land use is dense forest with some manmade grassland locally present,

Fig. 14 Soil profile 2

natural beech forest (*Fagus sylv*atica) with dwarf masterwort (*Hacquetia epipactis*) and balm-leaved archangel (*Lamium orvala*) (Fig. 15, Table 3).

Morphology of the profile 3 is as follows: A_1 horizon is 18 cm thick (gradual and smoot boundary), mollic horizon; texture class is silty clay loam; dark reddish brown (7.5YR 3/3, moist) colour; slightly moist water content; strong subangular blocky structure; slightly hard consistency; common angular coarse fragments; few fine roots. A_2B horizon is 9 cm thick, mollic horizon; texture class is clay loam; dark reddish brown (2.5YR 3/3, moist) colour; slightly moist water content; strong subangular blocky structure; slightly hard consistency; gradual and smoot boundary; many angular coarse fragments; few fine roots. A_3C horizon is 6 cm thick, mollic horizon; texture class is clay loam; brown (7.5YR 4/2, moist) colour; slightly moist water content; strong subangular blocky structure; slightly hard consistency; clear and wavy and smoot boundary; abundant angular coarse fragments; few fine roots. R horizon is 27 cm thick, compact Triassic dolomitic limestone, slightly cracked at the contact.

According to the WRB soil classification (IUSS Working Group WRB 2015), the name of the soil is Dolomitic Leptic Rendzic PHAEOZEM (Hyperhumic, Loamic) (Fig. 16).

Fig. 15 Location of the soil profile 3

Table 3 Selected soil properties and sequence of horizons of profile 3

Horizon	Depth [cm]	Textural class	$CaCO_3$ [%]	pH [KCl]	Organic matter [%]	Base saturation [%]
A_1	0–18	Silt clay Loam	4.2	7	19.6	94.0
A_2B	18–27	Clay Loam	4.5	7.5	19.1	95.8
A_3C	27–33	Clay Loam	5.3	7.5	18.5	94.6
R	33–(60)	–	–	–	–	–

3.4 *Soil Profile 4—Luvisol*

The fourth soil profile is located (N 46°04′57.1" E 14°26′02.0") on the karstic, levelled part of the upper, higher plateau, undulating (slope inclination 0°, the altitude of 370 m.a.s.l.), on hard Triassic dolomite, partly covered by colluvic material. Land use is dense forest, beech forest (*Fagus sylvatica*) with dwarf masterwort (*Hacquetia epipactis*) and balm-leaved archangel (*Lamium orvala*) (Fig. 17, Table 4).

Morphology of the profile 4 is as follows: O 2–0 cm, decomposed and dried beech leaves (Fagus sylvatica). A horizon is 10 cm thick, ochric (clear and smooth boundary); texture class is silty loam; brown (7.5YR 4/3, moist) colour; slightly moist water content; very few medium gravel angular-subrounded slightly weathered coarse fragments; moderate nutty subangular blocky structure; soft consistency; fine roots are common. B_w horizon is 29 cm thick gradual and smooth boundary); texture

Fig. 16 Soil profile 3

Fig. 17 Location of the soil profile 4

Table 4 Selected soil properties and sequence of horizons of profile 4

Horizon	Depth [cm]	Textural class	$CaCO_3$ [%]	pH [KCl]	Organic matter [%]	Base saturation [%]
O	2–0	–	–	–	–	–
A	0–10	Silt loam	0.9	6.2	5.4	46.1
B_w	10–31	Silt loam	0.2	6.3	1.2	40.7
B_t	31–41	Silt clay Loam	0.8	6.9	0.3	68.3
B_tC	41–44	Silt clay Loam	1.5	7.4	–	73.3
R	44–(60)	–	–	–	–	–

class is silty loam; yellowish brown (10YR 5/6, moist) colour; slightly moist water content; very few fine gravel; angular-subrounded slightly weathered coarse fragments; moderate subangular blocky structure; friable consistency; fine and medium roots are common. B_t horizon is 10 cm thick, argic horizon (gradual and smooth boundary); texture class is silty clay loam; strong brown (7.5YR 5/6, moist) colour; slightly moist water content; very few fine gravel; angular-subrounded slightly weathered coarse fragments; strong subangular blocky structure; firm consistency; fine and medium roots are common. B_tC horizon 3 cm thick still argic horizon (clear and wavy boundary); texture class is silty clay loam; strong brown (7.5YR 5/6, moist) colour; slightly moist water content; many fine gravel; angular-subrounded slightly weathered coarse fragments; strong subangular blocky structure; firm consistency; few medium roots. R horizon is 16 cm thick, compact Triassic dolomite, moderately weathered at the contact (dolomitic sand and silt) and slightly cracked.

According to the WRB soil classification (IUSS Working Group WRB 2015), the name of the soil is Endodolomitic Epileptic LUVISOL (Neocambic, Cutanic, Epidystric, Anoloamic, Ochric) (Fig. 18).

4 Soil Sequence on Hard and Compact Carbonate Parent Material

The basic interaction between slope processes, landforms in fluvial geomorphic systems and soil formation or other soil parameters is more or less known (Gerrard et al. 1992). The question that remains however is what impact the relief and its elements on soils in karst geomorphic system has where slope processes are different. In karst, vertical drainage is dominating with almost no surface runoff. Chemical weathering usually predominates over physical and denudation is mostly in form of a solution. Accumulation of once weathered material (bedrock) on the other hand is negligible (Bergant et al. 2011).

Soil genesis in the research area shows modified sequence, already described by Stritar (Stritar 1990) in his book Landscapes and landscape systems (Fig. 19). Among which, he also described soil sequences or using his terminology pedosequences on

Fig. 18 Soil profile 4

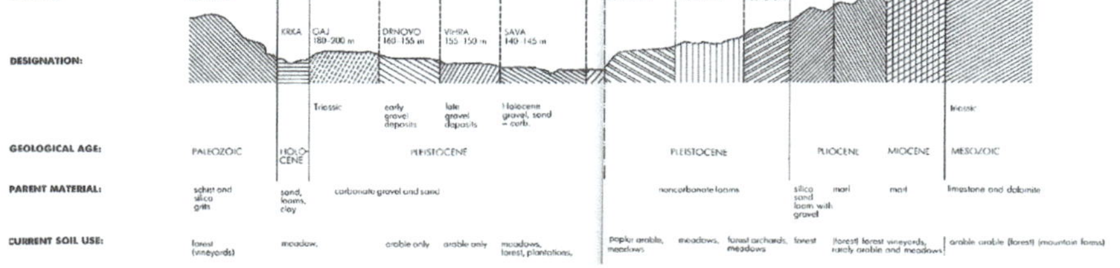

Fig. 19 Connection between chronological base and pedosequence (an example of Krško Polje and its surroundings) (Stritar, 1991)

hard carbonate rocks (limestones and dolomites). He states this is the most important soil sequence in Slovenia, because it appears in all-natural regions (high and low mountains and hills and especially in karstic areas). The developing stages of soil on hard carbonate rocks show considerable regularity in appearance (limestones and dolomites of different periods) throughout the regions of Slovenia. The variety of soil groups on limestones and dolomites is closely connected or conditioned by the specific geomorphology of these surfaces, which are modulated by irregularly distributed sinkholes, smaller valleys and precipitous, larger plateaus, terraces and undulating valleys. This variegated relief by itself would not be a cause for such sudden changes and variations in soil forms where it is not for the associated micro-relief of the rock. It dissects the surface: numerous pockets, cracks, crevices and hollows, as well as larger rocks and boulders, which have broken through the soil cover to the surface and hinder or even prevent the use of this land for agricultural purposes. Only a dolomite base does not show this variegation, rocks and boulders are very rarely on the surface and the surface is equally covered with soil. Humic Leptosols are moderately deep (around 20 cm) and equally thick. More developed decalcified soils (Cambisols and Luvisols) have a larger proportion of clays (Sušin 1964).

In the chosen area hard, Triassic, carbonate rocks prevail and in this special case dolomites. Comparing to the pure limestone areas there exist certain differences (Repe 2018):

The area is distinctively karstic and undulating but it has little rock outcrops.

Dolomite is more prone to mechanically weathering, which results in larger share of coarse rock fragments and somewhat higher soil clay content. Nevertheless, soil is still very shallow.

The contact between soil and hard parent rock is irregular and broken. It depends on micro-scale, local fractures and corrosiveness of the rock. Soil shows extreme mosaic micro-pattern that cannot be effectively predicted by general parameters of lithology, topography or hydrology. By mosaic distribution, we can observe swift changes in the distribution and appearance of rock outcrops, Leptosols, Phaeozems, Cambisols, Luvisols and sometimes even Regosols (Fig. 20).

Due to the very good water permeability soils, karstic areas show very typical automorphic soil genesis. However, dolomite is less permeable than limestone therefore on some locations, Stagnic properties can be developed in Bt horizons.

In Slovenia, soils on dolomite can have extreme carbonate contents. However, this parameter is difficult to estimate in the field (not clearly recognizable), since warm HCl solution is required.

5 Conclusion

Due to the mosaic soil patterns (Fig. 20), soil sequence on hard and compact carbonate parent material is very difficult to estimate, model or predict. The sequence follows some general rule, which mainly depends on slope inclination,

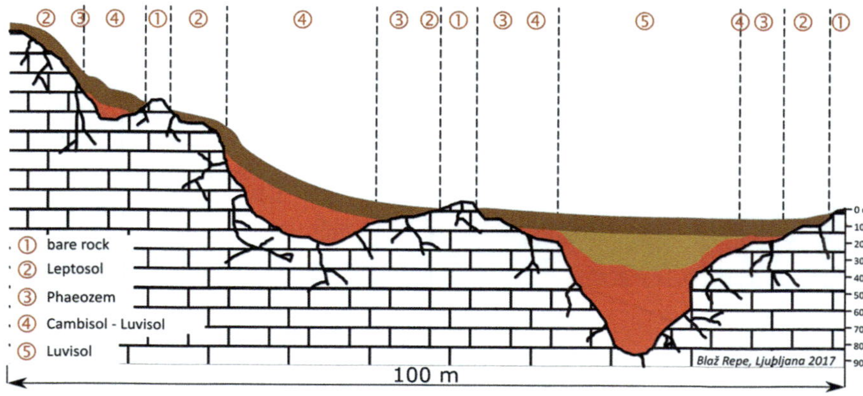

Fig. 20 Mosaic pattern of soil groups on carbonate rocks

i.e. Regosol—Leptosol—Phaeozem—Cambisol/Luvisol with increasing inclination. Bergant (Bergant et al. 2011) made different graphical and statistical analyses based on measurable topography parameters in the karstic area near town of Postojna. As he expected the Rendzic Leptosols and Chromic Cambisols were two predominant soil types found during the sampling. Soil samples were distinguished by their depth and thickness of horizons, organic matter content, water content, form of structural aggregates, texture, etc. An extreme mosaic structure in soil types and depth distribution was found, however, a detailed analysis of sampled transects proved the relationship between topography and soil horizon incidence, thickness and other soil morphological properties. The relationship between measurable relief elements and soil depth was calculated using the Pearson correlation coefficient and multiple regression analysis (soil depth/slope inclination $r = -0.54$; soil depth/curvature $r = -0.27$; $log\ (soil.depth) = 1.577 - 0.015 \times inclination - 0.041 \times curvature \pm 0.14$).

Using the recorded slope data and total curvature derived from DEM 12.5 as input variables a weak ($R = 56.2$, only 31.6% of variability can be explained), but still statistically significant correlation between dependent and independent variable (soil depth) was proved. Similar results were obtained by Repe (Repe 2006a) in the study of digital soil mapping in the Polhograjsko hills where 39.4% of variability had been explained on carbonate parent material. Higher slope inclinations have proportionally shallower soils, larger coarse fragment and lesser clay content, higher pH value, more carbonates and are showing less soil development. Conditions are vice versa is on gentler slopes with heavy leaching of carbonates. In addition to the slope inclination, most of the soil variability can be accounted to the micro-distribution of cracks, impurities and destructions of parent material that can only be determined exactly on the field (Repe 2018).

Theoretical suppositions have been partially confirmed in this field research of soil sequence (Fig. 21):

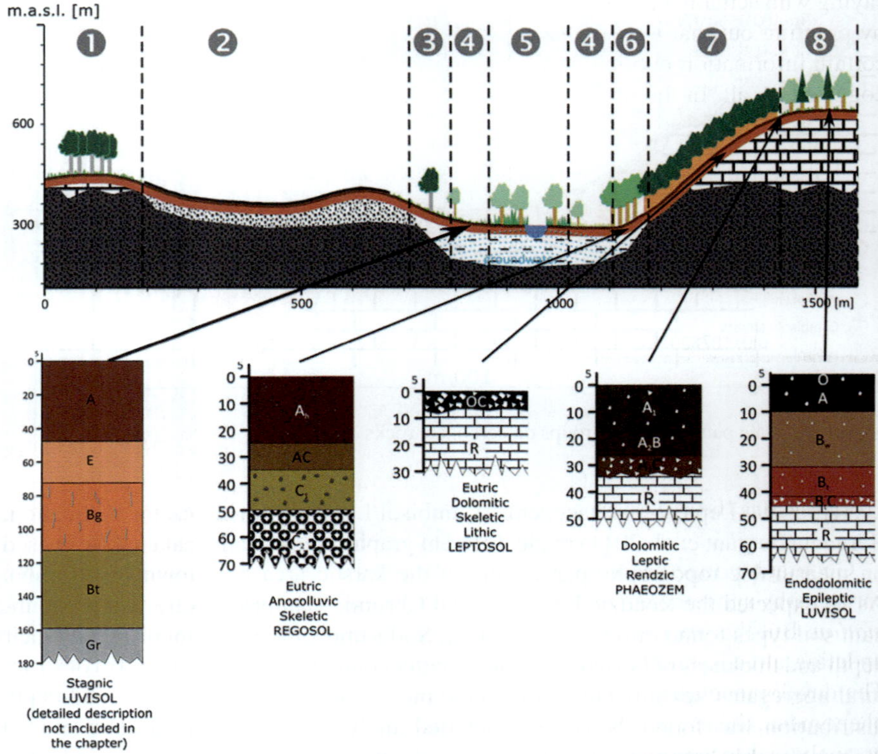

Legend:

	parent material	topography	soil group (part of the soil sequence)	land use
❶	conglomerate	undulating	deep, acid Luvisols	acidophilous *Fagus sylvatica* or *Pinus sylvestris* forest
❷	carbonate gravel	flat, river terraces	Skeletic Eutric Leptosol	fileds
❸	fine, clay and silt sediments	flat	deep, acid Stagnic Luvisols	meadows or *Pinus sylvestris* forest
❹	fine, clay and silt sediments, with deeper ground water	alluvial plain	Stagnosols and Gleysols	meadows and poor fields
❺	fine, clay and silt sediments, with shallow ground water	flood plain	Gleysols, Fluvisols	wet forest (*Salix Alba*, *Fraxinus excelsior* and *Alnus sp.*) and meadows
❻	colluvium	foot slope	Regosols and Stagnosols	moist forest (*Acer sp.*, *Fraxinus excelsior*, *Ulmus sp.*)
❼	dolomite / limestone	irregular slope	mosaic pattern of Rendzic Leptosols, Chromic Cambisols and occasionally Haplic Luvisols	beech forest (*Fagus sylvatica* with *Hacquetia epipactis* or *Lamium orvala*)
❽	dolomite / limestone	karstic, undulating	mosaic pattern of Haplic Luvisols, Chromic Cambisols, Phaeozems and occasionally Rendzic Leptosols	beech forest (*Fagus sylvatica* with *Acer pseudoplatanus* and *Abies alba*)

Fig. 21 Soil sequences in the foot of the Polhograjsko hills

1. The transition between flat and inclined surface, on slope ravines and former river terraces is indicated by an accumulation of soil particles (*Regosols*). They are of colluvial and delluvial origin, where water loses its erosive power and starts to accumulate (unsorted material, moister soils, hygrophilous plant species, poor meadows, fields are rare) (Repe 2007).
2. Slopes are marked with shallow and extremely shallow soils (*Leptosols*) . Soil genesis is extremely slow and is additionally held back by the heavy sheet erosion processes and denudation. By cutting trees in the past, this process of soil cover thinning had been accelerated by man. The only significant processes (besides erosion) are accumulation of humus and very slow mechanical weathering of bedrock (accumulation of rock fragments).
3. The transition between inclined and undulating surface is indicated by the increase of the fine particles, soils are thicker and move towards *Phaeozem/Cambisol* group. Coarse fragments move down to the lithic contact. The water movement is still downslope; therefore, there is no noticeable leaching or argic properties development. On these soils, there is a considerable amount of meadows, pastures or orchards.
4. Undulating karstic topography is generally marked by *Cambisols* and especially *Luvisols*, since water movement is mostly vertical, through the profile and cracks into the parent material. The clay content rises considerably (especially in the concave cracks, fissures and karst dolines) and can be accounted mainly to the residium of former, less resistant bedrock cover (e.g. flysch), colluviation processes from higher positions and the remnants from corrosion of carbonate rocks (iron oxides and other impurities). The percentage of fields and meadows rises considerably (Repe 2008a).
5. Soil genesis on hard carbonate parent material is very slow, 2000–3000 years for 15–20 cm of soil (Vidic 1998). On slopes, it is greatly exceeded by erosion.
6. In spite of the favourable physical and chemical properties of all soils in this sequence, they represent a serious restriction for agriculture (Gams 1974). This can mainly be contributed to shallow soils with many rock outcrops and lack of surface waters (running or stagnant). The percentage of forest cover in land use is the highest in Slovenia and can be up to 70 or even 90%. On northern, slope expositions away from the sun, it is an exclusive land use type. These soil sequences are typical areas of beech forest association (*Hacquetio-Fagetum*, *Lamio orvalae-Fagetum* and even *Omphalodo-Fagetum*) (Vrščaj et al. 2017).

Soils hiding beneath our feet or in nature below a thick forest or grassland cover heavily influence our lives. The very frequent appearance of soil types with quite different properties provides a sound basis only for forest. Arable land appears only in karstic valleys into which over the course of geological time, soil has been washed from the slopes (colluvial non-carbonate soils), as well as on larger plateaus. On steep positions and in general in mountain and high mountain karst, younger and shallower soil forms predominate and primarily a number of Leptosol groups which are in places sometimes covered by grasslands and more often by forest. Rural development in these regions is relatively slow because of the problematic supply water and problems of wastewaters. Sunny slope aspects suitable for building are in closed valley systems, which can communicate with more developed areas. This is also relatively stable land according to the earthquakes. In this soil sequence, it would be necessary clearly to delimit urban space from rural. Unregulated settlement usually uses good arable land, which is not overabundant in this soil sequence. There are usually more suitable sunny less steep positions with shallow soil (Leptosol, Phaeozem, eroded Cambisols soil, etc.) which gives a good ecological base for settlement (Stritar 1990).

This is especially the case of the researched Polhograjsko hills foot slopes. In the past, people understood the landscape and knew the value of hidden soils. The uplifted parts with relatively poorer and shallow soils were left under forest. Levelled land, with much better and deeper soil cover, was used for agriculture. The residential areas or rather residential spots were squeezed over transitional foot or toe-slopes (number 6 on Fig. 21). In such manner, the best soils had been preserved and shallow soils on slopes had been protected by tree roots and canopies against erosion.

It is somehow sad; today's generations have completely forgotten this hidden aspect of the landscape. That soils are (also) an integral part of land management and spatial planning.

Poor forest management (favouriting of spruce, *Picea abies*) leads to severe damages due to the glaze frost, wind/snow/storm falls and bark beetle attack which furthermore leads to clear cutting of forest (Fig. 22). On the other hand, nowadays, the best soils became one of the most desirable, the most eminent, the most expensive and the fastest growing suburban residential areas around the capital (Fig. 23).

Fig. 22 Cleared forest due to the catastrophic glaze frost in 2014

Fig. 23 Rapidly growing suburban residential areas at foot of the Polhograjsko hills (*photo* Aleš Jesenšek)

Revealing the hidden connections beneath our feet largely broadens the knowledge of a given landscape. In order to successfully preserve important landscape features and protect our soils, hidden concept should be obligatory considered, while conducting a holistic, geographical research.

References

ARSO. 2005. *Klimatografija Slovenije, Povprečna višina padavin, obdobje 1961–1990*. Ljubljana: Agencija Republike Slovenije za okolje.

Bergant, J., B. Repe, and B. Vrščaj. 2011. *Relief kot pedogenetski dejavnik na krasu*. Ljubljana: University of Ljubljana, Faculty of Arts.

Bushnell, T.M. 1942. *Some aspects of the soil catena concept*. Ann Arbor: Michigan.

Erhartič, B. 2010. Geomorphosite assessment. *Acta Geographica Slovenica* 50: 295–319.

Gabrovec, M. 1989. *Influence of the relief on geographical view of the Polhograjsko hills*. Department of Geography: Faculty of Arts, University of Ljubljana.

Gams, I. 1974. *Kras: zgodovinski, naravoslovni in geografski oris*. Ljubljana: Slovenska matica.

Gerrard, J., J.G. Gerrard, and A.J. Gerrard. 1992. *Soil geomorphology: an integration of pedology and geomorphology*. 1993 edition. London; New York: Spring

Grad, K., and L. Ferjančič. 1976. *Basic geological map, 1:100.000, section Kranj L 33–65 (interpretation)*. Beograd: Zvezni geološki zavod.

Hall, G.F. 1983. Pedology and geomorphology. In *Developments in soil science*, ed. L.P. Smeck, N.E. Hall, G.F. Wilding, 11:117–140. Elsevier Ltd. https://doi.org/10.1016/S0166-2481(08)70600-7.

IUSS Working Group WRB. 2015. *World reference base for soil resources 2014, update 2015 international soil classification system for naming soils and creating legends for soil maps. World soil resources reports No. 106*, 1st ed. Rome: FAO.

Jenny, H. 1941. *Factors of soil formation: A system of quantitative pedology*, vol. 1941. New York and London: Dover Publications.

Kočevar, H., and N.J. Vidic. 1998. *Izbrana poglavja iz osnov geologije*. Ljubljana: Biotehniška fakulteta, Oddelek za agronomijo.

Lovrenčak, F. 1994. *Pedogeografija*. Ljubljana: Oddelek za geografijo, Filozofska fakulteta, Univerza v Ljubljani.

Marinček, L. 1987. *Bukovi gozdovi na Slovenskem*, 1st ed. Ljubljana: Delavska enotnost.

Milne, G. 1935. Some suggested units of classification and mapping particularly for East African soils. *Soil Research* 4: 183–198.

Ministry of Agriculture and the Environment. 1999. *Digital soil map of slovenia*. Ljubljana: Ministry of Agriculture and the Environment.

Ogrin, D. 1996. Podnebni tipi v Sloveniji. *Geografski Vestnik* 68: 39–56.

Ogrin, D., Bobovnik, N., Cigale, D., Koželj, T., Natek, K., Ogrin, M., Prašnikar, D. et al. 2017. *Kamniška Bistrica - geografska podoba gorske doline*, ed. Darko Ogrin, 1. izd. Vol. GeograFF 22 . Ljubljana: Znanstvena založba Filozofske fakultete.

Ogrin, D., and D. Plut. 2009. *Aplikativna fizična geografija Slovenije*. Ljubljana: Znanstvena založba Filozofske fakultete.

Prus, T. 2000. *Klasifikacija tal slovenije—študijsko gradivo*. Ljubljana: Center za pedologijo in varstvo okolja.

Radinja, D. 1996. Regionalnogeografska monografija Slovenije. In *Regionalnogeografska monografija Slovenije. 2. del, Zahodni predalpski svet*, ed. Marja Bat, Andrej Černe, Jurij Kunaver, Darko Radinja, Mirko Pak, and Drago Perko. Vol. II. Ljubljana: Znanstvenoraziskovalni center SAZU, Geografski inštitut.

Repe, B. 2004. Soils of slovenia. In *Slovenia: A geographical overview*, ed. Milan Orožen Adamič, 51–56. Ljubljana, Slovenia: Association of the Geographical Societies of Slovenia, Založba ZRC.

Repe, B. 2006a. *Pedogeografska karta in njena uporabnost v geografiji (doktorska disertacija)*. Ljubljana: Univerza v Ljubljani, Filozofska fakulteta, Oddelek za geografijo.

Repe, B. 2006b. Novi pristopi pri proučevanju prsti v pokrajini. *Dela*.

Repe, B. 2007. Voda v prsti in ugotavljanje njenega razporejanja v odvisnosti od reliefa. *Dela*: 91–106.

Repe, B. 2008a. Ugotavljanje razprostranjenosti tipov prsti v pokrajini z uporabo klasifikacijskih metod daljinskega zaznavanja. In *Geografski informacijski sistemi v Sloveniji 2007–2008*, ed. Drago Perko, Matija Zorn, Nika Razpotnik, Marjan Čeh, David Hladnik, Marko Krevs, Tomaž Podobnikar, Blaž Repe, and Radoš Šumrada, 195–206. Založba ZRC.

Repe, B. 2008b. Ali smo že izgubili tla pod nogami?: Prst—zanemarjena vrednota preteklosti. *Delo (Polet)* January 24.

Repe, B. 2008c. Nezanimiva problematika. *Delo (polet)*, October 16.

Repe, B. 2009. Nezanimiva problematika II. *Delo (Polet)*, August 20.

Repe, B. 2010. Daleč od oči, daleč od srca. *Delo (Polet)*, March 4.

Repe, B. 2018. Soils of the Ljubljana basin western edge (foot slopes of the Polhograjsko hills). In *Soil sequences atlas II*, ed. Marcin Switoniak and Przemysław Charzyński, 1st ed., 2:201–2017. Torun, Poland: Machina Druku.

Smrekar, A., M. Breg, and P. Frantar. 2006. From drawing cognitive maps to knowing the protection zones for drinking water resources. *Acta Geographica Slovenica* 46: 7–31.

Smrekar, A., Šmid Hribar M., Tiran, J., Erhartič, B., Bole, D., Kladnik, D., Breg Valjavec, M. et al. 2014. *Interpretacija okolja na primeru Ljubljanskega barja*, ed. David Bole and Drago Kladnik. Ljubljana: Geoografski inštitut Antona Melika ZRC SAZU.

Stritar, A. 1990. *Krajina in krajinski sistemi*. Ljubljana: Partizanska knjiga.

Stritar, A. 1991. *Pedologija: kompendij*. Ljubljana: Biotehniška fakulteta, Agronomski oddelek - Ljubljana.

Sušin, J. 1964. *Doprinos k poznavanju terre rosse v Slovenskem Primorju*. Ljubljana: Univerza v Ljubljani, Biotehniška fakulteta.

Vrščaj, B., Repe, B., and P. Simončič. 2017. *The soils of Slovenia*. Dordrecht: Springer.

Vidic, N.J. 1998. Soil-age relationships and correlations: Comparison of chronosequences in the Ljubljana Basin, Slovenia and USA. *CATENA* 34: 113–129. https://doi.org/10.1016/S0341-8162(98)00085-X.

Zupančič, M., Marinček, L., Prešeren, M., Puncer, I., Tregubov, V., and V. Žagar. 1972. *Vegetacijska karta Slovenije, čistorisi–originali. Skenogram vegetacijske karte M 1:50.000 list Ljubljana-3.* Ljubljana: Biološki inštitut Jovana Hadžija, Znanstvenoraziskovalni center SAZU.

Zupančič, M., Prešeren, M., Puncer, I., Tregubov, V., and V. Žagar. 1973. *Vegetacijska karta Slovenije, čistorisi–originali. Skenogram vegetacijske karte M 1:50.000 list Bled-4.* Ljubljana: Biološki inštitut Jovana Hadžija, Znanstvenoraziskovalni center SAZU.

Uncovering Spatio-temporal Air Pollution Exposure Patterns During Commutes to Create an Open-Data Endpoint for Routing Purposes

Bram Vandeninden, Charlotte Vanpoucke, Olav Peeters, Jelle Hofman, Christophe Stroobants, Sam De Craemer, Hans Hooyberghs, Evi Dons, Martine Van Poppel, Luc Int Panis, and Frans Fierens

Abstract Air pollution is difficult to detect with human senses. It is to a large extent out of sight and out of sense, while causing a burden on our economy, our health and our environment. A relevant illustration of this is the exposure to air pollution during commutes. The air pollution commuters are exposed to remains to a considerable

---

B. Vandeninden (✉) · C. Vanpoucke · O. Peeters · F. Fierens
Belgian Interregional Environment Agency, Schaarbeek, Belgium
e-mail: bramvandeninden@gmail.com

C. Vanpoucke
e-mail: vanpoucke@irceline.be

O. Peeters
e-mail: opeeters@gmail.com

F. Fierens
e-mail: fierens@irceline.be

J. Hofman
Holst Centre, Eindhoven, The Netherlands
e-mail: Jelle.Hofman@imec.be

C. Stroobants
Flanders Environment Agency, Antwerp, Belgium
e-mail: c.stroobants@vmm.be

S. De Craemer
Department of Biology, University of Antwerp, Wilrijk (Antwerpen), Belgium
e-mail: sam.decraemer@gmail.com

H. Hooyberghs
VITO Environmental Modelling Unit, VITO, Mol, Belgium
e-mail: hans.hooyberghs@vito.be

E. Dons
Universiteit Hasselt, Hasselt, Belgium
e-mail: evi.dons@uhasselt.be

M. Van Poppel
VITO, Flemish Institute for Technological Research, Mol, Belgium
e-mail: martine.vanpoppel@vito.be

M. Krevs (ed.), *Hidden Geographies*, Key Challenges in Geography,
https://doi.org/10.1007/978-3-030-74590-5_6

extent a hidden geography, with, for example, a lack of available reliable information regarding the on-the-road concentrations of several air pollutants. This research aims to unravel, to the best possible extent, spatio-temporal air pollution patterns (active) commuters are exposed to. Cyclists and pedestrians can be unaware that they commute in polluted air. They often travel close to motorised traffic, resulting in high exposure to several air pollutants, which have elevated levels on the road due to vehicular emissions. Significantly higher concentrations of particulate matter (<2.5 μm), black carbon and nitrogen dioxide were found on roads with high-traffic intensities than on roads with less traffic, cycling highways or separated cycle lanes. The amplitude of the concentration differences between routes depends on both temporal factors, such as the season, the day of the week, or the time of day, and spatial factors, such as the traffic's density, the footpath or cycle lane's location, the architectural makeup (e.g. street canyons) and the meteorological conditions. Using high-resolution air pollution models, it is possible to distinguish between routes of higher and lower air pollution concentrations, allowing active road users to choose an alternative route to lower their air pollution exposure. However, on-the-road concentrations displayed by the Belgian ATMO-Street model are often considerably underestimated, especially for routes with high levels of motorised traffic. In general, for air pollution models to distinguish between routes, a minimum spatial-model resolution of 10 m^2 including street configuration effects (e.g. street canyons) is desired. For temporal resolution, static seasonal-hourly raster model data, calculated from a previous year's hourly data, are sufficient to make a scientifically sound distinction between alternative routes regarding exposure to air pollution. Those tools are a great help in uncovering the spatio-temporal pollution patterns (active) commuters are exposed to and also provide relevant insights to reduce the health and economic burden of air pollution, which is unseen to a large extent and of which most people are not aware. Additional research using microscale measurement setups to further unravel gradients in air pollutant concentrations and further reveal reliable estimates of on-the-road concentrations of those pollutants is recommended.

Keywords Air pollution · Exposure · Commute · Bicycling · Walking · Open data · Routing · Spatio-temporal patterns

L. I. Panis
Universiteit Hasselt, Diepenbeek, Belgium
e-mail: luc.intpanis@uhasselt.be

1 Introduction

Air pollution causes harmful effects on our health and the environment with its presence and effects often difficult to detect with human senses. Despite available resources such as measurement stations of air pollution and high-resolution modelisation of air pollution, spatio-temporal patterns of air pollution largely remain a hidden geography. When people cycle or walk for either commuting or recreational purposes, they are often exposed to extremely elevated concentrations of (invisible) Air Pollution (AP), and are frequently unaware of this exposure. In general, exposure to air pollution is often highest during commutes (Karanasiou et al. 2013; Rivas et al. 2016). The proximity of commuters to road traffic—one of the major pollution sources of air pollution—is an important explanation of this phenomenon (Dons et al. 2012, 2013), given that near-road horizontal gradients of air pollution can be very steep in the proximity of traffic (Sharma et al. 2009; Apte et al. 2017). Nitrogen dioxide (NO_2) concentrations and particulate matter concentrations can decrease 30–50% within the first 250 m from the traffic emission source while Black Carbon (BC) concentrations (and nitrogen oxide (NO) concentrations) decrease at an even steeper gradient, decreasing in some cases more than 50% within the first 200 m from the traffic emission source (Apte et al. 2017). Fisher and Metcalfe (2008) found a decrease of more than 40% in concentrations within the first 25 m from the road. Karner et al. (2010) found that concentrations of Ultrafine Particulate (UFP) matter and Electric Carbon (EC) more than halved within the first 100 m from the edge of roads. For air pollutants with a low background concentration, such as polycyclic aromatic hydrocarbons, some studies have even observed a difference of a factor ten in terms of concentrations on roads compared to 200 m from the edge of the road (Kim et al. 2015). Horizontal gradients of concentrations of air pollutants can also be influenced by meteorological conditions such as wind direction, wind speed and temperature (Padró-Martinez et al. 2012; Choi et al. 2013) and physical barriers such as landscape configuration (Gallagher et al. 2015). These gradients can be even steeper during peak hours. A study by Lonati et al. (2016) found that during peak hours, concentrations of BC and UFP are already 10–50% lower on a number of roads where cycling lanes are separated at least 2.5 m from traffic compared to on-road cycling lanes. Furthermore, a study conducted in Antwerp showed that distance to traffic and street typology are the dominant factors that determine the UFP and BC concentrations (Peters et al. 2014).

The air pollution to which cyclists and pedestrians are exposed can have serious effects on their health. Acute effects of exposure to peak concentrations can take the form of decreased pulmonary function and cardiovascular effects, nasal irritation (Cole-Hunter et al. 2013; Int Panis et al. 2017; Laeremans et al. 2018), as well as long-term effects from repeated or continuous exposure to high concentrations such as an increased risk of lung cancer, negative neurological effects, heart failure and asthma (Atkinson et al. 2013; Rückerl et al. 2011; Harris et al. 2016). Traffic-related exposure has also been linked to the development of allergic diseases while childhood allergies have been associated with exposure to traffic-related AP during pregnancy

(Deng et al. 2016). A study by Alvarez-Pederol et al. (2017) found that exposure to traffic-related BC and $PM_{2.5}$ during a commute to school is significantly associated with cognitive decline and a reduction in the growth of the working memory of affected children. Another study (Sunyer et al. 2017) found a significant link between short-term exposure to traffic-related AP and the attention level of children in school. Similarly, a significant association between exposure to traffic-related AP and amongst others memory span length was found, indicating that long-term exposure to traffic-related AP can adversely affect cognitive capacities (van Kempen et al. 2012). Because of the health effects of such exposure, it looks plausible that choosing a route with less air pollution can be beneficial for your personal health. Commuting via routes with lower AP levels whenever possible could hence reduce societal health costs. Car users are exposed to similar amounts of air pollution—depending on the study, slightly less/more compared to active road users, although active commuters (pedestrians, cyclists, etc.) inhale far more air pollution as a result of accelerated and deeper breathing during physical activity (Int Panis et al. 2010). While research indicates that the adverse health effects of air pollution do not outweigh the positive effects of increased physical activity among active commuters (Rojas-Rueda et al. 2011), a significant reduction in air pollution exposure (for example, switching daily commute to a route of similar distance and on average lower pollution levels) would increase the positive health effects of physical activity among pedestrians and cyclists. Travel time is likely to be increased on roads featuring considerable amounts of traffic (Lee and Sener 2019). Imagine, for example, traffic lights or places where it is difficult to cross the roads. Consequently, the exposure on traffic-dominated roads might be underestimated when using methods that calculate average concentrations based on distance. On the other hand, factors that cause cyclists to lower their speed or to wait at traffic lights decrease breath frequency and depths compared to when actively moving, implying reduced inhalation during those moments.

Decisions about which commute or recreational jogging routes to follow and the development of cycling infrastructure are usually based on criteria such as time, distance and safety, rather than the possibility of exposure of air pollutants (Giménez-Gaydou et al. 2019). Providing air quality data suitable for a routing context can serve various use-cases including helping users to identify cleaner routes, reducing exposure to air pollution and, as a consequence, reducing health and economic impacts. These health impacts and economic impacts are also a hidden geography. Another potential use-case is to combine the AP-data-endpoint with traffic data and detect mismatches between pollution and the use of the traffic infrastructure. Such open-AP-data can raise awareness about the causes and spatio-temporal patterns of AP-exposure during commutes and can empower citizens and/or local governments to take measures to reduce AP-exposure. Another measure is to construct cycling and pedestrian paths that are separated by at least some metres from road traffic, while reductions in motorised traffic designed to effectuate a model shift can also significantly reduce exposure to air pollution during commuting (Lonati et al. 2016; Pankow et al. 2014; Karanasiou et al. 2013). The benefits of creating cycling infrastructure and cycling lanes far outweigh the initial expenditure thanks to amongst others reduced healthcare costs (increased physical activity, improved air quality), diminished traffic congestion and potentially reductions in the number of traffic accidents (Buekers et al. 2015).

Providing a scientifically sound open-data endpoint, with modelled air pollution data displaying information about spatio-temporal patterns of air pollution during commute, can be useful for various purposes such as (1) making visible the invisible. That is, revealing spatio-temporal patterns and uncovering air pollutant concentrations commuters are exposed to (2) behavioural change in terms of routing choices for both pedestrians and cyclists in order to lower their personal exposure to air pollution and reduce the negative impact of air pollution on their health; (3) a policy-tool to support decisions as to where to implement new cycling infrastructure or other relevant facilities or where to take measures to reduce pollution; (4) an educational tool to raise awareness about the spatial patterns and causes of air pollution and (5) to stimulate more people to use active modes of transport. This could be realised through amongst others including AP-data in (smartphone) apps that promote physical activity (Boccupther et al. 2018) and a potential contribution to a modal shift through increased awareness regarding the harmful effects of motorised traffic on fellow citizens' health.

This research highlights the results of the following points:

1. A comparison of available AP modelled data with available measurements. From this, a conclusion is drawn regarding the suitability of using AP model data in a routing context (making visible spatio-temporal exposure patterns active commuters are exposed too). Various available models and model resolutions are compared;
2. A comparison of the different air pollutants and a discussion of the suitability of certain pollutants in a routing context;
3. A comparison of air pollution on high-traffic and traffic-calmed roads (modelled concentrations): case-studies for Antwerp, Leuven, Ghent and Orléans;
4. Unocvering Spatio-temporal exposure patterns of AP during commute including daily-hourly, monthly, seasonal and annual variations.
5. Guidelines 1–4 develop a scientifically sound open-data endpoint suitable in a routing context.

2 Data and Methods

2.1 Data

Various sources of data are employed in this paper including (1) modelled air pollution data from ATMO-Street models; (2) trajectory measurements of BC in the Antwerp, Leuven and Mechelen areas; and (3) NO_2 measurements from the CurieuzeNeuzen project (curieuzeneuzen.be).

2.1.1 Modelled Data

The modelled data used in this study originate from an integrated model chain called ATMO-Street (Lefebvre et al. 2013), which is a combination of the RIO, IFDM and OSPM models, which assess air quality on a regional (RIO), local (IFDM) and street canyon (OSPM) scale.

The regional contribution to the air pollutant concentrations is calculated on a $4 \times 4\ km^2$ resolution by the RIO land-use regression model (Janssen et al. 2008), using hourly measured concentrations reported by the official monitoring stations. Local increments due to road transport, shipping emissions and point sources are modelled using the IFDM Gaussian dispersion model (Lefebvre et al. 2011a, b). The IFDM departs from emission sources and models its distribution by means of meteorological parameters. Point source emissions (e.g. industrial sources) and line source emissions (road traffic, shipping) are added as an input to the IFDM, while other potential sources of air pollution such as households (e.g. wood-burning stoves) and agriculture are included with the calculation of RIO background concentrations. The model is a so-called receptor model: it computes the pollutant concentrations caused by the provided emission sources at specific receptor location. In the current setup, the nearest grid points are located at 15 m from the centre of the road, thus omitting the peaks on the actual roads. The IFDM is an open-street model, meaning that pollutants are dispersed in the air as if there are no obstacles, and as a result, it is not suitable for estimating air pollution within street canyons (small streets that frequently have high buildings). To tackle this shortcoming, the street canyon module OSPM (Berkowicz 2000) was used. OSPM models façade concentrations at both sides of a street canyon using a combination of a Gaussian plume model for the direct contribution and a box model for the re-circulating part of the pollutants in the street. Note that the actual peaks on the roads are also omitted by the OSPM. This model including the OSPM-module is used for all analyses in Belgium and the model is renamed recently to ATMO-Street. There is also one small analysis for the city of Orléans in France based on the annual map of 2018. For this, the model SIRANE is used, a high-resolution model that takes street configuration effects into account (Soulhac et al. 2011).

2.1.2 Measurement Data

Black Carbon Trajectory Measurements Several mobile measurement campaigns were carried out in the Antwerp, Mechelen and Leuven areas. BC was measured along different trajectories with a microAeth type AE51 (AethLabs, San Francisco, CA, USA). Results of the average measured BC concentrations (repeated measurements, conducted via a bike, for several routes) over the considered time period are used (for the Antwerp area, in addition to an analysis on a higher temporal resolution). These periods were as follows:

- For the Antwerp area: January 13, 2017–March 24, 2017 (Hofman et al. 2018);

- For the Leuven area: May 11, 2017–June 19, 2017 (http://www.airqmap.com/leuven.html); and
- For the Mechelen area: October 2017 and November 2017 (https://mechelen.meetmee.be/).

All these trajectory measurements were performed using an optimised noise-reduction algorithm (Hagler et al. 2011) with an ATN threshold of 0.05 (Hofman et al. 2018; http://airqmap.com, https://mechelen.meetmee.be), which has been also applied in similar studies in comparable cities (Peters et al. 2014; Van den Bossche et al. 2015). The trajectory measurements in the Antwerp area are further corrected using the loading correction algorithm of (Virkkula et al. 2007). For the Mechelen and Leuven measurements, filter tickets are frequently replaced in order to reduce the impact of loading corrections.

Measurement uncertainties related to those measurements are relatively small, with (Hofman et al. 2018) showing a Pearson-correlation of 0.77 for the 1-min level, 0.94 for the 5-min level and 0.97 for the half-hour level when the AE51 is compared with a more reliable non-portable measuring device (AE33) in one stationary measurement station in the Antwerp area. Furthermore, absolute deviations seem to be small with an RMSE of 0.49 for the 1-min level, 0.23 for the 5-min level and 0.15 for a half-hourly temporal aggregation (Hofman et al. 2018).

Curieuze Neuzen NO_2 data In the CurieuzeNeuzen project (Meysman and De Craemer 2018), NO_2 was measured via Palmes passive samplers at 20.000 building facades in Flanders from April 28 to May 26, 2018. On each site, there was one sampler. Based on this, the average NO_2 concentration over the investigated period was calculated for each sampler. There was an on-site calibration of the samplers at 20 reference measurement stations. The one-sigma uncertainty for the samples is 2.2 μg, implying that 63.8% of the samples lie within this one-sigma uncertainty (Meysman and De Craemer 2018). Sigma (for the entire dataset of 20.000 points) was calculated based on a comparison of a selection samplers at 20 reference measurement stations. Statistical models (based on regression analysis) were applied for the calculation of the uncertainty. A uniform correction constant was applied to all samplers—an increase of 3.63 $\mu g/m^3$ to the raw data result for each sampler because the samplers compared with the reference measurements stations delivered on average a result 3.63 $\mu g/m^3$ lower compared to those reference measurement stations (Meysman and De Craemer 2018).

For this analysis, only data results of samplers located near city-ringroads were analysed. This is a total of 57 locations, from ringroads of middle-sized to large Belgian cities such as Ghent, Mechelen, Leuven, Aalst, Bruges and Roeselare.

2.1.3 Data Preparation

Model post-processing The hourly air pollution model results calculated with both RIO-IFDM and ATMO-Street for BC and NO_2 were aggregated at various temporal scales and post-processed into suitable formats for further analysis:

a. Annual data (2016, 2017 and 2018);
b. Selected specific hourly maps (2016, 2017): Hourly maps of 36 randomly selected days for RIO-IFDM (2016, 2017) and 18 randomly selected days for ATMO-Street (2017 only). Specific intervals (7–8 UTC, 12–13 UTC and 16–17 h UTC) are investigated from these randomly selected days;
c. Annual-hourly maps (2017): Hourly concentrations averaged per hour of the day over a calendar year; e.g. the mean modelled concentration between 8 and 9 a.m. throughout 2017;
d. Seasonal-hourly (2017): Concentrations averaged per hour of the day over a meteorological season; and
e. Monthly-hourly maps (2017): Hourly concentrations averaged per hour of the day over a calendar month.

Route selection and GPX file conversion Several routes using the regular start-end locations of the routes used by individual commuters were tested (trajectories that are frequently cycled by commuters, e.g. from the railway station of a city to daily activity hotspots such as universities, large companies or from residential hotspots to the railway station, etc.). Based on these start-end locations, several routes were generated in the routing engine of Google Maps, resulting mostly in two or three route alternatives for each origin-destination route. To create these routes, the cycling transport mode was chosen as the default transport mode. These routes were converted to GPX-files using a GoogleMaps-GPX converter (https://mapstogpx.com).

2.2 *Methodology*

2.2.1 GPX-File: Equidistant Points Method

The framework of this research embodies a conversion of the automatically generated GPX-files from non-equidistant waypoints to GPX-files with equidistant points every 5 m along the considered trajectory (Fig. 1). The use of equidistant points every 5 m ensures a maximum uncertainty of 1% that is added to the uncertainty concerning the mean air pollution concentration calculated for a route with a distance >1 km. If repeated samples are taken, the 95% confidence interval of the mean air pollution concentration of a sample with a specified spatial resolution (1 sample point every

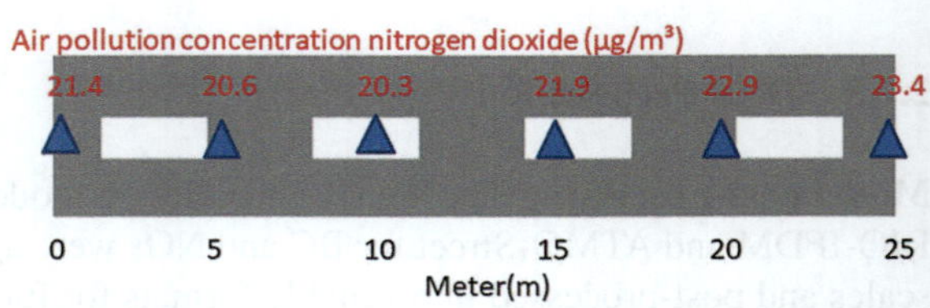

Fig. 1 Example calculation average NO_2 concentrations over a very short route of 25 m: = (21.4 + 20.6 + 20.3 + 21.9 + 22.9 + 23.4)/6 = 21.8 μg/m^3

× metres) should always fall within the 99% confidence interval of the mean air pollution concentration of this sample (±1%) using an infinite resolution (in this calculation, the assumption is that an infinite resolution is equal to 1 point every 1 cm, which implies for example 800,000 points for a route of 8 km). The conclusion using this method is that, for routes >1 km, one equidistant point every 5 m is suitable for further analysis. From a qualitative point of view, this makes sense because the model resolution (incl. street canyon effects) is 10 × 10 m. In our analysis, only distance is used, which means that the average concentration over a route is also calculated based upon distance, and does not take into account the speed of cyclists or attributes such as slopes and traffic lights.

2.2.2 Data Analysis: Modelled Exposure Along Alternative Routes

Average modelled air pollution concentrations for points along various route alternatives for selected real-life cases (real-life origins and destinations) are analysed, using the equidistant 5 m points method explained in the previous section. Different analyses are executed including:

1. An investigation as to whether there is a significant difference in mean air pollution concentrations between two routes. Shapiro-Wilk demonstrated that the distribution of the concentrations of the investigated routes significantly differs from a normal distribution ($p < 0.05$). For this reason, the test for a significant difference in concentrations between the mean of two different routes is determined via the Wilcoxon rank sum test, with continuity correction for the points with the maximum spatial resolution (equidistant points every 5 m). The null hypothesis, that there is no difference in air pollution concentrations between the two routes, is rejected if $p < 0.05$. No additional model uncertainty is added during this analysis;
2. The variation in the air pollution concentration from the origin to the destination of the route;
3. Descriptive statistics;
4. An analysis of spatio-temporal patterns (annual, seasonal, monthly, daily, hourly).

2.2.3 Data Analysis: Model and Measurements Comparison

BC trajectory measurements The results of the trajectory measurements are used to evaluate both the RIO-IFDM and the ATMO-Street model results for BC for the mentioned measurement campaigns in Leuven, Mechelen and Antwerp. For the Antwerp area, in addition the trajectory measurements of BC are also analysed at a higher temporal resolution, e.g. for specific days where both measurement and modelled data are available, the modelled data are compared with the measurement data for the same days and times (aggregated over the entire period available) as the modelled data (at 1-h temporal resolution). An additional analysis is performed on the effect of meteorological conditions on the model results for this dataset, including temperature, relative humidity, wind speed and wind direction.

For all analyses, the modelled values of both RIO-IFDM and ATMO-Street (1-h temporal resolution) are compared with measured values (aggregation of instantaneous measurements during monitoring, corrected for noise). Various error statistics are applied including:

- Coefficient of determination or R squared (R^2): Indicates the proportion of variance of the measured values that can be explained by the modelled values.
- Mean Bias: indicates whether there is, on average, a net model overestimation or underestimation of measured air pollution concentrations. Because underestimations and overestimations have the potential to cancel each other out, it tells us little about the absolute model deviations from measurements.
- FAC2: percentage of modelled values that are within 2 times and 0.5 times the measured concentration.
- The Index of Agreement (Wilmott et al. 2011; Carslaw 2015). Provides an indication for the sum of the magnitudes of the differences between the model-predicted and measured deviations about the measured mean relative to the sum of the magnitudes of the perfect model, in which the trajectory measurements are assumed to be a 'perfect model'.

2.2.4 Data Analysis: CurieuzeNeuzen NO_2 Data

The CurieuzeNeuzen results along the investigated routes in this study were compared to the modelled values over the same period (May 2018). Specific attention was given to analysis in an urban context and especially ring roads. For this analysis, the same error metrics are calculated in the same way as in the above section on BC trajectory measurements.

2.2.5 Measurement Stations

Measurement stations are used to compare correlations between pollutants in measurement stations. Correlations found in the measurement stations are compared to the correlations found in the modelled data (for the available pollutants).

3 Results

3.1 Comparison of Measurements and Models

See Fig. 2.

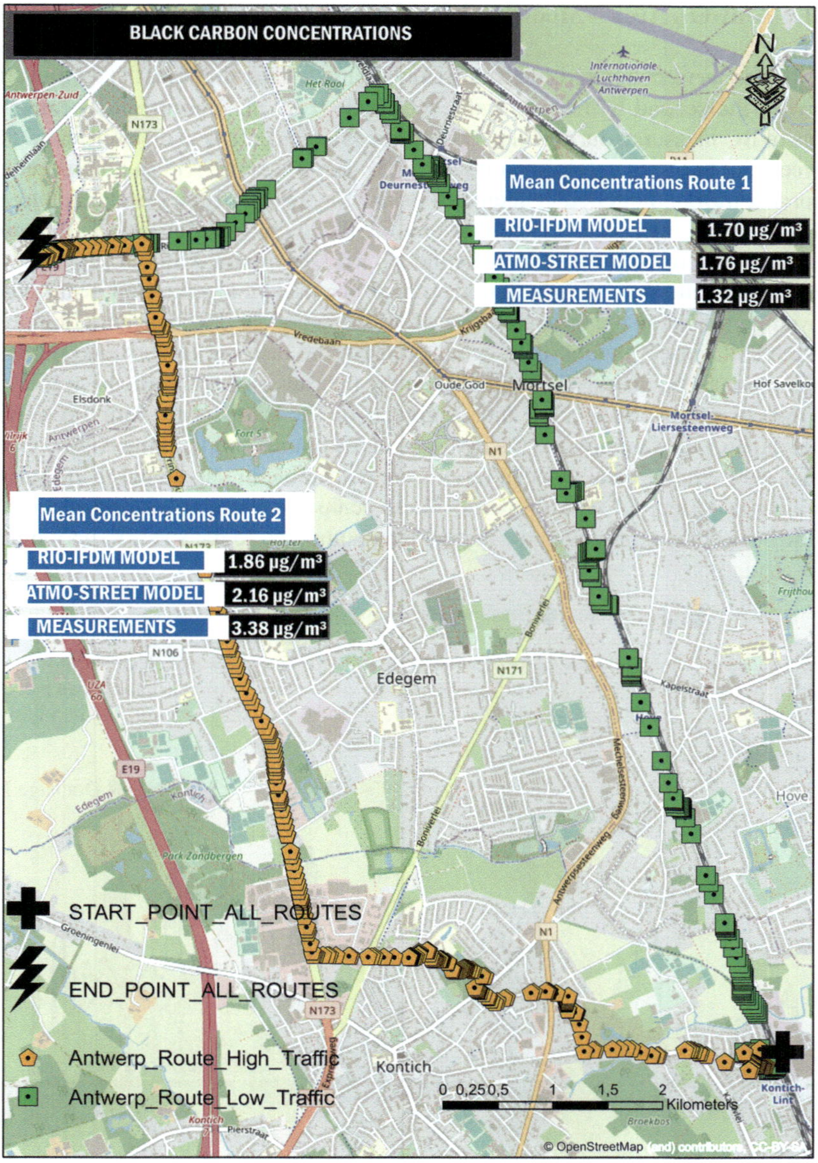

Fig. 2 Mean concentrations (modelled and measured) of BC for two routes in the Antwerp area. For the calculation of mean air pollution values per route, the original routes are transformed into routes with equidistant points each 5 meter as explained in Fig. 1

3.1.1 Antwerp Area—Modelled Versus Measured BC Concentrations

Two routes were compared, the first being characterised dominantly by a cycling highway, the second by heavily trafficked roads (Fig. 2). The trajectory measurements of BC measured during bike travel showed a concentration difference of 256% between the two routes (Hofman et al. 2018). For the route with high amounts of motorised traffic, modelled concentrations are for ±90% of the distance along the route below the measured concentrations for the ATMO-Street model (including street canyon effects), even ±95% for the RIO-IFDM model (without street canyon effects). The latter model fails to detect local increased concentration peaks and is on average far below the measured concentrations (Fig. 3). The concentration peaks in the ATMO-Street model often occur on the same or close to the same locations on the route as the observed local peaks in the measured concentrations (Fig. 3).

Figure 4 compares the same measurements on the traffic route with the ATMO-Street data for the same period (as in the previous figure) and with the ATMO-Street annual map. Similar peak concentrations are detectable in the annual map as in the specific-period map of ATMO-Street. These peaks are lower in the annual map, as are the base concentrations in the annual map compared to the specific-period map.

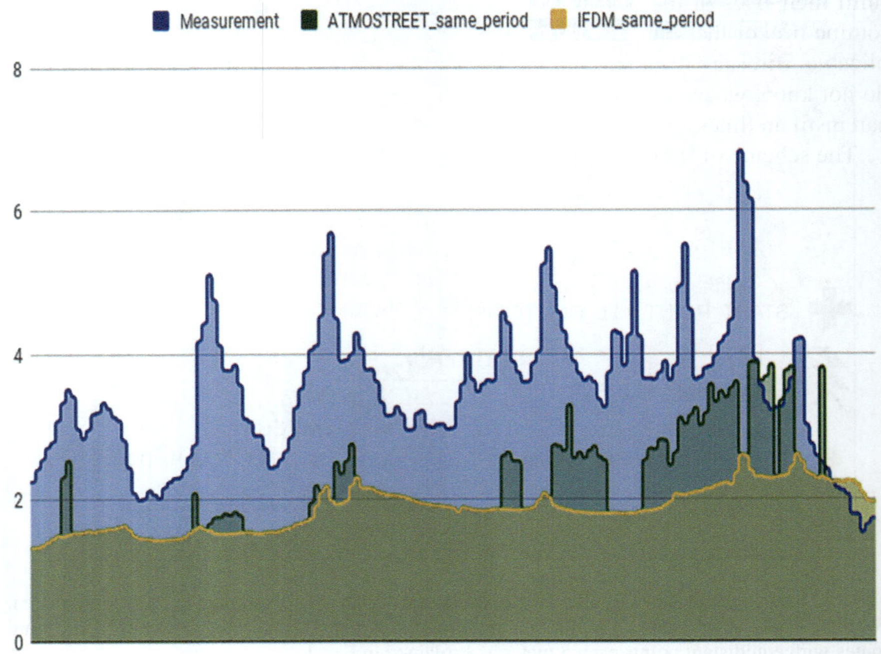

Fig. 3 Average BC concentrations for the traffic-dominated route (8.4 km in total): comparison of trajectory measurements with RIO-IFDM and ATMO-street model data for the same period. X-axis: distance over the route

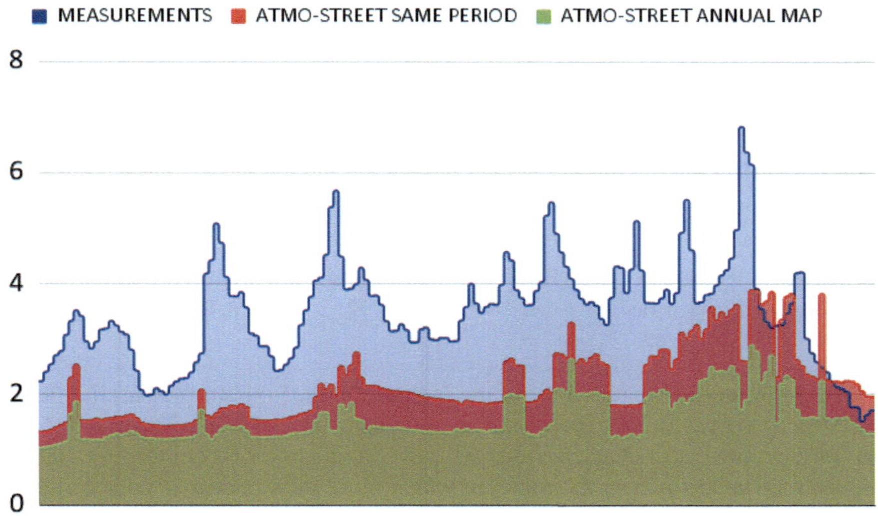

Fig. 4 BC measurements for the traffic-dominated route (8.4 km in total), comparing ATMO-street-data from the same period and with the ATMO-street annual data. X-axis: distance over the route

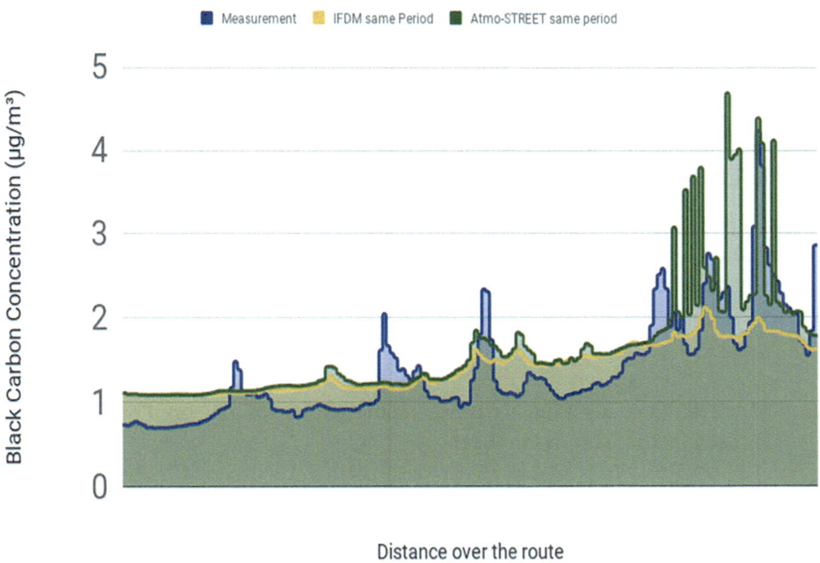

Fig. 5 Comparison of BC measurements on the cycling highway dominated route (9.0 km in total) with ATMO-street and RIO-IFM modelled values for the same period. X-axis: distance over the route

Table 1 Error metrics to quantify the deviation between BC measurements and BC modelled concentrations

R^2 values (−1, 1) **IOA VALUES (−1, 1)** *FAC2 VALUES (0–1)* MEAN BIAS (µg/m³)	High-traffic route	Low-traffic route
IFDM annual map	0.28; **−0.33**; *0.17*; −2.15	0.75; **0.66**; *0.98*; −0.15
ATMO-street annual	0.39; **−0.25**; *0.38*; −1.92	0.61; **0.65**; 0.99; −0.09
IFDM specific period	0.37; **−0.13**; *0.62*; −1.62	0.80; **0.52**; *1.00*; 0.34
ATMO-street specific period	0.42; **0.04**; *0.77*; −1.32	0.72; **0.48**; *0.98*; 0.39

For the route that featured a cycling highway (Fig. 5), both the baseline RIO-IFDM and ATMO-Street concentrations are higher than the measurements for the part of the route that is characterised by the cycling highway (about the first 70% of the route). The last part of the route, in which more traffic occurs, is detectable in the RIO-IFDM chart (some small upwards peaks indicating traffic emissions) while the peaks in the ATMO-Street model are above the RIO-IFDM peaks because of the street canyon effect. Some of the ATMO-Street peaks coexist well with the measured peaks; however, in some locations the model chain overestimates the influence of the buildings on the concentrations yielding large overestimations of the concentrations (false peaks). On the other hand, at some locations the effect of the buildings is underestimated, yielding peaks that are not captured in the ATMO-Street results (Fig. 5). The average error statistics over the entire route are listed in Table 1.

Correlation between the deviation of modelled and measured values versus meteorological characteristics.

No significant ($p < 0.05$) correlation was found between the deviation between the modelled and measured values on the one hand and meteorological characteristics on the other hand, including temperature, relative humidity, wind direction and scalar wind speed.

3.1.2 Modelled Versus Measured Concentrations for BC on Two Routes in the Mechelen Area

Two routes with the same origin and destination were compared (Fig. 6), one that runs along the ring road of Mechelen and the other through the city centre of Mechelen. For the route along the ring road, the modelled values are far below the values of the BC trajectory measurements. The ATMO-Street model gave a net mean bias of −3.36 µg/m³, when the model concentrations compared to the measured trajectory values. For the route that runs through the city centre, the deviation is much smaller, with a net mean bias of −1.37 µg/m³ (Table 2). Other error metrics (such as R^2, IOA and FAC2) gave better values for the route through the city centre compared to

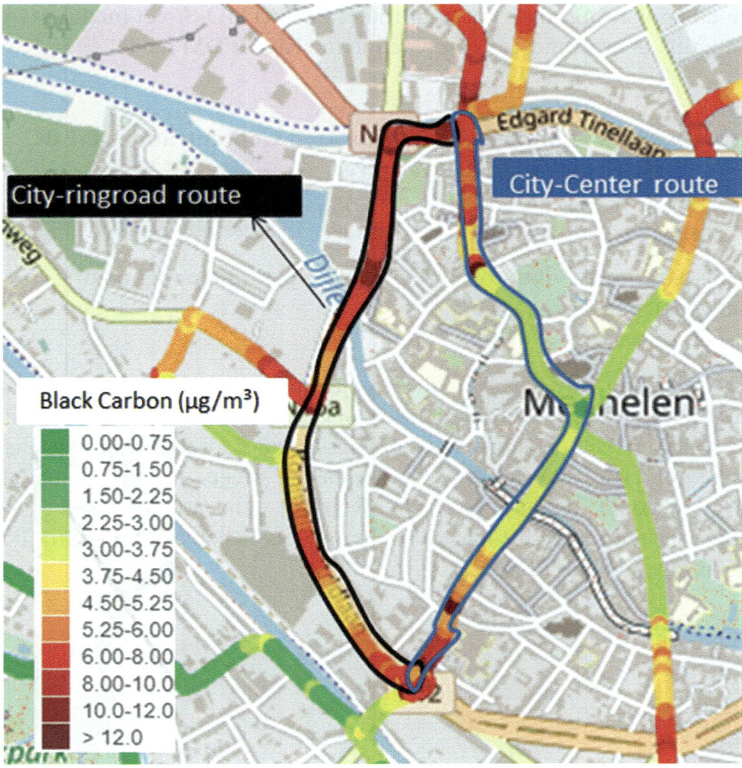

Fig. 6 Route comparison with concentration maps for the Mechelen area (adapted from https://mechelen.meetmee.be/kaart; background: © OpenStreetMap contributors)

Table 2 Error metrics to quantify the deviation between BC measurements and BC modelled concentrations

R^2 values (−1, 1) **IOA VALUES (−1, 1)** *FAC2 VALUES (0–1)* MEAN BIAS (µg/m³)	City centre route	Ring road route
IFDM annual map	0.40; **0.21**; *0.00*; −3.60	−0.62; **−0.38**; *0.00*; −5.67
ATMO-street annual map	0.29; **0.33**; *0.45*; −3.06	−0.57; **−0.31**; 0.18; −5.12
IFDM specific period map	0.34; **0.47**; *0.71*; −2.41	−0.66; **−0.15**; *0.25*; −4.18
ATMO-street specific-period map	0.37; **0.58**; *0.86*; −1.37	−0.48; **−0.03**; *0.52*; −3.36
ATMO-street annual hourly map—rush hours	0.37; **0.53**; *0.77*; −1.99	−0.09; **−0.04**; *0.55*; −3.67

the route along the ring road. For both routes, the ATMO-Street model gave better results in terms of error metrics compared to the RIO-IFDM model.

3.1.3 Comparison of Modelled and Measured Concentrations for BC for Three Roads in the Leuven Area

Three road sections with available BC measurements were compared (Fig. 7):

1. Naamsestraat (part close to the ring road) + part of the ring road
2. Bondgenotenlaan
3. Janseniussstraat.

The comparison only involved the annual map of 2016, because the higher temporal resolution ATMO-Street data were not available for 2016. A comparison for complete routes in the Leuven area was not possible because the trajectory measurements were only available for a limited number of streets, limiting potential route alternatives we could compare. Just as in Mechelen, concentrations on the ring road are the most heavily underestimated by the models, whereas the low-traffic route shows an acceptable bias (Table 3).

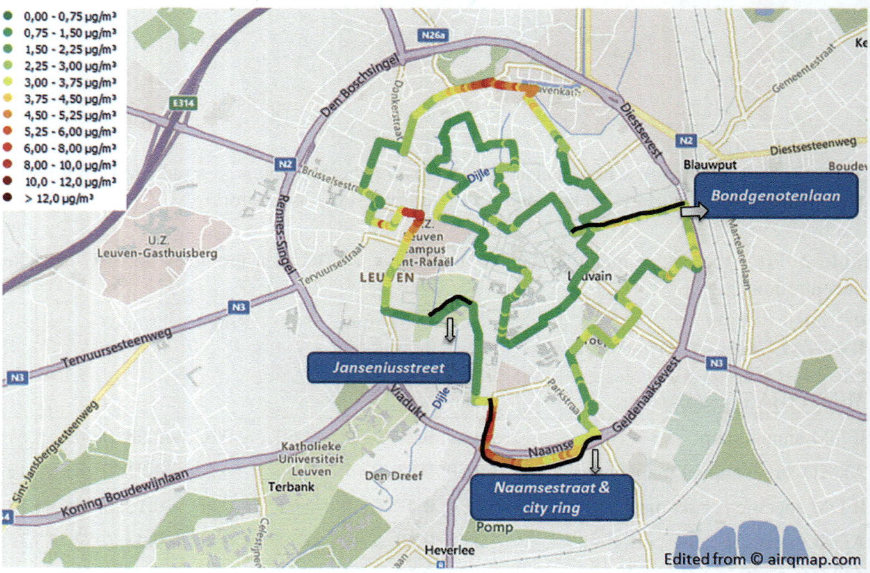

Fig. 7 Three road segments (highlighted black) analysed in the city of Leuven

Table 3 Error metrics to quantify the deviation between BC measurements and BC modelled concentrations

R^2 values (−1, 1) **IOA VALUES (−1, 1)** *FAC2 VALUES (0–1)* <u>MEAN BIAS (μg/m³)</u>	Naamsestraat and ring road	Bondgenotenlaan	Janseniusstraat Road section with limited amount of traffic
IFDM annual map	0.02; **−0.47**; *0.00*; <u>−3.50</u>	−0.11; **−0.51**; *0.18*; <u>−1.89</u>	0.35; **−0.50**; *1.00*; <u>−0.28</u>
ATMO-street annual	0.53; **−0.42**; *0.13*; <u>−3.34</u>	−0.07; **−0.26**; *0.91*; <u>−1.25</u>	0.49; **−0.42**; *1.00*; <u>−0.21</u>
ATMO-Street annual-hourly Rush hours	0.53; **−0.31**; *0.27*; <u>−2.67</u>	0.35; **0.03**; *0.94*; <u>−0.53</u>	0.51; **−0.66**; *0.93*; <u>+0.15</u>

3.1.4 CurieuzeNeuzen NO_2 Data: Comparison in an Urban Context

The results of the CurieuzeNeuzen project (Meysman and De Craemer 2018) made a direct comparison of measured NO_2 concentrations with modelled ATMO-Street NO_2 concentrations possible. This study performed a detailed comparison of model performances in an urban context, which resulted in the detection of a structural underestimation of NO_2 concentrations on ring roads. Previously, trajectory BC measurements revealed the largest model-measurements deviation on high-traffic roads and the largest deviations on city ring roads. As can be seen in the regression plot (Fig. 8), the average measured NO_2 of 40 $\mu g/m^3$ on city ring roads coexists with an average modelled concentration of between 20 and 30 $\mu g/m^3$.

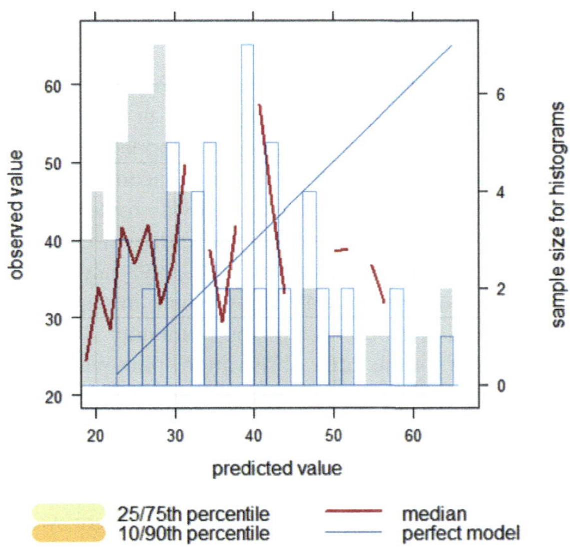

Fig. 8 Observed versus modelled values for city rings, taking into account all CurieuzeNeuzen points close to city rings in Leuven, Mechelen, Ghent, Bruges, Alost, Roeselare

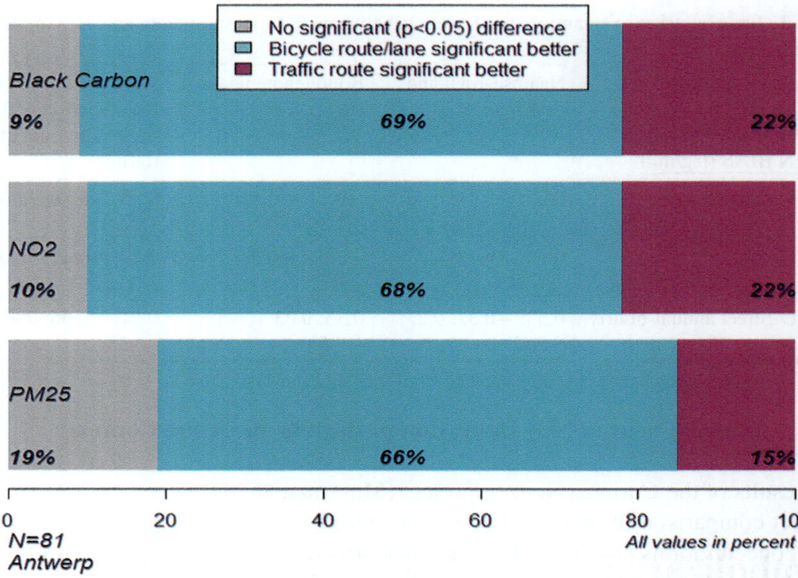

Fig. 9 Significant differences in concentrations on 36 randomly selected days (2016, 2017). For each day analysed at 07-08 UTC, 12-13 UTC, 16-17 UTC (RIO-IFDM model)

3.2 *Comparison of Pollutants*

3.2.1 Different Concentration Fluctuations Between Routes for Various Pollutants

Two previously analysed routes in the Antwerp area are compared for several randomly selected days, as explained in the methodology section. A comparison demonstrated that the ATMO-Street models (Fig. 10) concluded for all (100%) of the cases a significant difference in concentration of NO_2 and BC between the two routes (in favour of the cycling highway route), and in 90% of cases for $PM_{2.5}$. The RIO-IFDM (10 m) model (which excludes street canyon effects). Figure 9 indicated that for all pollutants considered (BC, NO_2 and $PM_{2.5}$) the bicycle highway as being less polluted than the traffic-busy route in approximately 70% of the cases, while the opposite was observed in approximately 20% of the cases (15% for $PM_{2.5}$). The explanation for the latter observation (pink part of the figure) has to do with model limitations and spatio-temporal patterns of air pollution and will be further elaborated in the discussion session. In approximately 10% of cases (20% for $PM_{2.5}$), no statistically significant difference between the considered cycling routes could be observed, and the null hypothesis that there is no difference could not be rejected.

On the same randomly selected days, there was on average a difference in the $PM_{2.5}$ concentration between the two routes of 3.1% according to RIO-IFDM and

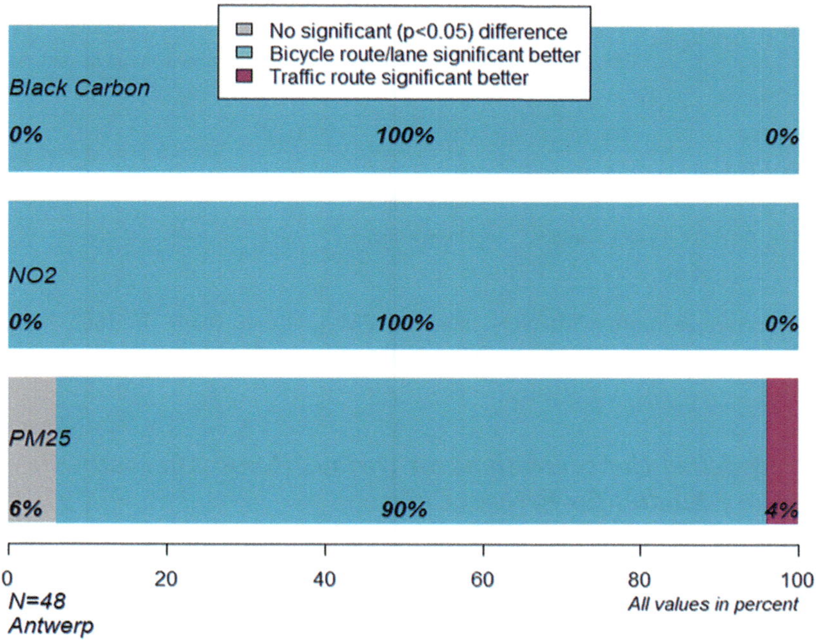

Fig. 10 Significant differences in concentrations on 18 randomly selected days (2017). For each day analysed at 07-08 UTC, 12-13 UTC, 16-17 UTC (ATMO-street model)

6.6% according to ATMO-Street, for NO_2, there was a difference of 13.1% according to RIO-IFDM and 26.4% according to ATMO-Street; and for BC concentrations a difference of 9.5% according to RIO-IFDM and a difference of 27.1% according to ATMO-Street.[1]

3.2.2 Correlation Between Pollutants

Considering hourly air pollution concentrations for the entirety of 2017, the model correlations between pollutants tended to be higher than correlations at the measuring stations. The RIO-IFDM model showed a nearly perfect correlation ($R^2 = 0.93$) between BC and NO_2. In the ATMO-Street model, this correlation was somewhat lower, but still strong ($R^2 = 0.78$). Correlations between firstly $PM_{2.5}$, and secondly NO_2 and BC seemed to be overestimated in the air pollution models, when the measuring stations were taken as reference. According to the measuring stations,

[1] It should be noted that models simulate near-the road concentrations and not on-the road concentrations, which are often—and especially in the case of a significant amount of motorised traffic on the road—far higher. This is examined in detail in the discussion section.

Table 4 Correlation between pollutants in models and reference measurement stations

	IFDM	ATMO-street	Sample of measurement stations
R^2 (BC-NO_2)	0.93	0.78	0.62
R^2 (BC-$PM_{2.5}$)	0.60	0.66	0.28
R^2 (NO_2-$PM_{2.5}$)	0.58	0.65	0.24
R^2 (NO_2-PM1)	Not available	Not available	0.39
R^2 (NO_2-C6H6)	Not available	Not Available	0.27

$PM_{1.0}$ correlates more with NO_2 than did $PM_{2.5}$ ($R^2 = 0.39$). Further, benzene correlates as strong as $PM_{2.5}$ with NO_2. Full results are listed in Table 4.

3.3 Modelled Concentrations on Various Route Alternatives: Case-Studies for Several Cities

In this section, all analyses are based on modelled air pollution values. For Leuven and Ghent, the difference in concentrations is analysed based on the ATMO-Street model while the SIRANE model is used for Orléans (France).

3.3.1 Leuven

Three routes were considered, each with the Leuven railway station as an origin and the Leuven University science campus as a destination. A considerable number of people commute from the origin to the destination each day by bike. The three route alternatives were generated by Google Maps. An analysis of the average air pollution concentration on each of those routes revealed that Route 2, which follows the ring road for a significant distance, has the highest concentrations (Fig. 11). By repeating the same analyses as for Antwerp in the previous section (using the same randomly selected days), the ATMO-Street model shows that Route 1 and Route 3 have significantly lower concentrations compared to Route 2 > 90% of the time, while Route 1 has significantly lower concentrations than Route 3 about 80% of the time (Fig. 12).

3.3.2 Ghent

Three routes were considered, each with the R4 cross section as a destination and Drongen railway station as an origin (Fig. 13). Based on the annual ATMO-Street map of 2017, the route along the railway line had the lowest NO_2 concentrations. The route through the city centre of Drongen had slightly higher concentrations, and the route along the R4 and N465 has by far the highest concentrations.

Fig. 11 Three routes analysed in the Leuven area. Before calculation of mean air pollution values, the original routes were transformed into routes with equidistant points each 5 meter as explained in Fig. 1

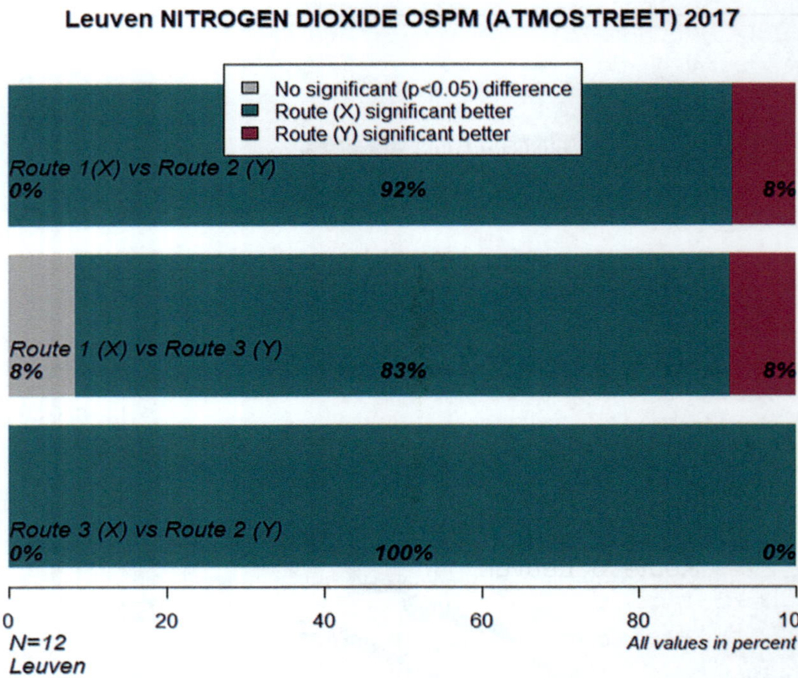

Fig. 12 Significant differences ATMO-street between three routes in the Leuven area for NO_2. Analyses based on a sample of 18 randomly selected days in 2017 (for each day, analysed at 07-08 UTC, 12-13 UTC and 16-17 UTC)

3.3.3 Orléans

Three routes were investigated, each with the Gare de Fleury Les Aubrais as an origin and Parc du Poutyl as a destination (Fig. 14). This time, the analysis was based on the SIRANE model annual map from 2018. Route 1 goes through the city centre, Route 2 uses main roads with a higher level of motorised traffic and Route 3 commences on main roads, soon heading to more quiet areas and smaller roads outside the inner city. Route 2 reveals the highest average concentration (20.4 $\mu g/m^3$), while Route 1 and Route 3 have similar concentrations of 18.0 $\mu g/m^3$ and 18.3 $\mu g/m^3$, respectively.

3.4 *Spatio-temporal Patterns of Exposure During Commute*

This section will look further at the previously analysed routes in the Antwerp and Leuven areas, digging somewhat deeper into the spatio-temporal characteristics of concentrations and consecutively exposure to air pollution.

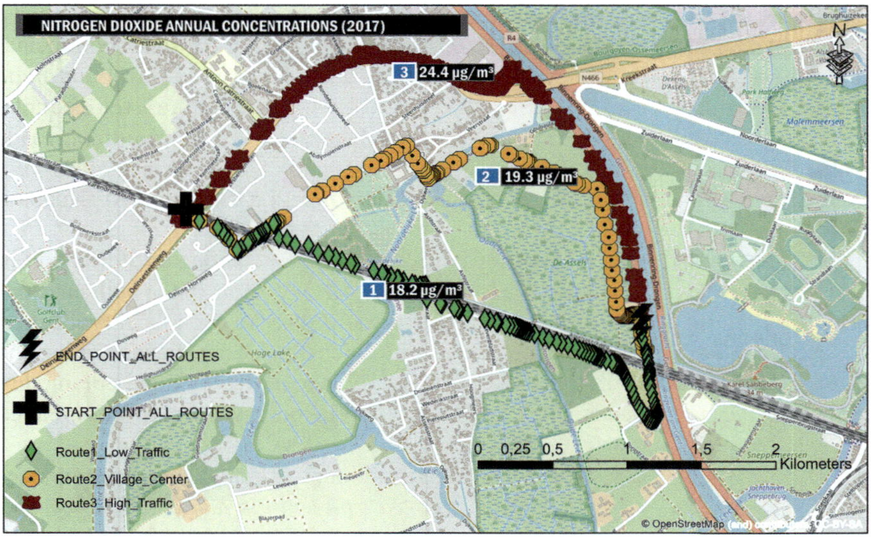

Fig. 13 Three routes in Ghent, route 1: along railway, route 2: through centre of village 'Drongen', route 3: main traffic roads

3.4.1 Antwerp

Daily patterns: When one considers the whole year, differences in concentrations between the two routes are seen to be the largest during morning peak hours. For both routes, concentrations are consistently higher than the spatial average for the whole of Flanders (Fig. 15). During night-time hours, concentration differences between different routes were smaller or even absent, compared to the daytime hours (Figs. 15 and 16). During the morning peak hours, late morning and early afternoon, the modelled concentration differences between the two routes in the Antwerp area are around 25% (Fig. 15). The time at which relative concentration differences between routes become smaller coincides approximately with sunset (Fig. 16).

Seasonal patterns: At all hours of the day, concentrations are clearly higher in winter compared to summer (Fig. 17). In summer, morning peak hours have higher NO_2 concentrations (compared to late-afternoon peak hours), while in winter, late-afternoon peak hours have higher NO_2 concentrations. In summer, relative and absolute differences between concentrations of different routes are larger than those in winter. For all analysed hours, the autumn and spring resulted in findings that fall somewhere in between summer and winter. For all seasons, concentration differences between routes are largest during the day, and the time at which the difference in concentration between the two routes disappears coincides approximately with sunset (Fig. 17).

Fig. 14 Three routes in Orléans (France)

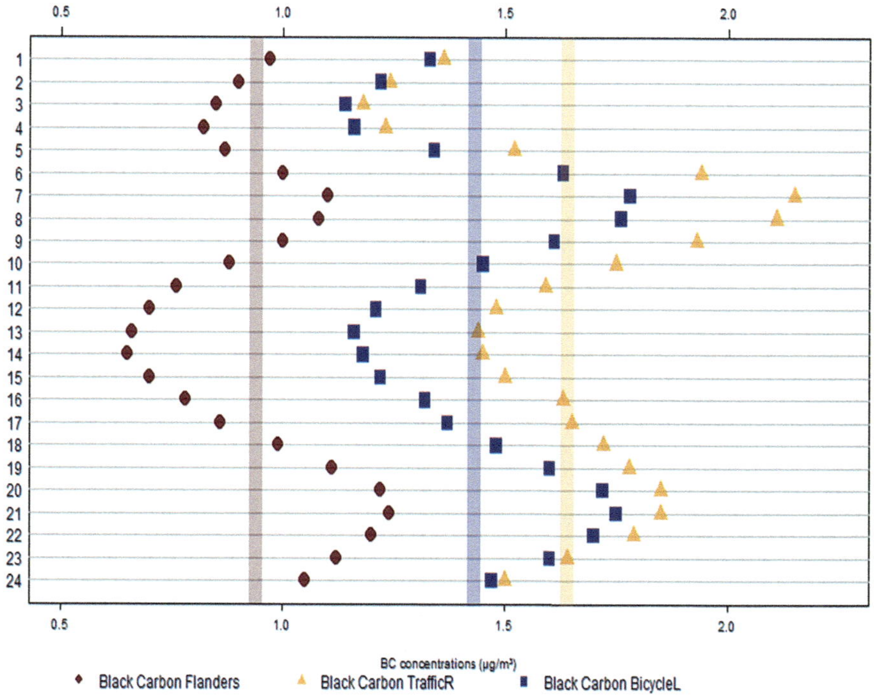

Fig. 15 Hourly patterns and averages of BC concentrations for Flanders and route 1 and route 2 in the Antwerp area (based on the 2017 annual-hourly map of ATMO-street)

3.4.2 Leuven

In Fig. 18, it can be seen that in summer, morning peak hours have a higher NO_2 concentration (than late-afternoon peak hours), while in winter, late-afternoon peak hours had higher NO_2 concentrations than in the morning. All findings (daily patterns, seasonal patterns, monthly patterns) are identical to the findings for the two routes in the Antwerp area.

Intra-seasonal variations in concentration differences between the two routes are small compared to inter-seasonal differences. This was found to be the same for absolute concentrations: there are large differences when compared season to season, and small differences, as well as an interchangeability of months, within season (Fig. 18). The impact of time of day on concentrations is of similar magnitude as the seasonal difference.

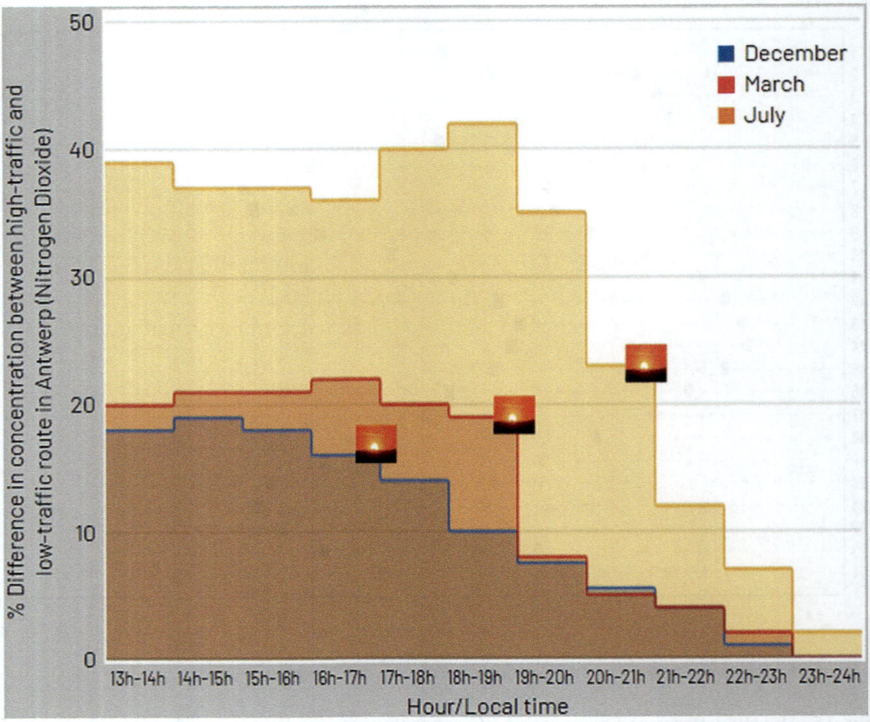

Fig. 16 The time at which difference in concentrations becomes smaller strongly depends on meteorological and daylight conditions

4 Discussion

4.1 Suitability of Model Data for Applications in Uncovering Spatio-temporal Air Pollution Exposure Patterns

It is clear that the measured BC values are on average higher, compared to the modelled BC values for streets with a considerable amount of traffic. One plausible explanation for this is the very steep gradient of BC away from the emission source, at an even more pronounced gradient than NO_2, which also has a pronounced gradient (Apte et al. 2017). Since the closest model receptor grid points are located at 15 m from the edge of the road, implying the models are not able to simulate correctly on-the-road concentrations, an underestimation of on-the-road concentrations can be expected. A second explanation is inaccuracies in the traffic emission data, which are often outdated, of limited availability and do not take traffic dynamics (amongst others congestion) into account. A third explanation might be inaccuracies in the BC measurements due to an instrumental measurement error (probably limited given

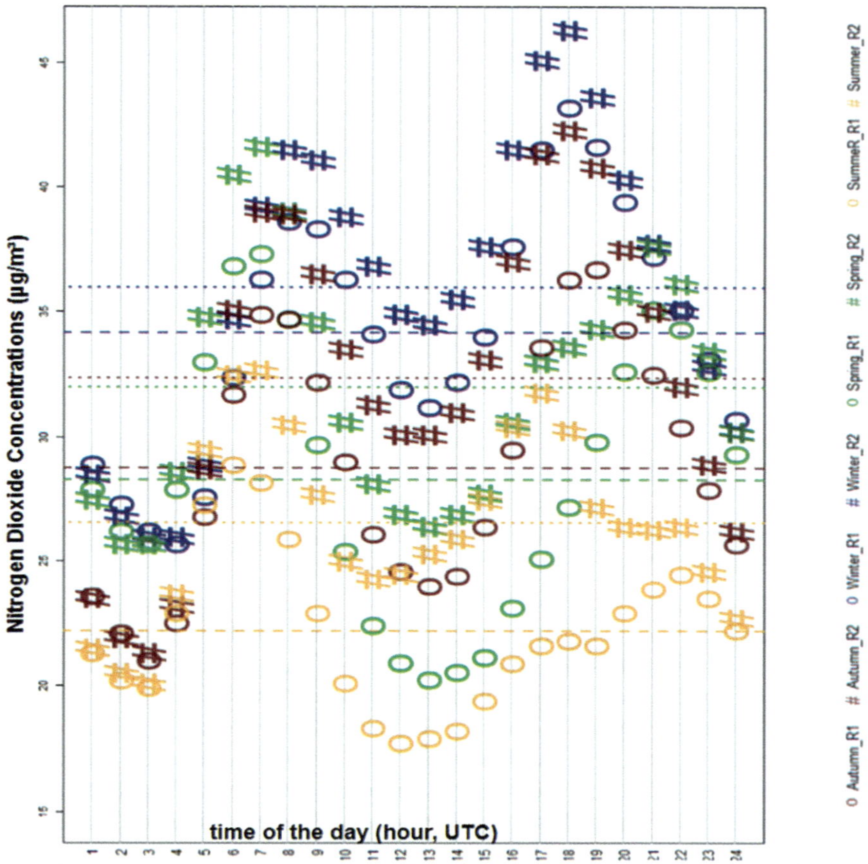

Fig. 17 Difference in concentrations between the two different routes in the Antwerp area: daily and seasonal variations

the good results when the monitor was validated with a more reliable device) and due to extreme and not representative conditions during mobile sampling during the measurement of instantaneous values (e.g. high polluting vehicle or extreme traffic conditions): mobile measurements provide an instantaneous value while the modelled data on the other hand calculates hourly mean concentrations and hence instantaneous peaks are equalled out. Those effects are partially levelled out because the data are used in an aggregated format (aggregation of repeated measurements on multiple days) and those effects are further reduced by the use of algorithms correcting for this noise. The data sets from all three cities indicate the same conclusions. As also highlighted in the introduction, also existing studies found remarkable differences between average concentrations on different routes. For example, Boniardi et al. (2019) found large differences in BC-exposure between home-to-school routes in

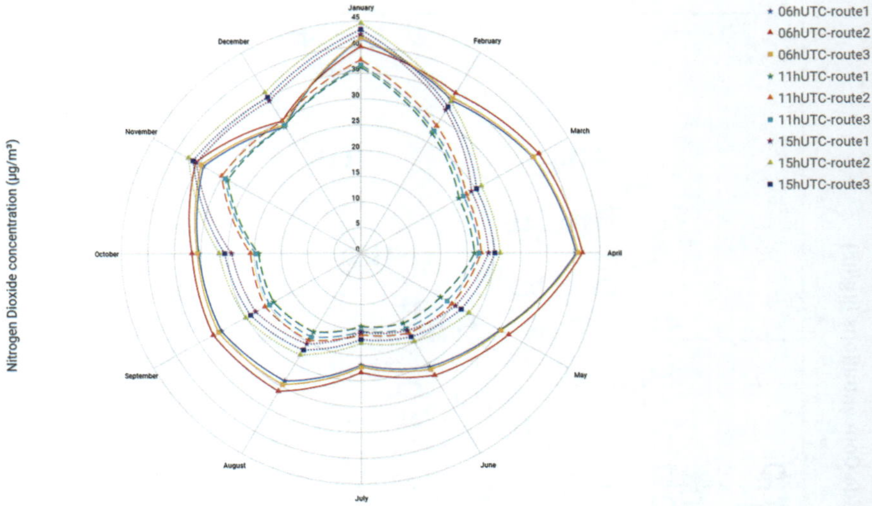

Fig. 18 Concentrations of NO_2 for the three routes in Leuven split up by month for 06-07h UTC, 11-12h UTC and 15-16h UTC

Milan. They also compared BC measurements with a model, in this case a Land Use Regression (LUR) model, and concluded that there is good agreement ($r = 0.74$) between model and measurement while there is an 30% underestimation of measurement concentrations by their LUR model (Boniardi et al. 2019). Likely, the phenomenon that even high-resolution air pollution models have the tendency to underestimate on-the-road concentrations on roads with a considerable amount of traffic, can be widespread in Europe, and even beyond, and not limited to the Belgian context.

The validation makes it clear that the BC monitoring device used for the measurements is reliable. Also, existing studies both in Belgium and elsewhere in Europe (e.g. Ljubljana) demonstrated BC is dominantly originating from traffic. A previous study in Ljubljana, where similar concentrations of BC as in Belgium were observed, found that in the Ljubljana area 70–85% of the BC originates from traffic (Ogrin et al. 2016). BC concentrations from biomass burning were more or less similar in all measuring sites in Ljubljana (between 0.9 and 1.2 $\mu g/m^3$) while the traffic contribution varied from 2.4 to 5.2 $\mu g/m^3$. Similarly, a study in Belgium found about between 25 and 28% of BC in winter originating from wood burning and around 70% of BC originating from traffic in winter in an non-urban area, with expectations for an even considerable higher percentage contribution from traffic in cities where there is often a high-traffic impact, as well as everywhere in summer, when wood-burning activities are low (VMM 2017). The measurements in Belgium showed very low background concentrations (low concentrations on roads with no or little traffic, even in residential areas, for example, <0.75 $\mu g/m^3$ in the city park of Leuven). Modelled background concentrations of BC are remarkably higher compared to measured background concentrations in car-free and residential areas, indicating a possible overestimation of BC concentrations by the models, possibly explained by the low number of measurement stations used to interpolate the background concentrations of the models. Measurement uncertainties with the used devices can also contribute to a deviation of modelled and measured values. On the other hand, the modelled concentrations are strikingly lower compared to the BC measurements on roads with a high amount of traffic, which can be explained by near-road (model) versus on-road (measurements) concentrations, and underestimated traffic emissions on roads with a lot of traffic.

The comparison between the ATMO-Street model with CurieuzeNeuzen results in an urban context clearly indicates that there exists a structural underestimation of NO_2 concentrations with regard to main traffic roads, such as a 20–30% structural underestimation on a number of analysed city ring roads in the Flanders region. In contrast to the BC trajectory measurements, the NO_2 measurements within the CurieuzeNeuzen project were gathered at a larger distance from the emission sources (cycling lane BC versus street-facing first-floor window NO_2), implying the deviations between measurements and models found in this analyses are for a large part attributable to shortcomings of the available traffic emission data. The model-measurement deviations in this analysis (20–30%) are smaller compared to the analysis of the BC trajectory measurements, where measurement concentrations on busy roads are often more than double the modelled concentrations.

However, despite this underestimation of the differences between a route with a low versus a route with a high amount of traffic due to the reasons discussed in the previous paragraphs, the models are capable of making a correct distinction between routes and correctly identifying the routes with significantly lower concentrations. This implies those model data are useful to include in applications such as route planners to make a reliable distinction between different routes, while falling short in displaying absolute concentrations active commuters are exposed to.

Error metrics show that the ATMO-Street model, which includes street canyon effects, leads to smaller model-measurement deviations compared to the RIO-IFDM model, especially on routes with a high amount of motorised traffic. The ATMO-Street does have a clear advantage over the RIO-IFDM model in terms of the ability to distinguish between routes. Several case studies in various cities (Ghent, Leuven, and Orléans) confirm that concentrations are—in an urban context of mid-sized cities—the highest on city ring roads and in car-busy street canyons, confirmed by the BC trajectory measurements (Leuven, Mechelen) and CurieuzeNeuzen data.

To conclude, high-resolution air pollution modelling using the ATMO-Street model is a reliable and essential tool to uncover and display the spatio-temporal patterns. Moreover, the models are excellently suitable in making a reliable binary distinction between the most polluted and least polluted route in the most situations. High-resolution air pollution models of NO_2 can reveal the often-hidden patterns of traffic-related air pollution, not only for NO_2, but also for strongly correlating pollutants such as BC, Benzene and UFP. Due to numerous reasons discussed in the above sections, the absolute concentration differences between the 'clean' and 'polluted' routes are often underestimated by the model.

4.2 Comparison of Pollutants

In Belgium, high-resolution models including street canyon effects are available for PM_{10}, $PM_{2.5}$, NO_2 and BC. $PM_{2.5}$, NO_2 and BC all show significant differences between concentrations on different routes. NO_2 and BC are clearly more correlated with traffic, compared to particulate matter. Therefore, they are the best to use in a routing context, where people are very close to the emission source and often exposed to concentrations far higher than what is visible from the modelled data and the reference measurement station data. Reliable data for on-the-road air pollutant concentrations is sparse, with pollutant level concentrations not only higher as what pedestrians and cyclists may assume or be aware of when travelling or recreating without searching for information, but also higher compared to what they are aware of when they consult information as, for roads with motorised traffic, on-the road concentrations are only seldom covered on air pollution monitoring websites. The extremely high on-the-road concentrations for several pollutants remain mostly a hidden phenomenon. Providing air pollution model data with a high resolution to use in (routing) applications, including open and transparent communication about the shortcomings, can be a first important step to raise more awareness about spatio-temporal patterns of air pollution exposure during commute.

For the choice of pollutants, attention should be paid to the possible decoupling of the correlation between NO_2 and BC concentrations. Because of the severe health effects of BC and ultrafine particulate matter (PM0.1), because of this decoupling and declining correlation with NO_2 due to the use of soot filters, it can be more suitable to use both NO_2 and BC to determine the cleanest route. The Belgian Interregional Environment Agency provides seasonal-hourly NO_2 maps as open data for amongst others routing applications for the primary reason that better validated air pollution maps exist for NO_2 compared to BC. The NO_2 data have been validated more than BC data and BC modelled data are only based on the interpolation of a very small number of reference measurement stations, potentially resulting in unreliable background concentrations. This is because the interpolation of the background is based on a limited number of measurement stations (± 25 in Belgium in 2018 for BC versus ± 80 measurements stations for NO_2) The urban background concentrations are calculated based on the interpolation of air pollution measurements at reference measurement stations. After that, point emissions and line emissions (such as traffic) are added to the model (see the methodology section). In the case of BC, it seems that the urban background concentrations are overestimated by the model in an order of magnitude between 0.25 and 0.5 $\mu g/m^3$ (as it was for the example visible in Fig. 5 with in the first part (first 75% of the distance from left to right) of the route, characterised by the cycling highway, higher model concentrations compared to ATMO-street modelled values are displayed most of the time.

Our research is bound by various limitations such as the availability of data. From our literature review, it was clear that, certainly UFP, and potentially also Benzene (or other volatile organic compounds) may also be of relevance in this context, because their presence very strongly depends on the amount of motorised traffic on a route.

However, the insufficient availability of data and a lack of model availability of those pollutants in Belgium made it impossible to investigate the importance of those pollutants. For UFP, various studies point to very serious health effects (respiratory, cardiovascular and neurological) while exposure to UFP is to a large extent linked to commuting (Knibbs et al. 2011). Benzene exposure also results in an elevated risk for the development of leukaemia. Annual mean concentrations of Benzene of 0.17 $\mu g/m^3$, 1.7 $\mu g/m^3$, respectively, are associated with an excess lifetime risk of leukaemia of 1/1 000 000 and 1/100 000, respectively (World Health Organization 2019).

4.3 Spatio-temporal Patterns of Exposure During Commute

While, without having reliable air pollution information, it already is challenging to distinguish between clean and polluted routes with human senses, spatio-temporal characteristics of the pollution patterns are an even more hidden geography, which is not or barely detectable with the human eye. This is not only relevant for the issue of air pollution itself: reliable pollution exposure patterns and estimates are needed to research and quantify the relationship between air pollution exposure and our health, healthcare costs and their economic burden. Those related phenomena only become revealed in a reliable way when the air pollution exposure patterns and concentrations are uncovered to a good extent.

It is important to consider daily patterns and to take into account larger concentration differences between sunrise and sunset during the daytime compared to night-time hours. As a result, annual-hourly maps enjoy a clear advantage over annual maps because of the important weight of night-time concentrations in annual maps when concentration differences between routes are very small. It should also be noted that there are very few active transport users on the road during the night, as confirmed by traffic counting campaign in Belgium.

Daily patterns also differ from season to season. In winter, the highest concentrations occur during late-afternoon peak hours (and concentrations remain high afterwards) while, in the other seasons, the highest concentrations are observed during morning peak hours (a more synchronised rush hour peak) and decrease quickly after those peak hours. The atmospheric mixing layer height descends sharply after sunset, reducing the vertical mixing of air pollutants, thus spreading the pollution that originates from the emissions over larger areas. In addition, wind speeds are lower which can also reduce the dispersion and dissolution of pollutants. Furthermore, there is more ozone formation in summer during evening peak hours that occur in the late afternoon (relative to the winter). Because, in summer, there is still light in the evening, UV-radiation and warmth during the evening peak hours enable the formation of ozone decreasing NO_2 during evening peak hours, especially on days with abundant sunshine. This is also visible looking to reference measurement stations (e.g. reference measurement stations Aarschot, Sint-Stevens-Woluwe, … data accessible through https://www.irceline.be/en/air-quality/measurements/monitoring-stations/interactive-viewer): it is visible that on warm summer days with abundant sunshine, NO_2 concentrations decrease during day and consecutively rise again during evening—around sunset—when ozone concentrations stagger and decline.

Another important observation is that the percentage concentration differences between various routes become smaller around sunset, implying the concentration differences become smaller much more early in winter (between 16h and 18h CET) compared to mid-summer (between 20h and 22h CET). The daily pattern of the height of the atmospheric mixing layer, what strongly depends on daylight, is probably the most important explanation for this. In addition, percentage concentration differences between routes are during the entirety of the day larger in summer compared to percentage concentration differences between routes in winter, with spring and autumn located in-between again.

Noteworthy is that, according to the model results, seasonal and daily concentration differences for one route are significantly larger compared to concentration differences between different routes in the same season and time of the day. In part this can be explained by the shortcoming of the models, namely the tendency to underestimate on-the-road concentrations on roads featuring a considerable amount of traffic. Models are perfectly able to distinguish between routes, but concentration differences between various routes will in many cases be larger as displayed by the models, with concentrations displayed for routes with traffic likely underestimated.

Due to the significant differences between seasons, it is recommended to choose seasonal-hourly maps over annual-hourly maps for routing. Inter-seasonal variations are more significant compared to intra-seasonal differences. Intra-seasonal differences are small, not significant and can vary from year to year. Hence, monthly-hourly maps don't have an advantage over seasonal-hourly maps, especially because the most recent available seasonal-hourly maps are often data from the year before.

The primary goal of this research is to uncover the often-hidden pattern of spatio-temporal exposure patterns during the commuting process and to enable people to change their typical commute routes to reduce their long-term exposure to harmful pollutants. Taking these aspects into consideration, it can be said that static seasonal-hourly maps are the best to use to determine the cleanest route.

5 Conclusion

In an urban context, the analysed air pollution models are generally able to make a reliable distinction between routes with lower and higher air pollution, making the hidden geography of air pollution exposure during commute visible to a considerable extent.

Despite that absolute on-road pollutant concentrations often being underestimated by the models on busy traffic network sections. Model inaccuracies can be both related to how the models are calculated (e.g. distance of the closest receptor points to the edge of the road resulting in an underestimation of on-street concentrations on streets featuring motorised traffic) and deficiencies in the traffic emission data (outdated traffic counts, no accurate reflection dynamic fluctuations like traffic jams or network changes, etc.). In general, exposure to air pollutants (BC, NO_2, $PM_{2.5}$, UFP) is highest on roads with a high amount of motorised traffic, such as ring roads and car-busy street canyons. In the Belgian context, NO_2 data are most suitable for the routing context since there are more quality-assured monitoring stations and that the NO_2 model results have been more extensively validated compared to BC. NO_2 can still serve as a proxy for concentrations of other pollutants such as UFP and BC and to a lesser extent also $PM_{2.5}$ in a routing context. It is important to consider daily and seasonal variations, given that we observed important differences between both the amplitude of absolute concentrations and relative concentration differences between routes, depending on the time of the day and season of the year.

The best results achieved in our research were by using static seasonal-hourly model data in a routing context to reduce pedestrians' and cyclists' exposure to air pollution during their commutes. To use air pollution models in routing apps or software, it is strongly recommended to use methodologies which are scientifically sound, such as the equidistant 5-metre point method for calculating average concentrations since this has a significant outcome on the result and a high spatial-model resolution of at least 10×10 m.

The results of our research suggest that further research is needed with respect to measuring and modelling on-the-road air pollution concentrations of pollutants linked to traffic. This can be useful for various purposes including an improved quantification of health effects (and related economic costs), calculating AP-exposure and dose-response relationships of air pollutants and making the public aware of extremely elevated pollution levels on roads with motorised traffic. This kind of pollution remains to a certain extent hidden for now, and the same is largely true for related health outcomes and economic costs. Moreover, to enable a reliable quantification of the health and economic impacts, a reliable quantification of on-the-road concentrations of air pollutants should exist. However implementing currently available air pollution models in applications can certainly already be an important step to make the hidden more visible, and enable citizens using active commute to switch to another route, for example, for daily commute or recreation, thereby significantly reducing their long-term exposure to air pollution.

Acknowledgements Be-good project Some of the research conducted was triggered by the ongoing Interreg project Be-Good and further solutions (open-data endpoint, demo-application) developed based upon this research are developed within the Be-good project. We acknowledge the Be-good project (https://www.nweurope.eu/projects/project-search/begood/) for their contributions and thank the project and all project partners for providing the opportunity to convert the results of this scientific research in real-world solutions and applications.

Black Carbon Measurement Data Mobile BC data in Mechelen and Leuven were collected as part of airQmap projects of VITO (www.airQmap.com). We thank Martine van Poppel, Jan Peeters and Stijn Vranckx for sharing these data. We are grateful to the volunteers of the Citizen Observatory Meet Mee Mechelen for their assistance in the collection of the mobile BC data in Mechelen (https://mechelen.meetmee.be/kaart). This data collection in Mechelen was partly funded by the European Union's Horizon 2020 project Ground Truth2.0, grand agreement No. 689744 and partly funded by the FLAMENCO project, a research project financially supported by Agentschap

Innoveren & Ondernemen (VLAIO) contract number IWT-SBO 150044. We thank the volunteers of the citizens platform Straten Vol Leuven and the ngo Leuven 2030 for their assistance in the BC data collection in Leuven. We are also grateful to the Provincie Vlaams-Brabant for funding the measurement campaigns of 2016, 2017 and 2019 within the framework of Duurzame Klimaat Project.

CurieuzeNeuzen data We thank all 20.000 citizens scientists in CurieuzeNeuzen Vlaanderen for their enthusiasm and efforts, and all team members of University of Antwerp, De Standaard, the Flanders Environment Agency (VMM), VITO, HIVA KUleuven and Kariboo! for their contribution to planning, logistics and communication. We additionally acknowledge the citizen movement Ringland and Stad Antwerpen for their contribution to the CurieuzeNeuzen Antwerpen project, which acted as a pilot for CurieuzeNeuzen Vlaanderen.

References

Alvarez-Pederol, M., I. Rivas, M. Lopez-Vicente, E. Suades-González, D. Donaire-Gonzalez, M. Cirach, M. de Castro, M. Esnaola, X. Basagana, P. Dadvand, M. Nieuwenhuijsen, and J. Sunyer. 2017. Impact of commuting exposure to traffic-related air pollution on cognitive development in children walking to school. *Environmental Pollution* 231: 837–844.

Apte, J.S., K.P. Messier, S. Gani, M. Brauer, T. Kirchstetter, M. Lunden, J.D. Marshall, C.J. Portier, R. Verrmeulen, and S.P. Hamburg. 2017. High-resolution air pollution mapping with google street view cars exploiting big data. *Environmental Science and Technology* 51 (12): 6999–7008.

Atkinson, R.W., I.M. Carey, T. van Staa, R.H. Anderson, and D.G. Cock. 2013. Long-term exposure to outdoor air pollution and incidence of cardiovascular diseases. *Epidemiology* 24: 44–53.

Berkowicz, R. 2000. OSPM—A parameterised street pollution model. *Environmental Monitoring and Assessment* 65: 323–331.

Boccupther, A.R., S. Edney, R. Plotnikoff, R. Curtis, J. Ryan, I. Sanders, A. Crozier, and C. Maher. 2018. Can smartphone apps increase physical activity? Systematic review and meta-analysis. *Journal of Medical Internet Research* 21: 3.

Boniardi, L., E. Dons, L. Campo, M. Van Poppel, L. Int Panis, and S. Fustinoni. 2019. Is a land use regression model capable of predicting the cleanest route to school? *Environments* 6: 90.

Buekers, J., E. Dons, B. Elen, and L. Int Panis. 2015. Health impact model for modal shift from car users to cycling or walking in Flanders: application to two bicycle highways. *Journal of Transport & Health* 2 (4): 549–562.

Carslaw, D.C. 2015. The openair manual—open source tools for analysing air pollution data. Manual for version 1.1-4. King's College London. http://www.openair-project.org/PDF/OpenAir_Manual.pdf. Accessed 23 Mar 2020.

Choi, W., M. He, K. Kozawa, S. Mara, A. Winer, and S. Paulson. 2013. Local scale spatial and temporal air pollution gradients in the Los Angeles area. https://ww3.arb.ca.gov/research/seminars/paulson3/paulson3.pdf. Accessed 9 Mar 2020.

Cole-Hunter, T., R. Jayaratne, I.B. Stewart, M. Hadaway, L. Morawska, and C. Solomon. 2013. Utility of an alternative bicycle commute route of lower proximity to motorised traffic in decreasing exposure to ultra-fine particles, respiratory symptoms and airway inflammation—A structured exposure experiment. *Environmental Health* 2013: 12–29.

Deng, Q., C. Lu, Y. Li, J. Sundell, and D. Norbäck. 2016. Exposure to outdoor air pollution during trimesters of pregnancy and childhood asthma, allergic rhinitis, and eczema. *Environmental Research* 150: 119–127.

Dons, E., L. Int Panis, M. Van Poppel, J. Theunis, and G. Wets. 2012. Personal exposure to Black Carbon in transport microenvironments. *Atmospheric Environment* 55: 392–398.

Dons, E., P. Temmerman, M. Van Poppel, T. Bellemans, G. Wets, and L. Int Panis. 2013. Street characteristics and traffic factors determining road users' exposure to black carbon. *Science of the Total Environment* 447: 72–79.

Fisher, G., and J. Metcalfe. 2008. *Good practice guide to assessing air emissions from land transport*. Technical report. Ministry for the Environment, Wellington, New Zealand.

Gallagher, J., R. Baldauf, C.H. Fuller, P. Kumar, L.W. Gill, and A. McNabola. 2015. Passive methods for improving air quality in the built environment: A review of porous and solid barriers. *Atmospheric Environment* 120: 61–70.

Giménez-Gaydou, D.A., A.C. dos Santos, G. Mendes, I. Frade, and A.S.N. Ribeiro. 2019. Energy consumption and pollutant exposure estimation for cyclist routes in urban areas. *Transportation Research Part D* 72: 1–16.

Hagler, G.S., T.L. Yelverton, R. Vedantham, A.D. Hansen, and J.R. Turner. 2011. Post-processing method to reduce noise while preserving high time resolution in aethalometer real-time black carbon data. *Aerosol and Air Quality Research* 11: 539–546. https://doi.org/10.4209/aaqr.2011.05.0055.

Harris, M.H., D.R. Gold, S.L. Shiman, S.J. Melly, A. Zanobetti, B.A. Coull, J. Schwartz, A. Gryparis, I. Kloog, P. Koutrakis, D.C. Bellinger, M.B. Belfort, T. Webster, R.F. White, S.K. Sagiv, and E. Oken. 2016. Prenatal and childhood traffic-related air pollution exposure and childhood executive function and behavior. *Neurotoxicology and Teratology* 57: 60–70.

Hofman, J., R. Samson, S. Joosen, R. Blust, and S. Lenaerts. 2018. Cyclist exposure to black carbon, ultrafine particles and heavy metals: An experimental study along two commuting routes near Antwerp, Belgium. *Environmental Research* 164: 530–538.

Int Panis, L., B. de Geus, G. Vandenbulcke, H. Willems, B. Degraeuwe, N. Bleux, V. Mishra, I. Thomas, and R. Meeusen. 2010. Exposure to particulate matter in traffic: A comparison of cyclists and car passengers. *Atmospheric Environment* 44 (19): 2263–2270.

Int Panis, L., E.B. Provost, B. Cox, T. Louwies, M. Laeremans, A.R.L. Standaert, E. Dons, L. Holmstock, T.S. Nawrot, and P. De Boever. 2017. Short-term air pollution exposure decreases lung function: a repeated measures study in healthy adults. *Environmental Health* 1: 60. https://doi.org/10.1186/s12940-017-0271-z.

Janssen, S., G. Dumont, F. Fierens, and C. Mensink. 2008. Spatial interpolation of air pollution measurements using CORINE land cover data. *Atmospheric Environment* 42 (20): 4484–4903.

Karanasiou, A., M. Viana, X. Querol, T. Moreno, and F. De Leeuw. 2013. Assessment of population exposure to air pollution during commuting in European cities. *ETC/ACM Technical Paper 2013/2*. European Topic Centre on Air Pollution and Climate Change Mitigation.

Karner, A.A., D.S. Eisinger, and D.A. Niemeier. 2010. Near-roadway air quality: synthesizing the findings from real-world data. *Environmental Science & Technology*.

Kim, K.H., S.-B. Lee, D. Woo, and G.-N. Bae. 2015. Influence of wind direction and speed on the transport of particle-bound PAHs in a roadway environment. *Atmospheric Pollution Research* 6 (6): 1024–1034.

Knibbs, L.D., T. Cole-Hunter, and L. Morawska. 2011. A review of commuter exposure to ultrafine particles and its health effects. *Atmospheric Environment* 45 (16): 2611–2622.

Laeremans, M., E. Dons, I. Avila-Palencia, G. Carrasco-Turigas, J.P. Orjuela, E. Anaya, T. Cole-Hunter, A. de Nazelle, M. Nieuwenhuijsen, A. Standaert, M. Van Poppel, P. De Boever, and L. Int Panis. 2018. Short-term effects of physical activity, air pollution and their interaction on the cardiovascular and respiratory system. *Environment International* 117: 82–90.

Lee, K., and I.N. Sener. 2019. Understanding potential exposure of bicyclists on roadways to traffic-related air pollution: Findings from El Paso, Texas, Using Strava Metro data. *International Journal of Environmental Research and Publich Health* 16: 371.

Lefebvre, W., F. Fierens, E. Trimpeneers, S. Janssen, K. Van de Vel, F. Deutsch, J. Viaene, G. Vankerkom, C. Dumont, C. Vanpoucke, W. Mensink, and J. Peelaerts. 2011a. Modelling the effects of a speed limit reduction on traffic-related elemental carbon (EC) concentrations and population exposure to EC. *Atmospheric Environment* 45: 197–207.

Lefebvre, W., J. Vercauteren, L. Schrooten, S. Janssen, B. Degraeuwe, W. Maenhaut, I. De Vlieger, J. Vankerkom, G. Cosemans, C. Mensink, N. Veldeman, F. Deutsch, S. Van Looy, W. Peelaerts, and F. Lefebvre. 2011b. Validation of the MIMOSA-AURORA-IFDM model chain for policy support: Modelling concentrations of elemental carbon in Flanders. *Atmospheric Environment* 45/37: 6705–6713.

Lefebvre, W., B. Degraeuwe, C. Beckx, M. Vanhulsel, B. Kochan, T. Bellemans, D. Janssens, G. Wets, S. Janssen, I. de Vlieger, L. Int Panis, and S. Dhondt. 2013. Presentation and evaluation of an integrated model chain to respond to traffic-and health-related policy questions. *Environmental Modelling and Software* 40: 160–170.

Lonati, G., S. Ozgen, G. Ripamonti, and S. Signorini. 2016. Effects of bike lane features on cyclists' exposure to black carbon and ultrafine particles. *Journal of Earth Sciences and Geotechnical Engineering* 6 (4): 317–330.

Meysman, F.J.R., S. De Craemer. 2018. CurieuzeNeuzen Vlaanderen, Het cijferrapport. Universiteit Antwerpen.

Ogrin, M., K.V. Mally, A. Planinšek, A. Gergorič, L. Drinovec, and G. Močnik. 2016. Nitrogen dioxide and black carbon concentrations in Ljubljana. *Geograff* 18.

Padró-Martinez, L.T., A.P. Patton, J.B. Trull, W. Zamore, D. Brugge, and J.-L. Durant. 2012. Mobile monitoring of particle number concentration and other traffic-related air pollutants in a near-highway neighborhood over the course of a year. *Atmospheric Environment* 61: 253–264.

Pankow, J.F., M.A. Figliozzi, and A.Y. Bigazzi. 2014. *Evaluation of bicyclists exposure to traffic-related air pollution along distinct facility types.* Report, Portland State University.

Peters, J., J. Van den Bossche, M. Reggente, M. Van Poppel, B. De Baets, and J. Theunis. 2014. Cyclist exposure to UFP and BC on urban routes in Antwerp, Belgium. *Atmospheric Environment* 92: 31–43.

Rivas, I., D. Donaire-Gonzalez, L. Bouso, M. Esnaola, M. Pandolfi, M.P. de Castro, M. Viana, M. Alvarez-Pederol, M. Nieuwenhuijsen, A. Alastuey, J. Sunyer, and X. Querol. 2016. Spatiotemporally resolved black carbon concentration, schoolchildren's exposure and dose in Barcelona. *Indoor Air* 26: 391–402.

Rückerl, R., A. Schneider, S. Breitner, J. Cyrys, and A. Peters. 2011. Health effects of particulate air pollution: A review of epidemiological evidence. *Inhalation Toxicology* 23: 555–592.

Rojas-Rueda, D., A. de Nazelle, M. Tainio, and M. Nieuwenhuijsen. 2011. The health risks and benefits of cycling in urban environments compared with car use: health impact assessment study. *BMJ* 343: 5306.

Sharma, A., D. Massey, and A. Taenja. 2009. Horizontal gradients of traffic related air pollutants near a major highway in Agra, India. *Indian Journal of Radio & Space Physics* 38: 338–346.

Sunyer, J., E. Suades-González, R. Garcia-Esteban, I. Rivas, J. Pujol, M. Alvarez-Pederol, J. Forns, X. Querol, and X. Basagaña. 2017. Traffic-related air pollution and attention in primary school children. Short-term association. *Epidemiology* 28 (2): 181–189.

Soulhac, L., P. Salizzoni, F.-X. Cierco, and R. Perkins. 2011. The model SIRANE for atmospheric urban pollutant dispersion: part I, presentation of the model. *Atmospheric Environment* 45: 7379–7395.

Van den Bossche, J., J. Peters, J. Verwaeren, D. Botteldooren, J. Theunis, and B. De Baets. 2015. Mobile monitoring for mapping spatial variation in urban air quality: Development and validation of a methodology based on an extensive dataset. *Atmospheric Environment* 105: 148–161.

Van Kempen, E., P. Fischer, N. Janssen, D. Houthuijs, I. van Kamp, S. Stansfeld, and F. Cassee. 2012. Neurobehavioral effects of exposure to traffic-related air pollution and transportation noise in primary school children. *Environmental Research* 115: 18–25.

Virkkula, A., T. Mäkelä, and R. Hillamo. 2007. A simple procedure for correcting loading effects of aethalometer data. *Journal of the Air and Waste Management Association* 57: 1214–1222.

VMM. 2017. *Contribution of wood smoke to air pollution.* Flanders Environment Agency. https://www.vmm.be/publicaties/inschatting-van-de-bijdrage-van-houtverbranding-door-burgers-aan-luchtverontreiniging-in-vlaanderen. Accessed 27 June 2020.

Wilmott, C.J., S.M. Robeson, and K. Matsuura. 2011. A refined index of model performance. *International Journal of Climatology* 32: 2088–2094.

World Health Organization. 2019. Preventing disease through health environment—exposure to benzene: A major public health concern. https://www.who.int/ipcs/features/benzene.pdf. Accessed 27 June 2020.

Burden of the Coastal Area with Solid Waste in Kornati National Park (Croatia)

Natalija Špeh, Robert Lončarić, Kristjan Breznik, and Maša Surić

Abstract The problem of waste disposal in the Kornati archipelago is clearly an issue of hidden geography since it is often neglected in the scientific literature and media alike. This paper summarises a realised joint preparation of the methodology for the inventory of solid waste in the coastal zone. The methodology was used and tested for the inventory of solid waste in Kornati National Park which encompasses a large part of the Kornati archipelago located in the central part of the Croatian Adriatic coast. Inventory of solid waste is the first and crucial step in the management of this often-overlooked issue. Approximately two-thirds of the land and the waters of the Kornati archipelago (217 km^2) in Central Dalmatia were proclaimed as a national park in 1980. It is a place of exceptional natural values, cultural heritage and occasional population, which is at the same time very attractive for visitors. The purpose of the National Park's management plan for the area was to maintain high aesthetics of the landscape and the extremely rich marine ecosystem. The observed natural conditions (winds and currents) and socio-economic features (environmental pressures during the peak tourist season) are causing specific problems within the park in terms of waste pollution. Human activities and their influence seem to be concentrated in the coastal zone: (a) areas with (occasional) settlements and infrastructure intended for the predominant nautical tourism, and (b) uninhabited areas mostly with bays and coves accessible only from the sea. Data show different origins of the waste according to the country of their production: Croatia, Turkey, Albania, Poland, Italy, Bosnia and Herzegovina, Montenegro, Germany, France, Greece, Hungary and North

N. Špeh (✉) · K. Breznik
Visoka šola za varstvo okolja, Trg mladosti 2, 3320 Velenje, Slovenia
e-mail: natalija.speh@vsvo.si

K. Breznik
e-mail: kristjan.breznik@mfdps.si

R. Lončarić · M. Surić
Department of Geography, University of Zadar, Ulica dr. Franje Tuđamana 24 i, 23000 Zadar, Croatia
e-mail: rloncar@unizd.hr

M. Surić
e-mail: msuric@unizd.hr

M. Krevs (ed.), *Hidden Geographies*, Key Challenges in Geography,
https://doi.org/10.1007/978-3-030-74590-5_7

Macedonia with prevailing household plastic waste and solid wood waste. Evidently, waste has not stopped accumulating there.

Keywords Marine pollution · Coastal area · Solid waste · Cluster analysis · Kornati National Park · Croatia

1 Introduction

Marine waste is a complex cultural and multi-sectoral problem that imposes tremendous ecological, economic and social effects, and costs to communities around the world. One of the substantial barriers to addressing marine waste is the absence of adequate scientific research, assessment and monitoring. There is a gap in the information necessary to evaluate the impact of marine debris on coastal and marine species, habitats, economic development, human health and safety and social values. More information is also needed to understand the status and trends in amount, distribution and types of marine debris. There is also a gap in capacity in the form of new technologies and methods to detect and remove accumulations of marine debris (United Nations 2016).

The case of the environmental (waste) burden of the oceans was recognised pretty much as a disparity. The key findings determined already in the 1990s are five global ocean gyres where plastic gathers due to the current circulation. Only about 20% of the ocean plastic comes from marine sources, such as discarded fishing equipment or cargo ship mishaps. About 80% of it washes out to the sea from the beach litter or is carried downstream in rivers (Parker 2014). As for most of the beaches, the major debris is plastic. The spatial distribution of plastic debris is affected by multiple factors, including land uses, human population, fishing activity and oceanic current systems (Ribic et al. 2010).

The interaction between humans and the environment has caused intense landscape transformations which often led to landscape degradation. Working together, scientific and professional knowledge are able to keep the natural and cultural environment in sustainable, self-regenerating conditions. Self-sustainability is a necessity also in the field of waste treatment.

The issue of sea waste pollution represents an even greater threat when its long-term impact is considered. Since the sea waste does not care for the borders, the sea waste management, on the opposite side, should. The spatial distribution of sea waste led by natural arrangements (prevailing winds, dominant sea currents) is undoubtedly a matter of hidden geography.

The Kornati archipelago in the central part of the Adriatic Sea recognised as a landscape of natural values is threatened by the sea waste pollution. The islands are promoted as a preserved and pristine natural environment, which is therefore attracting a growing number of visitors each year. However, mostly hidden from the public eye lies the problem of an increased burden of the sea waste along the shores of the Kornati. This issue is mostly omitted in both scientific literature and media coverage, which often focus on more positive topics (e.g. natural and cultural heritage, positive impacts of tourism, and nature protection) since emphasising the waste management issues in the archipelago could possibly divert visitors from the islands. Besides the sea waste along the coast, there is also a problem of inland waste disposal. In most cases, these sites are hard to detect as the locally produced waste is intentionally illegally disposed of in more secluded locations, and the local population is sometimes reluctant to discuss this topic. On the other hand, sea waste washed ashore can easily be spotted even from the distance, but most of this waste is not locally produced and its origin can only be determined by the detailed analysis which is the main topic of this paper.

Here, the first analysis of the solid waste accumulation in the Adriatic coastal area is presented. Based on the social–environmental approach, no economic indicators had been included so far. Since our life and economies strongly depend on the oceans (50% of oxygen is produced in the ocean), we explored one view of that global issue at the regional level of Kornati National Park. A real impact of (plastic) waste introduction to the marine ecosystem (e.g. in biota and sea-floor sediments) is far from being globally controlled thus presenting a very unpredictable issue and in the case of the Kornati archipelago, as long as no organised international action is taken in the Adriatic Sea, the sea waste will remain a hidden geography topic.

Visible anthropogenic influences in the form of various types of waste distributed along the island' coasts should have already been studied. However, we have not found evidence of any long-term study regarding the surveyed topic of different waste types and materials.

2 Study Area and Environmental Settings

The Kornati archipelago is situated in the Croatian part of the Middle Adriatic (Fig. 1) covering an area of ca. 320 km^2, and includes ca. 150 karstified land entities (either permanently or occasionally above the sea level) arranged in four island chains in NW–SE direction. Kornati National Park, where the research was conducted, encompasses 68% of the Kornati archipelago (217 km^2), i.e. 89 islands (Fig. 1). In geological terms, they are part of >8,000 m thick carbonate succession deposited in several episodes from Upper Palaeozoic to Palaeogene (Vlahović et al. 2005) reworked by intensive tectonics mostly related to Alpine orogeny. Subsequent karstification and Late Pleistocene–Holocene transgression finally shaped the archipelago.

Presently, environmental features and processes strongly depend on the regional oceanographic and local climatic settings. Namely, Adriatic is a semi-enclosed sea connected to the Mediterranean Sea through the relatively narrow Otranto Straight, and is deeply indented into the European continent. Its surface circulation is generally cyclonic (Zore 1956), characterised by the northwest current along the eastern Adriatic coast, and southeast West Adriatic Current (WAC) along the western Adriatic coast (Fig. 2). Along with cyclonic circulation, a short-term phenomenon manifested as transitional currents occurs on the scale from days to weeks. These currents

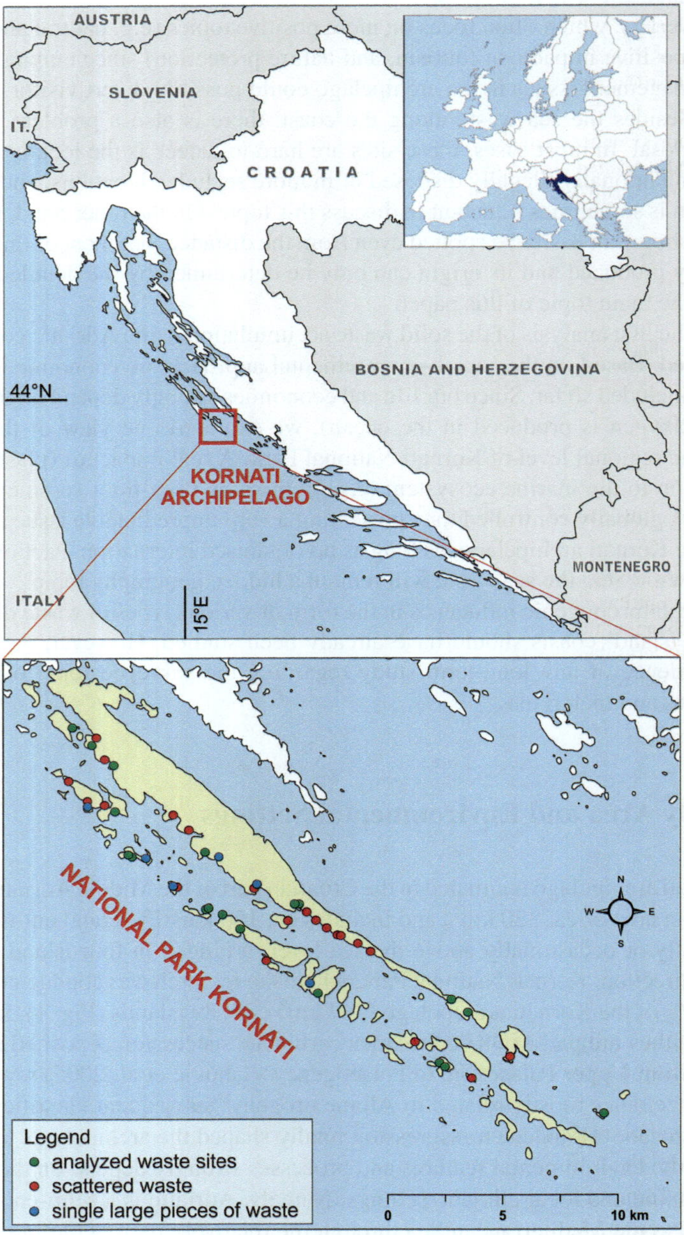

Fig. 1 Location of Kornati National Park and study sites within

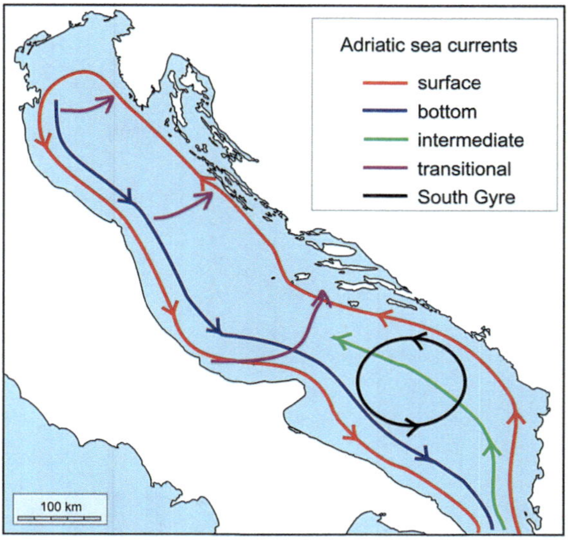

Fig. 2 Generalised Adriatic Sea circulations (after Morović et al. 2018)

are caused by a strong synoptic disturbance when strong bora and sirocco winds force WAC shift towards the Croatian coast, particularly at the Palagruža Sill and off Ancona (Vilibić et al. 2009; Morović et al. 2018). Astronomical tides have an average range of 25–30 cm, but occasionally meteorological influences, such as storm surges and tsunami-like sea-level oscillations induced by air pressure oscillations (Šepić and Vilibić 2011), can enhance the common range.

The climate of the study area is Mediterranean with hot and dry summers (Csa according to Köppen classification). Prevailing weather types are determined by the contact of the continental air masses from Central and Eastern Europe, and maritime air masses originating from the Atlantic Ocean and the Mediterranean Sea (Surić et al. 2018). During the summer, there is a dominant influence of the Azores High which supports stable, dry and warm weather with minimal, mostly local air circulation. During the autumn the high-pressure zone weakens, allowing the inflow of the Atlantic air masses within the low-pressure systems which are frequently moving across the Adriatic Sea bringing wet and windy weather throughout the autumn and early winter. During the winter, the weather often stabilises under the influence of the Siberian High, which induces incursions of the cold air over the Adriatic. These incursions, combined with the transition of low-pressure systems over the Mediterranean and the Adriatic Sea, induce the occurrence of the NE *bora* wind, the most frequent wind in the Kornati Islands (Fig. 3), which often reaches gale force. The second most frequent wind is the sirocco, SE wind, which is formed on the front side of the low-pressure systems and is typical for the autumn (CHMS 2020). Although it is not as strong as the bora wind, it produces the highest waves in the Adriatic Sea due to a larger fetch and a more constant speed. In November 2019, during a particularly strong sirocco episode, a record-breaking wave was recorded off

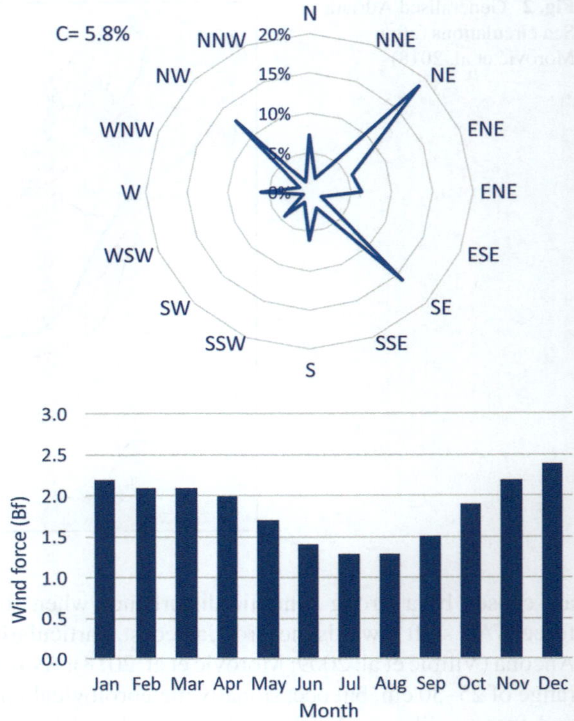

Fig. 3 Wind frequency and wind force recorded in Vela Sestrica meteorological station (NW part of the archipelago) for the 1990–2018 period

Dubrovnik, measuring 10.87 m (Hydrographic Institute of the Republic of Croatia 2019).

3 Historical Socio-economic and Landscape Changes

Although the Kornati Islands form an archipelago of around 150 islands, none of those are permanently inhabited. Scientists still debate the reasons for the absence of permanent settlements without reaching any consensus. The most probable cause is a combination of narrow-shaped islands which can only support small patches of arable land with sparse water sources and the distance from both the mainland and the settlements on the neighbouring islands. Additionally, these islands lack natural harbours, so only smaller vessels could anchor in few sheltered coves (Skračić 2013). Throughout history, the islands were inhabited only seasonally as the two cornerstones of the local economy—fishery and livestock breeding—were also seasonal. The settlers originated from the adjacent Dugi otok Island and from the late 1800s from the more distant Murter Island. The number of seasonally present population was always relatively small, usually between 200 and 300 inhabitants. Official census data cover the period from the mid-1850s onward. The first two censuses (in 1857 and 1869) showed no population on the Kornati Islands, while the 1880 census revealed 37 people living on the islands, and 23 of them on Kornat Island, the largest in the archipelago. Population peak of 313 inhabitants was recorded in the 1931 census, after which population rapidly declined, particularly after the end of the Second World War, so the 1953 census found no people living on the islands. During the last two decades, a small rebound of the population has occurred with 21 inhabitants recorded in the 2011 census (Magaš 2013). It is worth noticing that the census methodology has been changing, which prevents reliable comparison of the population numbers. Some censuses only recorded permanently inhabited population, while others recorded population which was present on the islands at the time of the census.

In their traditional economy, fishing prevailed as the main source of income. The target species were sardines, mackerels and bonitos with an occasional catch of tuna. Fishermen rarely ventured offshore and the school of fish was typically trapped with the net in bays and coves. Livestock breeding (mostly sheep) was extensive so animals were often left to fend for themselves and were occasionally gathered for milking or wool shearing. Other activities included olive growing, small-scale salt extraction and lime production which was—together with livestock grazing—responsible for the destruction of the islands' forests (Fig. 4).

The shift in the local economy came during the second half of the twentieth century. The 1960s saw the onset of mass tourism along the Croatian coast and a growing number of tourists visited the Kornati Islands. Tourism, although also seasonal and limited only to summer months, offered a far better perspective for the local population. Many people migrated to the mainland in order to work in tourism,

Fig. 4 Patches of arable land protected from livestock in the past on an otherwise barren island

while some of the old homes and storages on the islands were refurbished and offered to tourists. Along with the restoration of old houses the new ones were built, and at such rate that the number of newly built homes in the late twentieth and early twenty-first century surpassed the number of houses built in the entire history of these islands. For example, at the beginning of the 1800s, there were only around 20 houses in the whole of the archipelago. By the 1960s, this number increased to over 270, and today the number of houses there stands at over 800 houses (Magaš 2013). Furthermore, the islands have become very popular as a destination for nautical tourism which prompted the opening of two marinas and other tourism infrastructure (Fig. 5) in the archipelago. One of the marinas is located within the borders of the national park. The number of tourists was constantly rising until the early 1990s when tourism was halted by the outbreak of the Croatian War for Independence. The revival of their tourism began shortly after the end of the war in 1995, and today the national park is annually visited by more than 100,000 people. The real number is probably significantly greater because official statistics do not include visitors who do not enter the park. However, alongside evident economic benefits, tourism also carries numerous threats: increased traffic congestion, air and water contamination, increased quantities of waste, increased noise, a possible loss of the local identity and non-reinvestment of revenues achieved in the protected areas.

Such pressure on the environment necessitated some level of legal protection. The first initiatives appeared during the 1960s when it became apparent that the number of visitors was on the rise and the landscape began to suffer. The final result of the

Fig. 5 Restaurant and mooring area on Smokvica island

environment protection initiatives came in 1980 when a large part of the archipelago became a national park—the second highest level of protected areas in the Croatian legal system.

4 Environmental Starting Point

On the global level, the factors contributing most to the environmental risks are the fishing industry, chemical pollution and eutrophication, physical changes of the ecosystem (invasions of allochthones species) and global climate changes (National Research Council 1995). Kornati NP is usually considered an unpolluted reference area in environmental and ecological studies, but with growing tourism and increasing exposure to anthropogenic inputs. Some parts of the national park are still well preserved and mostly intact. On the other hand, there are some areas where human activity has already made an evident impact and has consequently contributed to the changes in environmentent. The pilot study showed that areas closer to marinas and seasonally inhabited villages (Fig. 6) are more likely to be affected by anthropogenic pollution. Yet, more distant areas remain anthropogenically unaltered (Ilenič et al. 2018).

5 Methodology and Fieldwork

The vessel-based pre-research of Kornati National Park directed to the systematic survey of the sea waste in the islands' area has been conducted as early as 2002. We recognised the exploration as reasoned and quite sensible, although we speak of the

Fig. 6 Seasonal settlement Vrulje, the largest on the islands, located in the cove sheltered from the dominant winds and near a favourable agricultural zone. The name of the settlement indicates the presence of freshwater (*vrulje* = submarine springs)

protected area of many natural and cultural values. In addition, Kornati has had the reputation of the tourist (nautical) attraction since the 1960s. Additionally, we used Sustainable Tourism Development Strategy for the Broader Kornati National Park Area (2015), which was produced following the Kornati National Park Management Plan (2014–2023). The mentioned documents have been confirmed also to keep the Kornati archipelago ecosystems' goods and services as a natural, self-regeneration environment.

The fieldwork was carried out in May 2018 as a cooperation of the Environmental Protection College Velenje and the Department of Geography, University of Zadar. To assess the protection regime in Kornati National Park area, we altered the method of indicators as the main methodological approach. For the purpose of fieldwork, we also adapted the questionnaire of landfill indicators. Sea kayak was used as the means of transport, since we had to directly approach the waste locations to create the database of ashore waste.

Firstly, in order to introduce principles of solid waste accumulation ashore in the archipelago of Kornati islands, we made a model based on evidence of 11 indicators. It helped to evaluate the waste locations: (1) geomorphological (2) coast characteristics, orientation, (3) openness of the bay/cove to the sea, (4) vegetation conditions, (5) distance of waste from the sea, (6) waste distribution, (7) landscape exposure of the waste area, (8) the amount of waste (in m^3), (9) origin of waste (information obtained from the packaging), (10) type and percentage of the waste type, and (11) share of plastic waste (in %).

5.1 *Cluster Analysis*

We applied the method of hierarchical agglomerative clustering to define statistically important characteristics for the waste deposited ashore. Cluster analysis itself can be understood as a method or a task to group a set of units in such a way that units inside the same group (also called a cluster) are as similar as possible and as different as possible compared to units from other groups. As a technique, cluster analysis was already introduced in the first half of the previous century by different authors (Driver 1932; Tryon 1939; Cattell 1943). There are several clustering algorithms, one of the most popular being connectivity-based clustering also called hierarchical clustering. Based on distances calculated between units, we can produce a dendrogram when applying hierarchical clustering. It provides a hierarchy of clusters that merge with each other at different distances. Based on the dendrogram, the number of clusters can be determined.

For the purpose of this research, we applied the method of hierarchical agglomerative clustering with the Ward method and squared Euclidian distance (Everitt et al. 2011) to define statistically important characteristics for the ashore waste accumulation. Six indicators (out of 11) were normalised and included in the clustering analysis. Based on the dendrogram in Fig. 13, the location's (19) features were joined into four clusters.

6 Results

The research encompassed the most indented west coast of Kornati NP islands in order to include as many locations as possible. The collected fieldwork data enabled the following categorisation of waste sites: (A) 19 locations were recorded, where the list of indicators could be completely filled in and statistically analysed (Cluster) (*analysed waste sites* in Fig. 1), (B) 21 locations were redefined as the waste spread along the shore area (category *scattered waste* in Fig. 1) and (C) 6 locations characterised with the large individual waste pieces, e.g. abandoned fishing vessel (category *single large pieces of waste* in Fig. 1).

6.1 Evaluation of the Natural Settings

Wind conditions are discussed in relation to the waste locations in bays' and coves' exposition to the prevailing wind directions (Fig. 7). Coasts with south and northwest exposure of the bays/coves proved to be the most affected by the waste was accumulation and coasts with southeast and southwest exposure followed. No waste has been recorded in the bays/coves with the northeast exposure.

Although the absence of waste in the bays/coves exposed to the predominant bora wind (NE) would imply that the wind is not the main contributor of the sea waste, detailed analysis of the prevailing winds and the origin of the waste suggests otherwise. Namely, most of the sea waste can be tracked down to the illegal landfills along the Albanian coast (Tudor and Janeković 2011) which is then washed into the sea, particularly during the autumn when gale-force SE wind (*scirocco*) often occurs. During such events, huge amounts of this sea waste can be found along the coast of the mainland and the islands in southern Croatia. When gale-force winds occur over longer periods of time, the sea waste can be carried further towards the central and

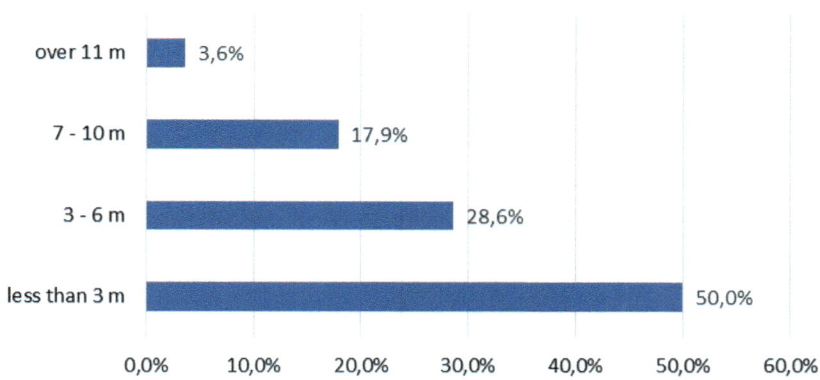

Fig. 7 Exposure of the bays and coves to the prevailing winds

even northern part of the Adriatic Sea. The effect of the scirocco wind on the spatial distribution of the sea waste is amplified by the fact that it coincides in direction with the dominant sea current in the Adriatic Sea. In the case of the Kornati Islands, since the scirocco is the second most frequent wind, it can be assumed that the wind in combination with the prevailing sea currents is the main contributor of the sea waste which is supported by the fact that the coasts exposed towards SE are usually the most affected by sea waste pollution (Fig. 7).

The paper also presents data about the origin of the waste (Fig. 8) which is mostly determined from the "made in" labels (e.g. on the food packages). Such determination is not reliable because the country of origin written on a package does not tell us in which country this item has been disposed of. For example, Italian-made automobile tyre can be washed to sea from the illegal waste site on Albanian shore, so the origin of that tyre is Albania and not Italy. Based on the "made in" labels, we determined that in most cases waste had a domestic origin (16 cases), followed by the waste originating from Italy (9 cases), Turkey (9) and Greece (8).

The prevailing orientation of the coast with waste sites was northwest–southeast (Fig. 9). This is in accordance with the prevailing NW–SE orientation of the eastern Adriatic coast.

According to the distance of the accumulated waste from the sea, the most frequent category of the waste distribution was the closest to the sea; in 50% of the locations, waste was deposited less than 3 m from the sea. The distance from the waste density was decreasing (Fig. 10).

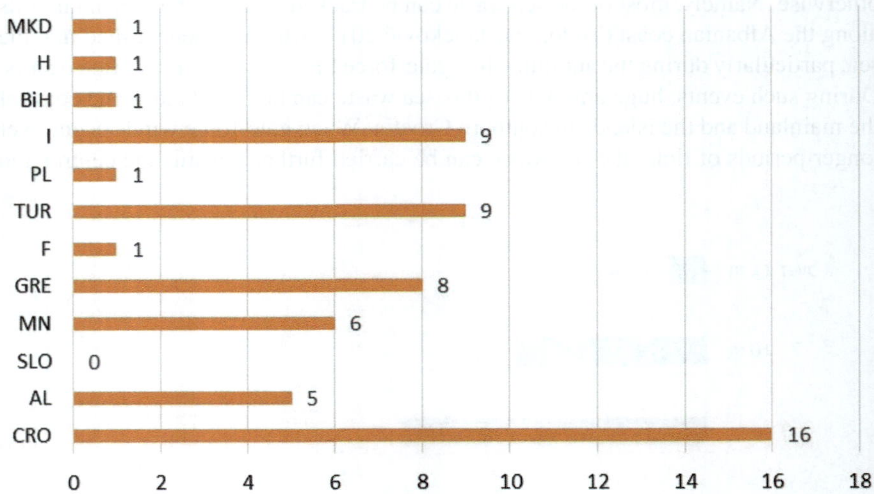

Fig. 8 Origin of the waste. Key: MKD—North Macedonia, H—Hungary, BIH—Bosnia and Herzegovina, I—Italy, PL—Poland, TUR—Turkey, F—France, GRE—Greece, MN—Montenegro, SLO—Slovenia, AL—Albania, CRO—Croatia

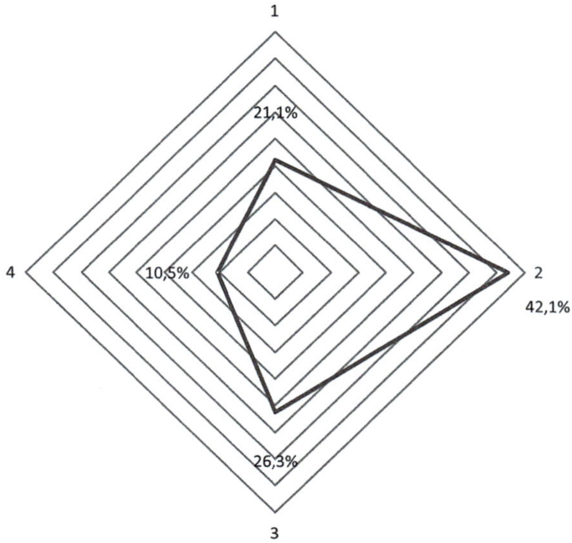

Fig. 9 Coast orientation. Key: 1 (N-S); 2 (NW–SE); 3 (NE-SW); 4 (E-W)

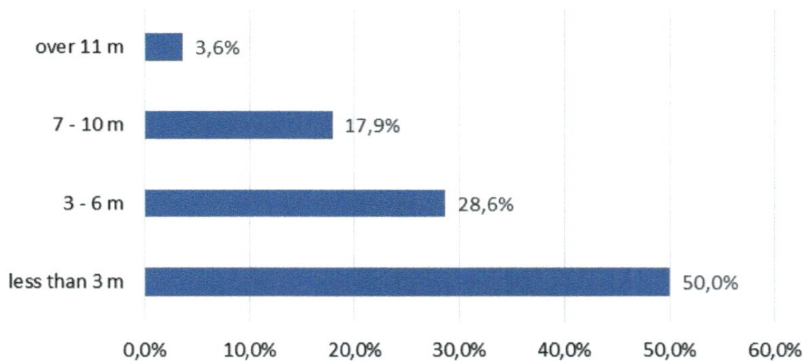

Fig. 10 Distance of the waste from the sea

6.2 Social Features of the Waste Locations

The following data on the anthropogenic settings were considered: (1) the access to the location and (2) the type and the share of the waste. The share of waste type indicator was favourable to the human impact (the maximum share represented household waste, which accounted for 32.1%; Fig. 12). All the waste locations were accessible by the sea (100%) and some also by (worse) path (26.3%)—Fig. 11. We reasoned the result from the occasional, tourism (nautical) activities causing overpopulation of the researched aquatory from May to October.

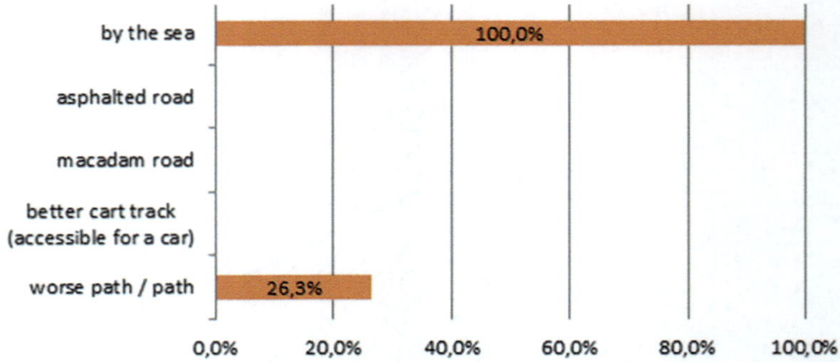

Fig. 11 Access to the waste location

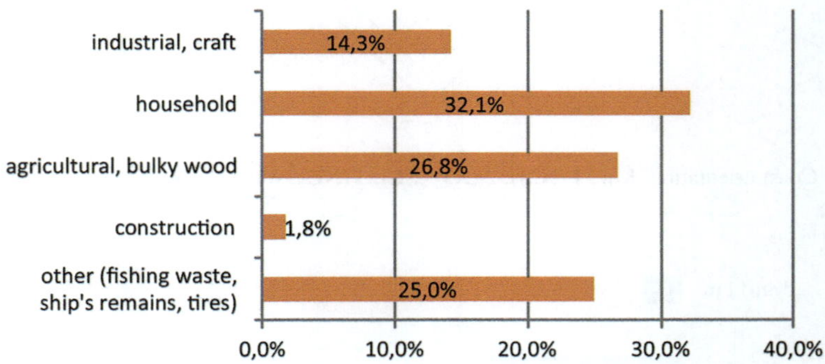

Fig. 12 The type and the share of the waste

6.3 *Cluster Analysis*

Following the method of hierarchical agglomerative clustering to define statistically important characteristics for waste accumulation ashore, the following six indicators (out of 11) were normalised and included in the clustering analysis: (1) the amount of waste, (2) the type and the share (in %) of waste accumulated, (3) the share of plastic waste, (4) the distance of waste from the sea, (5) the distance from a settlement area and (6) the origin of waste production. Based on the dendrogram in Fig. 13, locations' characteristics (19) were joined into four clusters (groups).

Four groups and their descriptions were formed taking also the rest of the five indicators surveyed into account:

1. Group 1: The highest amount of waste, biomass prevailing, SE openness of the bays/coves to the sea (Fig. 14);
2. Group 2: West openness of the bays/coves to the sea, mixed waste structure, the lowest amount of waste;
3. Group 3: The most frequent plastic, household waste;

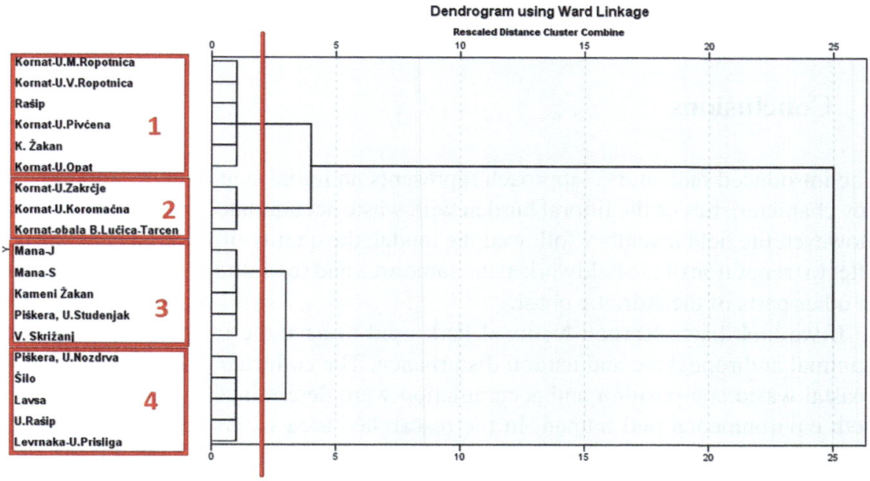

Fig. 13 Dendrogram of agglomerative hierarchical clustering of 19 waste locations in Kornati region

Fig. 14 The deposit of sea waste ashore, biomass prevailed (cluster group 1)

4. Group 4: Miscellaneous.

7 Conclusions

The introduced indicators'' approach represents an initial step in understanding the key characteristics of the littoral burden with waste accumulated in the Adriatic Sea. However, the field inventory followed the model, the quality of which will be proven after (a) repetition of the fieldwork at the same area and (b) a study of another island(s) in other parts of the Adriatic coast.

If we look back, Kornati National Park used to have the image of an area with minimal anthropogenic and natural disturbance. The collected data showed that the coastal waste composition and accumulation were determined by various impacts, both environmental and human. In the researched area we listed (A) 19 locations where it was possible to fill in the questionnaires completely; the waste material lay ashore in various condensed forms (e.g. Cluster Group 1, Fig. 14), (B) category were locations (21 cases) with the dispersed of waste ashore and (C) individual pieces of the waste (6 locations).

Apparently, the observed environment of Kornati NP should be discussed as an intertwined identity of (A) their applicable value based on the (B) well preserved natural values.

Following the essential pollution aspect of the research, we numbered the following findings of the inquiry:

(1) Considering the diversity of the sea waste type, we estimated 26.8% out of 52.8 m^3 of total waste (biomass; agricultural, bulky wood) as not risky for the environment. The rest should be treated as a potential for reuse or recycle.
(2) Waste in this area was distributed in layers (0.3 m average thickness) in 47.7% cases, and 8.4% in piles. In 47.9% of cases, we found waste scattered.
(3) In 73.3% of cases, waste locations presented the distance from the nearby village over 1000 m and an individual house was over 1000 m away in 57.9% of cases.
(4) In 78.9% of cases, the waste sites were assessed as exposed.
(5) The average amount of waste per location summed 2.8 m^3 (of a total of 52.8 m^3).
(6) In 84.2% of cases, the sites exhibited a state of occasionally washed ashore waste.
(7) The average share of plastic waste accounted for 65.4% of total waste.

To upgrade the above-concluded results, we composed some recommendations:

1. Regarding the waste structure, three types of potential sanitation measures are recommended: (a) disposal of waste to a municipal landfill for further waste management, (b) composting and (c) waste incineration.
2. Considering the human–nature relationship, a coordinated environmental action and engagement of the public should perform (a) a coordination of the private

sector and within it and (b) a multidisciplinary approach of land-based and sea-based activities including (human, material) sources to establish the actual and potential effects of (sea waste) plastics.
3. To cope with protected natural values of national importance and their use, some of the issues (conflicts) would be prevented if we gave top priority to public education about the sea and the sea waste issue and the awareness of its importance prior to the legislation and short-term economic gains.

The obtained results can be considered as an initial step in the waste management in the archipelago or as a signal to the public and the authorities that there is an issue of the sea waste pollution and that there is a need for organised action to deal with this issue on the local, national and international (Adriatic region) level. This research can also help to move the problem of the sea waste in the Kornati Islands from "hidden" to "visible" geography.

Acknowledgements The development of the methodological approach was inspired and carried out during the first author's Erasmus mobility.

Lecturers of the Environmental Protection College Velenje and the University of Zadar, Department of Geography, carried out their research activities together with the students of Zadar institution.

References

Cattell, R.B. 1943. *The description of personality: Basic traits resolved into clusters.* American Psychological Association.
Croatian Meteorological and Hydrological Service. 2020. https://meteo.hr/index_en.php. Accessed 20 Feb 2020.
Driver, H.E. 1932. *Quantitative expression of cultural relationships.* University of California Press
Everitt, B.S., S. Landau, M. Leese, and D. Stahl. 2011 *Cluster analysis*, 5th edn. Wiley.
Hydrographic Institute of the Republic of Croatia. 2019. http://www.hhi.hr/en. Accessed 20 Feb 2020.
Ilenič, A., S. Lojen, I. Župan, et al. 2018. Is the Kornati National Park still an acceptable reference area for environmental studies. *Geosciences* 8: 385. https://doi.org/10.3390/geosciences8110385.
Magaš, D. 2013. Zemljopisna obilježja kornatskih otoka u funkciji upoznavanja njihove toponimije. In *Toponimija kornatskog otočja*, ed. V. Skračić, 19–65.
Management Plan of Kornati National Park 2014–2023. 2014. http://www.np-kornati.hr/images/plan_upravljanja/Plan_upravljanja_NP_Kornati_tekst_%20i_zone.pdf. Accessed 11 Feb 2020.
Morović, M., A. Ivanov, and M. Oluić. 2018. Mapping of oil slicks in the Adriatic Sea: Croatia case study. In *Oil pollution in the Mediterranean sea: Part II*, eds. A. Carpenter, and G. Kostianoy. Springer International Publishing.
National Research Council. 1995. Understanding marine biodiversity. https://doi.org/10.17226/4923. Accessed 4 Feb 2020.
Parker, L. 2014. The best way to deal with ocean trash. National Geographic. https://www.nationalgeographic.com/news/2014/4/140414-ocean-garbage-patch-plastic-pacific-debris/. Accessed 2 Oct 2019.
Ribic, C.A., S.B. Sheavly, D.J. Rugg, and E.S. Erdmann. 2010. Trends and drivers of marine debris on the Atlantic coast of the United States 1997–2007. *Marine Pollution Bulletin* 60: 1231–1242. https://doi.org/10.1016/j.marpolbul.2010.03.021.

Skračić, V. 2013. Kurnatari u Kornatima. In *Toponimija kornatskog otočja*, ed. V. Skračić, 7–18.
Surić, M., R. Lončarić, N. Bočić, N. Lončar, and N. Buzjak. 2018. Monitoring of selected caves as a prerequisite for the speleothem-based reconstruction of the quaternary environment in Croatia. *Quaternary International* 494: 263–274. https://doi.org/10.1016/j.quaint.2017.06.042.
Sustainable Tourism Development Strategy for the Broader Kornati National Park Area. 2015. http://np-kornati.hr/images/novosti/sustainable-tourism-development.pdf. Accessed 10 Feb 2020.
Šepić, J., and I. Vilibić. 2011. The development and implementation of a real-time meteotsunami warning network for the Adriatic Sea. *Natural Hazards and Earth System* 11: 83–91. https://doi.org/10.5194/nhess-11-83-2011.
Tryon, R.C. 1939. *Cluster analysis; correlation profile and orthometric (factor) analysis for the isolation of unities in mind and personality*. Lithoprinters and Publishers: Edwards Brother Inc.
Tudor M., and I. Janeković. 2011. Modelling the origin and fate of waste materials on the south-eastern Adriatic coast (Croatia). In *Geofizički izazovi 21. stoljeća (proceedings)* 29.
United Nations. 2016. The first global integrated marine assessment world ocean assessment I. https://www.un.org/Depts/los/global_reporting/WOA_RPROC/WOACompilation.pdf. Accessed 11 Feb 2020.
Vilibić, I., J.W. Book, G. Beg Paklar, M. Orlić, V. Dadić, M. Tudor, P.J. Martin, M. Pasarić, B. Grbec, F. Matić, H. Mihanović, and M. Morović. 2009. West Adriatic coastal water summertime excursion into the East Adriatic. *Journal of Marine Systems* 78: 132–156.
Vlahović, I., J. Tišljar, I. Velić, and D. Matičec. 2005. Evolution of the Adriatic carbonate platform: Palaeogeography, main events and depositional dynamics. *Palaeogeography, Palaeoclimatology, Palaeoecology* 220 (3–4): 333–360. https://doi.org/10.1016/j.palaeo.2005.01.011.
Zore, M. 1956. On gradient currents in the Adriatic Sea. *Acta Adriatica* 8: 1–38.

Potential for the Development of Astronomical Tourism in Kumano City, Mie Prefecture, Japan

Takumi Isono and Nobunari Itoh

Abstract This study discusses the potential for the development of astronomical tourism in Kumano City. An analysis is conducted not only of tourism development efforts undertaken by the local government and the tourist association, but also of astronomy-related tourist behaviors and tourists' opinions about stargazing activities. Astronomical tourism in Kumano City is regarded as the creation of new tourism consumption opportunities, given the lack of existing night tourist attractions. However, it is necessary to differentiate the contents of stargazing activities from other areas to attract tourists because Kumano City has not been recognized as a popular astronomy-related destination yet. Overall, astronomical tourism tends to focus strongly on features of educational tourism, such as explanations of constellations by specialists. On the other hand, tourists who are interested in astronomy in Kumano City are attracted by leisure or comfort in the context of stargazing activities. Therefore, to promote astronomical tourism, it is necessary to not only improve human resource development, by providing interpreters for stargazing, but also create warm spaces and environments where one can comfortably watch the starry sky in Kumano City and engage in academic activities. Isono (The Tourism Studies 31:5–18, 2019) explains that such lighthearted stargazing products can produce child-friendly opportunities to enjoy peaceful night-time activities, even though the starry sky itself is unlikely to be a major tourist attraction. Thus, it is necessary to create stargazing programs related to the strengths of Kumano City such as the scenic natural landscape or warm winter climate.

Keywords Astronomical tourism · Regional revitalization · Peripheral area · Hiddentourism geography · Kumano city

---

T. Isono (✉) · N. Itoh
Faculty of Education, Mie University, Tsu, Japan
e-mail: takumi.i_freedom@hotmail.com

N. Itoh
e-mail: nitoh@edu.mie-u.ac.jp

M. Krevs (ed.), *Hidden Geographies*, Key Challenges in Geography,
https://doi.org/10.1007/978-3-030-74590-5_8

1 Introduction

1.1 Research Background

Light pollution (increased night brightness) is a major obstruction to astronomical observation. Satellite images show that the night sky in developed countries and BRICs (Brazil, Russia, India, and China) is steadily increasing in brightness (Itoh et al. 2018). Falchi et al (2016) noted that 80% of Americans and 70% of Japanese live in areas where the Milky Way cannot be seen. Thus, the opportunity to watch the sparkle sky is becoming a rare experience for urban residents (Longcore and Rich 2004).

Under these circumstances, discussions on tourism development in relation to the dark night sky have become more active in Japan and abroad. The International Dark-Sky Association, established in 1988, designates International Dark-Sky Places where one can experience educational activities addressing light pollution, such as stargazing and astronomical observation tours. Collison and Poe (2013) explained that astronomical tourism offers attractions for tourists to visit places where the dark sky is free from light pollution. National parks and reserve authorities have been collaborating to create stargazing programs for tourists, especially in Western countries (Rodrigues et al. 2015).

Due to the rapid decline of traditional industries, many peripheral cities and towns have sought to revitalize their regions by tourism development. Depopulated regions without notable tourist attractions have joined the astronomical tourism movement as a means of regional development (Narita and Ueda 2013). Local governments have attempted to create unique attractions to draw tourists (Murakami 2001), as regional night-time resources present excellent options to differentiate one tourist location from another. Peripheral areas in particular offer abundant natural elements that are limited in urban areas, and thus can offer exciting one-of-a-kind experiences such as night safaris (e.g., Kawanami 2016; Hosaka et al. 2017) and stargazing activities (e.g., Uda and Isono 2019). In addition, astronomical tourism is an effective strategy for regional development as it can create a demand for long-term stays where perhaps only day trips have been common. Thus, astronomical tourism is likely to develop as a new type of sustainable tourism and has been drawn attention as one of the hidden tourism geography topics such as the above studies.

The promotion of astronomical tourism has been steadily improving the environment since 1988. The Ministry of the Environment has held events to not only increase awareness for the environmental protections about light and air pollution, but also to promote the utilization of starry sky for education and tourism. Astronomers have organized a group that trains and certifies astronomical guides to lead tours at night-time attractions. (Tomita and Okyudo 2009). Furthermore, in 2008, the Japan Tourism Agency declared that *sora tourism*[1] (which includes astronomical tourism) is a successful and increasingly popular facet of the country's tourism industry.

---

[1] For more information about sora tourism: https://soratourism.com/.

Some peripheral areas that have been successful in astronomical tourism were studied as examples of destination-based tourism. Nagai (2016) explored the background and development process of astronomical tourism in a Japanese rural area which had suffered from a decreasing number of tourists to its hot springs. Uda and Isono (2019) described the characteristics of business operators who conduct stargazing activities for tourists in Ishigaki City, which is famous for nature-based tourism. However, few studies have focused on the tourists' perspectives. A demand assessment for astronomical tourism is needed. Collison and Poe (2013) noted that astronomy-related tourists are often overnight tourists; however, they analyzed only the very basic attributes of astronomy-related tourists. Therefore, the specific characteristics of tourists' behaviors, motivations, and needs remain unclear. It is necessary to form a process for selecting astronomy-related tourist destinations to ensure an astronomical tourism industry with a stable supply and demand balance.

1.2 Objective and Methodology

This study discusses the potential for the development of astronomical tourism in Kumano City (Mie Prefecture in Japan). An analysis is conducted not only of tourism development efforts undertaken by the local government and the tourist association, but also of astronomy-related tourist behaviors and tourists' opinions about stargazing activities. To achieve the purpose of this study, fieldwork based on interviews was conducted during March and April 2019. The staffs of the Kumano City Office and the Kumano City Tourist Association were interviewed about the current situation of astronomical tourism promotion. To understand the characteristics of astronomy-related tourist behaviors and their opinions, a web questionnaire survey which the authors commissioned to Macromill, Inc. was conducted for 209 overnight tourists who had stayed at least one night in Kumano City since 2014 (see Sect. 4 for the detailed information).

2 Study Area

2.1 Overview of Kumano City

Mie Prefecture is divided into five regions (Hokusei, Chunansei, Ise-Shima, Iga, and Higashi-Kishu), and Kumano City is located in the Higashi-Kishu region (Fig. 1). Kumano City is blessed with a rich natural environment, but it has a low habitable area ratio (12.3%) due to the overall mountainous terrain. The warm current (Kuroshio) flows off the coast of the Higashi-Kishu region, making it a mild climate in Mie Prefecture, especially in winter (Fig. 2).

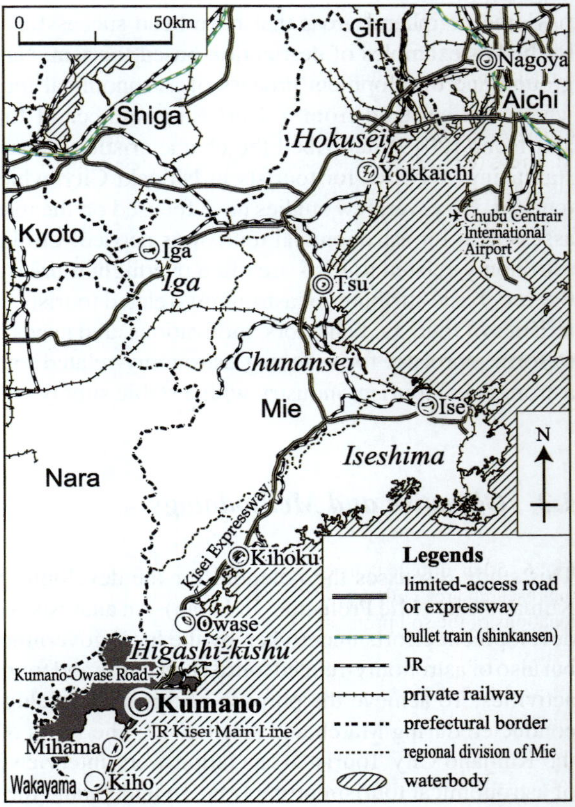

Fig. 1 Study area. *Source* Author

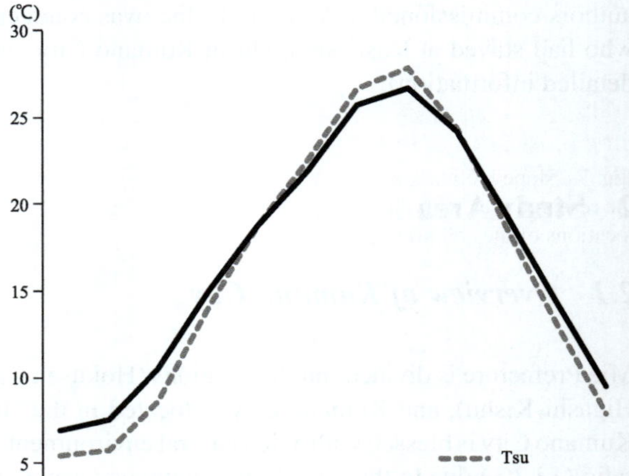

Fig. 2 Average temperature per month in Tsu and Kumano City (1985–2019). *Source* Japan Meteorological Agency

The Higashi-Kishu region, including Kumano City, is far from urban areas, such as Nagoya, Osaka, Yokkaichi (Hokusei), and Tsu (Chunansei) cities, which caused delays in the development of transportation infrastructure. As for road traffic, accessibility from urban areas has been improved since 2006 because of the gradual development and expansion of the Kisei Expressway and the Kumano-Owase Road. This led to a declining population and limited employment opportunities. According to the national census (2015), the population of Kumano City greatly decreased from 1960 (39,150) to 2015 (17,322). The elderly population ratio rose substantially from 1960 (7%) to 2015 (42%). This shows the current situation of fewer children and aging society in Kumano City. Regarding the ratio of employed persons to occupation types in Kumano City, primary industries account for 7.9% of the workers, secondary industries for 17.3%, and tertiary industries for 74.8%. Therefore, it can be pointed out that Kumano City has a high proportion of primary and tertiary industries because the average ratios of Mie Prefecture are 4.0% for primary industries, 33.6% for secondary industries, and 62.4% for tertiary.

2.2 *Kumano City as a Tourist Destination*

The Higashi-Kishu region is situated in a part of the World Heritage sites known as the Sacred Sites and Pilgrimage Routes in the Kii Mountain Range, so designated in 2014, and in the Yoshino-Kumano National Park. Thus, there are a variety of natural and cultural tourist resources in the Higashi-Kishu region. The Kumano-Kodo, a network of pilgrimage trails partially recognized as a World Heritage site, is one of the most famous and popular tourist attractions in the Higashi-Kishu region, and Kumano City has various local resources related to the Kumano-Kodo (Fig. 3). After the World Heritage certification, the number of tourists to the Kumano-Kodo is increasing on a continuing basis (Fig. 4). However, the Higashi-Kishu region is located far from major urban areas and, as such, has experienced a delay in public transportation development. Thus, compared to other regions, the number of tourists in the entire Higashi-Kishu region has been very small despite being certified as a World Heritage site (Fig. 5). In addition, half the tourists to the Higashi-Kishu region are from Mie Prefecture, and the remaining 40% are from Aichi or Osaka Prefectures. On the other hand, the number of tourists from Tokyo is extremely small. Thus, it can be pointed out that the tourists' residential area in the Higashi-Kishu region is limited to the small, that is, prefectural area.

As for the tourist attractions in Kumano City, cultural resources related to the Kumano-Kodo and natural resources like the Yoshino-Kumano National Park are widely distributed. The number of tourists and tourist attractions (e.g., Onigajo and Hanano-iwaya Roadside Station), which are located in a coastal area, are larger than those in mountainous areas because they are more easily accessible by road traffic (Fig. 6). In addition, Maruyama Senmaida, Yunoguchi Spa, Seiryu-so (Ryokan), and Dorokyo Gorge, which are distributed along national roads 311 and 169 in mountainous areas, are regarded as main tourist attractions for good accessibility by

Fig. 3 Matsumoto Toge (one of the Kumano-Kodo sites related the World Heritage) (*Source* Taken by the author, June 2019)

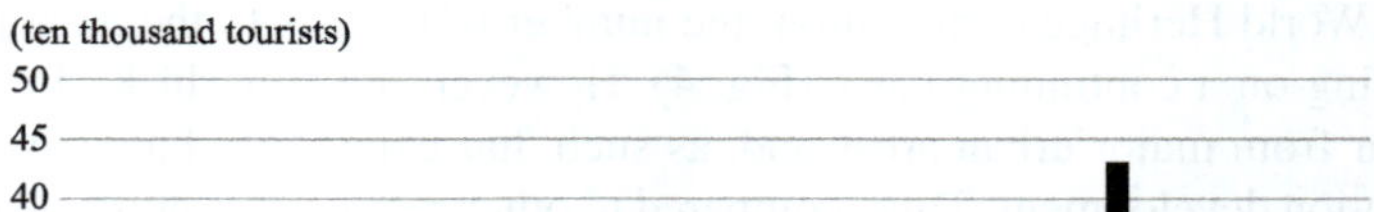

Fig. 4 The number of tourists to visit the Kumano-Kodo in Mie Prefecture (2000–2017) (*Source* Higashi-Kishu Chiikishinko Kousha)

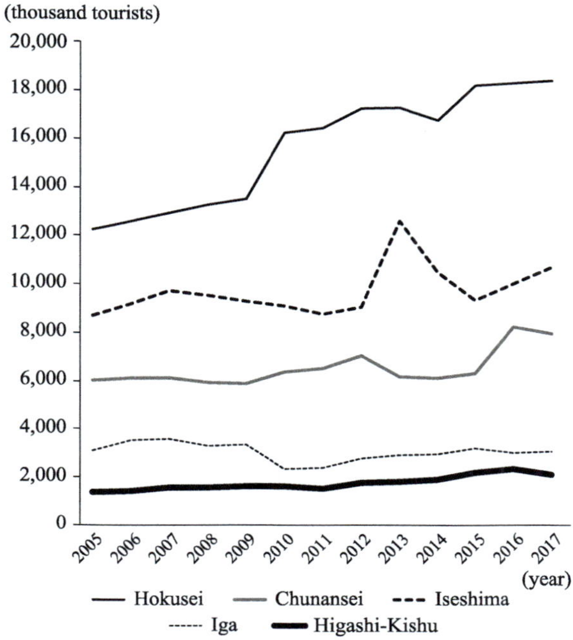

Fig. 5 The number of tourists of each region in Mie Prefecture (2005–2017) (*Source* Mie Prefecture)

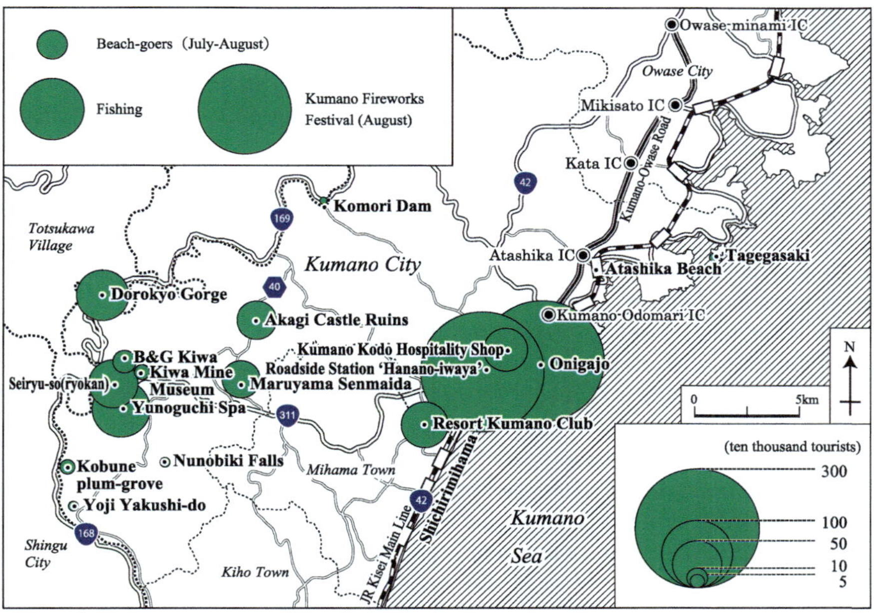

Fig. 6 The number of tourists at the major tourist attractions in Kumano City (2018) (*Source* Kumano City)

Fig. 7 A scene of the Kumano fireworks festival (2017) (*Source* Kumano City Tourist Association)

car. However, the absolute number of tourists in mountainous areas is much lower than that in coastal areas with good access from urban regions.

As for night-time activities, the Kumano Fireworks Festival, held every August 17th, is one of the biggest firework events in Japan with 10,000 fireworks being launched (Fig. 7). The famous fireworks capitalize on Kumano's rich nature and are launched into the night sky over the Kumano Sea, and include such notable displays as the one at Onigajo and the self-destructing fireworks on the water's surface. Figure 8 shows that the number of tourists is by far the largest in August. In addition, the Maruyama Senmaida Torch Procession is held every June as a night event in Kumano City (Fig. 9). The torch procession was held every year to drive away crop-eating insects until 1953; it was then revived in 2004 when the Kumano-Kodo was recognized as a World Heritage site. Thus, summer is regarded as the most important tourist season in Kumano City.

However, the Higashi-Kishu region, including Kumano City, needs to develop and promote new tourist attractions in the context of the World Heritage site and national park. This is because it is relatively less competitive than other regions, such as Wakayama Prefecture. However, it has been suggested that night tourist attractions in Kumano City need to be improved. Itoh et al. (2018) clarified that Kumano City has favorable conditions for starwatching, as the night sky as seen in the mountainous areas of Kumano City is as dark as that of Mauna Kea (the highest point in the state of Hawaii, America). In response to this, promoting astronomical tourism increased and an industry–government–academia collaboration project was established by the Kumano City Office, the Kumano City Tourist Association, and Mie University in 2018. The following section will describe how dark the night is in Kumano City and how the industry–government–academia collaboration has evolved.

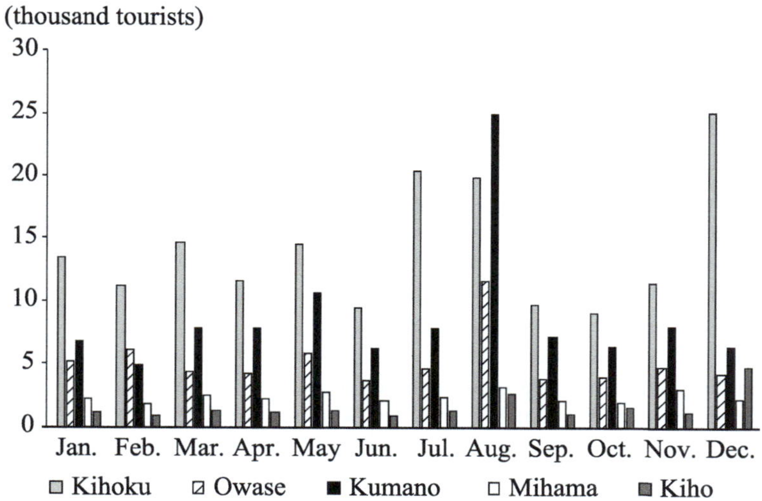

Fig. 8 The number of tourists in each city or town per month in the Higashi-Kishu region (2018) (*Source* Mie Prefecture)

Fig. 9 A scene of the Maruyama Senmaida torch procession (2019) (*Source* Kumano City Tourist Association)

3 Night-Time Environment in Kumano City

3.1 Night Sky Brightness in Kumano City and the Surrounding Area

The visibility of a starry sky depends on various conditions, even in the same region, such as starwatching time, local weather conditions, the altitude of the observation point, and the moon phase. The most essential condition for a star-filled night sky is the brightness of the sky. In this section, the results of the measurement of night sky brightness around Kumano City are discussed.

It is necessary to use common indices to compare sky brightness at various observation points. There are two main methods to evaluate sky brightness. The first is the observation from space using satellites, which make it possible to continuously observe light pollution around the world. The second method is from the ground. Falchi et al. (2016) summarized recent results regarding light pollution observed by satellites as the light pollution map (Fig. 10), and people can easily know observed values on arbitrary points in the world. The values in the map indicate radiance, emitted energy per unit time, area, and solid angle (Wcm^{-2} str^{-1}) from the ground,

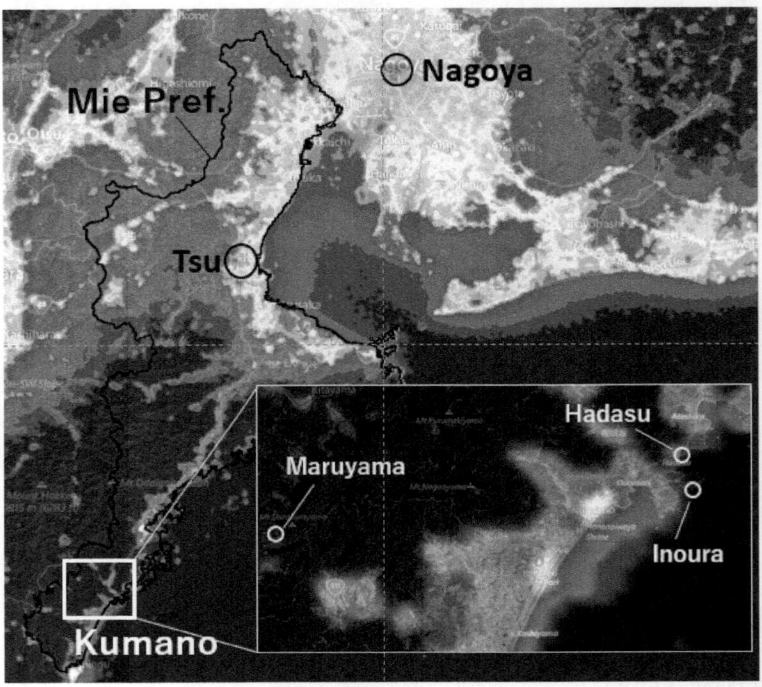

Fig. 10 Light pollution map (*Source* https://www.lightpollutionmap.info/ [Cited: 2020/04/08])

Table 1 Comparison of radiances

Location	Radiance ($Wcm^{-2}\ str^{-1}$)
Ginza/Tokyo	153.92
Mie University/Tsu City	7.21
Kumano Station/Kumano City	5.21
Ichiura/Kumano City	0.20
Maruyama Senmaida/Kumano City	0.23
Mauna Kea/Hawaii	0.08

and the map is very useful to determine dark areas around the world. For example, Table 1 shows the situation of light pollution at several representative locations. It is clear that Ginza, the central area of Tokyo, is approximately 2,200 times brighter than the Mauna Kea Summit, where many large telescopes are located for astronomical research, and it is also found that we can get the dark sky as well as in Hawaii within 10-min driving distance from the city center in Kumano.

On the other hand, night sky brightness observed from the ground could differ from values observed by satellites, because night sky brightness on the ground is composed of stellar light, zodiacal light, airglow, scattered light from airglow, and zodiacal city light, while radiance observed by satellites is mainly due to misdirected city light. Therefore, it is necessary to investigate night sky brightness from the ground to access the visibility of a starry sky.

In general, there are three ways to measure the night sky brightness, that is, by the naked eye, dedicated devices such as a Sky Quality Meter (SQM), and a digital camera. The night sky images taken by digital cameras help us to investigate not only night sky brightness but also night sky color, which cannot be measured by both the naked eye and SQM, because digital cameras take images with RGB colors. In addition, people can calibrate the characteristics of a camera at any time by capturing images of the same stars in every shot. The detailed procedure to measure night sky colors can be found in the study by Itoh et al. (2018).

Night sky brightness is generally expressed by the magnitude per unit square arcsecond, and the measurement results of night sky brightness in Kumano City are reported in Table 2. The results of other areas are also listed for comparison. It is necessary to note that sky brightness becomes darker as numerical values in the table increase. The measurements were taken at an effective wavelength of 545 nm, which is almost the same wavelength as the most sensitive range of human eyes (Bessell 2005). The number of visible stars decreases as night sky brightness increases. Our rough estimate indicates that the number of visible stars in Kumano City is more than 30 times (in order of magnitude) larger than in Nagoya. It was also found that night sky brightness at Hadasu-jinja Shrine and Inoura Park is almost at the same value as in the Mauna Kea, one of the most preferred sites for astronomical observations. These results show that the quality of the starry sky in Kumano City is superior to that of Hawaii.

Table 2 Measured night sky brightness

Location		Night sky brightness (mag arcsec^{-2})
Maruyama Senmaida	Kumano City, Mie, Japan	20.4
Hadasu-jinja Shrine		21.1
Ichiura Park		21.2
MieUniversity	Tsu City, Mie, Japan	18.2
Shirakawa Park	Nagoya City, Japan	16.3
Mauna Kea	Hawaii, USA	21.1

3.2 Industry–Academic–Government Project

The development of astronomical tourism has been started for promoting regional activation in several communities of Japan. Although the main drivers of the projects are local governments and private sectors, academic organizations participate in the projects in certain cases. Mie University, a national university in Mie Prefecture, has a mission to contribute to regional development via research outcomes and provides financial support to promote cooperation between businesses, local governments, and the university. The project to promote public relations activities relating to the starry sky in Kumano City was started in cooperation with Kumano City Office and Mie University in 2018. The academic staff of this project majors in astronomy, and the main bodies associated with the project from Kumano City are the tourism division of the City Office and the Kumano Tourist Association.

The staffs conducted site surveys to evaluate the quality of the starry sky in Kumano City and its surrounding areas, as aforementioned in the first project year. In the second year, the staffs started to investigate tourist demand for night-time tours, including astronomical tourism. They also created a flyer titled Kumap (a combination of Kumano and map) which listed preferable sites to watch skies with beautiful starry photographs (Fig. 11). On the other hand, it is necessary to promote private companies' participation in this project to maintain continuous activities. To this end, it is indispensable to cultivate human resources in addition to the demand generation for astronomical tourism.

4 Analysis of Tourists' Motivations for Visiting and Behaviors

This chapter explains the characteristics of tourists' motivation for visiting and behaviors to evaluate the demand for astronomical tourism in Kumano City through an analysis of a web questionnaire commissioned by Macromill, Inc. in March 2019.

Fig. 11 Kumap (*Source* Kumano City Tourist Association)

The respondents, 209 residents of Kumano City since 2014, were asked about basic attributes such as gender, age, origin, and visit frequency to Kumano City (responses are detailed in Table 3). This questionnaire also included questions about tourist motivations and behaviors, such as places visited, willingness to revisit Kumano City, the way to spend the night, intention to participate in night activities, and so on, to evaluate the demand for astronomical tourism in Kumano City.

Figure 12 reveals that many tourists visited the World Heritage site of Onigajo (51.2%), Kumano-Kodo (48.3%), and Shichirimihama (24.9%), which were easy to access by car. In addition, Maruyama Senmaida (20.6%), Yunoguchi Spa (22.0%),

Table 3 Basic attributes of the web questionnaire respondents (2019) (*Source* Web questionnaire)

Gender	Male: 57.9% Female: 42.1%
Age	20–24: 4.3% 25–29: 7.2% 30–34: 8.1% 35–39: 9.1% 40–44: 8.1% 45–49: 12.0% 50–54: 15.3% 55–59: 10.5% 60–: 25.4%
Origin	Mie: 45.5% Aichi: 20.1% Osaka: 7.2% Chubu (Gifu, Shizuoka): 8.1% Kinki (Nara, Wakayama, Shiga, Kyoto): 19.1%
Visit Frequency	1st: 42.6% 2nd: 31.1% 3rd: 9.1% 4th: 2.9% 5th: 1.4% 6th–: 12.9%

Fig. 12 Visited or experienced ratio and hope to visit or experience ratio it next time in Kumano City (2019) (*Source* Web questionnaire)

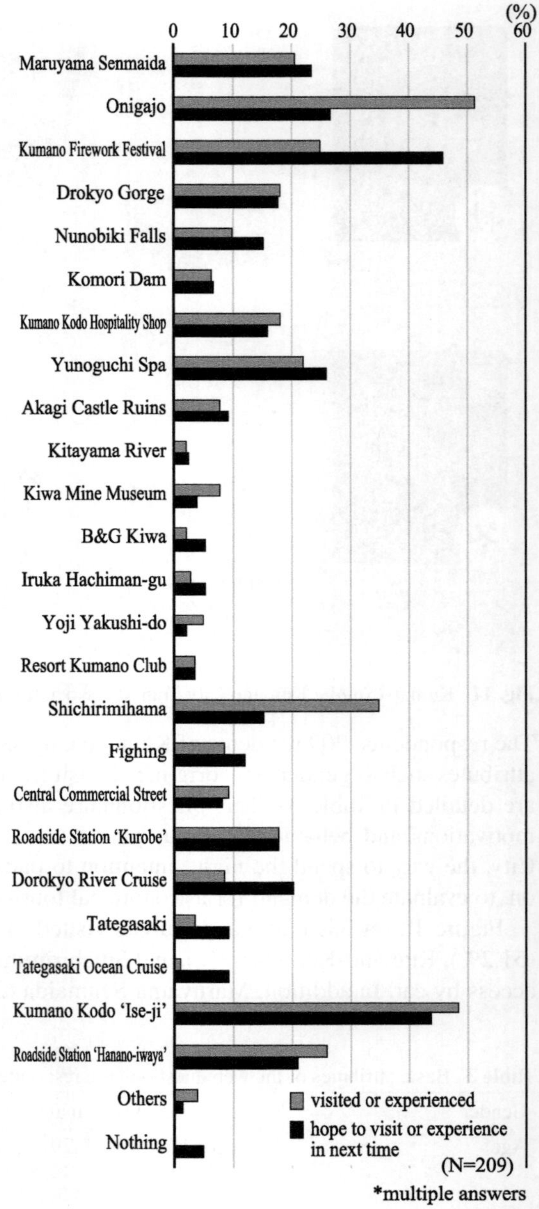

and Dorokyo Gorge (18.2%) emerged as popular tourist attractions symbolizing Kumano City. Similar to results reported in Fig. 6, it was found that most tourists visited Kumano City by car and stayed at roadside stations such as Hanano-iwaya (25.8%) and Kurobe (17.7%). In terms of places or experiences that the tourists hoped to visit the next time, a majority of responses included the Kumano Firework Festival (45.9%) and Kumano-Kodo (44.0%). Compared to the actual visit rate, tourists were more interested in the Dorokyo River Cruise (experienced: 8.6%, hope to experience it next time: 20.1%) and the Tategasaki Ocean Cruise (experienced: 1.0%, hope to experience it next time: 9.1%), which are both activity-based tourist attractions, and highlight the need to improve consumption opportunities to acquire repeat visitors.

As seen in Fig. 13, which describes night tourist behaviors, 55.0% of tourists did not go out at night and spent time relaxing in their accommodations as they conversed with companions or watched TV. Although some tourists visited the baths outside in Kumano City (24.4%) or walked around and ate dinner near their accommodation (18.2%), a few enjoyed outdoor activities that involved observing the night life and the night sky. Furthermore, only 1.4% of tourists participated in night-time activities, indicating that very few tourists enjoyed experiential consumption in Kumano City at night.

Figure 14 shows tourists' intentions to join stargazing activities if conducted. Most tourists indicated an interest in stargazing activities wherein they could leisurely see the starry sky, such as while lying on a reclining chair (42.1%), relaxing at a foot spa (34.0%), or enjoying dinner under the stars (34.0%). Activity-based tourist experiences such as stargazing on night cruises (28.2%) were also found to be attractive.

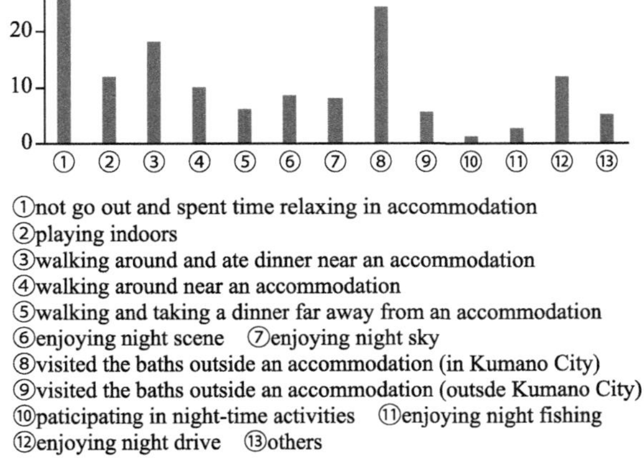

Fig. 13 The way of spending night-time (after 17) of respondents (2019) (*Source* Web questionnaire)

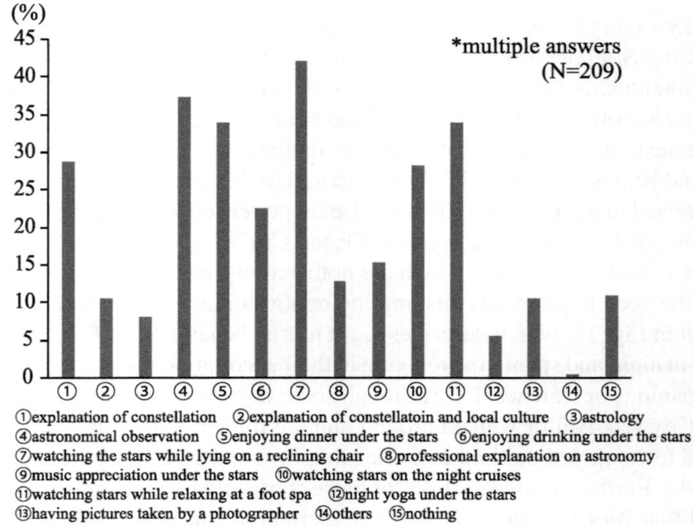

Fig. 14 Respondents' intentions to join stargazing activities if conducted (2019) (*Source* Web questionnaire)

Moreover, respondents indicated the need for astronomical observation (37.3%) and explanation of constellation (28.7%). On the other hand, there was low demand for academic content such as explanations of relationships between astronomy and local culture (10.5%) like the legend and stories of Xu Fu or some anthology of waka poems and professional explanations of astronomy (12.9%).

To summarize the results, the astronomy-related tourists tended to be interested in leisure-oriented stargazing activities, such as looking up at the starry sky while sitting on reclining chairs or taking footbath and dining out under the stars. On the other hand, it did not show such high expectations of the experts' explanation of stars or the relationship between astronomy and the local culture, although there was a certain demand for astronomical tourism. Most tourists visited Kumano City for vacations and enjoyed natural tourist attractions, such as the World Heritage sites, the national park sites, and hot spring facilities by car. In addition, although almost all the tourists were interested in stargazing activities, they rarely participated in these as they tended to not engage in any activities and relax after dinner.

5 Discussion and Conclusion

Tourists tend to visit the various attractions related to the World Heritage or the national park that are broadly distributed throughout Kumano City. However, according to the web questionnaire, few tourists enjoy experiential consumption in

Kumano City. To be conducive to revisits, it is necessary to prepare new tourism contents. Moreover, Kumano City is located far from the urban area, and tourists are apt to stay there. Hence, the Kumano City Office was faced with the need to enhance certain night tourist attractions and opportunities. In the context of night tourist attractions, there are famous events such as the Maruyama Senmaida Torch Procession in June and the Kumano Firework Festival in August, held on a particular day each summer. However, it has been suggested that night tourist attractions in Kumano City be improved.

In these circumstances, motivation to promote astronomical tourism has increased, and an industry–government–academia collaboration project was established by the Kumano City Office, the Kumano City Tourist Association, and Mie University in 2018. The Kumano City Office and Kumano City Tourist Association later undertook the development of astronomy-related materials for tourism such as a night map for stargazing, while Mie University organized stargazing parties. Tomita and Okyudo (2009) explain the importance of industry–government–academia collaborations as well as human resource development for the promotion of astronomical tourism such as astronomy-related guides. In the future, human resource development through industry–government–academia collaboration will be necessary for Kumano City.

Astronomical tourism in Kumano City is regarded as the creation of new tourism consumption opportunities, given the lack of existing night tourist attractions. However, it is necessary to differentiate the contents of stargazing activities from other areas to attract tourists because Kumano City has not been recognized as a popular astronomy-related destination yet. Overall, astronomical tourism tends to focus strongly on features of educational tourism, such as explanations of constellations by specialists. On the other hand, tourists who are interested in astronomy in Kumano City are attracted by leisure or comfort in the context of stargazing activities. Therefore, to promote astronomical tourism, it is necessary to not only improve human resource development, by providing interpreters for stargazing, but also create warm spaces and environments where one can comfortably watch the starry sky in Kumano City and engage in academic activities. Isono (2019) explains that such lighthearted stargazing products can produce child-friendly opportunities to enjoy peaceful night-time activities, even though the starry sky itself is unlikely to be a major tourist attraction. Thus, it is necessary to create stargazing programs related to the strengths of Kumano City such as the scenic natural landscape or warm winter climate.

Recently, interest in night-life tourism and night-time economy promotion has been increasing in Japan because it leads to new large economic impacts through long-stay or accommodation demands. Certain large cities, especially Tokyo and Osaka, are immensely involved in developing night-life tourism because of increasing inbound demand. On the other hand, some peripheral areas also work positively to enrich night tourism contents such as astronomical tourism as one of the regional revitalization measures. Therefore, it is highly possible that tourist destinations based on the starry sky will face stiff competition in the future. Thus, it is necessary for some peripheral areas seeking to promote the dark sky as a tourist attraction to discuss how to discover and capitalize on regionality in the overall concept and strategy of astronomical tourism.

As for the previous geographical studies in Japan, astronomical tourism has not been much controversial ever because most tourism-related activities are implemented in the daytime. However, as mentioned above, some depopulated regions without notable tourist resources have paid attention to the starry sky as one of the regional attractions. Therefore, it can be pointed out that astronomical tourism will become a noteworthy research topic as the hidden tourism geography in the future.

Acknowledgements We would like to be grateful to the Kumano City Office and the Kumano City Tourist Association staffs for supporting our field survey. A part of this study was presented at the EUROGEO 2019 (at Ljubljana, Slovenia). This research is financially supported by the Grant-in-Aid for Young Scientists (No. 19K20566, 2019–2023, Representative: Takumi ISONO) and the Mie University Regional Contribution Activities (2019–2021, Representative: Nobunari ITOH).

References

Bessell, M.S. 2005. Standard photometry system. *Annual Review of Astronomy and Astrophysics* 43: 293–336.

Collison, F.M., and K. Poe. 2013. Astronomical tourism: The astronomy and dark sky program at Bryce Canyon National Park. *Tourism Management Perspectives* 7: 1–15.

Falchi, F., P. Cinzano, D. Duriscoe, C.C.M. Kyba, and C.D. Elvidge. 2016. The new world atlas of artificial night sky brightness. *Science Advances* 2 (6): e1600377. https://doi.org/10.1126/sciadv.1600377.

Hosaka, T., M. Kurimoto, and S. Numata. 2017. Entomological culture and insect-related tourism in Japan. *The International Journal of Tourism Science* 10: 57–64.

Isono, T. 2019. Development of guided night tours for foreign tourists in Shibuya City, Tokyo. *The Tourism Studies* 31 (1): 5–18.

Itoh, N., M. Koshimura, T. Hagihara, and A. Kato. 2018. Night sky brightness measurement in Kumano: Relation between night sky brightness and sky color. *Bulletin of the Faculty of Education Mie University* 69: 31–37.

Kawanami, T. 2016. Changes in tourists in the Ogasawara Islands after registration as a world heritage site. *Geographical Review of Japan Series A* 89: 118–135.

Longcore, T., and C. Rich. 2004. Ecological light pollution. *Frontiers in Ecology and the Environment* 2 (4): 191–198.

Murakami, K. 2001. Tourism and local community. In *Introduction to tourism*, ed. N. Okamoto, 287–308. Tokyo: Yuhikaku Publishing Co., Ltd. (Title etc. translated by the author)

Nagai, T. 2016. *Let's sell the starry sky*. Tokyo: KADOKAWA (Title etc. translated by the author)

Narita, K., and H. Ueda. 2013. Production of the "Starry Sky Video Camera" for the awareness campaign of light pollution. *Memoirs of Faculty of Education and Human Studies, Akita University* 68: 17–22.

Rodrigues, A.L.O., A. Rodrigues, and D.M. Peroff. 2015. The sky and sustainable tourism development: A case study of a dark sky reserve implementation in Alqueva. *International Journal of Tourism Research* 17: 292–302.

Tomita, A., and M. Okyudo. 2009. Starting the training course for astronomy guide at Wakayama University. *Faculty of Education Wakayama University, Bulletin of the Center for Educational Research and Training* 19: 99–104.
Uda, T., and T. Isono. 2019. A study on constructing process of starry sky as tourism resources in Ishigaki Island, Okinawa Prefecture. *Geographical Space* 12: 277–294.

Hidden Historical and Sacred Places

Visualising the Subterranean: Tunnels and Flows Beneath a Welsh Lead Mining Landscape

Mark Nuttall

Abstract This chapter discusses controversies that surrounded the construction of drainage tunnels beneath the lead mining landscape of Halkyn Mountain in North East Wales in the nineteenth and early twentieth centuries. Lead mining had a dramatic effect on the surficial and interior environments of the region, shaped social relations, and created distinct communities. As miners probed deeper to reach new veins, water levels rose in shafts and passages. Mine owners and syndicates embarked on drainage schemes. Mining and tunnelling sculpted the underlands of North East Wales, but provoked debates about power and privilege, the occupation and ownership of the subterranean, the nature of strata, and concerns over the flow, disruption and contamination of underground watercourses and local water supplies. Drawing on archival material, the chapter discusses the visualisation of Halkyn Mountain's hidden geographies, particularly through debates over the construction of the Milwr Tunnel. A socio-technical assemblage, the Milwr Tunnel was subject to a boosterism about "splendid" and "promising" discoveries of riches by unwatering abandoned mines and probing further underground to extract ore from new lodes. Engineers and miners were celebrated as pioneers and adventurers in revealing and shaping hidden worlds and unlocking mineral wealth, allowing, in the process, companies to map and control the subterranean. Above ground, there were local anxieties over the spread of mine shafts below and tunnels were criticised as monopoly schemes. Tunnels and underground channels raised questions over purpose, design and social and environmental impacts, and demands were placed on drainage companies to meet local social and economic obligations. The formations, structures, materialities and aesthetics of the underground, geological maps, and the flow of carboniferous limestone water, were essential to testimony presented at parliamentary hearings that visualised the shape and rupture of subterranean depths.

Keywords Lead mining · Drainage tunnels · Subsurface visualisation · Volumetric geographies · Halkyn Mountain · Flintshire · Wales

M. Nuttall (✉)
Department of Anthropology, University of Alberta, Alberta, Canada
e-mail: mark.nuttall@ualberta.ca

M. Krevs (ed.), *Hidden Geographies*, Key Challenges in Geography,
https://doi.org/10.1007/978-3-030-74590-5_9

1 Introduction

Recent approaches to territory, political geology, and elemental geographies have drawn attention to how thinking volumetrically about space and place enlivens our understanding of subterranean environments (e.g. Elden 2013; Bobbette and Donovan 2019; Hawkins 2020a). Volumetric perspectives have much to offer historical studies of mining, especially with respect to how the subsurface has been imagined, made legible, visible and knowable, and controlled, and how resource potential is anticipated, assessed and calculated (e.g. Endfield and van Lieshout 2018; Kama and Kuchler 2019; Kuchler 2017). As a contribution to scholarship on this and on the representations and politics of encounters with earthly materials and geological forces, the liveliness, restlessness and motion of rock and stone (e.g. Cohen 2015), and with the ways human activities are entangled with the earth's geological and temporal depths (cf. Clark 2017), this chapter discusses controversies that surrounded the construction of drainage tunnels to dewater the lead mines of North East Wales in the late nineteenth and early twentieth centuries. Venturing into the hidden geographies of Halkyn Mountain in the county of Flintshire, it explores how geological imaginaries and knowledge about the shape of the underground, the nature of strata and rock and theories about the flow and direction of water through Carboniferous limestone were essential to the visualisation of subterranean depths, the calculation of resource potential, and the social and environmental effects of tunnel construction. Halkyn Mountain was a busy working landscape on the surface as well as below. Local lives and economies were bound up with the formations, structures and materialities of the underground. Lead mining shaped the landscape and distinct communities grew around it. For Tilley and Cameron-Daum (2017: 5), landscape "has a material form with textures and surfaces, wet and dry places, scents and sounds, diurnal and seasonal rhythms, places and paths and cultural forms and built architecture that, through differential experience, is constitutive of different identities," and people and landscapes "are entangled in a network of material and social relations" (ibid.: 6). In the Halkyn Mountain area this entanglement involves the spaces, forms, substances, materials, pressures, forces and flows of what lies above and below ground.

Specifically, the chapter's focus sharpens on debates over the construction of the Milwr Tunnel, a socio-technical assemblage, which began in 1897 and was extended several times over the next sixty years. Tunnelling drained lucrative mines and allowed the discovery of new lodes, but concerns were expressed over the disruption and contamination of underground watercourses and local water supplies. Anxieties about the effects of the Milwr Tunnel on the water of a nearby holy well—a site of pilgrimage—were expressed by opponents to the project. The chapter draws on current research that is examining historical and archival material, including documents, engineering plans, maps and schemata, letters, newspaper reports, and the transcripts of parliamentary hearings to discuss how these debates centred on limestone, fissures and cracks, the nature of strata and mineral veins, the location of cross-courses and the direction of water flow. Geology and hydrology were essential to both the promotion and opposition of tunnel construction and for how Halkyn's subsurface was imagined, represented and given form. The tunnel was subject to a boosterism about "splendid" and "promising" discoveries of riches by unwatering abandoned mines and probing further underground to extract ore from new lodes. Engineers and miners were celebrated as pioneers and adventurers in revealing and shaping hidden worlds and unlocking mineral wealth. This also involved expressions of an aesthetics of the underground (cf. Hawkins 2020b). Vertical imaginaries and the geo-logics and geo-optics of understanding, measuring, ordering and peering into subterranean depths sharpened ways of sensing and visualising Halkyn Mountain's interior geographies. Indeed, newspapers at the time reported on the parliamentary hearings as a battle between geological experts as they argued about the nature of the known, the unknown and the hidden.

2 The Lead Mining Landscape of Halkyn Mountain

Its topographical designation suggests a lofty peak, but Halkyn Mountain (Mynydd Helygain, in Welsh) is an upland plateau of mainly common land on a spine of Carboniferous limestone. Reaching a height of 293 m (961 ft.), it covers an area of 604 ha (1,492 acres) in Flintshire in North East Wales. Flintshire adjoins the English county of Cheshire and the Halkyn Mountain area is in many ways connected to and embedded economically and socially in the nearby Deeside conurbation. This region stretches across the Flintshire-Cheshire border and includes the city of Chester (Cheshire's county town) and several towns near the canalised length of the River Dee (Afon Dyfrdwy) which flows into the Dee Estuary (Aber Dyfrdwy). The Welsh–English borderlands shift, dissolve and merge here—as does much of the social and familial landscape—and this chapter arises from anthropological research (involving ethnographic and historical methods) that is part of a larger project concerned with the historical and contemporary nature of the region's rural-urban fringes, edgelands, in-between places, and post-industrial landscapes and waters. Along with lead mining, coal mining, brickmaking, copper smelting and steel production were also major industries in the nineteenth and twentieth centuries in Flintshire. Much of the region is filled with evidence of old lead mines, disused collieries, claypits and brickworks and this draws one's attention to the underground as much as the surficial, to reflect on what it means to live, work and deal with rock, clay, coal, lead, sand, water, and so on.

Looking east from Halkyn Mountain, farmland underlain by coal measures slopes down and—cut through by the A55 expressway (the main road across North Wales, from Chester to Holyhead)—reaches to the towns of Flint and Bagillt and the salt-marsh, mud and sands of the Dee Estuary, which Les Roberts describes as liminal "'empty' spaces in-between the coastlines of Wirral (in England) and Flintshire" (Roberts 2012: 104). An industrial landscape dominates the view of much of the Dee Estuary and Deeside. Major sites include the Connah's Quay Power Station, Dee Power Station, the Tata Steelworks, a paper mill and, a little further to the southeast, the Airbus aircraft wing factory at Broughton—while the Clwydian Range (Bryniau

Clwyd) of hills frames the view to the west. Its lowlier geography than the Clwydians and the higher, rugged mountains further west in the Ordovician landscape of Snowdonia belies Halkyn Mountain's historical importance as one of Britain's most valuable sites for the extraction of non-ferrous metals (lead and zinc ore) and quarried stone (mainly for different varieties of limestone and chert) over several centuries (Fig. 1).

The population around Halkyn Mountain is concentrated in four main villages—Halkyn (with around 2,800 people), Pentre Halkyn (approximately 1,100 people), Rhosesmor (around 560) and Rhes-y-Cae (some 350). A number of scattered houses, smallholdings and farms on the edges and slopes also make up the wider area. The town of Holywell lies near the plateau's northeastern edge. With a population of just under 10,000 Holywell is known for St. Winefride's Well, a site of pilgrimage renowned for having miraculous curative and healing properties. As this chapter relates, the well, which is in the care of St. Winefride's Church, a Roman Catholic parish church, was also a central actant in the debates concerned with tunnelling and in the narratives of Halkyn's hidden geographies

Halkyn Mountain is surrounded by farms but its common land has never been enclosed. The common land is owned largely by the Grosvenor Estate, which is under the proprietorship of the Duke of Westminster (also known as Earl Grosvenor) and whose seat is Eaton Hall, just over the border near Chester. Around 200 properties are registered as having ancient commoners' rights and some 1,500 sheep graze the rough, unploughed and unfenced pasture. Quarrying remains a major activity on Halkyn Mountain today, with three operations supplying a range of aggregates and

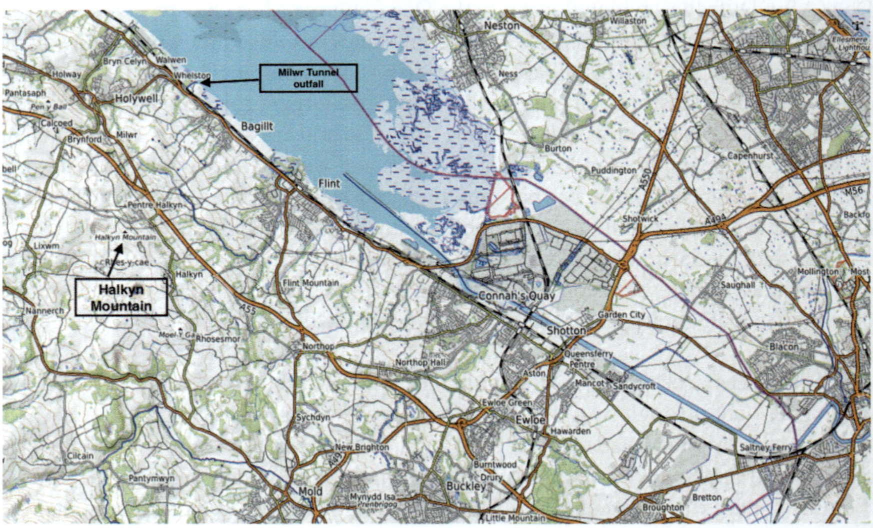

Fig. 1 Map of Flintshire showing Halkyn Mountain and the Milwr Tunnel outfall (author's annotations). Citation: Map courtesy of OpenTopoMap https://opentopomap.org/#map=12/53.2279/-3.0600 Map data: ©OpenStreetMap contributors, SRTM | map style: © OpenTopoMap (CC-BY-SA)

asphalt. The earliest mining of metal ores may have taken place in the Halkyn region during both the Bronze and Iron Ages. While precise archaeological evidence for this is lacking, extraction would have focused on shallow surface workings, with open cuts along the line of an ore vein. Mining for lead and silver in the area followed the Roman occupation of Britain, with operations supplying the frontier legionary garrison of Deva (the origin of Chester, established in the AD 70s on the banks of a navigable stretch of the River Dee). Extraction was also significant during the Middle Ages (Ebbs 2008; Frost and Jones 2004).

In the eighteenth and nineteenth centuries, the area became one of Britain's major lead mining zones (Frost and Jones 2004; Williams and Williams 2012). Landed families such as the Grosvenors began acquiring mines and mining licences in the early 1600s, and were given rights by the Crown in 1614, which meant they had effective control over much of the industry on Halkyn. A number of Flintshire and Cheshire families accrued wealth from lead mining, often by leasing mining rights from the Grosvenors. Many were also involved in smelting and other ancillary industries, such as silver production and silversmith work (Williams 2012; Williams and Williams 2012), which were established in nearby towns and villages, especially along the Flintshire shores of the Dee Estuary. The Quaker London Lead Mining Company also acquired mining rights in Flintshire in 1692 and was a major player in the region's lead and silver production for the next one hundred years, before abandoning its Welsh interests to concentrate on its mines in northwest England (Armour 1956). The Port of Chester, on the Welsh–English border, and small harbours along the Dee Estuary were key to the export of lead, and Chester shipping agents and merchants were prominent figures in the lead mining industry that developed not just around Halkyn Mountain, but in other parts of Flintshire and elsewhere in North East Wales. Lead products were transported by ships from estuarine ports to other parts of Britain, to Ireland and global markets, and the region prospered.

In the early eighteenth century, a lead mining boom brought incomers from Derbyshire, Cornwall and other parts of England. Skilled as lead miners in their home regions, they were often recruited by mining companies who required different kinds of extractive labour and subsurface knowledge than found in the communities around the nearby Welsh coal fields. They settled in the Halkyn district, some learned Welsh, and married into local families. English surnames such as Bagshaw, Nuttall, Redfern and Wagstaff became local names (Rhodes 1968; Ellis 1998) and remain part of today's social landscape. Many lead miners and their families lived in poor conditions on smallholdings and kept sheep and other livestock. This was a stark contrast to those who owned the mines and smelters, were engaged in trade, commerce and the silversmith industry, and who often built and lived in fine houses in rural Flintshire, Chester and other parts of Cheshire. Extraction, commerce and other industrial ventures contributed to the formation and shaping of the landscape and to the social and economic makeup of a region in which people and their livelihoods, and the fortunes of landowners, mining companies, entrepreneurs, and merchants, were linked together by a connection to Carboniferous strata.

The hidden geographies of Flintshire's underlands were revealed as miners dug into the earth, sunk shafts and carved out passages in a search for ore bodies. Mine shafts were excavated vertically and horizontal passages—or levels—were driven off to give access to a mineral vein and to cross-courses (which are the intersections of east to west veins with north to south faults). When an ore body was discovered, it would be worked until the entire deposit was removed. With this burrowing, excavation and tunnelling, miners discovered interior chambers, caves, caverns, vertical pots, rift passages, watercourses and lakes that became part of a subsurface extractive zone transected by levels and other mine workings. In mapping the underground as extractive territory, shafts, lodes, veins and cross-courses were named—for example, Long Rake, Old Rake, Silver Rake, Deep Level, Pant-y-pydew, Pant-y-pwll-dwr, Chwarel Las, Powell's Lode, China Rake, Pant-y-ffrith and Caleb Bell. Knowledge of subsurface topography and strata was essential when miners navigated their way below ground. Many subsurface names also conjured images of geological riches deep in the earth and diagrammes and maps of veins and cross-courses were used to make the Halkyn underlands visible to potential investors. At the end of the nineteenth century, it was believed that an immense subterranean world of streams, rivers, lakes and caverns—and lodes—was still to be explored. Halkyn's interior was becoming known at a time when the scientific interest in caves was taking shape. This was also a time of growing popular interest in the elemental nature of the underworld, in descent narratives, and in adventurous subterranean travel (McCausland 2018). This was a different kind of frontier to imagine—there was no open space of air and light, but only darkness and walls of rock—requiring a vertical, rather than horizontal, perspective in making subsurface representation.

The last lead mine on Halkyn Mountain was closed in 1987. Around 250 major mining sites, with thousands of shafts, have been identified (Coyle 2010). Shafts, trial pit workings, and clusters of spoil heaps stretch along mineral veins and survive as evidence of lead mining's impact on the landscape and the subsurface. Many shafts have collapsed and their locations are indicated by grassy depressions and hollows, often surrounded by humps and mounds of spoil and mine waste (Fig. 2). Derelict land reclamation schemes have also altered the surface since the 1970s. Large numbers of deep shafts, some of which were also dug for ventilation and drainage, have been infilled and capped by stone and concrete, or have been covered with iron grates and fenced off. Apart from a few houses and mine offices—some of which have been converted to modern dwellings—most of the buildings and structures associated with past mining activities have disappeared, although ruined limekilns connected to quarrying and the production of hydraulic lime are now historic sites. Halkyn Mountain has designation as a landscape of historic interest—the area also includes prehistoric sites such as Moel y Gaer, an Iron Age hillfort overlooking Rhosesmor. As well as the shaft sites, Halkyn Mountain's surface is pock-marked by the foundation stones of buildings and boundary walls between mining concessions, horse whim circles and leats (artificial water channels), and is criss-crossed by miners' tracks, paths, and traces of old roads and wagonways along which horse-drawn carts carrying mined materials and lime would rattle. All this gives a tantalising glimpse of the activity that took place underground.

Fig. 2 Old lead mine workings on Halkyn Mountain. Photograph: Mark Nuttall

3 Watery Underground Spaces, Tunnels and the Geo-optics of Halkyn's Hidden Geographies

Lead mining was an encounter with limestone and damp, wet spaces. Flooding and drowning posed a significant risk. As Younger (2005) points out, mines not only produce minerals they "make water" and this was a considerable problem in the Flintshire mines. Moving and working within the watery underground required hydrogeological knowledge and demanded quick thinking about the best way to remove the water that flooded the spaces being hollowed. When miners probed deeper within Halkyn Mountain, water levels rose in shafts and passages, hindering production and bringing more hazards to an already dangerous occupation. Before steam pumps were first put to use after the invention of the Newcomen engine in 1712 (the London Lead Company first installed an engine house on Halkyn Mountain in 1729), underground water was taken to the surface in buckets or pumped out using horse-drawn machinery. As with lead mining elsewhere in Britain, such as Derbyshire (e.g. Endfield and van Lieshout 2018), Halkyn's water problem intensified after the 1750s once the more easily accessible veins had been worked. There were often violent, dangerous inrushes (cf. Younger ibid.), flooding passages, filling shafts and closing active workings. Adits were constructed, but mines flooded again when workings reached below them. The rising cost of coal to keep the steam pump engines in operation was prohibitive for some smaller mining operations, but the powerful flow of water also made effective steam pumping difficult. Attempting to solve the problem, mine owners and syndicates embarked on grander drainage schemes and tunnel construction. In 1774, the Holywell level was commenced by the proprietors of the Holway Mine near Holywell—some three metres above St. Winefride's Well and for a length of 1500 m—and was used as a canal to barge ore from the mine (Appleton 1989). In 1818 Robert Grosvenor, the 2nd Earl Grosvenor, initiated the construction of the first major tunnel—known as the Halkyn Deep Tunnel—from Nant-y-Flint to drain water from the mines. It was abandoned in 1822, but work resumed in 1838 and the tunnel enabled the development of the Deep Level Lode (Appleton ibid.). Throughout the nineteenth century, small mining ventures were taken over by large companies and Flintshire was one of Britain's major lead mining areas, second only to the North Pennines orefield in England (Coyle ibid.; Frost and Jones ibid.). Yet the more easily accessible ore had been worked out and miners went deeper and below the water table. By the 1870s, some Halkyn mines had closed—water was certainly a problem, but many suffered from a fall in ore prices and competition from imported ore. Drainage tunnels were considered vital for rejuvenating the industry.

In 1875, an Act of Parliament empowered the Halkyn District Mines Drainage Company to take over and extend the Deep Level tunnel work that had been initiated by Robert Grosvenor. The company was also empowered to levy royalties from every occupier of a mine within the area it drained. Work commenced on the Halkyn Deep Level tunnel in 1876 (also known as the Halkyn Tunnel, Old Drainage or 1875 Tunnel). In August 1879, the company announced that the drainage scheme was successful—a strong current of water had been tapped and was flowing along the channel. By the late 1880s, the company reported to its shareholders that the tunnel had affected the drainage of a large tract of the country, led to "a wonderful discovery of lead ore", and had improved the condition of the lead trade, and so creating a far more hopeful feeling in the Halkyn district than had existed for many years. By 1901, the tunnel had reached the South Llyn-y-pandy Mine, nearly five miles (eight kilometres) from the portal.

A second tunnel—known as the Milwr Tunnel, or Sea-Level Tunnel—was begun in 1897 at sea level near Bagillt on the Dee Estuary. This was driven by the Holywell-Halkyn Mining and Tunnel Company, which was formed by a group of mining companies and empowered by an Act of 1896 to drain mines north of the Halkyn District Mines Drainage Company area (Frost and Jones ibid.). Geoscientific knowledge and data on the appraisal of subsurface resources supported promotional material about the region's promise. In 1896, the company published a prospectus for private circulation to potential investors, promoting Halkyn Mountain as containing "many marketable and valuable minerals" within "the richest metalliferous zone in Flintshire."[1] It gave details of reports from the company's engineers on the mines in the area and the estimated revenue and probable profits after completion of the tunnel "from the minerals proved in the cutting of the Tunnel in new lodes." It came with a fold-out map showing the planned route of the Milwr Tunnel, the mines and lodes it would drain and the ground it would open up, with a cross-cut of the tunnel stretching from sea level and the shafts running down into the Halkyn Mountain interior. The prospectus proclaimed that

[1] 'Milwr Tunnel Scheme. Prospectus of the Holywell-Halkyn Mining and Tunnel Company Limited (1896),' Flintshire Record Office, Document D/PG/303.

> Halkyn mountain contains an immense quantity of minerals, not merely lead ore and blende, but also cement, stone, chert (largely used in the Staffordshire Potteries), limestone (hydraulic and ordinary), superior fireclay. Owing to the present cost of high transport, the output of these minerals is limited, but with the cheaper transport afforded by the Tunnel, this output will greatly increase.[2]

In enticing investors, the company emphasised the advantages of location, transportation and harbours. The tunnel was necessary not only for draining water from the mines, but for placing the Halkyn district in immediate contact with the main line of the London and North-Western Railway and the River Dee, and through the latter with the River Mersey, the Manchester Ship Canal and various seaport towns.

These tunnels enabled the flow of water away from the mines and allowed access to the ore. Mining and tunnelling sculpted the underlands of North East Wales, and tunnel construction was subject to a boosterism—company reports to shareholders and the minutes of annual meetings are replete with claims of "splendid" and "promising" discoveries of riches to be made by unwatering abandoned mines and probing further into the uncharted underground to extract ore from vast lodes. For example, in March 1896, *The Flintshire Observer* reported on "substantial progress" made in preparations for the beginning of the Milwr Tunnel venture, announcing that it would "have undoubtedly great industrial bearing in this district." When progress with the tunnel was reported on during the second ordinary general meeting of the Holywell-Halkyn Mining and Tunnel Company, which was held in Chester in December 1897, it was remarked that the company had "materially increased its area" and that prospects were good for the development of a rich lead-bearing zone.[3] Immense quantities of lead were said by the company's executives and engineers to be buried in water and, in March 1898, further progress with the Milwr Tunnel was proclaimed as opening up "an exceedingly rich district." It was reported how "company after company have attempted to follow this great wealth to deeper workings, but each company succumbed to the water difficulty".[4] The Holywell-Halkyn Mining and Tunnel Company argued it was breaking down frontiers.

Above ground, however, there were many, including parish councillors, local business owners and farmers who were concerned about the spread of mine shafts, levels and tunnels. Debates raged about power and privilege, the occupation and ownership of the subterranean, and the nature of strata. Concerns were expressed that tunnelling not only diverted and disrupted the flow of underground watercourses, but that mining contaminated local water supplies. It was argued that the water draining from the Milwr Tunnel had the same source as that for St. Winefride's Well (also known as St. Winifred's Well and the variation in spelling is reflected in some of the documents, and in the words of key actors, referred to in this chapter). The water entering the well also feeds the Holywell Stream, which empties into the Dee Estuary near Greenfield, some two miles (three kms) away. A number of industries and businesses depended on the water, but it was also the source for Holywell's domestic

---

[2] Ibid.

[3] https://papuraunewydd.llyfrgell.cymru/view/3753823/3753828/45/Milwr%20Tunnel.

[4] https://papuraunewydd.llyfrgell.cymru/view/3753913/3753921/102/Milwr%20Drainage.

water supply. Opposition to further tunnel construction focused on concerns and anxieties that the flow of water to the well—which was said to come from the south in the area of tunnel construction—would be cut. Arguments were put forward that local businesses and even pilgrims to the well should be compensated for the effects of any disruption to the underground stream that provided local water supplies. For example, in September 1896, the *South Wales Echo* had this to say:

> It is generally held that the principal source of supply to the well is a prolific subterranean stream which runs from Minera, near Wrexham, to Holywell, unwatering in its course a vast number of lead mines and the district generally through which it passes. The stream or "cross-course" in question is known as the "Caleb Bell" so named from the man who discovered it; and the fear is that the tunnel proposed to be constructed might tap this stream and so divert the principal source of supply to the well. The various local authorities have been asked by the promoters of the scheme for their sanction to it; but before doing so they have decided to get the best expert opinion obtainable on the matter. The diminution of the effluent would be a very serious matter not only for Holywell town and trade, but also for the manufacturers on the stream, who at present derive the motive power for their machinery from the overflow of the water...."[5]

However, there were many geologists and mining engineers who disagreed. The source of the water feeding St. Winefride's Well was disputed by the advocates of the tunnel—and the geologists and hydrologists who were brought into support this position at parliamentary hearings in 1904 were cited that the water entered the well from the northwest, an area that would be unaffected by any tunnelling.

These controversies over the alteration of surface landscapes and the subsurface, and especially to the flow and contamination of water, dominated the parliamentary committee meetings and hearings over tunnel extension. In a seminal article, Martin Rudwick (1976) argues that geology did not develop fully as a science until it produced diagrammes and maps of strata and representations of interior geographies. The production of geological maps also became essential for understanding the location and nature of mineral resources. For example, Eric Nystrom (2014) explores how mining engineers in the late nineteenth and early twentieth centuries in America produced maps and models of mines that allowed the visualisation of subsurface geologies, tunnels and shafts. Similarly, for Halkyn Mountain, geologists and mining engineers conceptualised and gave order to the hidden geographies of the interior. Their maps and schemata, and other technical representations of strata and underground workings, provided further impetus to the exploration of these underlands by envisioning them as resource spaces of great potential. Through these geo-logics and geo-optics—techniques of classifying, ordering, seeing, imaging, mapping, making and knowing—spectacular representations of deep, dark places (that were sublime and beautiful) as regions of abundance intimated great returns to those who invested in subsurface zones of promise. The formation and shape of the underground also became essential in the parliamentary debates about the tunnels.

[5] "St. Winefride's Well: Feared tapping of the source," *South Wales Echo*, 2nd September 1896.

4 Locating the Watershed

Expansion of the Milwr Tunnel began in the early years of the twentieth century. Legislation to allow it was set out in the Milwr and District Mines Drainage Bill, which was debated in the House of Commons and before a House of Lords Committee in 1904. The Bill was promoted by the Duke of Westminster, the Earl of Derby, other landowners and mine owners, and those with interests in the tunnel company. The scheme consisted of a system of tunnels capable of carrying off 10,000,000 gallons of water per day; it was argued, however, that the tunnels already draining the mines ran through a formation of porous Carboniferous limestone from which the Holywell Stream drew its supply. Sir Thomas Esmonde, who was Member of Parliament for the Irish constituency of North Wexford at the time, moved the rejection of the Bill in support of local Flintshire interests. Holywell District Council, for example, was opposed to the tunnel, and many others in the area who depended on the Holywell Stream—including farmers, local business owners, mill owners, and riparian owners—worried about the environmental and social effects of interference with Holywell's water supply, as well as the impact on St. Winefride's Well. Concerns were expressed that the scheme was being promoted to enable a company to earn dividends and mine owners to make a profit at the risk of depriving the inhabitants of Holywell of their water supply and ruining the industries on which they depended for their livelihoods. The motion for the rejection of the Bill was seconded by Sir Charles McLaren (MP for Bosworth in Leicestershire, but also a Baronet of Bodnant in North Wales), who argued that if all the water "was to be sucked from the land" many Flintshire communities would be seriously threatened. The damage which would be done by the proposed underground tunnelling was irreparable, it was argued, and the town and industries would be ruined forever. Among the Bill's supporters was Samuel Moss, MP for Denbighshire Eastern in North Wales, who rejected the views of the opponents and argued that up to 1874 many mines in the neighbourhood of the well were worked at a level below it yet the flow of water remained unaffected.[6]

At the heart of discussion was the nature of limestone and the direction of subsurface water flow. McLaren argued that "the peculiarity of the water was that it did not flow on the surface of the ground, but went down fissures in the limestone strata, and at a certain level the water rose up again in great volumes. It was quite obvious that if tunnels were driven down below the existing stream the subterranean reserves would be drained to such an extent as to make the water flow not in its accustomed channel."[7] In written submission given on 7 July 1904, W. Fitzherbert Brockholes stated that

> The weight of evidence points to the probability, if not absolute certainty, that the volume of water from St. Winifred's Well will be greatly reduced even if the whole underground

---

[6] House of Commons Debate, Second Reading of the Milwr and District Drainage Bill, 9th May 1904, Hansard vol. 134, cc836.

[7] House of Commons Debate, Second Reading of the Milwr and District Drainage Bill, 9th May 1904, Hansard vol. 134, cc836.

> stream be diverted, and in the face of the evidence given to that effect I am at a loss to understand how Parliament could refuse in common equity the reasonable appeal for the provision of due compensation in the event of injury to the property of others being caused by the proposed drainage works.[8]

Supporters of the Bill pointed out that the area covered by the drainage scheme was acknowledged by numerous experts to be very rich in minerals. While Holywell District Council was opposed to the tunnel, Flint town council had unanimously adopted a resolution in favour of it on the grounds that it would be of advantage to the development of the area. The Bill's supporters also pointed out that other tunnels in the Halkyn area had not caused any harm to water flow, and disputed that Holywell derived its entire water supply from the underground stream diverted by the Milwr Tunnel. Robert Yerbergh, MP for Chester, said that while he could quite understand the feeling of those who held that St. Winefride's Well was a holy well sacred to the use of the sick and the lame, "who resorted to it for cure", those fears were "without justification," and that, given that many mines had closed because of flooding, the scheme put forward was essential to the development of the district.

The Bill proceeded through the House of Commons and the tunnel project was considered by a House of Lords Committee on the Halkyn drainage scheme. The parliamentary hearings were filled with discussion about the nature and shape of the subterranean. Limestone was described in incredible detail by geologists and debate raged as to the exact source of the watershed that supplied the well. The Caleb Bell cross-course and belief in the existence of a major subterranean channel became key elements in the narrative. The starting point for Caleb Bell is south of Rhes-y-Cae, and it terminates at the centre of Holywell High Street. It traverses the limestone beds of the district, and its course is indicated on the surface by numerous shafts. Experts argued that if tunnels were to be driven towards Caleb Bell they would interfere with the flow of water into the well. For example, Sir Aubrey Strahan from the Geological Survey (and who was later the Survey's director from 1914–1920) had authored a memoir on the geology of North East Wales which included his extensive survey of the Halkyn district (Strahan 1890), and he expressed his opinion that the tunnelling would pose a great risk of taking a large part of the water away from the well.

William Boyd Dawkins, another renowned geologist, was one of the key experts called in support of the opponents. He argued at the House of Commons hearings that a large underground reservoir lay below Halkyn Mountain, and that the subsurface geology was characterised by irregular cracks and fissures that influenced water flow. In testimony that was tinged by a sense of wonder, he described Halkyn's hidden underground geography as follows:

> The limestone is riddled naturally with joints; so much so that when it is exposed it takes up all the rainfall, less the evaporation. When the rain gets into the limestone it dissolves passages for itself and it forms clearly defined caverns, underground channels, which sometimes one can traverse for very considerable distances. There are sometimes great subterranean halls, and there are subterranean waterfalls and all the phenomena which you get, say in a hilly

---

[8] Opposition in the House of Commons Session 1904, Milwr and District Mines Drainage Bill, Document D/MT/1078, Flintshire Record Office.

district were the water is tumbling on the surface of the ground from a high level to a low level.[9]

Giving testimony to the committee in March 1904, Dawkins said of the drainage scheme that

> In my opinion it must lower the level of the water in this area, which is, some of it at all events, on its way to St. Winifred's Spring;….it must deplete St. Winifred's Spring; and as to the further question – as to whether it would destroy St. Winifred's Spring altogether – that is a matter which is really a matter of experience; but by judging from other cases in which similar depletion has taken place (the Severn Tunnel) for example, I am of opinion that the effect of this long system of tunnels will be to destroy the head of water from which at all events a considerable portion of the water supply is derived, to such an extent that St. Winifred's Well may be absolutely dried up.[10]

James Mansergh, an English civil engineer also provided evidence in support of the Bill's opponents. He informed the committee that he had made a personal inspection of St. Winefride's Well, the local area, and the local industries that depended on the stream. He gave a description of Carboniferous limestone overlain by millstone grit and spoke about how the limestone was extremely permeable, with large fissures and caves, faults, and mineral lodes. The nature of limestone was central to his testimony, which described how water passed from limestone to millstone grit, and how exposed areas of dry limestone indicated the rapidity with which water finds its way into the rocks by way of fissures and faults. Mansergh talked about how it had been established that there was a fairly steep gradient in the subsoil water from south to north, towards St. Winefride's Well. He argued that "it is clear that a very considerable lowering of the subsoil water level will ensure, with the inevitable result of so flattening the water gradient summit to St. Winifred's Well that the discharge at that point will be materially reduced and may be altogether stopped."[11]

Nathanial M. Griffith, a mining engineer from Wrexham, argued that the Caleb Bell cross-course was key to an understanding of whether or not the tunnel would affect the water supply to St. Winefride's Well and the Holywell Stream. Griffith produced a drawing showing the position of the Milwr Tunnel and the five proposed tunnels—and he showed on geological maps that it was crossed by a large fault, which threw the strata down to the south, with the beds to the east of Caleb Bell dipping towards it for some distance north. He argued that this would bring underground water to Caleb Bell at the point of its junction with the fault, which offered greater flow for the passage of water. Griffiths argued that if Caleb Bell was tapped at a level lower than the well its feed would suffer.

Arthur Caradoc Williams, chief assistant to Henry Enfield Taylor, a mining engineer from Hawarden, Flintshire and who practised in Chester, spoke about an inspection he had made of the district and of the Caleb Bell vein, the depths of shafts and the

---

[9] Opposition in the House of Commons Session 1904, Milwr and District Mines Drainage Bill, Document D/MT/1078, Flintshire Record Office.

[10] Opposition in the House of Commons Session 1904, Milwr and District Mines Drainage Bill, Document D/MT/1078, Flintshire Record Office.

[11] Opposition in the House of Commons Session 1904, Milwr and District Mines Drainage Bill, Document D/MT/1078, Flintshire Record Office, pp. 2–3.

water level. He showed the committee detailed geological maps with the locations of east, west, north and south running veins. In particular, he pointed to how Caleb Bell, for nearly the whole of its length, traversed the limestone beds of the district and explained how its course was indicated on the surface by numerous shafts. He described how limestone strata are traversed and cut up by cracks, joints, veins and faults, and that all these form channels for the water falling on the surface and carry it away underground. The parliamentary committees heard that this was a dry landscape—how rain finds its way quickly into the ground, little runs off, and how Caleb Bell extended across the limestone belt and provided a necessary main channel along which the water could find its way.

Material submitted for consideration by the committee not only included geological maps and models, but also statistics on the yield of water at the well. For example, the committee heard about a report produced in 1895 by Major-General Charles Scrope Hutchison, who was an inspector for the government's Board of Trade, which concluded that the average yield of St. Winefride's Well was reduced by nearly 50% after the Halkyn Mines Drainage Tunnel was driven and brought into use in 1882. At the time, however, and despite Strahan's detailed survey and the evidence presented, other geologists argued that the exact nature and extent of Caleb Bell was not known, and the tunnel proponents drew on this when claiming the water would not be affected. The evidence in support of the proponents was put forward to argue that the water flowed from the west or northwest and not from the south. For example, Honoratus Lloyd put forward the case to the House of Commons committee that the water came from the west or possibly the south-west. He referred to a diagramme of lodes, veins and faults and argued that no one could say where the water flowed and that there was no known or defined underground channel. The House of Lords committee heard from leading geologists William Whitaker and Richard Hill Tiddeman that the well was supplied by underground water that flowed from the west. Tiddeman said that he thought there to be an impervious barrier somewhere to the south of the well which would make any water flowing from the south to turn north somewhere to the west of it, and which would prevent it from reaching the shrine.

Despite the opposition, the Bill passed into law. The committees were not convinced of the arguments put forward that the water feeding the well came from the south. Notable supporters in the House of Lords included those landowners with lead mining interests in Flintshire, such as the Duke of Westminster and the hearings, and the debates focused on the Halkyn underlands as spaces of possibility. At the Holywell-Halkyn Mining and Tunnel Company Annual Meeting in May 1905, consulting engineer Captain Matthew Francis said that

> The company would not only have the safest, but the straightest and most capacious, and in its drainage effects the most exhaustive main drainage tunnel that to my knowledge has even been made to unlock the mineral wealth of any district and, it should prove a source of profit to the company as well as a boon to the metalliferous miner for many years to come.

5 Further Tunnel Expansion and St. Winefride's Well Runs Dry

Plans for further expansion of the Milwr Tunnel were put forward almost a decade later. In 1913, the Halkyn District Mines Drainage Bill was presented in the House of Commons to empower the Halkyn District Mines Drainage Company through an Act of Parliament that would allow it to extend the tunnel into their drainage area, as well as several other tunnels under Halkyn Mountain. A powerful economic argument was put forward by the tunnel company and by mine owners. In particular, Flint County Council was supportive of the Bill and argued that drainage of the mines would restore the prosperity of the lead mining industry, increase employment in the area, and benefit other trades and industries. The council argued that unless the Bill was passed and the tunnel constructed the mining industry in the district would languish and gradually come to an end. The flow of water into St. Winefride's Well was once more central to the parliamentary enquiry that debated the Bill that preceded the Act and maps and models were used as necessary evidence.[12]

A petition against the Bill was brought forward by Lady Anna Maria Mostyn and John McKean. Lady Mostyn was the widow of Sir Pyers Mostyn, who had died in 1912, and who had opposed the 1904 Bill. McKean, an Irish nationalist politician and MP for South Monaghan from 1902 to 1918, was a frequent visitor to Holywell and desired to represent the interests of visitors and pilgrims to St. Winefride's Well. McKean had also opposed the 1904 Bill and had argued then that the flow of water to the well had been diminished because of the Milwr Tunnel. The petition outlined how the Mostyn family was the owner in fee of lands, mills and other buildings on the Holywell Stream, with a legal right to use the water which was indispensable for those lands, industries and businesses. The petition stated

> The Holywell Stream flows from St. Winifred's Well into the Dee near Greenfield, a distance of about two miles. It derives its water almost entirely from the Well and the Well is fed by a stream flowing through what experts believe to be a known and defined subterranean channel which they can trace for a distance of several miles inland across and beyond the district in which the works proposed by the Bill and the works which can, under the Bill, be connected with them are situate. But whether the Well is fed by one or more subterranean streams and whether their course can be certainly defined or not, the injury which could be caused to your first-named Petitioner by the interruption of the flow of the stream and the abstraction of its water, would be equally great. The Well is, moreover, the chief and the only wholesome and unfailing source of supply of water for domestic purposes in the town of Holywell.[13]

Mostyn and McKean argued that large numbers of pilgrims and others visited the shrine every year and the use of its waters for bathing was an important source of revenue for Holywell. The petition also stated that

---

[12] D/MT/1084 min of Evidence taken before the Select Committee of the House of Commons on the Halkyn District Mines Drainage Bill, July 1913.

[13] D/MT/1084 min of Evidence taken before the Select Committee of the House of Commons on the Halkyn District Mines Drainage Bill, July 1913.

> As far back as any records exist the flow of water from St. Winifred's Well was remarkably constant and never fell below six million gallons per diem. But about 25 years ago in consequence mainly in not wholly (as your Petitioner Dame Anna Maria Mostyn believes) of the construction of the Halkyn Tunnel, a deep level drainage tunnel similar and adjacent to the tunnels proposed by the Bill, the flow of water in St. Winifred's spring suddenly fell to one-half of its former volume and at that point has since remained.[14]

In the evidence she presented to the 1913 House of Commons committee, Lady Mostyn argued that St. Winefride's Well was fed from a distant and proved source, which the Halkyn Company tapped to a considerable extent in 1882 in their first operations, and that the deeper level of many large tunnels, precisely in the same direction, would withdraw a further supply of water from the well. She spoke eloquently about Carboniferous limestone and that after the heaviest storms the well remained clear for two-and-a-half days before discolouring. Lady Mostyn referred to local tradition in Flintshire, and particularly around Holywell, that held that a great river flowed underground—if a network of tunnels were completed and linked together, she argued, it would disrupt the flow of water and seriously affect the well. McKean also argued that the length of time it took for the water in the well to become discoloured following heavy rainfall showed that the water originated from a considerable distance to the south. However, William Whitaker appeared as a witness to discuss the nature of strata. He said that it was difficult to determine where the watershed of the well was located, but felt that a large outcrop of limestone to the north-west was a likely place, and that the well would collect water from a variety of sources, not just from water flowing from the south and south-west.[15] Other geological experts and mining engineers produced maps to explain that the water came from the north and west—evidence being presented that south of the well there were numerous faults in the shape of lodes, which dipped to the south, which meant that the geological structure precluded the theory that water from the south flowed to the well.[16]

Captain Matthew Francis was also examined as a witness. He told the committee that he had lived in the area all his life, had more than 45 years experience of mining in the region and had also tunnelled for many miles. He argued that the geological structure precluded the theory that water from the south—that is, from the district that would be drained by the tunnel extensions—went to the well. Instead, pointing to a map, he said that it was his opinion that the watershed feeding the well was situated to the west-north-west and south-west of Holywell. The committee heard evidence from Francis and others that a number of mines south of Holywell had been drained—in some cases below sea level—without affecting the well. The argument that no sudden diminution in the flow of water to the well occurred when the Halkyn Tunnel that was first driven was key to the expert witness statements. At the same time, Francis and other experts argued that the existing Milwr Tunnel was nearer the

[14] D/MT/1084 min of Evidence taken before the Select Committee of the House of Commons on the Halkyn District Mines Drainage Bill, July 1913.

[15] D/MT/1084 min of Evidence taken before the Select Committee of the House of Commons on the Halkyn District Mines Drainage Bill, July 1913.

[16] D/MT/1084, Minutes of Evidence taken before the Select Committee of the House of Commons (Group E) on the Halkyn District Mines Drainage Bill, Tuesday July 1st, 1913, p. 5.

well—about 199 feet below the level of the well at a point 1 ¾ miles (about 2.8 kms) distant—than any part of the proposed tunnel network would be and that it had not affected the flow of water.

Geologist and mining engineer Frederick Henry Hatch was also called as expert witness and declared that he felt there was very little probability of damage being done to the well. Cross-examined by several committee members about the origins of water for the well, Hatch pointed to the geological map of the area and argued that the supply was from the limestone areas to the west. He said the limestone overlay Silurian shales on the west and dipped under the coal measures on the eastern side of the Halkyn area. Indicating the area of limestone and the location of cross-courses on the map, Hatch said that

> The blue is the area of carboniferous limestone from which the water of the district is supplied. In my opinion, the water that comes from the well is supplied from this area, which lies to the west and north-west….The dip of the limestone is from the west to the east, and the natural flow of the water would be along the dip…..That is well known to every geologist, and of course is shown on the Geological Survey map.[17]

He explained how a limestone formation has fissures in it, but that the fissures in the Halkyn district have been filled by the deposit of mineral matter and sealed, thereby producing the veins and the cross-courses:

> in this area, there are a great number of these cross-courses which run north to south and veins which run from east to west which act as barriers to the flow of water and the result of it is to divide this particular area of limestone up into practically water-tight compartments, and that is borne out by the irregular level at which the water stands in these shafts.[18]

For Hatch, cross-courses acted as natural dams, ponding the water, not carrying it, and he did not see the tunnel as bringing any new source of danger. Supporters of the Bill argued that there was no evidence that St. Winefride's Well was fed by a defined, hidden subterranean channel which could be traced for a distance of several miles. Geologists were also called as expert witnesses to present this view. They also argued that very extensive pumping of underground water had been going on very much nearer the well and which had not affected it. The argument that mills were dependent on the stream was also disputed and that local industry and businesses had another significant source of water supply that was independent of the well and which could not be affected by the tunnel and drainage works. Figures from gauging studies were presented to show that the water in St. Winefride's Well was not constant and varied considerably with rainfall and that no relation between pumping from the mines and the rise and fall in the well could be proved. Evidence was also heard that draining mines was a matter of safety for the miners, would prevent drowning, and provide air for those underground. Also, more men could be employed if mines were

---

[17] D/MT/1084, Minutes of Evidence taken before the Select Committee of the House of Commons (Group E) on the Halkyn District Mines Drainage Bill, Wednesday July 2nd 1913, p.30

[18] D/MT/1084, Minutes of Evidence taken before the Select Committee of the House of Commons (Group E) on the Halkyn District Mines Drainage Bill, Wednesday July 2nd 1913, p. 30
p. 32

not waterlogged. And so, the proponents and supporters of the Bill put forward a persuasive case that St. Winefride's Well would not be affected and that the Holywell Stream would be able to supply the same amount of water to the mills. The Bill passed into law and the Halkyn District Mines Drainage Company was empowered to extend the Milwr Tunnel and construction work began.

In May 1916, however, there was a notable reduction in the flow of water to St. Winefride's Well. An investigation was commissioned by the Mostyn Estate and the report, completed by mining engineers W.J. Davies and Cyril Davies the following September, concluded that it was down from 6,000,000 gallons per day to as low as 86,000 gallons a day and to a high of 900,000 gallons a day. Cyril Davies had inspected the Milwr Tunnel extensions on Friday 1 September and found that a number of fissures had been struck within the area from which the evidence presented in 1904 and 1913 contended the supply to the well derived. Concerned over the impacts on the town's water supply and the industries in the area, Holywell Urban Council also requested inspection reports and gaugings of the flow of water to the well. These reports concluded that the quantity of water issuing from the fissures crossed in the Milwr Tunnel extensions was equal to the decreases registered at the well between April and August 1916, and accounted for the diversion of the source of supply.[19] By January 1917, the well had dried up—although flow eventually resumed it was not enough to provide water for local industry and so a level was driven from an abandoned mine shaft—the Roskell Shaft—to replenish the flow into the Holywell Stream.

6 Conclusions

From the late nineteenth century, the construction of tunnels in the Flintshire lead mining areas became central to the realisation of ideas about promising discoveries in an expanding resource zone. As socio-technical assemblages, the need for tunnels under Halkyn Mountain was argued as essential for the opening up of a subterranean frontier—and there are parallels in how oil and gas pipelines are promoted today around the world (e.g. see Nuttall 2014 for a discussion of how this is so in north-west Canada). Miners and engineers were celebrated as pioneers and adventurers as their digging, extraction and tunnelling revealed hidden worlds and unlocked the mineral and metal wealth of North East Wales. However, as miners probed deeper to reach new veins, water levels rose in shafts and passages. Drainage schemes were attempted to ease the problem and dewater the mines. For mine owners and their investors, construction of the Milwr Tunnel was considered a triumph of engineering interventions and of human progress over nature. This allowed mining and tunnel

[19] These reports along with a series of letters and details of water gauges are archived at the Flintshire Record Office, Document D/MT/1085.

companies to map, mark out and stake claims to rich lodes, and control the subterranean. However, tunnels provoked debates about power and privilege, the occupation and ownership of the subterranean through the placing of infrastructure, and concerns over the flow, disruption and contamination of underground watercourses and local water supplies. Yet, as the transcripts of the parliamentary committees and hearings show, the lives and experiences of people who ventured in and out of the depths, who engaged with the underground, worked with rock and extracted the ore, were not taken into consideration. The words of the miners, as well as local people who lived with water, are not inscribed in the official record. It was geologists and mining engineers who were considered the experts—and this chapter's narrative shows that it was their knowledge and theories about the formations, structures, materialities and aesthetics of the underground, along with geological maps, inventories of rock, stone, metals and minerals, and the flow of Carboniferous limestone water, that became to the testimony at the 1904 and 1913 parliamentary committee hearings. Geoscientific knowledge was essential for describing and envisioning what was known about the underground and the hidden (cf. Kuchler 2019). Just as imperial expeditions across oceans and terrestrial surfaces used images "as key techniques in the processes of investigating, ordering, explaining, and possessing – or attempting to possess – nature" (Bleichmar 2012: 7), maps and models of mines, the depiction of strata, and hydrogeological knowledge were essential for the visualisation, possession and engineering of Halkyn's interior spaces (see also, Nystrom 2014).

The Milwr Tunnel expanded. Throughout the 1930s, Halkyn District United Mines, which was formed in 1928, continued driving the tunnel and with mining ore from the veins accessible from it. Some 650 men were employed. In 1938, however, the low price of lead forced the laying-off of most of the workforce. No mining was carried out during the Second World War, but quantities of TNT explosive were stored underground on Halkyn Mountain for the Ministry of Supply in specially constructed stores. In 1948, work on driving the Milwr Tunnel was restarted. This continued until 1957, when the tunnel reached the Cathole Vein, near the Mold-Ruthin road, over ten miles (sixteen kms) from the portal. It had been intended to drive the tunnel into the mines to the south, in the Maeshafn-Llanarmon area of Denbighshire, but this plan was abandoned and mineral leases in this area were given up in 1960. Lead ore was mined at intervals up to 1977, but most extractive industry on Halkyn was focused on quarrying high-grade limestone for Pilkingtons, a glass manufacturer. The Milwr Tunnel was maintained because it had become an important water supply for the Courtaulds rayon mills at Holywell and Flint (the last mill closed in 1989). The last lead mine on Halkyn Mountain may have closed in 1987, but each day 23,000,000 gallons of water flow into the Dee Estuary from the Milwr Tunnel outfall, known locally as "The Holy," near Bagillt (Fig. 3). This flow is closely monitored and much of it is used by industry on Deeside.

Fig. 3 The Milwr Tunnel outfall into the Dee Estuary, near Bagillt, Flintshire. Photograph: Mark Nuttall

Acknowledgements This research is funded by the Henry Marshall Tory Chair research programme in the Department of Anthropology at the University of Alberta. Many thanks go to the staff at the Flintshire Record Office at Hawarden/Penarlâg in North East Wales for access to archival documents on lead mining and the Milwr Tunnel. I am also grateful to Marko Krevs for his comments and to the anonymous reviewers for their careful reading and suggestions on the first draft of this chapter.

References

Appleton, P. 1989. Limestones and caves of North Wales. In *Limestones and caves of Wales*, ed. T.D. Ford, 218–253. Cambridge: Cambridge University Press.

Armour, C. 1956. The trade of Chester and the state of the Dee navigation, 1600–1800. Unpublished Ph.D. thesis, University of London.

Bleichmar, D. 2012. *Visible empire: Botanical expeditions and visual culture in the hispanic enlightenment*. Chicago and London: The University of Chicago Press.

Bobbette, A., and A. Donovan (eds.). 2019. *Political geology: Active stratigraphies and the making of life*. Cham: Palgrave MacMillan.

Clark, N. 2017. Politics of strata. *Theory, Culture and Society* 34 (2–3): 211–231.

Cohen, J.J. 2015. *Stone: An ecology of the inhuman*. Minneapolis: University of Minnesota Press.

Coyle, G. 2010. *The riches beneath our feet: How mining shaped Britain*. Oxford: Oxford University Press.

Ebbs, C. 2008. *The Milwr Tunnel: Bagillt to Loggerheads*. Cris Ebbs: Llanarmon-yn-ial.

Elden, S. 2013. Secure the volume: vertical geopolitics and the depth of power. *Political Geography* 34: 35–51.

Ellis, B. 1998. *The history of Halkyn Mountain: The mountain with lead in its veins*. Holywell: Helygain.
Endfield, G.H., and C. van Lieshout. 2018. Water and vertical territory: the volatile and hidden historical geographies of Derbyshire's lead mining soughs, 1650s–1830s. *Geopolitics*. https://doi.org/10.1080/14650045.2018.1486299.
Frost, P., and N. Jones. 2004. Interpreting the mining landscapes of north-east Wales. In *Mountains and orefields: Metal mining landscapes of mid and north-east Wales*, ed. N. Jones, M. Walters and P. Frost 101–157. Council for British Archaeology Research Report 142. York: Council for British Archaeology.
Hawkins, H. 2020a. Underground imaginations: Environmental crisis and subterranean cultural geographies. *Cultural Geographies* 27 (1): 3–22.
Hawkins, H. 2020b. 'A volcanic incident': Towards a geopolitical aesthetics of the subterranean. *Geopolitics* 25 (1): 214–239.
Jones, N., Walters, M., and P. Frost. 2004. *Mountains and orefields: Metal mining landscapes of mid and north-east Wales*. CBA Research Report 142. York: Council for British Archaeology.
Kama, K., and M. Kuchler. 2019. In *Political geology: Active stratigraphies and the making of life*, eds. A. Bobbette and A. Donovan, 105–146. Cham: Palgrave MacMillan.
Kuchler, M. 2017. Post-conventional energy futures: Rendering Europe's shale gas resources governable. *Energy Research and Social Science* 31: 32–40.
McCausland, E. 2018. Caverns measureless to man: Subterranean rivers and adventurous masculinities in the Victorian lost world novel. *Journal of Victorian Culture* 23 (3): 350–365.
Nuttall, M. 2014. Pipeline politics in northwest Canada. In *Polar geopolitics? Knowledges, resources and legal regimes*, ed. R. Powell and K. Dodds, 227–283. Cheltenham: Edward Elgar.
Nystrom, E.C. 2014. *Seeing underground: Maps, models, and mining engineering in America*. Reno: University of Nevada Press.
Rhodes, J.N. 1968. Derbyshire influences on lead mining in North Wales in the 17th and 18th centuries. *Bulletin of the Peak District Mines Historical Society* 3 (6): 339–351.
Rudwick, M. 1976. The emergence of a visual language for geological science 1760–1840. *History of Science* 14 (3): 149–195.
Strahan, A. 1890. *The geology of the neighbourhoods of Flint, Mold, and Ruthin*. Memoirs of the Geological Survey. London: His Majesty's Stationary Office (HMSO).
Tilley, C., and K. Cameron-Daum. 2017. *An anthropology of landscape: The extraordinary in the ordinary*. London: UCL Press.
Williams, A.R. 2012. Hidden bullion: Silver production in north-east Wales. *Welsh Mines and Mining* 2: 33–44.
Williams, C.J., and R.A. Williams. 2012. Rediscovering the lead and zinc production of north-east Wales. *Welsh Mines and Mining* 2: 13–31.
Younger, P.L. 2005. 'Making water': The hydrogeological adventures of Britain's early mining engineers. In *200 years of British hydrogeology*, ed. J.D. Mather, 121–157. Geological Society, London, Special Publications 225. London: Geological Society of London.

Hidden Geographies of the Sixteenth Century Road Network and Trade in Georgia's Portion of the Silk Road

Roman Maisuradze and Tamar Khardziani

Abstract Over time, landscape elements are being transformed by natural, cultural and socio-political processes. There are occasions when important geographic events and objects become hidden, or traces of their existence become extinct. Often, the history of the landscape is hidden and can be revived based on historical sources, folk narratives and physical remains. The present paper deals with the sixteenth-century road network and import–export of goods in the Samtskhe-Javakheti region. The Ottoman census materials known as the Great Defter of Gurjistan Vilayet is the primary source of the research. In the Samtskhe-Javakheti region (Georgia), in the study period, the structure and function of the road network was different from today. Locally produced exports, as well as roads connecting settlements in each historic administrative unit, are described in the paper. The chapter tells about difficulties associated with travel, as well as the flows and main directions of imported and exported goods and the connection centres. The study shows the state of the mediaeval road network and trade patterns in this section of the South Caucasus Corridor.

Keywords Mediaeval road network · Import–export · Landscape · Historical geographic information system (HGIS) · Silk Road · Samtskhe-Javakheti · Georgia

1 Introduction

The work deals with the hidden elements of the landscape that played an important economic and cultural role several centuries ago. Because of political and socio-economic changes, the significance of some geographical objects or events has declined or disappeared over time. Road network and trade-economic relations are

R. Maisuradze (✉)
Faculty of Exact and Natural Sciences, Department of Geography, Ivane Javakhishvili Tbilisi State University, Tbilisi, Georgia
e-mail: romani.maisuradze@tsu.ge

T. Khardziani
Department of Physical Geography, Vakhushti Bagrationi Institute of Geography, Ivane Javakhishvili Tbilisi State University, Tbilisi, Georgia
e-mail: tamar.khardziani@tsu.ge

M. Krevs (ed.), *Hidden Geographies*, Key Challenges in Geography,
https://doi.org/10.1007/978-3-030-74590-5_10

the phenomena that are transformed by socio-political changes. Road network is a vital geographical object and a prerequisite for societal development. It determines the economic, social and cultural development of the regions that drive product circulation and human involvement in public activity locally and internationally. A road is a unit that for centuries and millennia has shaped the interdependence of the local environment and distant territories. Its role has not diminished to this day (Jacobson 1940). New roads often repeat the directions of ancient walking paths and ways, but some have changed considerably. Road directions and their socio-political importance change significantly throughout history, thus the remnants of old roads could retain the information accumulated over millennia (Vion 1989; Watteaux 2017). The focus of the work is to unhide hidden landscape elements that have been in operation for centuries, but their role and meaning have changed significantly over time. Such change was related to the geopolitical, social and economic developments in the study region.

Road network formation has been around for millennia, and the importance of some road sections periodically increases or decreases. Changing the function and importance of the road network is a dynamic process and, therefore, from ancient times, they have had travel, trade, commercial and military roles. The ancient Egyptian, Sumerian and Babylonian sources describe the direction of the oldest roads. Historic roads are known to link the East with the West and the inner continental areas with the sea. At the same time, they also made it easier to travel and trade between cities. Under the rule of city-states and centralized empires, for example, during Achaemenid Iran and the Roman Empire, there were ways of prioritizing roads (Braudel 1972; Astour 2000; Colburn 2013). There were examples where roads were given specific names by their direction and primary function (Sitwell 1981; Knapton 1996).

Our study region (Fig. 1), Samtskhe-Javakheti, is an interesting territorial unit in this respect, as the diverse natural landscapes and various forms of the built environment and land use are present there. For centuries, well-developed local production and active trade relations with its surrounding areas had been distinguishing the region. Our interest was in the study of the region's road network and commodity turnover on the background of the East-West trade relations and the corridor so-called Silk Road.

The Silk Road is an iconic East-West bridging corridor that encompasses numerous roads, highways and *caravans*, from Antioch-Constantinople to present-day Central China and India (Richthofen 1877; Saliba 2008; Harold 2008; Williams 2014; Jones 2009). The term combines many branches of the primary and auxiliary roads described in different literary and cartographic sources; however, some researchers believe that this road has not lost its significance for millennia and is still relevant to many countries today (Christian 2000; Karluk and Suleyman 2014; Williams 2014). Both historiographical sources as well as archaeological material illustrate the importance of these roads and the diversity of trade goods (Tomber et al. 2008; Spengler et al. 2014). Discussion of the road network and export–import of goods in Samtskhe-Javakheti is essential as an integral part of the Silk Road.

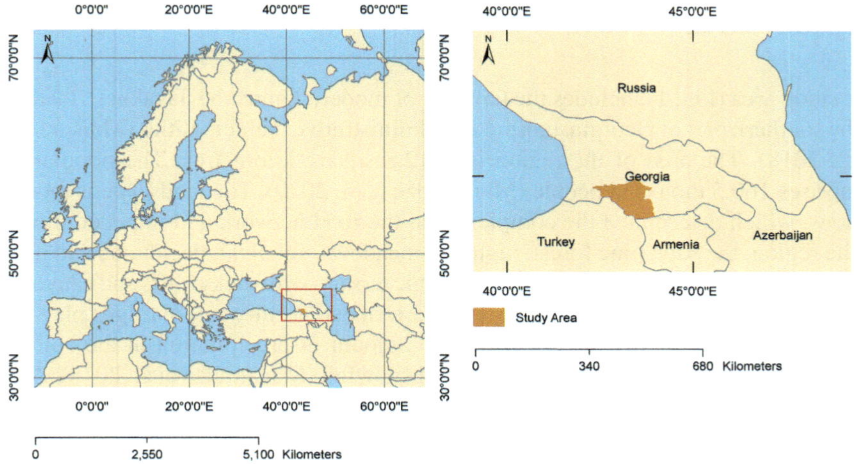

Fig. 1 Location of the study area

The roads and *caravanserai* here served to the travellers to and from Georgia, the Caucasus, the Middle East and European countries.

The fact that in some modern cartographic sources the Silk Road goes beyond the South Caucasus Corridor does not mean that Georgia and the South Caucasus had no transit function, as records of some European travellers can serve as the evidence of the carriageway existence of in the region (Kharebava 2018). The transit function of the South Caucasus Corridor is evidenced by the 1254 population census document. At that time, the Georgian population was around 8 million (Verulava 2019), which was more than 2% of the world population at that time (Klein et al. 2010). Accordingly, the population in the whole Caucasus region should have been even higher. Such a large population in the region is also emphasizing the importance of the trade and accompanying roads.

Mediaeval road network and trade were associated with population concentration and urban settlements. The document from the sixteenth century (The Great Defter of Gurjistan Vilayet) has described the settlements of the study area (Jikia 1941, 1958). In the twentieth century, academician Sergi Jikia published the Georgian version of this document in three volumes (Maisuradze et al. 2020). In the sixteenth century, the influence of the Ottomans intensified in the region. The demographic, economic and urban environments of this period determined the different character of the road network from that of today. In this work, we have surveyed the hidden elements of the environment based on the written historical sources, and the material remains preserved in the landscapes.

2 Study Area

Our study area (Fig. 1) includes the territory of modern Samtskhe-Javakheti, located in the southern part of Georgia, uniting six administrative municipalities (Maisuradze et al. 2018). The area of the region is 6412.8 square kilometres. The population comprises 160.5 thousand people (Maisuradze et al. 2020). The landscape is mountainous, which determined the complexity of the road network and impeded travel. In the region, the subalpine forest, highland meadows, subnival, deciduous and dark coniferous landscapes are present. The volcanic terrain, with rocky cliffs and canyons, characterizes the relief of the region. The western and north-western ridges play the role of oro-climatic barriers. The increased amount of precipitation in the coldest period of the year is causing an abundance of snowfall (Maisuradze et al. 2018). As a result, some of the roads were blocked and stopped operating; the landscape features of the study region significantly determined road functioning.

3 Materials and Methods

We used the following methods to study the subject: 1. Analysis of documentary and historiographical sources; 2. Analysis of existing literary and cartographic materials; 3. Field expeditions; 4. Development of the geo-information system by using ArcGIS, analysis of road network and import–export of goods using the Historical Geographic Information System (HGIS) technique.

The Great Defter of Gurjistan Vilayet was the main source of the research, upon which we developed GIS layers and attribute tables. The analysis of this document was particularly critical in identifying the essential centres of local production.

We also used secondary works based on various historical sources, which implied analytical research on the economic situation of mediaeval Samtskhe-Javakheti and trade relation issues, including major local markets (Makalatia 1938, 1957; Svanidze 1971; Lomsadze 1975). Besides, we also studied cartographic materials published in the last century, reflecting the state of the road network in the past (Javakhishvili 1926; Aslanikashvili 1953). They depict roads of various uses that have been active in the past centuries.

We conducted field surveys in the summer of 2017 and 2018. The study aimed to determine the direction of the sixteenth-century roads and their socio-economic significance. We studied the settlements and the state of the roads connecting them. We have put local roads on the map which are not part of the modern road network.

We used 1: 50,000 and 1: 25,000-scale Soviet topographic maps and ArcGIS base map satellite images. We identified currently active and non-active mountain passes. We identified the settlements that served as market centres and where both imported and local goods would sell. Then we developed a classification according to the direction of the imports. After that, we analysed the roads according to the directions of imported and exported goods.

4 Main Results

4.1 Hidden Geopolitics

Since the study period is the second half of the sixteenth century, we analysed the issue in terms of space-time. The centre of the region, just like today, was the town Akhaltsikhe. However, the road vector was not directed only to one dominant—Akhaltsikhe-Tbilisi direction, because Georgia at that time included several independent kingdoms. The kingdoms of Kartli and Imereti bordered Samtskhe-Javakheti, so the road network was not just one-way and was more multi-vector. Nowadays, in the region, only two major highways have international importance, during the sixteenth–seventeenth centuries many branches of roads existed with international function.

4.2 Unhidden Modern Road Network

It should be noted that currently, in the study region, we have identified the following categories of modern roads: 1. International highways; 2. Asphalted local; 3. the motor ground, covered with gravel; 4. Motor ground; 5. Non-motorized roads; 6. Railway; and 7. Narrow railway. The classification of the roads is based on their morphological nature, the type of cover and their function. We do not have paths in the classification because it was challenging to put them on the map due to the scale.

4.3 The Hidden Mediaeval Road Network

We have grouped the sixteenth-century roads into the following categories: 1. Main roads; 2. Interregional roads; and 3. Local roads (Fig. 2).

The classification of old roads is more general, as, at that time, there were no railways and solid-cover roads, and the most advanced modes of transportation were chariots and carts. The term international is symbolic here because, at that time, there was no border control, and a person who came to a foreign land acted following the local rules. Therefore, the status of the roads was not clearly defined, and their convenience and reliability determined their importance.

With the help of the ArcGIS, we have determined the length and directions of the communication units, which is shown in the Table 1.

The total length of the main roads comprised 855.61 km. Their direction partially repeated the direction of today's main highways. In addition to the above, there were relatively less congested roads that crossed the Arsiani, Adjara-Imereti, Trialeti and Samsar-Javakheti ridges, which are classified as interregional roads on the map

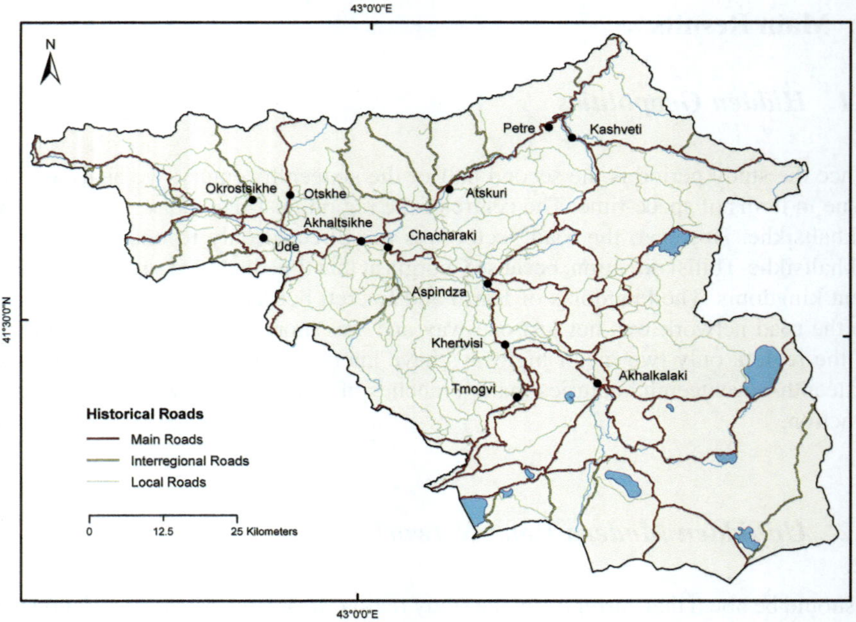

Fig. 2 Historic road network in the study region

Table 1 Length of the road network on the territory of Samtskhe-Javakheti, from the sixteenth century

Road type	Length (km)
Main roads	856
Interregional roads	307
Local roads	1382
Total	2545

(Fig. 2). Furthermore, the roads leading to the settlements, to the agricultural lands, and to the pastures were classified as local roads.

When classifying the road network, we excluded pedestrian and equestrian paths because they existed only temporarily, and their traces have not been preserved. The length of the road network is presented in the Table 2, which shows the development of road communications in the sixteenth century. Furthermore, what influence the road network condition had on the formation of the political-socio-economic picture of the region.

As can be seen from the data, the road network was quite well developed (Table 1). The length of the main roads exceeded the length of the interregional roads. Roads connecting local and rural areas and agricultural lands formed the most frequent network, as travel and commodity movement were frequent among rural settlements.

Table 2 Research region and road network, their length according to road types in administrative units

	Length of roads, according to their types, km			
Historic administrative units (*Liva* and *Nahia*)	Main roads	Interregional roads	Local roads	Total
Akhaltsikhe *Liva*	245	199	696	1140
Mzvare	12	18	46	76
Chrdili	20	0	55	75
Ude	51	37	83	171
Kvabliani	33	24	33	90
Otskhe	21	4	65	90
Atskuri	30	75	68	173
Okrostsikhe	20	9	25	54
Aspindza	49	0	155	205
Chacharaki	9	33	165	207
Khertvisi *Liva*	134	0	290	424
Khertvisi	37	0	104	141
Tke-Javakleti	97	0	128	225
Buzmareti	0	0	58	58
Akhalkalaki *Liva*	278	48	210	536
Akhalkalaki	229	48	123	401
Tmogvi	27	0	31	58
Nialiskuri	22	0	56	78
Childiri *Liva*	69	26	18	113
Kanarbeli	69	26	18	113
Potskhovi *Liva*	18	3	25	46
Chrdil	18	3	25	46
Petre *Liva*	113	13	130	257
Petre	55	0	62	117
Kashveti	58	13	69	140

The most frequent road networks from the administrative units were characteristic to the Akhaltsikhe, Khertvisi and Akhalkalaki *Livas* (Table 2).

In Akhaltsikhe *Liva*, by the main road network were distinguished Ude and Aspindza *Nahias*, while in terms of interregional roads, Atskuri *Nahia*, which may be due to the roads leading to the Adjara-Imereti ridge, through which Samtskhe-Javakheti would connect to the Kingdom of Imereti. The local roads in Aspindza, Udes, Otskhe and Atskuri have an unusually high rate, which is related to the abundance of villages and the population density.

We have identified two categories of roads in Khertvisi Liva: main and local roads. The local cart and the equestrian road network are also frequent, which indicates the movement between the rural settlements and the density of the settlement network. Buzmareti Valley is distinguished, where there are only local roads. The valley is located close to Niali plateau and occupies the northern direction from Erusheti-Karzaneti road. One direction of the main trade route was winding near the village Buzmareti. The second direction was connected to the road Shalosheti-Gundi-Alandza, which bypassed the Buzmareti *Nahia*.

In Akhalkalaki Liva, the total length of the main roads was longer than in other Livas. Part of the Javakheti plateau, which included the Akhalkalaki *Nahia*, was connected to the Kingdom of Kartli, Paravani-Chikiana, Tabatskuri and Javakheti ridge, which was a branch of the Silk Road and ran eastward in the lower reaches of the Mtkvari River to the present-day Azerbaijan and Iran.

In Kanarbeli *Nahia*, which was located next to Akhalkalaki, local and interregional roads were presented in smaller numbers. Part of the *Nahia* extended to the plateau that is now part of the Republic of Turkey. The trade routes to Artaan, Kars and Erzurum passed by Lake Chrdili. The second direction was connecting the region with the Armenian districts.

The Georgian part of the Kanarbeli *Nahia* occupied a small area, which explains the small total length of the local roads. In addition, part of the roads passed through the villages and then joined the road leading to Artaani, Erzurum and Kars. The road network was frequent in Potskhovi and Petre *Livas*, which were connected to the trade routes leading towards the rivers Potskhovi and Gujaratistskali, as well as to the Adjara-Imereti and Trialeti ridges, which connected the study region to the Imereti and Kartli kingdoms. Consequently, these roads were individual branches of the Silk Road (Fig. 2).

4.4 Historic Road Network as a Hidden Landscape Element

Road communications are elements of public importance in the landscape. They ensured the development of trade relations on the ground and the circulation of goods, and relations between settlements or territorial units. There are many historic roads in the region, the importance of which is currently diminished. Roads that were once a connecting artery between regions and kingdoms have become a means of communication at the local level. Some of them remain only as gravel roads, the function of which is only to connect the rural settlements or to reach the agricultural lands, or they are used for pastoralism. The direction of the roads also gives us information about which goods went in which direction and where the main centres of local production were located.

The road network left its mark on each landscape. Road network determined to some extent whether particular landscape was involved in the activities of the society or vice versa, or became uninhabited and remained beyond the active use. Those old road communications, which at first glance do not seem to be visible in the landscape,

Fig. 3 **a** The historic road leading to the Tmogvi-Vardzia canyon; **b** A mediaeval bridge over the river Paravnistskali, in the village Tontio; **c** The section of the old road connecting Kartli and Samtskhe near the village of Abuli; **d** Caravanserai ruins on the territory of Akhalkalaki Fortress (*Source* Photographed by Roman Maisuradze, 2017–2019)

are the hidden element of the public activity of the space, which has been so important for centuries.

Landscape categories were distinguished from each other by different road network densities (Fig. 4). The mountain valleys and plateaus were distinguished by the length and high density of the road network, as well as in the middle mountains, where the dark coniferous forest landscapes are represented. Interregional roads, which connected Samtskhe-Javakheti to the kingdoms of Imereti and Kartli, mostly overlapped with the dark coniferous forest landscapes. In addition to the main caravan routes, other roads were connecting the regions that crossed the Adjara-Imereti and Trialeti ridges. The landscapes of the highland meadow and meadow-steppes occupied a significant part of the region. The roads crossing the steppe landscapes of the Javakheti plateau as well as the Samsari and Javakheti ridges created one system. The high intensity characterizes the road network in the landscapes of the volcanic canyon, which includes the Aspindza-Tmogvi-Vardzia section. Although the length of the road network here lagged behind that of the valleys, the dark coniferous forest, the volcanic ridges and plateaus of the highlands and the landscapes of the volcanic canyon cover a much smaller area than the landscapes mentioned above. In conclusion, road communications were quite well developed in the study region.

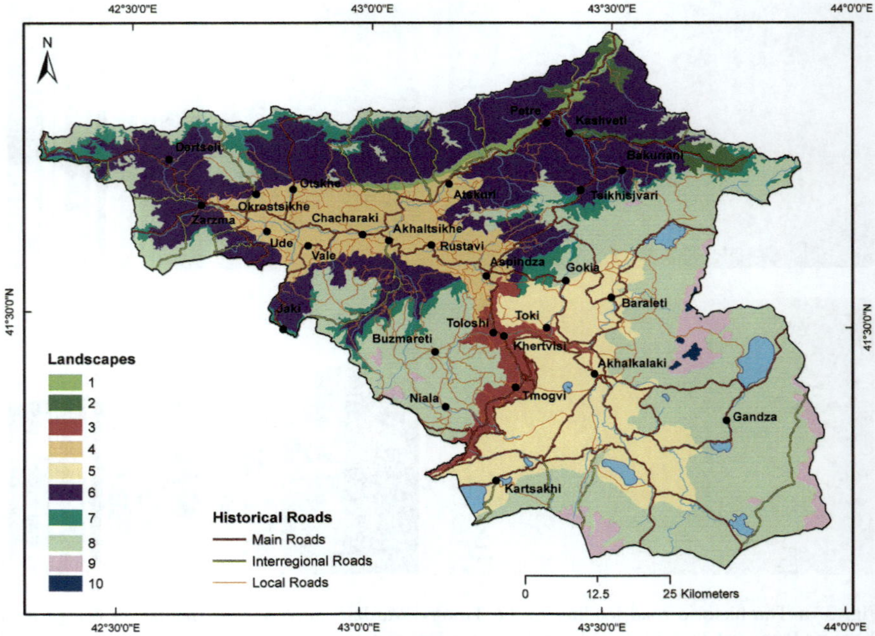

Fig. 4 Landscapes of Samtskhe-Javakheti: 1. Lower mountain landscape with oak and oak-hornbeam forests; 2. Middle mountain landscape with hornbeam-beech and chestnut-beech forests; 3. The landscape of the Middle Mountain Volcanic Canyon, with Hemi-xerophilic shrubland, steppe and open forests; 4. Meddle mountain valley landscape with steppe vegetation, rarely with forest and shrubbery; 5. The landscape of middle mountain plateaus with steppe vegetation; 6. Middle mountain landscape with beech, beech-dark coniferous and pine-beech forest; 7. Upper mountain landscape with birch and birch-pine forests; 8. The subalpine landscape of highlands, with meadows and rare forests; 9. Highland alpine landscape with meadows and alpine shrubs; 10. The subnival landscape of highlands

The data of the road network shows that the density in the landscapes of the volcanic canyons and Samtskhe Valley is particularly high and equals 0.8–0.9 km/km^2, which is significantly higher than in the landscapes of lowland forests and volcanic plateaus, where the density is 0.35–0.5 km/km^2 (Fig. 4).

The high density of the road network along the Samtskhe Kvabuli and Mtkvari Valley is connected with the concentration of the leading agricultural and cultural centres. A significant number of the fortresses and towns were concentrated here, and the population density was higher than in other landscapes.

The road network density is relatively low in the middle mountain dark coniferous forest landscapes. For highland subalpine meadows and meadow-steppe landscapes, this figure drops to 0.26 km/km^2.

Part of the roads was periodically closed due to precipitation, mostly heavy snow, or natural hazards (Table 3). Some closures have been linked to frost and unfavourable travel conditions. Travelling with the modes of transportation at that period required

Table 3 The length of seasonally operating roads

	Road length (km)	The length of roads that were closed seasonally due to climatic and relief factors	%
Main roads	856	132	15
Interregional roads	307	195	63
Local roads	1382	302	21
Total	2545	628	100

more time than today. The modes of transportation in the region were horse, donkey, mule and camel. The means of transport of the cargo were carts, carriages and waggons. The roads crossing the Adjara-Imereti and Arsiani ridges are characterized by heavy snow and avalanches, which significantly hindered travel. The passes of the Trialeti ridge were less snowy but hard to cross due to the steepness. The peaks of the Samsari and Javakheti ridges were covered with snow during the cold period of a year, but the snow cover lagged behind the precipitation on the Arsiani and Adjara-Imereti ridges. However, the snow cover here was somewhat behind the Arsiani and Adjara-Imereti ridges. On rough, snowy, icy roads, one could cover 5–10 km by foot in a short day, while carts and waggons with load could cover 10–15 km with great difficulty. During bad weather, the roads were closed.

Almost a quarter of the roads were periodically closed, and travel was challenging, particularly to the directions of Adjara, Imereti, Guria and Kvemo Kartli. The case concerned military movements as well as domestic travel and trade relations.

4.5 *The Road as a Hidden Economic and Geopolitical Unit*

From the military-defensive perspective of that time, the roads may have a positive effect on the socio-political situation. This circumstance allowed the people to be better prepared and to use the seasonal closure of roads for defence. From the historical events of the study period, the multiple invasions of the Ottomans and the Iranians are noteworthy. In addition, significant threats were attacks by Dagestani gangs, followed by looting and human trafficking (Alimbarashvili 2013). This situation greatly affected Samtskhe-Javakheti and the Georgian kingdoms. This situation was exacerbated by trade in the Ottoman Empire and the existence of slave markets, where captives were sold at very high prices at the time (Fisher 1978; Pamuk 2000; Özmucurand Pamuk 2002). This circumstance prompted the Dagestani warriors living in the closed valleys to find an easy and at the same time ruthless way to earn an income. This event was preceded by the fact that Dagestani people lost lands in the north, which they had been using for centuries for grazing. The event was followed by the proliferation of Russian *Stanits* and their entry into the

Russian Kingdom. Poverty and land scarcity forced them to find alternative livelihoods, such as slave markets in the cities of the Ottoman Empire, which stimulated human trafficking. Such a situation posed a threat to travellers and traders.

Interregional travel routes have played an essential role in the lives of locals. Road safety and security should be of particular importance. If we consider both domestic and foreign trade, the various branches of the East-West connecting roads, we can assume that the trade and commerce were both local, regional and international, as *caravans* and traders were transported mainly by main roads. The steady movement of *caravans* on these roads was associated with both the existing infrastructure of the roads (*caravanserai*, *Funduki*, settlements) as well as the difficulties encountered here (Fig. 3).

The Table 4 shows that wine, spices, fruit and fabrics had to go the longest route. Destinations of imported goods mainly were located around the centres of *Livas* and the principal settlements of the *Nahias*. The places where the fairs were held were located in the area of the castle Rabats and crossroads. In this respect, the Rabats of Akhaltsikhe, Ude, Otskhe and Atskuri fortresses stood out. Also important were the Rabats of Khertvisi, Tmogvi, Aspindza and Akhalkalaki, located near the caravanserais of Nakalakari and Paravani (Svanidze 1971; Lomsadze 1975).

Wine and spices have travelled the longest in the region, followed by salt, textile, glue, paints and clothing. These goods were also transported from various points, and their delivery to their destinations meant a covering of considerable distance. Various goods passed on the other roads. Sometimes goods of a similar category

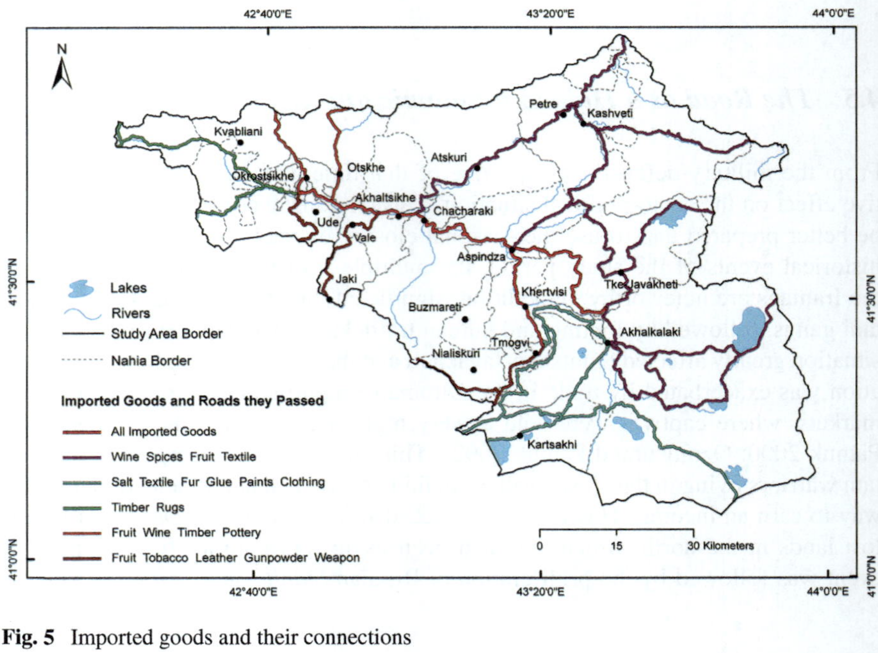

Fig. 5 Imported goods and their connections

Table 4 Roads and imported goods

Names of the goods passing through the main roads	Road length	%
Salt, textile, fur, glue, paints, clothing	190.1	22.22
Wine, spices, fruit, textile	292.11	34.14
Wood, rugs, dairy products	82.92	9.69
Fruit, mash, wood, clay products	58.92	6.89
Fruit, tobacco, processed leather, gunpowder, guns	101.52	11.87
All imported goods passed through the destination	130.02	15.20
Total	855.61	100

Table 5 Directions of import and export

Direction	Length, km	Share, %
Towards Artaan-Kars-Erzurum	39.23	4.58
Towards Adjara-Guria and Trabzon	70.02	8.18
Towards Potskhovi-Artaani-Shavsheti	62.49	7.3
Towards Imereti	55.86	6.53
Towards Kartli and Kakheti	149.13	17.43
Towards Armenia	58.65	6.86
Central roads, where several roads leading to different regions were connected	420.23	49.11
Total	855.61	100

are found in different combinations because they depended on many factors. Export goods took the same route, and these roads served to take local products out of the region (Table 5).

The goods, which were produced in Samtskhe-Javakheti (Table 4), can be considered as export goods for this period. Since there is a small amount of data on production, we will rely on all the written and digital data we have in the census materials for local settlements.

Roads crossing the region were the means of connecting the East and the West. In addition, passing *caravans* were buying and transporting products from local markets. To some extent, local traders also exported them (Fig. 6). Traders from neighbouring countries also occupied some sectors.

The region was famous for its grain crops and was considered Georgia's granary. We can conclude that wheat, barley, rye and millets were the products that were exported from the region (Maisuradze 2019b). Honey, wax and propolis are products that were made in abundance in the region. In this regard, Kvabliani gorge, the middle part of Uraveli, Javakheti plateau and Atskuri area, as well as Borjomi gorge and Sadgeri plateau are distinguished. Exports should have included a significant amount of local wine and grape products, Churchkhela and raisins (Maisuradze et al. 2020). Flaxseed oil should have also been exported (Maisuradze 2019a), which was made

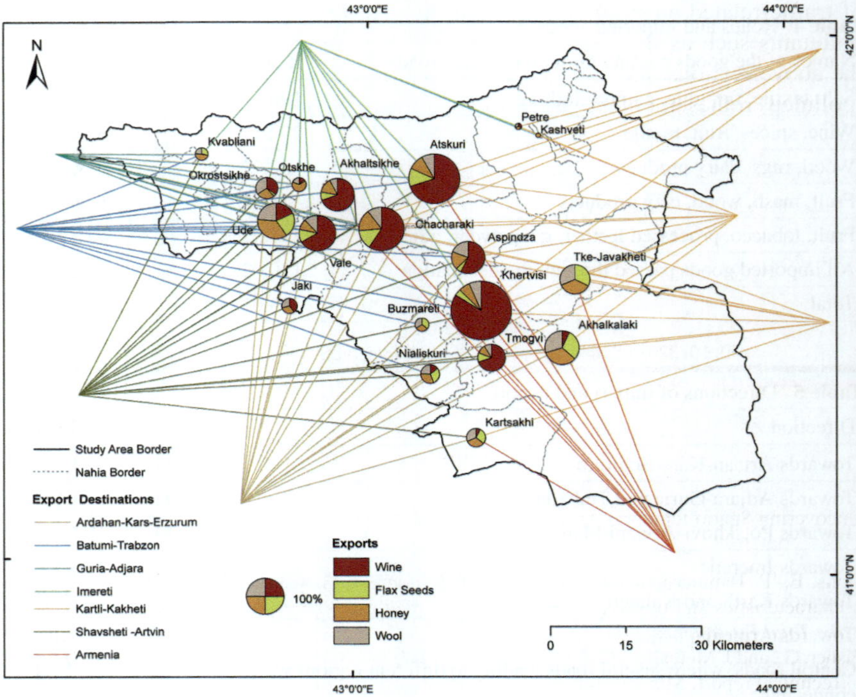

Fig. 6 Road network analysis according to the primary export goods and their directions

in the region for this period, as evidenced by the presence of *Bezirkhanes*, flaxseed oil processing manufactures. Flaxseed oil was later produced on the territory of Kvemo Kartli, on the present-day Tsalka, Gomareti and Dmanisi plateaus. Paints and small items were made in settlements that were well fortified, for example, in Tmogvi, Khertvisi, Aspindza, Atskuri, and Akhaltsikhe. Swords and other weapons were produced in the same settlements. The exported goods were wool, felt and leather, as evidenced by the abundance of sheep, cattle and pigs in the region. The pork skin was the best material to make armour, shoes and saddles. Later, their export should have been weakened due to a large number of the Islamic population in the neighbouring regions. Exports of timber and wood products were mainly directed to the south, east of Anatolia and Armenia-Iran. In terms of exports of iron goods, the Tmogvi-Khertvisi strip was distinguished, where the best iron products were made.

Goods were imported from Kars-Erzurum, namely, weapons, shotguns and ammunition, as well as leather, paints and tobacco. Salt and spices, clothing, textiles, fur and dyes were imported by three different routes along the southeastern roads. Goods were imported from Kartli and Kakheti in four ways: 1. Borjomi gorge; 2. Gujaratistskali gorge and the pass over the Trialeti ridge; 3. via Tabatskuri; 4. Through the Faravani. In this way, they mainly imported wine, spices, fruit and textile. Different types and tastes of wine were in demand here; also, the types of fruit that did not grow in Meskheti. The same can be said about the fabric imported from the East. Clay products, timber and wine, which were different from the local ones, were imported from Imereti via Zekari Pass and Fersati plateau. From the west, through the Kvabliani gorge and Goderdzi pass, timber, dairy products, rugs and other local woven goods were imported. The central transport axis on the map represents the section of the road where almost all imported goods passed (Fig. 5).

Remains of *caravanserai* are still preserved in Rabats of Atskuri, Akhaltsikhe and Akhalkalaki (Fig. 3); besides, the sources mention the *caravanserais* of Khertvisi and Paravani (Lomsadze 1975). The village of Paravana was neither a regional centre nor a fortified fortress. The caravanserai existed here not because of the status of the settlement, but because of its location. Before Chikiani Pass, traders and travellers had to walk the long and challenging road and needed a break. The caravanserai remains preserved near the village Nakalakevi indicate that this section was crowded and in addition to Khertvisi-Tmogvi, another stopping point was needed.

The abundance of *caravanserais* and market places is a testament to the fact that the Samtskhe-Javakheti region was part of the East-West Corridor, and the roads here operated within the great road system that von Richthofen (1877) called the Silk Road.

In the sixteenth century, the function of the East-West Corridor was significantly reduced. However, this fact should be perceived in society as a temporary and unfavourable period. We think it would be hard to imagine that they would have to live in the conditions of military and predatory-robbery attacks for centuries to come. It would probably be more challenging to trade peacefully, move caravans and bring local produce to market.

Since the late sixteenth century, part of the roads acquired the function of transporting kidnapped captives. The slave trade was legal in the Ottoman Empire, but it was banned in the Georgian kingdoms, Kakheti, Kartli and Imereti. Therefore, the gang groups chose the well-hidden roads. Such roads ran mostly in forested landscapes. There was a market in Akhaltsikhe where captives were sold and then taken to various cities in the Ottoman Empire or neighbouring countries. Such a vicious form of trade must have been linked to the weakening of the role of the corridor, the isolation of the mountainous regions of the Northern Caucasus, and ongoing political changes on the North Caucasus plains, causing hardship and poverty in its eastern part. As an alternative to this, the Ottoman Empire indirectly offered them a market where, unfortunately, large-scale slave trade took place. Developments in the region have significantly weakened the Georgian kingdoms and led to a change in the negative direction of the *caravan* routes.

The study of hidden and forgotten elements once again confirms how essential they represent in determining historical road communications and import–export of goods and in restoring the historical picture. Presented research brings the socio-cultural significance of the region to the forefront, as well as revives the hidden information and knowledge about the space.

5 Conclusions

In Samtskhe-Javakheti, road communications were well developed in the sixteenth century. They only partially replicated the direction of modern roads. The region was connected in several ways with the Georgian kingdoms, as well as with modern Armenian territory and the central districts of the Ottoman Empire.

Roads were mainly concentrated in the centres of the administrative units (*Liva*, *Nahia*) and the settlements around the castles (*Rabati*).

According to the landscape analysis, the road network was particularly dense in the landscapes of the Samtskhe Valley, the Javakheti plateau and the Tmogvi-Vardzia volcanic canyon, so these areas were more populated and economically advanced.

Only part of the roads was permanent. Roads leading to Imereti and Adjara-Guria were closed for a long time. The roads from Borjomi gorge and Chikiani pass were the busiest. They were part of the Caucasus Branch of the Silk Road and the East-West Corridor.

In the sixteenth century, the East-West Corridor could no longer function as before, although the production of goods was still maintained. The primary export goods were wheat, barley, rye, wine, fruit, honey and propolis, wax, wool, felt, nuts, flaxseed oil, dyes and leather products, saddles, iron goods and weapons. Imported goods were weapons, salt, spices, tobacco, oriental and dried fruits, wine, textile, dyes, clay products, timber and precious items.

From the sixteenth century, the function and cargo turnover of the Samtskhe-Javakheti road network decreased. The region has become a relatively isolated region, with fewer ties to the Georgian kingdoms and a reduced role in the functioning of the East-West Corridor. The very uncomfortable function of the roads has been strengthened—the transfer of abducted people who were sold to the Ottoman Empire and its neighbouring countries. The travel and trade destination of the region's road network was renewed when Georgia gained independence. These road communications are of geopolitical importance to the territory of the South Caucasus and Eastern Anatolia. Therefore, they have both local and regional significance. Their prospects are enhanced by the rich historical experience of the existence of road communications.

Acknowledgements This research was supported by the Shota Rustaveli National Science Foundation of Georgia (SRNSFG) [grant number YS-2016-44].

References

Alimbarashvili, I. 2013. "Lekianoba" and military-defensive system of Kartl-Kahheti kingdom from the XVIII century to the 30s of XIX century. Ph.D. thesis, Tbilisi.

Aslanikashvili, A. 1953. *Gurjistan Vilaiet*. Tbilisi.

Astour, M. 2000. Overland trade routes in Ancient Western Asia. In *Civilizations of the ancient near East*, ed. Jack M. Sasson, 1401–1420. Peabody, MA: Hendrickson.

Braudel, F. 1972. *The Mediterranean and the Mediterranean world in the age of Philip II*. New York: Harper Collins.
Christian, D. 2000. Silk roads or steppe roads? The silk roads in world history. *Journal of World History* 11 (1): 1–26.
Colburn, P.H. 2013. Connectivity and communication in the Achaemenid Empire. *Journal of the Economic and Social History of the Orient* 56: 29–52.
Fisher, A.W. 1978. *The sale of slaves in the Ottoman Empire: Markets and state taxes on slave sales, some preliminary considerations*. Digital Library for International Research Archive. http://www.dlir.org/archive/items/show/2713. Accessed 8 April 2020.
Harold, F. 2008. Caravan routes of Iran. *The Silk Road* 6 (1): 17–28.
Jacobson, H.R. 1940. *A history of roads from ancient times to the motor age*. Atlanta: Georgia School of Technology.
Javakhishvili, A. 1926. *Map of Georgia*. Tbilisi: Tbilisi State University.
Jikia, S. 1941. *Great book of Gurjistan Vilaiet. Part II*. Tbilisi: Mnatobi.
Jikia, S. 1958. *Great book of Gurjistan Vilaiet. Part III*. Tbilisi: Publishing of the Georgian Academy of Sciences of Soviet Republic of Georgia.
Jones, A.R. 2009. Centaurs on the Silk Road: Recent discoveries of Hellenistic textiles in western China. *The Silk Road* 6 (2): 23–32.
Karluk, R.S., and K.C. Suleyman. 2014. Bridging civilizations from Asia to Europe: The Silk Road. *Chinese Business Review* 13 (12): 730–739.
Kharebava, N. 2018. *Regesten der zweisprachigen georgisch-persischen Urkunden der Safavidenzeit*. Bamberg: University of Bamberg Press.
Klein, G.K., A. Beusen, and P. Janssen. 2010. Long-term dynamic modeling of global population and built-up area in a spatially explicit way: HYDE 3.1. *The Holocene* 20 (4): 565–573.
Knapton, J. 1996. The Romans and their roads—The original small element pavement technologists. In *5th International Concrete Block Paving Conference*. Tel-Aviv, Israel.
Lomsadze, S. 1975. *Samtskhe-Javakheti: From the middle of XVIII to middle of XIX centuries*. Tbilisi: Metsniereba.
Maisuradze, R., M. Elizbarashvili, T. Khardziani, and M. Tatishvili. 2018. Landscapes of Samtskhe-javakheti region, transformation degree and forms. *International Multidisciplinary Scientific GeoConference: SGEM: Surveying Geology & Mining Ecology Management* 18: 487–494.
Maisuradze, R., T. Khardziani, and T. Eradze. 2020. Retrospective mapping of the XVI century Samtskhe-Javakheti viticulture and fruit farming. *Miscellanea Geographica* 24 (1): 5–15.
Maisuradze, R. 2019a. Flax culture in the XVI century Samtskhe-Javakheti. *Annals of Agrarian Science* 17 (1): 28–35.
Maisuradze, R. 2019b. Major agricultural crops in the XVI century Samtskhe-Javakheti. *Annals of Agrarian Science* 17 (1): 16–27.
Makalatia, S. 1938. *Meskhet-Javakheti*. Tbilisi: Federatsia.
Makalatia, S. 1957. *Borjomi Valley*. Tbilisi: Sakhelgami.
Özmucur, S., and Ş Pamuk. 2002. Real wages and standards of living in the Ottoman Empire, 1489–1914. *The Journal of Economic History* 62 (2): 293–321.
Pamuk, Ş. 2000. *A monetary history of the Ottoman Empire*. Cambridge: Cambridge University Press.
von Richthofen, F. 1877. Über die zentralasiatischen Seidenstrassen bis zum 2. Jahrhundert n. Chr. Verhandlungen der Gesellschaft für Erdkunde zu Berlin.
Saliba, G. 2008. China and Islamic civilization: Exchange of techniques and scientific ideas. *The Silk Road* 6 (1): 9–16.
Sitwell, H.H.N. 1981. *Roman roads of Europe*. NY: St. Martin's Press.
Spengler, R., M. Frachetti, P. Doumani, L. Rouse, B. Cerasetti, E. Bullion, and A. Mar'yashev. 2014. Early agriculture and crop transmission among Bronze Age mobile pastoralists of Central Eurasia. *Proceedings of the Royal Society B: Biological Sciences* 281: 20133382. https://doi.org/10.1098/rspb.2013.3382.

Svanidze, M. 1971. *From the history of Georgian-Ottoman relations in XVI-XVII centuries*. Tbilisi: Metsniereba.
Tomber, R., A. Simpson, and P. Copeland. 2008. *Indo-Roman trade: From pots to pepper*. London: Duckworth.
Verulava, T. 2019. Family support policy in Georgia. *Economics and Business* 3: 96–108.
Vion, É. 1989. L'analyse archéologique des réseaux routiers: une rupture méthodologique, des réponses nouvelles. *Paysages Découverts* 1: 67–99.
Watteaux, M. 2017. The road network in the Longue Durée: A reading key of the history of territories. *Open Archaeology* 3: 149–174.
Williams, T. 2014. *The Silk Roads: An ICOMOS thematic study*. Charenton-le-Pont: ICOMOS.

Boundary Stones and Their "Hidden" Legacy in Slovenia

Matija Zorn and Peter Mikša

Abstract A boundary stone, boundary marker, border marker or border stone is a robust physical marker that identifies a land boundary, especially a change in the direction of a boundary. Usually, it is a stone. Natural stone was used for boundary stones, which were later made of concrete or other materials. They were usually placed in a particularly visible spot. Many boundary stones feature information, such as an abbreviation identifying the holder of the border and a date. Boundary stones separating countries usually include abbreviations of countries they are separating, as well as the date when the border was delineated. We focused on boundary stones in the territory of modern-day Slovenia, which in the Habsburg Monarchy, before World War I, divided the Duchy of Styria and the Kingdom of Hungary and, in the interwar period, the Kingdom of Italy and the Kingdom of Yugoslavia. These boundary stones no longer serve their original purpose; however, as markers of the past, they are historical witnesses. They represent an administrative legacy that is today mostly hidden in the cognitive perception of these boundaries.

Keywords Historical geography · Political geography · Border studies · Boundary stone · Cultural heritage

1 Introduction

The border "*represents a relatively static model of territorial demarcation that can be expressed physically—whether through human-constructed border stones, walls, or fences—or through natural features such as rivers, mountain ranges, and even trees that became endowed with human-constructed meaning as border markers …*" (Lee and North 2016: 2). Borders and boundary markers are a basis for administrative

M. Zorn (✉)
Research Centre of the Slovenian Academy of Sciences and Arts, Anton Melik Geographical Institute, Ljubljana, Slovenia
e-mail: matija.zorn@zrc-sazu.si

P. Mikša
University of Ljubljana, Faculty of Arts, Department of History, Ljubljana, Slovenia
e-mail: peter.miksa@ff.uni-lj.si

M. Krevs (ed.), *Hidden Geographies*, Key Challenges in Geography,
https://doi.org/10.1007/978-3-030-74590-5_11

divisions or divisions of land, either between states or within them (Waldhäusl et al. 2013). On plans and maps, the border is represented by a drawn line, while on location different markers are used to mark out the demarcation line (Slak et al. 2019). Nowadays, these boundary markers are part of cultural heritage because they have a functional and aesthetic value. Additionally, they serve as a witness to the date of placement and the function of delimitation. Concurrently, they are the sole witness to land survey conducted on site (Kozorog 2008). In the case of changed boundaries, they represent a hidden legacy of past borderscapes.

Boundary markers were often placed in prominent spots, i.e. next to roads, gorges and watercourses; consequently, many boundary markers are not preserved due to the expansion of infrastructure or due to natural processes. The bulk of old boundary markers can thus be found in forests (Kozorog 2008).

Before the introduction of boundary markers, borders were approximate and ran along natural boundaries, such as gorges, forests, ridges, prominent trees or boulders. These were more or less border areas or strips than exact demarcation lines. Boundaries between seigneuries were recorded in urbaria. Borders were not defined in detail, which resulted in many border-related disputes. Somewhat more permanent boundary markers were introduced with geodetic land surveys and their depictions on maps. The area of modern-day Slovenia saw the first exact geodetic measurements taken in the early nineteenth century when a land survey (the so-called Franciscan cadastre) was produced in a scale of 1:2,880 (Kozorog 2008). The administrative division of the Slovene territory was first depicted in more detail on maps of the First Military Survey of the Habsburg Monarchy from the second half of the eighteenth century, in a scale of 1:28,800 (Zorn 2007). Exact boundaries are drawn in the Franciscan cadastre, which is not the case with boundary stones; in the First Military Survey of the Habsburg Monarchy, detailed geodetic measurements were yet to be taken, but boundary markers (such as boundary stones (Figs. 1 and 2) or boundary ditches (Rajšp and Kološa 2000)) are marked in some places.

This chapter deals with boundary stones, disregarding other types of boundary markers. A boundary stone (border stone) "… *is a physical marker that identifies the start of a land boundary or the change in a boundary, especially a change in direction* … [and] *have often been used to mark critical points on boundaries between countries, states or local administrations* …", as well as private land-holdings. Traditionally, they were often made of stone, but later also of concrete or a mixture of materials. "*They are typically placed at a notable or especially visible point. Many are inscribed with relevant information, such as abbreviation of the boundary holder and often a date.*" (Guo 2018: 56).

We focused on boundary stones in the territory of modern-day Slovenia, which in the Habsburg Monarchy, before World War I, divided the Duchy of Styria and the Kingdom of Hungary and, in the interwar period, the Kingdom of Italy and the Kingdom of Yugoslavia. Not only individual boundary stones are preserved but also in certain sections entire series of boundary stones are extant (Figs. 6 and 7). These boundary stones no longer serve their original purpose; in their respective locations, they act as historical witnesses, as markers of the past.

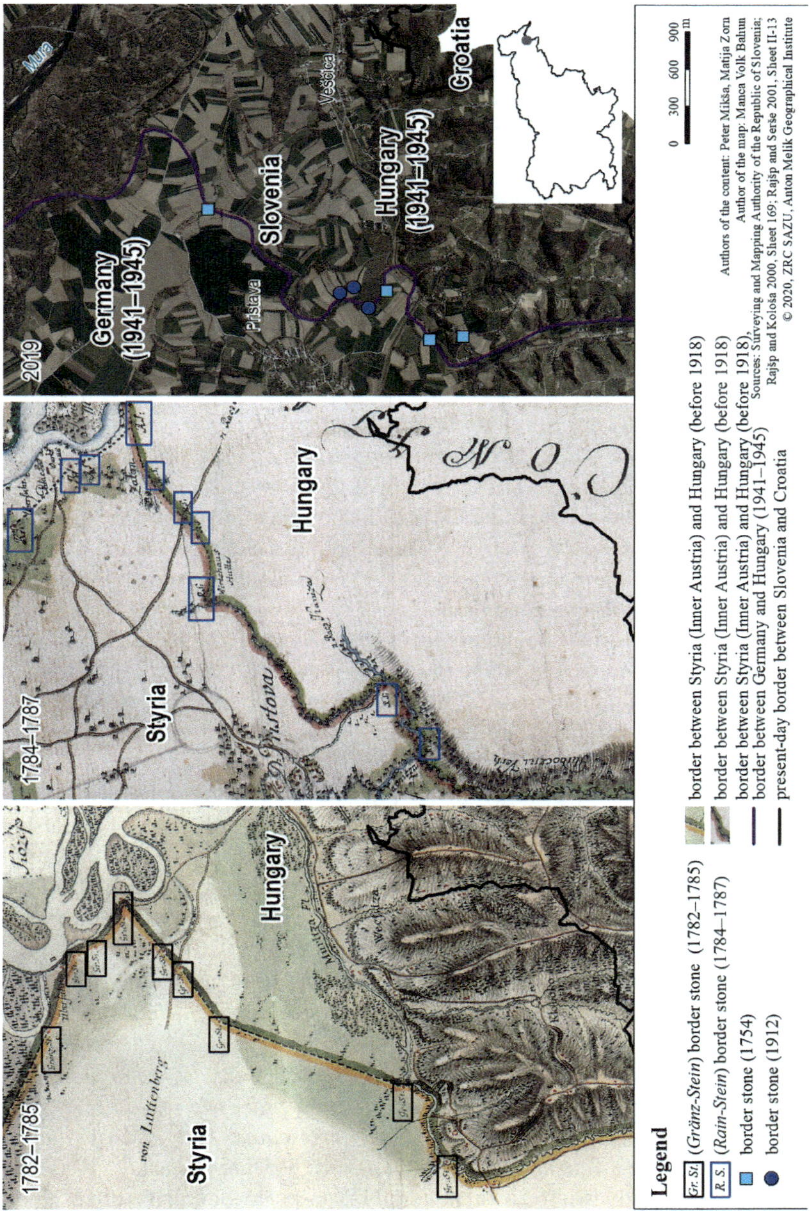
1782–1785
Styria
Hungary
von Luttenberg
1784–1787
Styria
Hungary
2019
Germany (1941–1945)
Slovenia
Hungary (1941–1945)
Croatia
Mura
Pristava
Veščica
Legend
Gr. St. (Gränz-Stein) border stone (1782–1785)
R. S. (Rain-Stein) border stone (1784–1787)
border stone (1754)
border stone (1912)
border between Styria (Inner Austria) and Hungary (before 1918)
border between Styria (Inner Austria) and Hungary (before 1918)
border between Styria (Inner Austria) and Hungary (before 1918), border between Germany and Hungary (1941–1945)
present-day border between Slovenia and Croatia
0 300 600 900 m
Authors of the content: Peter Mikša, Matija Zorn
Author of the map: Manca Volk Bahun
Sources: Surveying and Mapping Authority of the Republic of Slovenia; Rajšp and Kološa 2000, Sheet 169; Rajšp and Serše 2001, Sheet II-13
© 2020, ZRC SAZU, Anton Melik Geographical Institute

◂**Fig. 1** Boundary stones between the Duchy of Styria and the Kingdom of Hungary are recorded on maps of the First Military Survey of the Habsburg Monarchy in the second half of the eighteenth century. The map on the left-hand side was produced during the survey of the Kingdom of Hungary between the years 1782 and 1785 (Rajšp and Serše 2001), and the map in the centre was produced during the survey of Inner Austrian provinces between 1784 and 1787 (Rajšp and Kološa 2000). The orthophotograph on the right-hand side shows the border between Germany and Hungary during World War II, which was based on the demarcation of the Duchy of Styria and the Kingdom of Hungary before World War I, as well as preserved boundary stones. Modern-day border between Slovenia and Croatia is outlined on all three maps

As still standing boundary markers of the past, they are a visual representation of contemporary "phantom borders" (cf. Kolosov 2020) within modern-day Slovenia, i.e. "*political borders that once were, are no more, but–nevertheless–somehow still are*" (Zajc 2019: 298) or with other words "*political borders, which politically or legally do not exist anymore, but seem to appear in different forms and modes of social action and practices today*" (von Löwis 2015: 99). They can be regarded as "*scars of history,*" i.e. the boundaries that do not exist anymore but still have an impact on society (Kolosov 2020: 1). These boundaries and with them boundary stones represent an administrative legacy that is, in this study, mostly hidden in the cognitive perception of presented boundaries.

The concept that former boundaries still hold legacy in the landscape is not new to geography (e.g. von Löwis 2015, 2017; Kolosov 2020), as we can within geography itself find statements that the boundaries are "*history imprinted in space*", and that historical boundaries are important "*in strengthening or building territorial identity, contemporary cultural and political territorial patterns …*" (Kolosov 2020: 2). Former boundaries and their remains are for geography also important as they "*determine the cultural landscape of the borderlands,*" and they also hold significance as cultural heritage (Kolosov 2020: 4). As such they may be attractive to tourism and thus have development potential. There are many examples of monuments, museums, historical paths, etc. along former boundaries as they may have a high symbolic significance (Kolosov 2020; Kumer et al. 2020).

Regarding material remains along boundaries, it is important to note that "*geography at its most basic level promotes awareness of what's around you*" (O'Reilly 2020: 54), and around us are many archaeological/historical remnants that are embedded in borderscapes and hold value especially for cultural geography (cf. Hill 2015). "*Geographical literacy interpreting places in the present, must be cognizant of changes due to long and short time scales. This includes heritage–the cumulus of human inheritances both material and non-material …*" (O'Reilly 2020: 68). In this study, we focus on material heritage, i.e. boundary stones that may seem marginal but may still help us understand some hidden legacies of Slovenian landscapes.

Fig. 2 Boundary stones between the Duchy of Styria and the Kingdom of Hungary near Središče ob Dravi. Left: The map with locations of boundary stones, which was produced during the survey of Inner Austrian provinces between the years 1784 and 1787 (Rajšp and Serše 2001). Right: An orthophotograph with locations of preserved boundary stones. Modern-day border between Slovenia and Croatia follows the course of the former border of the Duchy of Styria and the Kingdom of Hungary in this section

2 Boundary Stones Between the Duchy of Styria and the Kingdom of Hungary

In the Habsburg Monarchy, whose part was the territory of modern-day Slovenia up to the end of World War I, one can locate many boundary stones that identified boundaries of seigneuries, dominical or ecclesiastical estates, as well as administrative and provincial boundaries (Lisec et al. 2020). The border between the Inner Austrian province Styria (Duchy of Styria) and the Kingdom of Hungary, whose south-western part ran in the territory of modern-day Slovenia, is an example of such a boundary. The border was more than 70 km long (Table 1); its northern part ran roughly along the watercourse Kučnica/Kutschnitza, which is nowadays border between Slovenia and Austria. Its central part ran on the river Mura, which is today entirely located in Slovenia, while the border's southern part ran across the low hills of Slovenske Gorice, in the location of the modern-day border between Slovenia and Croatia.

A large portion of the border ran on watercourses. "*Rivers are natural geographical dividers whose shifting courses hamper permanent administrative demarcation, especially in flat areas exposed to frequent flooding and meandering. Thus, rivers as borders may cause problems because they are not static. Moreover, they are very dynamic and tend to change their courses; for example, one country may claim that the 'old' course is the border and the other that the 'new' course is the border*" (Perko et al. 2019: 213). This holds true for the river Mura as well, where the border was fixed in the first half of the thirteenth century (Kos 1969) and on account of which disputes and lawsuits occurred as early as in the first half of the sixteenth century. In the early sixteenth century, a Hungarian nobleman changed Mura's river course; consequently, the river began to erode the Styrian side, wearing away fields and even posing a threat to houses. As a result, three villages disappeared on the Styrian side. The dispute escalated to the extent that both sides positioned armed forces on their riverbank (Zelko 1984, 1996). In an effort to establish peace, the king had to intervene and sent his army to the area. River-related disputes, in some of which blood was spilled, continued later as well. In the seventeenth and eighteenth centuries, Styrians complained repeatedly that Hungarians changed the border arbitrarily (Kovačić 1926; Zelko 1984, 1996; Hozjan 2013). Mura's riverbed changed continuously to the detriment of one or the other side also due to frequent floods. As a result, either the Styrian or the Hungarian side grew in size. In such instances, the land ended up at the hands of a landowner on the opposite bank. This altered border should have brought about a change in land tax as well; however, this was not the case, and Styrian peasants were still obliged to pay tax for land that was due to the changed riverbed situated on the opposite bank (Čuček 2016). As a consequence of Mura's newly formed riverbed during the great flood of 1676, Hungary obtained about 1,500 ha of land at the expense of Styria (Radovanovič 2007).

The border was finally defined between 20 May and 1 December 1755. To mark the border, serfs of seigneuries situated along the river Mura built embankments, and large boundary stones were placed in some places (Kovačić 1926). The oldest

Table 1 Boundary stones between the Duchy of Styria and the Kingdom of Hungary recorded on maps of the First Military Survey of the Habsburg Monarchy from the second half of the eighteenth century (Rajšp and Kološa 2000; Rajšp and Serše 2001) in the area of modern-day Slovenia (*entire border on the river Mura; **excluding the section on the river Mura; – no data). Two maps, the original and the copy, were produced for each section

The approximate area (from the north to the south)	Length (km)	Survey Inner Austria (1784–1787)			Survey Hungary (1782–1785)		
		Section number	Original (number of boundary stones)	Copy (number of boundary stones)	Section number	Original (number of boundary stones)	Copy (number of boundary stones)
The border running predominately on the watercourse Kučnica; modern-day border between Slovenia and Austria	26	143	7	7	I-6	0	0
The border running predominately on the watercourse Kučnica; modern-day border between Slovenia and Austria		144	11	7	I-7	1	0
The border on the river Mura; nowadays in Slovenia	30*	167	6	5	I-8	6	6

(continued)

Table 1 (continued)

The approximate area (from the north to the south)	Length (km)	Survey Inner Austria (1784–1787)			Survey Hungary (1782–1785)		
		Section number	Original (number of boundary stones)	Copy (number of boundary stones)	Section number	Original (number of boundary stones)	Copy (number of boundary stones)
The border on the river Mura; nowadays in Slovenia		168	5	5	II-12	1	1
The northern part: the border on the river Mura; nowadays in Slovenia The central and southern part; the low hills of Slovenske Gorice; modern-day border between Slovenia and Croatia	17.5**	169	12	9	II-13	11	10

(continued)

Table 1 (continued)

The approximate area (from the north to the south)	Length (km)	Survey Inner Austria (1784–1787)			Survey Hungary (1782–1785)		
		Section number	Original (number of boundary stones)	Copy (number of boundary stones)	Section number	Original (number of boundary stones)	Copy (number of boundary stones)
The low hills of Slovenske Gorice; modern-day border between Slovenia and Croatia		198	7	7	II-14	–	7
Total number of boundary stones			48	40		19	24

material proof associated with delineating the border is a boundary stone near the village of Veščica ("*pri Kregarju*", i.e. at Kregar's; Fig. 3a) that was placed during

Fig. 3 Preserved boundary stones between the Duchy of Styria and the Kingdom of Hungary in the proximity of Veščica (**a** and **b**; Fig. 1) and Središče ob Dravi (**c** and **d**; Fig. 2). **a** A boundary stone from 1674; **b** a boundary stone from 1754; **c** a boundary stone from 1912; **d** a boundary stone from 1912 and modern panel fencing (photography by Matija Zorn in January 2020) on the border between the Republic of Slovenia and the Republic of Croatia

the rule of Leopold I of Habsburg. The bulk of preserved boundary stones are from the mid-eighteenth century, from the period of rule of his granddaughter, Maria Theresa (Damjanovič et al. 2016). These boundary stones bear the date 1754 (Fig. 3b). Boundary stones were numbered and placed in a manner that made them optically connected (Pavličič 1995; Krnc 2013; Damjanovič et al. 2016). They were recorded on maps of the First Military Survey of the Habsburg Monarchy in the 1780s (Figs. 1 and 2, Table 1). The bulk of these boundary stones was washed away by the river Mura and its changing riverbed, a few of them were demolished due to tillage (Kovačić 1926), and some of them can be seen on site to this day (Pavličič 1995; Ratiznojnik 2013). When outlining the border, a few kilometres of the riverbed were canalized, which marks the beginning of the first large-scale regulation of the river Mura (Hozjan 2013).

As mentioned above, boundary stones were recorded on maps of the First Military Survey of the Habsburg Monarchy. The border between the Duchy of Styria and the Kingdom of Hungary was mapped twice in the scope of this survey, namely in the military survey of the Kingdom of Hungary between 1782 and 1785, as well as in the scope of the military survey of Inner Austrian provinces between 1784 and 1787. The mapped area was divided into sections. Two maps were produced for each section in both military surveys, the original and the copy (Rajšp and Kološa 2000; Rajšp and Serše 2001) which somewhat differed in details. Consequently, there are four maps available for each part of the border. The part of the border in question was divided into six sections in both surveys (Table 1).

Table 1 shows that there were approximately twice as many boundary stones recorded in maps produced in the scope of the survey in Inner Austrian provinces than in that of the Kingdom of Hungary and that there are differences between originals and copies. In the scope of a survey of Inner Austrian provinces, a boundary stone was recorded approximately every 1.5 km (on the original and just shy of 2 km on the copy) and in the case of that of the Kingdom of Hungary approximately every 3 km (on the copy).

The course of the riverbed differs on maps produced within both surveys as well; it is more exact and detailed on maps produced in the scope of survey of Inner Austrian provinces (Hozjan 2005, 2007). The demarcation line is also outlined differently. Figure 1 shows that the border is drawn quite rectilinearly in the survey of the Kingdom of Hungary, while in the case of that of Inner Austrian provinces, it follows relief features of the stream that represented the border. A few boundary stones bearing the date 1754 (Fig. 3b) are preserved next to the stream. The date indicates the period when the border was marked out meticulously for the first time. A few boundary stones are also marked with the date 1912, signifying the period when the border was also surveyed (Fig. 1).

On the maps of the First Military Survey of the Habsburg Monarchy, boundary stones are marked either on the demarcation line itself (Figs. 1 and 2) or somewhat away from it, particularly in places where the border followed the river Mura. Boundary stones are recorded on the map of Section 167 of the survey of Inner Austrian provinces on both banks of the river, as well as on river islets (Hozjan 2005), while on the map of Section 168, the border is outlined on the left (Hungarian) bank, and the bulk of boundary stones is marked on the right (Styrian) bank (Hozjan 2007). On the complementary map of Section II-12 of the survey of the Kingdom of Hungary, the border is delineated in the centre of Mura's riverbed (Hozjan 2007), and there are considerably fewer boundary stones recorded. Along with boundary stones, there is also a dry border ditch recorded on the map of Section 144 of the survey of Inner Austrian provinces (Rajšp and Kološa 2000; Hozjan 2005).

After World War I, boundary stones between the Duchy of Styria and the Kingdom of Hungary lost their purpose. They were replaced by new boundary stones between the newly established Republic of Austria and the Kingdom of Serbs, Croats and Slovenes (subsequently Kingdom of Yugoslavia) in the northern part, i.e. in the area of the river Kučnica as far as its confluence with the river Mura. At present, these boundary stones identify the border between the Republic of Austria and the Republic of Slovenia.

In the area of the river Mura, boundary stones were no longer needed because both banks became part of the Kingdom of Serbs, Croats and Slovenes (subsequently Kingdom of Yugoslavia) or its administrative unit, the Drava Banovina (Drava Banate; Slovene: *Dravska Banovina*). Although the boundary was gone, the river Mura remained until today an important regional identity divide between *Prekmurci* (literary 'people on the other side of the river Mura') on the left bank of the river, who were part of the former Kingdom of Hungary and *Štajerci* (Styrians) on the right bank of the river, who were part of the Duchy of Styria (Geršič 2020).

In the southern part, i.e. in the area where the border runs across the low hills of Slovenske Gorice, which also became part of the Kingdom of Serbs, Croats and Slovenes (subsequently Kingdom of Yugoslavia), the border was transformed into an administrative border between the Drava Banovina and the Sava Banovina (subsequently Banovina of Croatia), where no boundary stones were placed. Nor were new boundary stones placed in this section during World War II, when this line became the border between Germany and Hungary. After World War II, this line became the border between the Socialist Republic of Slovenia and the Socialist Republic of Croatia within the Federal People's Republic of Yugoslavia (subsequently Socialist Federal Republic of Yugoslavia), which were not delineated by boundary stones. Nowadays, the demarcation line in this section is the state border between the Republic of Slovenia and the Republic of Croatia and is yet to be marked with boundary stones because the border between these two states is yet to be determined (Zadeva … 2019). This section thus only featured boundary stones that delineated the Duchy of Styria and the Kingdom of Hungary and were until recently the only border markers between these two states (Figs. 2 and 3c). On account of the migrant (refugee) crisis (from the second half of 2015 onwards; Klemenčič and Verbič Koprivšek 2017), Slovenia installed barbed wire fences and subsequently, panel fencing (Fig. 3d) along the border, which along with old boundary stones serves as (an unofficial) border marker.

3 Boundary Stones Between the Kingdom of Italy and the Kingdom of Yugoslavia

The border between the Kingdom of Italy and the Kingdom of Serbs, Croats and Slovenes (subsequently Kingdom of Yugoslavia) in the territory of modern-day western Slovenia was established after World War I, after the decline of the Habsburg Monarchy, whose part it had been prior to that. The border was a result of a treaty signed by these two states in 1920 and was to a great extent in line with Italian aspirations for its eastern border that had been stated in the secret Treaty of London, signed by Italy and the Entente in April 1915. The treaty between both countries was signed in the Italian town of Santa Margherita Ligure near Rapallo, wherefore the border is also known as "the Rapallo border" (Cattaruzza 2011, 2017; Mikša and Zorn 2018; Zorn and Mikša 2018).

The new border ran approximately along the watershed of the Black Sea and of the Adriatic. Looking further back, we see that the watershed constituted the border between provinces in the period of the Roman Empire and that here ran the border between the Italian and the remaining part of the empire after the fall of *limes*. It was here that in Late antiquity defence structures were built to hinder other peoples to penetrate northern Italy from the east. The area served as a dividing line also later in history. In the Habsburg Monarchy, it represented the border between Inner Austrian provinces of Carniola and Gorizia (future Austrian Littoral) (Mikša and Zorn 2018; Zorn and Mikša 2018).

By signing the treaty, both kingdoms bound themselves to form a mixed delimitation committee that would outline the demarcation line, survey it and see to it that boundary stones would be placed. The committee commenced work in February 1921, finishing it in late 1926. According to the committee, in the territory of modern-day Slovenia and as far as Rijeka (modern-day Croatia), the demarcation line measured 244.5 km in length; following the Free State of Rijeka's integration into Italy, the border grew in length, totalling almost 264 km (Žorž 2016).

In the years 1920–1925, a mapping survey was produced for demarcation purposes by members of the Italian Military Geographic Institute (Italian: *Istituto Geografico Militare*) in Florence. There were 84 maps in a scale of 1:5,000 produced for the area extending from the Austrian-Italian-Yugoslav tripoint (Slovene: *Tromeja* or *Peč*; German: *Dreiländereck* or *Ofen*; Italian: *Monte Forno*) to Rijeka (Žorž 2016). Along with the demarcation line itself, these maps also include all boundary stones (Fig. 4). In a separate table, each boundary stone's altitude was stated, along with its distance from consecutive boundary stones, and the cumulative distance to the boundary stone located at the tripoint.

The delineation on paper was followed by outlining the border with boundary stones on location. Texts published in Slovene periodicals from the period when the border was marked out are interesting to read. For instance, the Slovene Mountaineering Society warned hikers not to remove boundary markers that were placed by the delimitation committee (Turistom 1922).

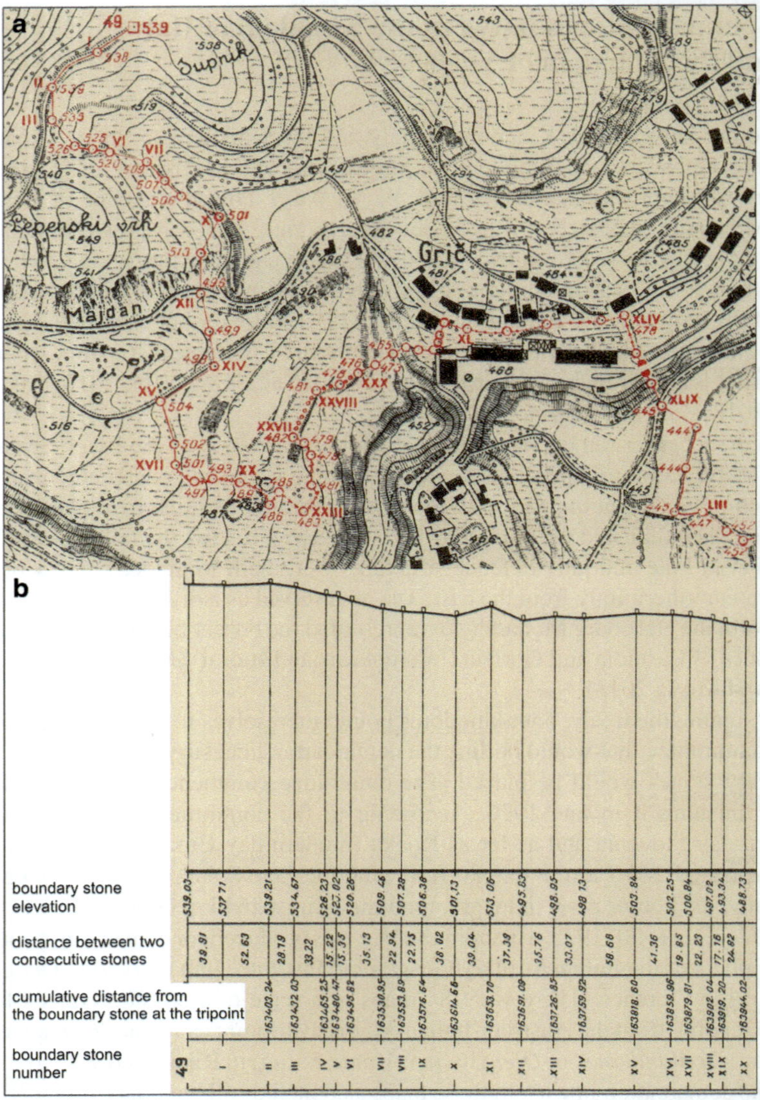

Fig. 4 **a** A detail of the map in a scale of 1:5,000, showing the course of the Rapallo border in the settlement of Planina (map no. 56). The red line indicates the course of the border (the Kingdom of Italy lies to the left (west) of the line, the Kingdom of Serbs, Croats and Slovenes (subsequently Kingdom of Yugoslavia) to the right (east) of it). Boundary stone no. 49, the main boundary stone, is marked with a square with red lines (top left on the map). Located at the international border crossing Planina, on the road from Logatec to Postojna, two special boundary stones are indicated by two solid red circles. Circles bounded with a red curve stand for secondary (intermediate) boundary stones inscribed with Roman numerals. The boundary stones' altitude is written in red Arabic numerals. **b** Data on the boundary stones' locations are an integral part of each map: their altitude, the distance between two consecutive boundary stones and the cumulative distance to the first boundary stone at the tripoint (ZRC SAZU Anton Melik Geographical Institute Archive)

The two delegations decided to use concrete border markers to mark out the demarcation line; they were prism-shaped, made of reinforced concrete and affixed to a rocky foundation (Žorž 2016). They agreed to use four types of boundary stones, whose shape would be suited for their intended use (Fig. 5; Žorž 2016):

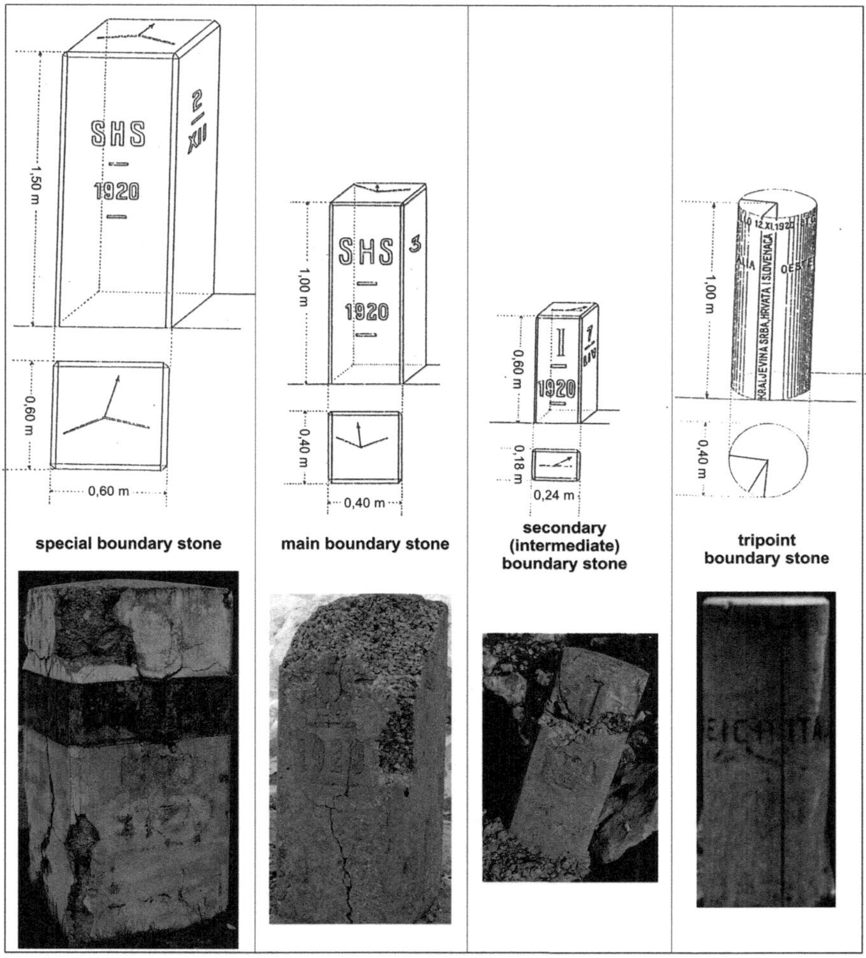

Fig. 5 Boundary stones' dimensions, inscriptions, and markings (top). The front and the rear side featured a code signifying the names of countries and the year when the Treaty of Rapallo was signed. An Arabic numeral denoting the number of the sector was featured on sides, and a Roman numeral indicated its sequence within a sector. Two lines on its top face pointed towards consecutive boundary stones, the arrow towards the north. Three types of preserved boundary stones can be seen on photographs (left to right; photography by Matija Zorn). The original tripoint boundary stone, which is depicted on the extreme right, is not preserved (the photograph dates back to 1939; Nani Poljanec Archive)

- Main boundary stones (Italian: *termini principali* or *cippo principale*) were placed on important, prominent, high-lying spots. Measuring one metre and marked with a consecutive number written in Arabic numerals, these boundary stones were the first border marker in each sector.
- Secondary (Italian: *termini secondari*) or intermediate boundary stones were placed between main boundary stones; their function was to outline the exact course of the demarcation line. They were 60 cm tall and featured the sector's consecutive number and a consecutive Roman numeral within the sector.
- Special boundary stones (Italian: *termini speciali*) were placed in spots where the demarcation line traversed important (road) communications or border crossings. Their markings corresponded to secondary boundary stones; however, they were taller, measuring 1.5 m in height.
- The tripoint boundary stone (Italian: *termine triconfinale*) stood on the summit of Mount Peč (Tromeja); it signifies the tri-border point of the Republic of Austria, the Kingdom of Italy, and the Kingdom of Serbs, Croats and Slovenes (subsequently Kingdom of Yugoslavia).

The border was divided into seventy sectors that were separated by main boundary stones featuring numbers 1–70 written in Arabic numerals (boundary stone no. 70 stood on the contact point with the Free State of Rijeka; there were 4,508 secondary boundary stones standing between main boundary stones) and the date 1920, as well as the letter "I" on the Italian and the abbreviation "SHS" (from 1929 onwards the letter "J") on the Yugoslav side. Additionally, a Roman numeral was added beneath the sector number. Their top face featured a line pointing towards the previous and the next boundary stone, along with a marking pointing northwards. Following the Free State of Rijeka's integration into Italy (1924), the number of main boundary stones increased to 79 and that of secondary boundary stones to 5,098; additionally, 38 special boundary stones were standing (totalling 5,215) (Žorž 2016). Nowadays, the share of preserved boundary stones in respective sectors varies between less than one-tenth and a half (Pečelin 2003). Figure 6 shows the state of preservation of boundary stones located in the Julian Alps, in Sectors 15 and 16, at about 1,800 and 2000 m above sea level. About three-fifths of locations of boundary stones were recorded, with about one-seventh of boundary stones still standing. Figure 7 shows the condition of boundary stones in the Prealps (Cerkno Hills), in Sectors 31 and 32, at an elevation of approximately 1,000 m. About one-eighth of boundary stones' locations were recorded, with less than one-tenth of boundary stones still standing. Their state of (non)preservation is partly associated with their planned and unplanned removal (due to tillage or expansion of traffic routes) after the "fall" of the border, as well as with weathering and other natural processes (Fig. 8; Mikša and Zorn 2018).

Officially (*de iure*), the Rapallo border (and boundary stones with it) existed up to 1947, when the border between Italy and Yugoslavia shifted somewhat westwards after World War II and after the Treaty of Peace with Italy was signed. However, the border's status changed already during World War II. Following the German occupation of parts of Slovenia, its northern part became the state border between Italy and Germany in 1941 (the letter "J" signifying Yugoslavia was replaced by

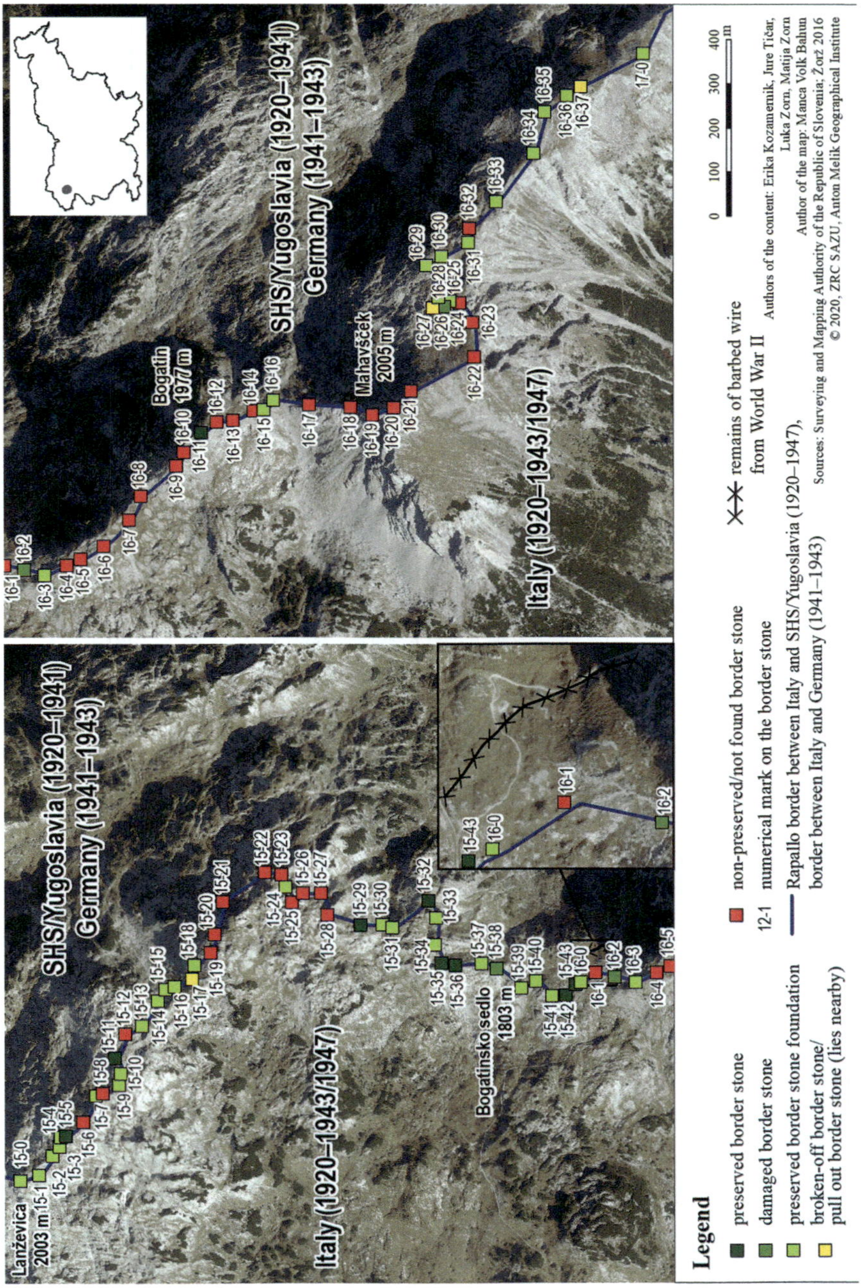

Fig. 6 Preserved boundary stones between the Kingdom of Italy and the Kingdom of Yugoslavia in Sectors 15 and 16 of the Julian Alps. In this area, boundary stones of the Rapallo border served as border markers between Italy and Germany during World War II

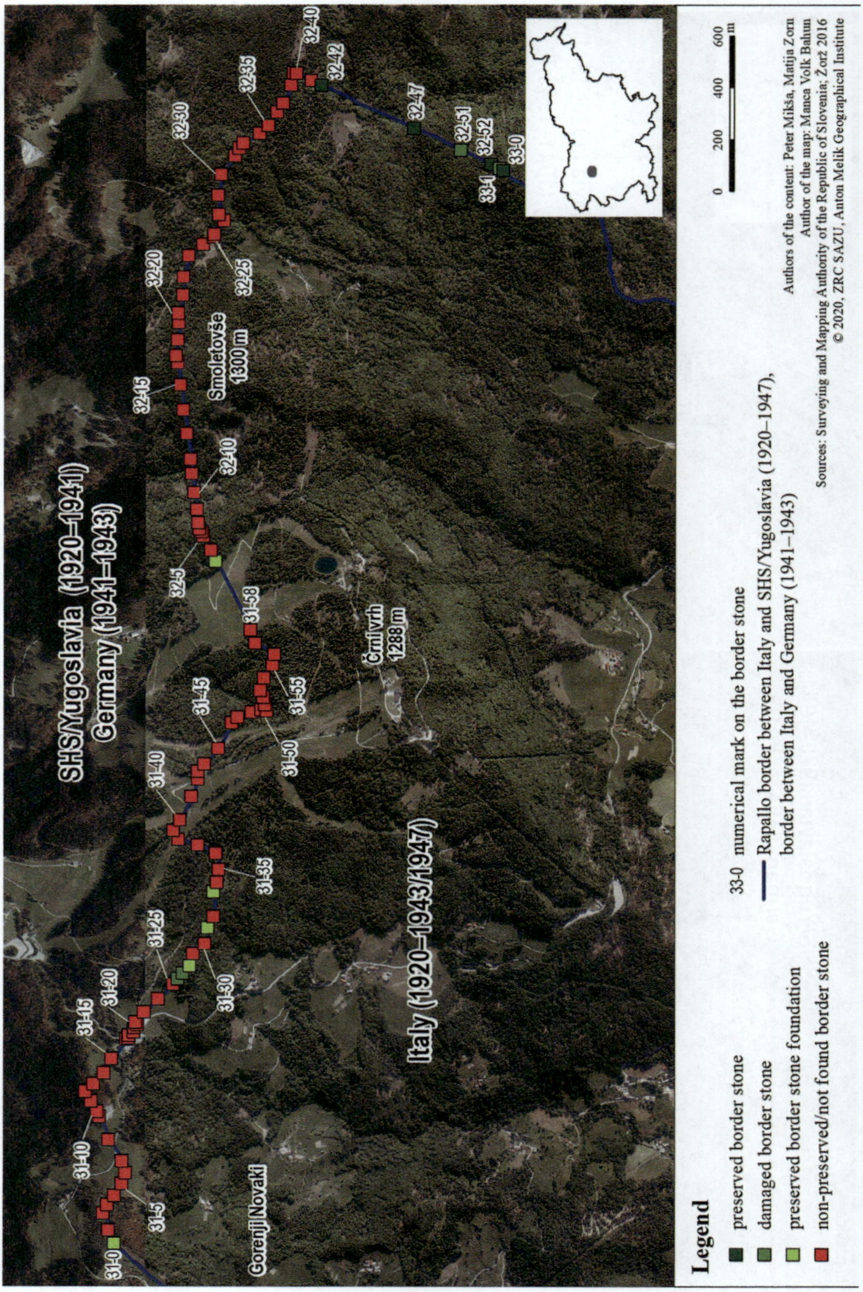

Fig. 7 Preserved boundary stones between the Kingdom of Italy and the Kingdom of Yugoslavia in Sectors 31 and 32 of the Prealps (Cerkno Hills). Boundary stones of the Rapallo border served as border markers between Italy and Germany during World War II in this area as well

Fig. 8 The condition of boundary stones of the Rapallo border, the border between the Kingdom of Italy and the Kingdom of Yugoslavia in Sectors 15 and 16 of the Julian Alps. A few boundary stones were broken; identifiable are solely their foundations (boundary stone no. 15–40), while others were knocked down and lie near their former location (15–17). Impacted by man, the alpine environment, and, first and foremost, intensive weathering, boundary stones fell into ruin (15–34, 15–37, 15–38, 16–01). However, a few boundary stones (15–36) are very well preserved (photography by Matija Zorn)

“D” for Germany; Fig. 8, boundary stone no. 15–36) and its southern part Italy’s internal, provincial border. Officially, this situation remained unchanged even after Italy’s capitulation in 1943, and the Italian part of modern-day Slovene territory came under German administration (the so-called Operational Zone of the Adriatic Littoral). After the end of World War II in 1945 and up to 1947, the territory to the west of the Rapallo border was under the Yugoslav military administration (Troha 2005; Mikša and Zorn 2018; Zorn and Mikša 2018).

Following the official discontinuation of the border, boundary markers were removed in more prominent spots (e.g. former border crossings); in places, this happened even sooner, when the territory to the west of the Rapallo border ended up under the Yugoslav military administration (Žorž 2016; slika 9a). There were even organized special events to remove boundary stones (Naglič 2005). Some boundary stones were removed already during World War II. Partisans removed the boundary stone on Mt. Triglav, Slovenia’s highest mountain and symbol of Sloveneness (Strojin 1980), in the summer of 1944 (Svetek 1985; Fig. 9b, c).

Three aspects can be highlighted in terms of the legacy of the Rapallo border: (1) the Rapallo border’s potential for tourism, (2) the Rapallo border as a creator of regional identity, and (3) its reflection in various administrative divisions. The first point includes numerous preserved boundary stones and other border structures constructed in the interwar period (mostly fortifications created by both sides for defence purposes). Several municipalities situated along the former border boast of thematic trails that raise their visitors’ awareness of the border’s existence; hikes along the Rapallo border are organized. IT technology plays an important role in “legacy tourism” of the Rapallo border (Mikša and Zorn 2018; Kumer et al. 2020); it allows for a meticulous visualization of the Rapallo border, which is freely available online. Based on the digitalization of delimitation maps of the course of the border in a scale of 1:5,000 (Fig. 4), a point and line vector data layer were produced that enable a detailed insight into the course of the Rapallo border in modern-day landscape via the Internet portal “Rapalska meja” (Fig. 10; Zgodovinsko … 2020).

The second aspect of the Rapallo border’s legacy is regional identity. Slovenes tend to define their regional identity based on Inner Austrian provinces from before World War I or, in the case of Carniola, based on its internal division into Upper (Slovene: *Gorenjska*), Lower (Slovene: *Dolenjska*) or Inner Carniola (Slovene: *Notranjska*). Thus, for instance, the Rapallo Border in Inner Carniola turned the former Inner Carniolans in the Idrija region, the Vipava Valley, as well as the Ilirska Bistrica area into *Primorci* or “people from the Littoral”. This implies that their identity originates in the former Italian-held territory (Mikša and Zorn 2018; Zorn and Mikša 2018; Geršič 2020).

The legacy of the Rapallo border is seen in various administrative divisions as well. Its northernmost part, located between the Upper Sava Valley (Slovene: *Zgornjesavska dolina*) and the Canale Valley (Italian: *Val Canale*), is still used as the border between Italy and Slovenia (Fig. 11). The border also continues to be used as the border between the Ljubljana and Koper dioceses (especially to the north of Žiri), as the border between the Northern Littoral (Slovene: *Severna Primorska*) and Upper Carniolan municipalities, and more to the south some of its sections continue to be

Fig. 9 Removal of boundary stones of the Rapallo border. **a** an extracted boundary stone at the Podlanišče border crossing in the summer of 1945 (Štefan Rutar Archive); **b** the boundary stone at the summit of Mt. Triglav removed by partisans in the summer of 1944 (Slovenian Alpine Museum Archive); **c** a memorial plaque was installed in the location of the former boundary stone in 2018 (bottom left) to commemorate the removal of the boundary stone in 1944 (photography by Matija Zorn)

used as the border between the municipalities of Logatec and Postojna, Cerknica and Postojna, Cerknica and Pivka, as well as Pivka and Loška Dolina. Additionally, between 1945 and 1947, the Rapallo border served as the demarcation line between Yugoslavia and Zone B of the Julian March (Italian: *Venezia Giulia*) and in the 1950s, as the border between the district of Nova Gorica and that of Kranj. It is still used as the border between the Gorizia (Slovene: *goriška statistična regija*) and Upper Carniolan (Slovene: *gorenjska statistična regija*) statistical regions, and in Slovenia's extreme south (to the southwest of Babno Polje), several kilometres are used as modern-day border between Slovenia and Croatia (Jarc 2002; Mikša and Zorn 2018; Zorn and Mikša 2018). In the area where the Rapallo border still serves as the border between Italy and Slovenia, boundary stones stand in the same locations

Fig. 10 The Rapallo border on the Internet portal "Rapalska meja" (Zgodovinsko … 2020). The figure shows the same detail of the border as Fig. 4. The figure is adapted—the blue line indicates the course of the border and dark green squares boundary stones featuring numbers written in Roman numerals, matching those on actual boundary stones. The number written in Arabic numerals marks the first boundary stone in Sector 49. By means of boundary stones' digitalized locations, one can look for them on site and see if they are still preserved

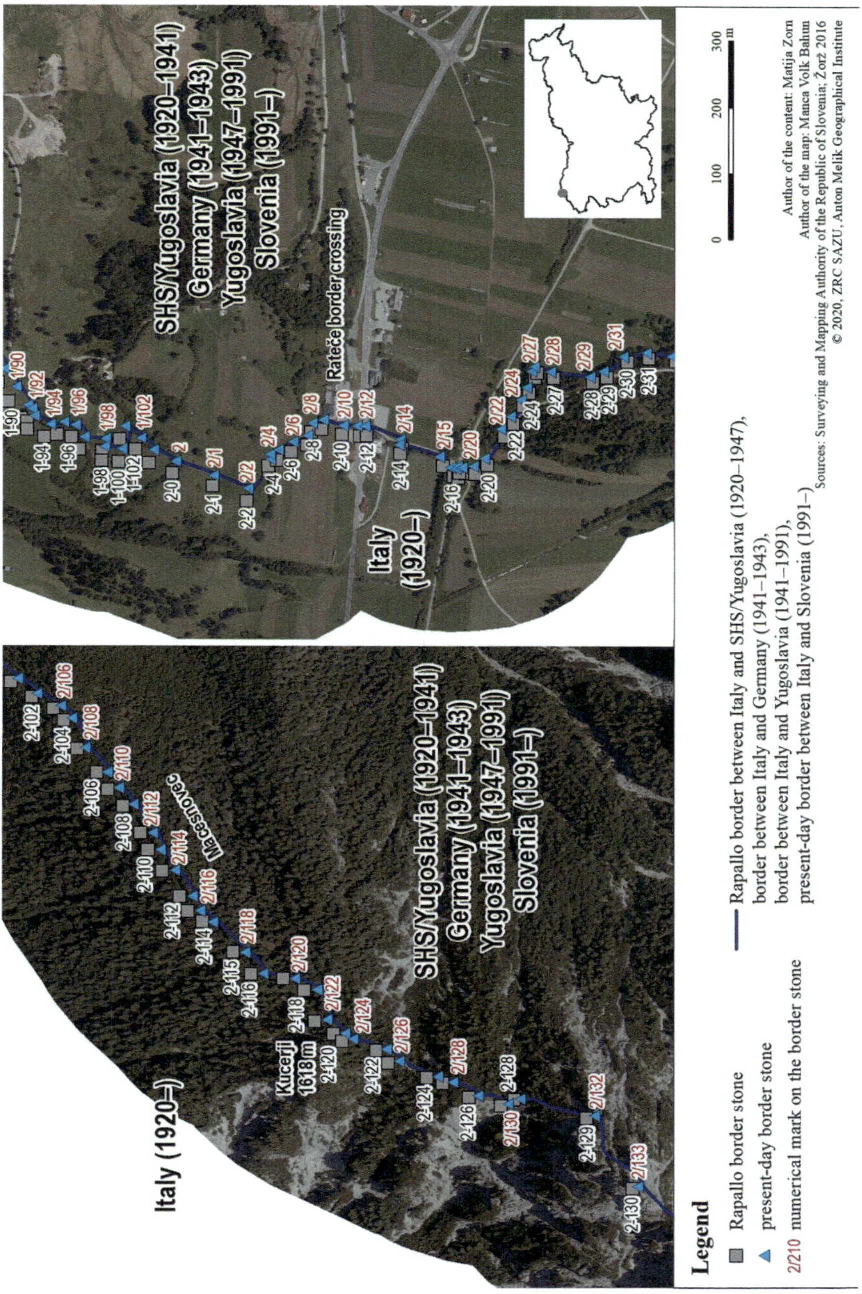

Fig. 11 Matching locations of boundary stones of the Rapallo border between the Kingdom of Italy and the Kingdom of Yugoslavia with those of contemporary boundary stones between Italy and Slovenia (near Rateče, in the proximity of the modern-day Italian–Slovene–Austrian tripoint)

as boundary stones of the Rapallo border (Fig. 11). However, these contemporary boundary stones bear a different date, i.e. 1947 or the date when the Treaty of Peace with Italy was signed after World War II, rather than the year 1920 that signifies the date when the Treaty of Rapallo was signed.

4 Conclusions

Borders transcend mere spatial delimitations; they are associated with legal, cultural, historical, social, economic and other entities (Waldhäusl et al. 2013). Boundary stones are generally considered to be of high proof value in international border disputes (Zadeva … 2019). A state's border must be clearly defined or measured in a manner that allows for its re-establishment at any given moment if boundary markers are either destroyed or moved. The position of boundary stones must be defined meticulously with coordinates in an agreed coordinate system (Mlakar 1993).

Old boundary markers that no longer serve their purpose are subject to natural processes (e.g. floods, weathering, hillslope processes, vegetation growth) and anthropogenic activities (intentional or unintentional removal) (Waldhäusl 2019). A mere 24 boundary stones are recorded in Slovenia's Register of Immovable Cultural Heritage, 3 of which are associated with the border between the Duchy of Styria and the Kingdom of Hungary (Fig. 3a, b) and none of which is associated with the Rapallo border (Register … 2020). Their non-protection contributes significantly to their difficult preservation. Despite their non-existing protection, a few local associations recognized the boundary stones' importance for the preservation of historical memory and its tourist potential; a few boundary stones were even granted the status of a cultural monument of local importance (Lisec et al. 2020). This is particularly the case with boundary stones on the Rapallo border, where many boundary stones were restored (Kozorog 2008; Slak et al. 2019; Škodič 2020).

The fact that, inter alia, boundary markers were nominated for inscription on the UNESCO World Heritage List speaks in favour of the growing awareness that boundary stones constitute important cultural heritage (Waldhäusl et al. 2014; Waldhäusl 2017; Lisec et al. 2020). As such they are interesting as intangible cultural heritage in terms of tradition of regulating rights in space and as tangible cultural heritage as archaeological remnants.

Boundary stones presented here no longer serve their original purpose; however, in their respective locations, they are historical witnesses as markers of the past. Today they act as a visual representation of "phantom borders" reflected in their cognitive perception. In both case studies, this is mostly reflected through regional identity—"*there is a dialectic interdependence between boundaries and identity*" (Kolosov 2020: 5). In the first case study, the river Mura still represents an important regional identity boundary between people on both sides of the river (Geršič 2020), although the boundary was gone more than a century ago. Similarly, the boundary presented in the second case study is today also strongly reflected in the regional identity (Geršič 2020), although the boundary was gone more than seven decades ago. These "phantom borders" may also be regarded as "hidden geographies" in past borderscapes (Kumar Rajaram and Grundy-Warr 2007), with boundary stones being their physical remnants.

Acknowledgements The work was supported by the Slovenian Research Agency through research programs "The Slovene History" (P6-0235) and "The Geography of Slovenia" (P6-0101) and research projects "Make This Land German... Italian... Hungarian... Croatian! The Role of the Occupation Border in the Denationalization Policy and the Lives of the Slovene Population" (J6-8248) and "Creating, Maintaining, Reusing: Border Commissions as a Key to Understanding Contemporary Borders" (J6-2574).

References

Cattaruzza, M. 2011. The making and remaking of a boundary—The redrafting of the eastern border of Italy after the two world wars. *Journal of Modern European History* 9 (1): 66–86.

Cattaruzza, M. 2017. *Italy and its eastern border, 1866–2016*. New York, London: Abingdon, Routledge.

Čuček, F. 2016. K problematiki štajersko-hrvaške dravske meje konec 18. stoletja [On the issue of the Styrian-Croatian border on the river Drava at the end of the 18th century]. *Prispevki za novejšo zgodovino* 56 (2): 116–128.

Damjanović, D., N. Kiš, M. Krivec, A. Magdič, E. Sapač, and S. Vešligaj. 2016. Mura in njena preteklost [River Mura and its past]. *Proteus* 78 (6–8): 266–273.

Geršič, M. 2020. *Pokrajinska imena kot dejavnik identitete* [The names of regions as an identity factor]. Geografija Slovenije 36. Ljubljana: Založba ZRC.

Guo, R. 2018. *Cross-border resource management*. Amsterdam: Elsevier.

Hill, L.J. 2015. Human geography and archaeology: Strange bedfellows? *Progress in Human Geography* 39 (4): 412–431. https://doi.org/10.1177/0309132514521482.

Hozjan, A. 2005. Gornje Prekmurje in del Porabja v jožefinski vojaški izmeri 1784 [The Upper Prekmurje and a part of the Porabje Region on the Josephinian Military Land Survey 1784]. *Zbornik soboškega muzeja* 8: 147–170.

Hozjan, A. 2007. Dolnje Prekmurje v jožefinski vojaški izmeri 1784 in pomen vira za raziskave tedanjega prostora med Rabo in Muro [The Lower Prekmurje on the Josephinian Military Land Survey 1784 and the source's importance for the area between the rivers Raba and Mura at the time]. *Zbornik soboškega muzeja* 9–10: 7–29.

Hozjan, A. 2013. Reka Mura na slovenskem v novem veku [The River Mura in Slovenia in Modern Times]. *Ekonomska i ekohistorija* 9: 16–27.

Jarc, R. 2002. *Rapalska meja: primer reliktne meje in njeni vplivi na družbeno prostorsko strukturo Zahodne Slovenije* [The Rapallo order: An example of a relict border and its impact on the social and spatial structure of Western Slovenia], B.A. Thesis. Ljubljana: Filozofska fakulteta.

Klemenčič, M., and M. Verbič Koprivšek. 2017. Response of parliamentary parties in the Republic of Slovenia to the mass arrivals of migrants. *Razprave in gradivo* 79: 145–163.

Kolosov, V. 2020. Phantom borders: The role in territorial identity and the impact on society. *Belgeo* 21 (2): 1–17. https://doi.org/10.4000/belgeo.38812.

Kos, M. 1969. Meja proti Ogrski in Hrvatski v štajerskem Podravju [The Border towards Hungary and Croatia in the Styria's Drava Region]. In *Poetovio – Ptuj 69-1969: zbornik razprav ob tisočdevetstoletnici*, ed. V. Bračič, 83–91. Maribor: Obzorja.

Kovačić, F. 1926. *Ljutomer: zgodovina trga in sreza* [Ljutomer: A history of the market town and of the Srez]. Maribor: Zgodovinsko društvo.

Kozorog, E. 2008. Mejna znamenja na Severnem Primorskem [Boundary markers in the Northern Primorska Region]. *Goriški letnik* 32: 77–90.

Krnc, J. 2013. Veržejske podobe: Umetnostna, arhitekturna in zgodovinska dediščina [Images of the Municipality of Veržej: Artistic, architectural and historical heritage]. In *Občina Veržej*, ed. F. Just, 49–51. Murska Sobota: Franc-Franc.

Kumar Rajaram, P., and C. Grundy-Warr (eds.). 2007. *Borderscapes: Hidden geographies and politics at territory's edge*. Minneapolis: University of Minnesota Press.

Kumer, P., M. Zorn, G. Žorž, and P. Gašperič. 2020. From the Great War to interwar fortifications: Changing narratives attached to the military landscape in Western Slovenia. In *Places of memory and legacies in an age of insecurities and globalization*, ed. G. O'Reilly, 247–263. Cham: Springer. https://doi.org/10.1007/978-3-030-60982-5_11.

Lee, J.W.I., and M. North. 2016. Introduction. In *Globalizing borderland studies in Europe and North America*, ed. J.W.I. Lee and M. North, 1–13. Lincoln: University of Nebraska Press.

Lisec, A., J. Dajnko, E. Flogie Dolinar, and M. Čeh. 2020. Mreža meja in mejnikov: nominacija za UNESCOvo svetovno dediščino [A network of boundaries and its monuments: UNESCO World Heritage nomination]. *Geodetski vestnik* 64 (3): 403–415.

Mikša, P., and M. Zorn. 2018. Rapalska meja: četrt stoletja obstoja in stoletje dediščine [The Rapallo border: a quarter century of existence and a century of heritage]. In *Nečakov zbornik: procesi, teme in dogodki iz 19. in 20. stoletja*, ed. K. Ajlec, B. Balkovec, and B. Repe, 605–641. Historia 25. Ljubljana: Znanstvena založba Filozofske fakultete.

Mlakar, G. 1993. Geodezija in planinstvo 2 [Geodesy and mountaineering 2]. *Planinski vestnik* 93 (11): 486–489.

Naglič, M. 2005. *Dediščina rapalske meje – Rupnikova linija in Alpski zid, življenje ob rapalski meji v letih 1918–43(47): poskus utemeljitve nove muzejske zbirke* [The legacy of the Rapallo Border—Rupnik Line and the Alpine Wall, life along the Rapallo border in the years 1918–43(47): Attempting to substantiate the new museum collection]. Žiri: Pegaz International.

O'Reilly, G. 2020. Geography: Place and memory nexus. In *Places of memory and legacies in an age of insecurities and globalization*, ed. G. O'Reilly, 43–69. Cham: Springer. https://doi.org/10.1007/978-3-030-60982-5_3.

Pavličič, S. 1995. Stari mejniki [Old boundary markers]. *Gea* 5 (8): 54–55.

Pečelin, V. 2003. *Življenje na Žirovskem v času rapalske meje: Žiri kot pomemben obmejni kraj* [Life in the area of Žiri in the period of the Rapallo border: Žiri as an important border town], B.A. Thesis. Ljubljana: Filozofska fakulteta.

Perko, D., M. Zorn, R. Ciglič, and M. Breg Valjavec. 2019. Changing river courses and border determination challenges: The case of the Slovenian–Croatian Border. In *Geospatial challenges in the 21st century*, ed. K. Koutsopoulos, R. de Miguel González, and K. Donert, 213–230. Cham: Springer. https://doi.org/10.1007/978-3-030-04750-4_11.

Radovanovič, A. 2007. Štajerski del Pomurja v 16. in 17. stoletju [The Styrian part of the Mura Region in the 16th and 17th century]. *Zbornik soboškega muzeja* 9–10: 31–55.

Rajšp, V., and V. Kološa (eds.). 2000. *Slovenija na vojaškem zemljevidu 1763–1787, Karte, 6. zvezek = Josephinische Landesaufnahme 1763–1787 für das Gebiet der Republik Slowenien, Karten, 6. Band*. Ljubljana: ZRC SAZU, Arhiv Republike Slovenije.

Rajšp, V., and A. Serše (eds.). 2001. *Slovenija na vojaškem zemljevidu 1763–1787, Karte, 7. zvezek = Josephinische Landesaufnahme 1763–1787 für das Gebiet der Republik Slowenien, Karten, 7. Band*. Ljubljana: ZRC SAZU, Arhiv Republike Slovenije.

Ratiznojnik, A. 2013. Na valovih Mure in zgodovine: zgodovinski razvoj občine Veržej [On the waves of the River Mura and history: Historical development of the Municipality of Veržej]. In *Občina Veržej*, ed. F. Just, 15–24. Murska Sobota: Franc-Franc.

Register kulturne dediščine [Cultural Heritage Register]. 2020. https://gisportal.gov.si/portal/. Accessed 20 Aug 2020.

Slak, J., J. Triglav, and D. Boldin. 2019. *Dediščina katastrov na Slovenskem: Digitalni arhiv zemljiškega katastra, katastra stavb in državnih prostorskih načrtov = The cadastral heritage*

of Slovenia: The digital archive of land cadastre, building cadastre and national spatial plans. Ljubljana: Surveying and Mapping Authority of the Republic of Slovenia.
Strojin, T. 1980. Triglav - simbol slovenstva in gorništva [Triglav—A symbol of Sloveneness and mountaineering]. In *Triglav, gora naših gora*, ed. T. Strojin, 5–15. Maribor: Obzorja.
Svetek, L. 1985. Na Triglavu med narodnoosvobodilnim bojem [On Triglav during the national liberation struggle]. *Planinski vestnik* 85 (9): 394–398.
Škodič, D. 2020. Stoletnica rapalske pogodbe: Meja v naših gorah [A centenary of the Rapallo border: The border in Slovene nountains]. *Planinski vestnik* 120 (5): 4–18.
Troha, N. 2005. Urejanje mejnih vprašanj [Addressing border-related questions]. In *Slovenska novejša zgodovina: Od programa Zedinjene Slovenije do mednarodnega priznanja Republike Slovenije 1848–1992, 2*, ed. J. Fischer, 908–915. Ljubljana: Mladinska knjiga.
Turistom [For tourists]. 1922. *Slovenec* 50 (174): 3.
Waldhäusl, P. 2017. The Network of boundaries and its monuments for World Heritage. In *Border areas—Encounter areas: Neighbourhood conflicts and neighbourhood co-operations in Europe*, 38–44. Berlin: ICOMOS.
Waldhäusl, P. 2019. Die Grenzen der Vermessung. In *20. Internationale Geodätische Woche Obergurgl 2019*, ed. K. Hanke, and T. Weinold, 222–231. Berlin: Wichmann Verlag.
Waldhäusl, P., A. Günther, K. Hanke, M. Hiermanseder, H. König, R. Mansberger, G. Navratil, G. Schennach, and C. Twaroch. 2013. Sind das landsweite Netzwerk von Grenzen und die Grenzsteine ein Weltkulturerbe? *DGPF Tagungsband* 22: 534–542.
Waldhäusl, P., C. Twaroch, G. Navratil, R. Mansberger, H. König, M. Hiermanseder, K. Hanke, G. Schennach, and A. Günther. 2014. Der Grenzstein als Symbol für das Grundeigentum soll UNESCO Welterbe werden. *Österreichische Zeitschrift für Vermessung und Geoinformation* 102 (1): 11–23.
von Löwis, S. 2015. Phantom borders in the political geography of east central Europe: An introduction. *Erdkunde* 69 (2): 99–106. https://doi.org/10.3112/erdkunde.2015.02.01.
von Löwis, S. 2017. Phantom borders and ambivalent spaces of identification in Ukraine. *L'Espace Géographique* 46 (2): 126–142.
Zadeva PCA št. 2012-04 v zvezi z arbitražo po arbitražnem sporazumu med Vlado Republike Hrvaške in Vlado Republike Slovenije, podpisanem 4. novembra 2009 [PCA Case No. 2012-04 in the matter of an arbitration under the arbitration agreement between the Government of the Republic of Croatia and the government of the Republic of Slovenia, signed on 4 November 2009]. 2019. *Uradni list Republike Slovenije* 18.
Zajc, M. 2019. Contemporary borders as 'phantom borders': An introduction. *Südosteuropa* 67 (3): 297–303. https://doi.org/10.1515/soeu-2019-0023.
Zelko, I. 1984. Stoletne razprtije zaradi reke Mure med Avstrijo in Madžarsko [Century-long disputes between Austria and Hungary on account of the River Mura]. *Kronika* 32 (1): 24–27.
Zelko, I. 1996. *Zgodovina Prekmurja* [A history of Prekmurje]. Murska Sobota: Pomurska založba.
Zgodovinsko društvo rapalska meja [Rapallo Border Historical Society]. 2020. https://www.rapalskameja.si/zemljevid/. Accessed 20 August 2020.
Zorn, M. 2007. Jožefinski vojaški zemljevid kot geografski vir [The Josephinian Military Land Survey as a geographical source]. *Geografski vestnik* 7 (2): 129–140.
Zorn, M., and P. Mikša. 2018. The Rapallo border between Italy and Yugoslavia after the First World War. *Histoire des Alpes* 23: 165–181.
Žorž, G. 2016. *Varovanje rapalske meje in vojaška navzočnost na območju XI. Armadnega zbora* [Protection of the Rapallo Border and Military Presence in the Area of XI Army Corps], M.A. Thesis. Ljubljana: Filozofska fakulteta.

The Monastic Carthusian Space

Michele Stoppa

Abstract The Carthusian monks spend their extraordinary existence in an atmosphere of great contemplative silence and total hiding, protected by an extremely rigid form of enclosure, so their monasteries are also called simply deserts. The Carthusian space constantly proposes a functional organization of the environments strictly dependent on the monks' peculiar kind of life. This includes spaces reserved for fathers, spaces dedicated to the times of community life and, finally, spaces reserved for brothers. This is specified in the Statutes of their Order, although from time to time it has been harmoniously adapted in creative terms to the morphology of the different sites. Undoubtedly the Carthusian monasteries constitute a paradigmatic example of hidden geographies. They can however be discovered to some extent through an accurate reconnaissance of the precious testimonies—more or less well preserved and valorized from a cultural point of view—that these monks have left in the places where they have lived in the past. Such testimonies can also be found in the territories of the Alpine-Adriatic region, i.e., within the current borders of Slovenia and Austria.

Keywords Monastic geography · Charterhouses · Alpine-adriatic region · Slovenia · Austria

1 Origin and Specificity of the Carthusian Order: The Peculiar Kind of Life

In Italy, in the last 20 years organized researches on the *Geography of Religion* have been developed in the frame of a Working Group of the Association of Italian Geographers (A.Ge.I) led by Graziella Galliano, University of Genoa. The results have been published in the journal Geotema (Galliano 2002, 2003). Nowadays these studies are currently continuing by the WG Geografie del Sacro, led by Gianfranco

M. Stoppa (✉)
Associate professor, Geography, Department of Mathematics and Geosciences, University of Trieste, Trieste, Italy
e-mail: mstoppa@units.it

M. Krevs (ed.), *Hidden Geographies*, Key Challenges in Geography,
https://doi.org/10.1007/978-3-030-74590-5_12

Battisti, University of Trieste. In this context a field of study dedicated to monastic geography is also promoted, with particular attention to the comparative analysis of the imprint made by the different expressions of Christian monasticism in the cultural landscape of Europe.

This contribution reports the first results of a prolonged research effort dedicated to Carthusian testimonies. For reasons of space it considers only those existing in the territories of the ancient Carthusian province of *Alemania superior* today included within the borders of Slovenia and Austria.

The results set out below derive from a fruitful intertwining that involved on the one hand an analysis of the Carthusian sources and on the other hand a prolonged field work. Beginning in 2008, it involved also north-eastern Italy, in addition to Slovenia and Austria.

Unlike the cultural inheritances left by other religious orders, the Carthusians did not develop a rigidly repetitive landscape architecture. With evident flexibility they preferred harmonizing each time the physiographic characteristics of the territories chosen for the new foundations with the austere demands of their kind of life.

The Carthusians are undoubtedly the most austere monastic order of the Catholic Church of the Latin rite (Padri certosini 1997, 2000; Kartuzijani in Kartuzija Pleterje; L'Ordine Certosino; San Bruno e i certosini; Thir 2018; Zadnikar 1995). Their kind of life was initially undertaken by Saint Bruno of Cologne (1030–1101) in 1084, when, with a small group of friends—the so-called seven stars of the Carthusian order, of which five priests and two lay faithful—he settled in an estate donated by the holy bishop Hugh of Grenoble. Located in a remote and inaccessible valley of the Chartreuse Pre-Alps, it began the original nucleus of the order's parent company.

Called to the papal court in Italy by Blessed Urban II, San Bruno soon obtained permission from the Pope to resume his solitary life, thus giving rise in 1090 to a second foundation in Serra San Bruno in Calabria, where he died (Moleres & Tassone 2009). After a short time (1197) this foundation passed to a Cistercian obedience to return to the Carthusian obedience only in 1514 (Hogg 2004).

San Bruno, however, did not write a rule of life; the Consuetudines Cartusiae—which represent the first draft of the Statutes of the order—were in fact codified only later by Guigo, fifth prior of the Grande Chartreuse, in response to the pressing requests by the new foundations which marked the expansion of the order all over our continent.

The new foundations initially flourished in France, gradually expanding into the rest of central and northern Europe starting from Switzerland (1146) and Slovenia (1160), but also entered the Iberian Peninsula (Hogg 2004). The maximum development of the Order took place in the fourteenth century. The subsequent historical events determined a significant contraction starting from the Protestant Reformation and later, starting from the age of enlightenment, owing to the suppressions imposed by the civil authorities of various countries (Hogg 2004; Fig. 2).

Currently the Carthusians are still present in Europe, with male monasteries in France, Germany, Great Britain, Italy, Portugal, Slovenia, Spain and Switzerland, as well as four recent foundations in North America (United States), South America

(Brazil and Argentina) and Asia (South Korea). There are also five female monasteries, four in Europe (France, Italy and Spain) and a new foundation in Asia (South Korea) (L'ordine certosino).

It should be noted that the Carthusian nuns led a religious life of a more distinctly cenobitic character, which are reflected in the architecture and organization of the spaces of the respective monasteries. An experiment to meet the desire of the Carthusian nuns, also aimed at adapting the female branch to the more austere kind of life led by the male one, was successfully undertaken in the Charterhouse of Vedana in Veneto at the end of the last century.

The Carthusians are essentially solitary (hermits) who, moreover, dissolve the harshness of hermitage by wisely harmonizing it with appropriate moments of cenobitic life, constituting the so-called Carthusian family, composed of fathers and brothers (converts and donated). The former, consecrated in the order of presbiters, leads a life predominantly devoted to prayer, *lectio divina* and meditation while the latter shows greater commitment to the work-related activities necessary for the existence of the community. The complementary kind of life led by fathers and brothers, respectively, is reflected on the characteristics of the spaces where these men lead their extraordinary existence, as will be illustrated in the following paragraphs.

2 The Carthusian Order and Its Spread in the Eastern Alpine-Adriatic Region

The territory of the eastern Alpine-Adriatic region was divided between different Carthusian provinces (see Fig. 1). The Venetian monasteries belonged, for example, to the Carthusian Province Tusciae and their foundation is more recent than in the German-speaking foundations (fourteenth century for the Charterhouse of Montello, fifteenth for all the others, see Fig. 2).

Looking at the present-day Veneto region, we see that the Carthusians established their monasteries in highly different environments, including a karstic hill (Montello), a lagoon (Venezia, where the Carthusians settled in an earlier Augustinian monastery and adapted it to their way of life), an urbanized plain (Padova), an Alpine foothill (Vedana) and a lowland one (Vigodarzere).

In this contribution, however, we privilege the study of the monasteries of the Carthusian Province *Alemaniae superioris*, currently included in the territories of Slovenia and Austria (see Fig. 2). The oldest foundation—dating back to the 12th century—is Seitz's (Žiče; see Cartusialover) near Celje in Štajerska, a remarkable cultural center and home to the oldest pharmacy in Slovenia. The Charterhouse of Seitz, among other things, gave rise to that of Mauerbach (Cartusialover; Huber; Knall-Brskovsky 1999) which, in turn, originated the Carthusian monasteries Gaming (Cartusialover; Hermann & Spandl 2007), Schnals—afterward incardinated in the Carthusian province Alemaniae inferioris—and Aggsbach (Cartusialover; Thir 2018).

Fig. 1 Eastern Alpine-Adriatic region: Carthusian provinces and their monasteries (borders adapted from Hogg 2004)

Excluding the monastery of Pleterje (Cartusialover; Kartuzijani in Kartuzija Pleterje; Zadnikar 1995), it can be said that the foundations located in the Slovenian territory today are older, dating back to the twelfth and thirteenth centuries, than those existing in Austrian territory, established in the fourteenth century. The

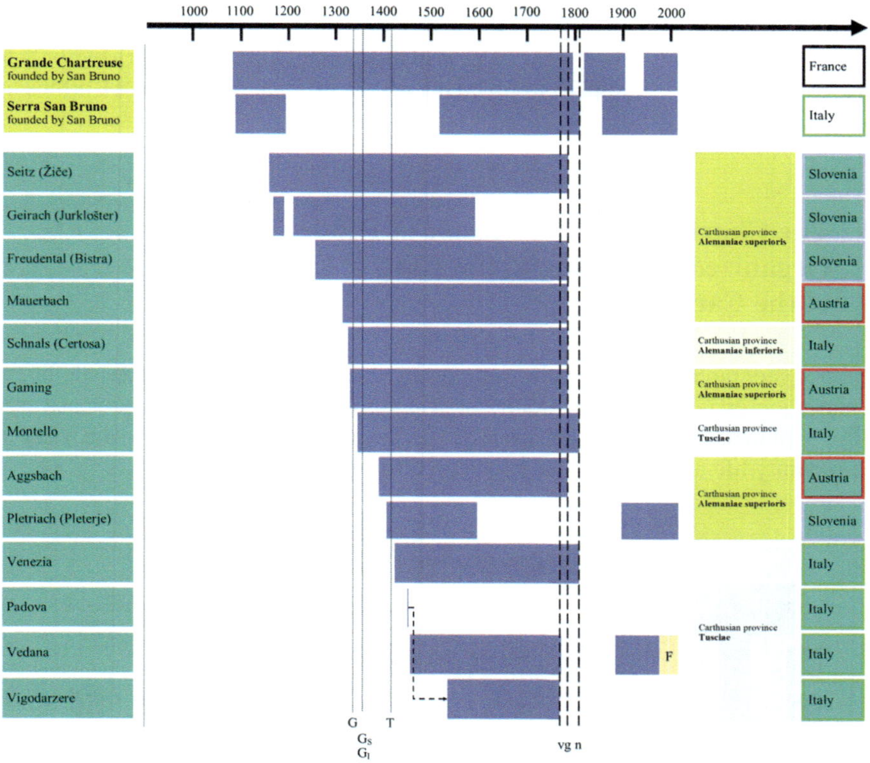

Legend

F = nuns (female branch of the Carthusian order)

G = erection of the Carthusian province **Alemaniae**
G_I = erection of the Carthusian province **Alemaniae inferioris**
G_S = erection of the Carthusian province **Alemaniae superioris**
T = erection of the Carthusian province **Tusciae**

v = suppression by the Serenissima Republic
g = Joseph II of Absburg suppression

Fig. 2 Foundation, persistence and suppression of the Charterhouses of the eastern Alpine-Adriatic region

suppressions (1782) were all imposed by Emperor Joseph II of Habsburg in all the territories of the Austro-Hungarian Empire.

The Charterhouse of Pleterje[1] has a unique history: it can be considered an example of a regenerated Charterhouse, as repopulated and, in fact, refounded, after having

[1] Pleterje is located in Dolenjska at the foot of the Gorjanci mountains near the Slovenian-Croatian political border.

been previously suppressed, with French monks forced to abandon their Carthusian monasteries. The only one—in the context of the Alpine-Adriatic-Eastern region—to preserve even today, despite its troubled history, its original function as a Carthusian hermitage.

3 The Charterhouses: A Mosaic of Hidden Spaces Organized According to the Kind of Life of the Carthusians

From a geographic point of view, we can say that the peculiar characteristic of the Charterhouses is without any doubt the tripartite division of the settled space (see Fig. 3). In fact, there are spaces reserved for fathers, spaces dedicated to the times of community life and, finally, spaces reserved for brothers.

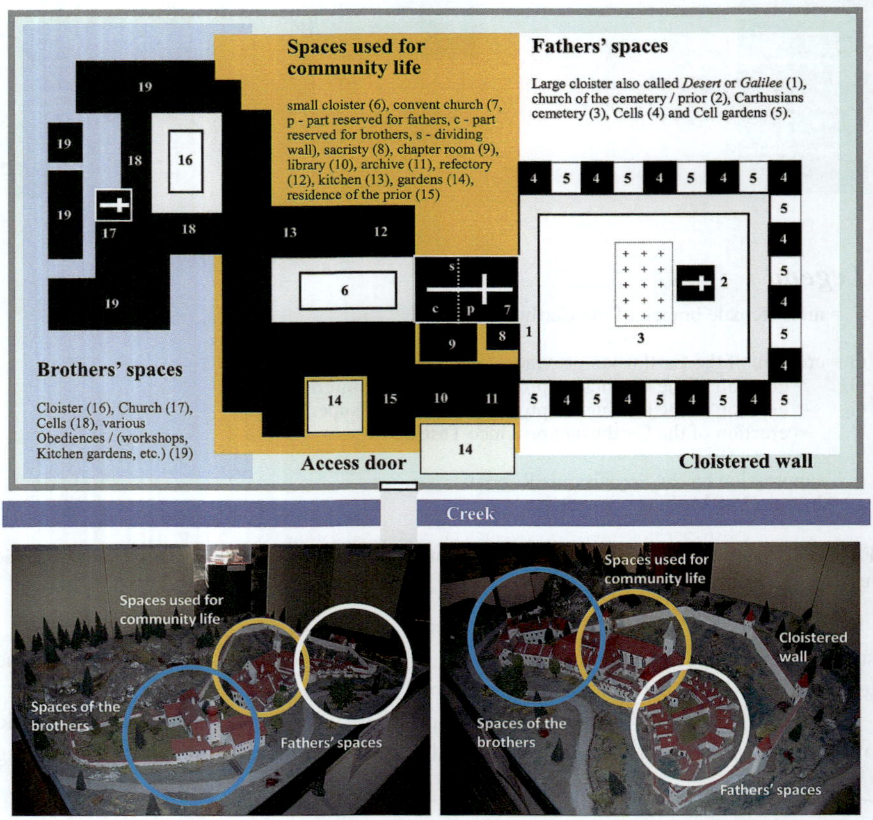

Fig. 3 Divisions of Carthusian space (Aggsbach Charterhouse: model in the cloister museum)

The space reserved for the fathers is represented by the large cloister (see Fig. 8), a sort of covered walkway (open or closed, depending on the rigidity of the climate), usually square or rectangular in shape, along whose sides are located the Carthusian cells, real huts with their own gardens (see Fig. 9). It as a link between the cells of the fathers, the Church and the spaces dedicated to common life (refectory, chapter room, library, etc.).

The brothers' spaces, on the other hand, include much smaller cells, because the brothers remain inside for a much shorter time. These, as a rule, are distributed around a minor cloister and in the vicinity of the *obediences*, or the workshops where the brothers practice various activities required by the regular development of daily life in the Charterhouse.

These three sets of spaces—which we can consider real spatial units—are unmistakably distinguishable from one another and are necessarily always present in the Carthusian monasteries, usually enclosed within the enclosure wall, as they are functional, indeed, essential to the genre of life of these monks.

In the case of the oldest Carthusian monasteries the spaces of the brothers were physically separated from those of the hermitage, coming to establish in a neighboring location placed downstream the lower house, also called Correrie (L'Ordine Certosino). This characterizes the Grande Chartreuse, the mother house of the Carthusian order, and within the Eastern Alpine-Adriatic region, the Seitz/Žiče Charterhouse originally populated with monks from the Great Chartreuse. In the case of Seitz, it was located in Špitalič, where the church of the brothers is still visible.

4 The Adaptation of Spaces According to the Topography of the Site

The reciprocal position of the three spatial units with respect to the building that serves as the center of gravity of the Carthusian complex—that is the Church of the monastery—in fact changes from Charterhouse to Charterhouse (see Figs. 4, 5, 6). This lack of uniformity depends to a large extent on the different geographical features of the site that houses the monastic settlement from time to time.

Likewise, if the geometry prevalent in the Carthusian complexes is based on the shape of the quadrilateral (and when possible, if the topography of the site allows it, of the square—see Charterhouses of Mauerbach and Pleterje—or of the rectangle, see Charterhouse of Gaming), this form, in turn, may undergo in each case the necessary deformations depending on the topographical layout of the site (see Aggsbach Charterhouse).

This in no way interferes with the life of Carthusian monks, given that, unlike other religious orders (see for example the Cistercians), they have not developed an architecture of their own that must necessarily be re-proposed in all foundations.

The monasteries of the eastern Alpine-Adriatic region, hinged in the Carthusian province *Alemaniae superioris*, from a topographical point of view are mostly

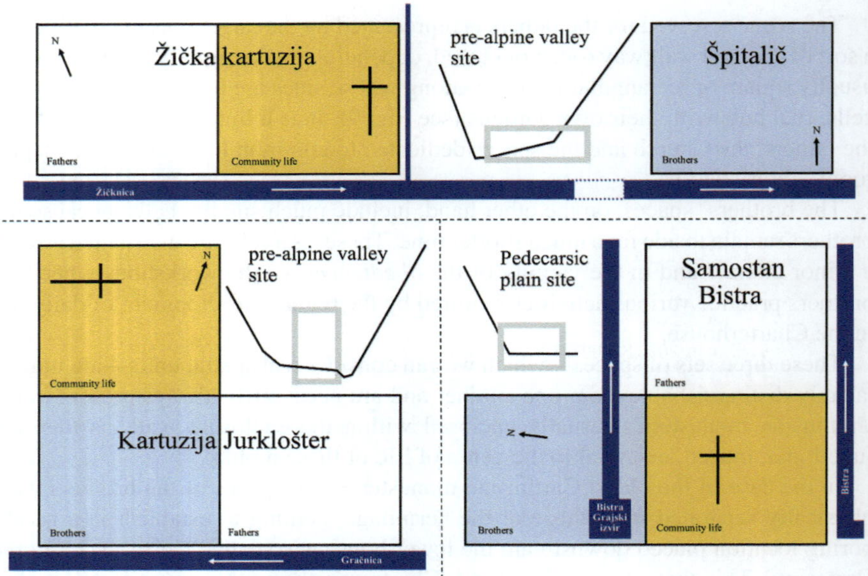

Fig. 4 Spatial units in the suppressed Slovenian charterhouses

Fig. 5 Spatial divisions of the Slovenian charterhouse still in use

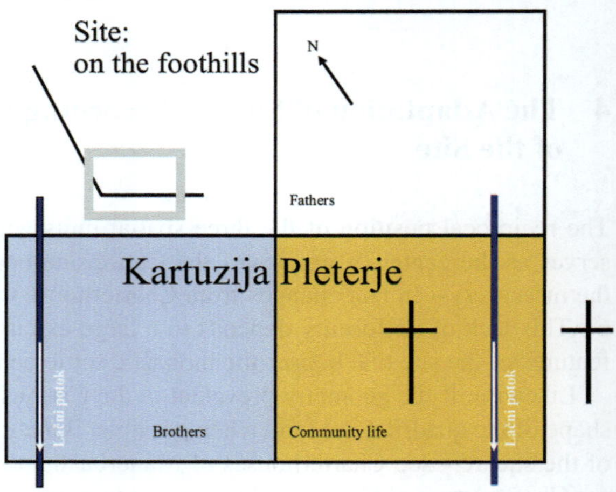

located in valley varices—and on the immediately facing valley slopes. In the presence of particularly narrow valleys this guarantees the vital space necessary for the construction of the Charterhouse and, in both cases can undoubtedly favor hiding.

These areas are characterized by soft foothills of the pre-Alpine type (Charterhouse of Žiče, Jurklošter and Mauerbach), alpine (Charterhouse of Gaming) or mountain (Charterhouse of Aggsbach in the Dunkelsteinerwald, transdanubian offshoot

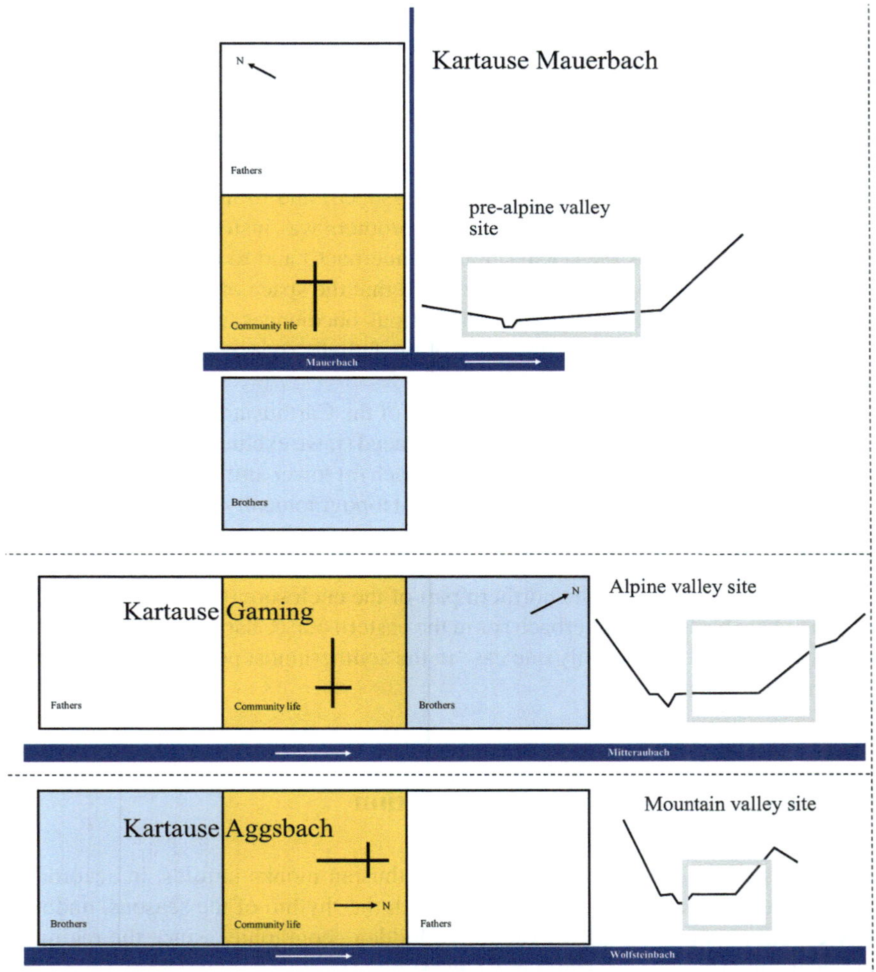

Fig. 6 Spatial units in Austrian charterhouses

of the Bohemian Massif) or in marginal foothills (Charterhouse of Pleterje) or Karst foothills (Charterhouse of Bistra; see Cartusialover; Mlinaric 2001).

It is interesting to note that, in the case of these Carthusian monasteries, at least one of the longer sides of the enclosure perimeter is almost always flanked by a stream, sometimes even two consecutive sides at a confluence of a tributary (Charterhouse of Žiče), while there are cases in which the waterways penetrate into the enclosure (monasteries of Pleterje and Mauerbach) or, finally, in one case the Charterhouse is located near copious karstic resurgences (Charterhouse of Bistra). For obvious reasons, it is the obediences that are mainly located in closer proximity to the banks of the waterways.

In the case of Slovenian Carthusians, the space of the fathers is (was, if dismantled) normally located behind the church of the monastery (Charterhouse of Jurklošter, Bistra and Pleterje) or to the left of the church (Carthusian monastery of Žiče), the space of the brothers was instead located to the right of the church (Charterhouse of Jurklošter) or to the left of the church (Charterhouse of Bistra and Pleterje). In the case of Austrian Carthusian monks, the space of the fathers is located behind the church (Charterhouse of Mauerbach and Aggsbach) and to the right of the church (Charterhouse of Gaming); the space of the brothers was instead located in front of the church (Carthusians of Aggsbach and Mauerbach) and to the left of the church (Charterhouse of Gaming). It should be noted that the space of the brothers, usually noisier due to the work carried out in the various obediences, is normally located in an area as far as possible from the space of the fathers, so as not to disturb the great silence of the great cloister.

It is also interesting to observe, in the case of the Carthusian monasteries in question, how the space of the brothers is always placed (if we exclude the Charterhouse of Bistra, located on the plain, and that of Aggsbach) at lower altitudes than the space of the fathers, reproducing, somehow, the original topographically more depressed location of the lower house (*Correrie*) compared to the upper house (*Grande Chartreuse*). Furthermore, in relation to the cardinal points, we note that the space of the fathers is more frequently located in the northern part of the enclosure (Charterhouse of Žiče, Pleterje, Aggsbach and Mauerbach) or in the eastern one (Charterhouse of Jurklošter and Bistra), while there is only one case in the southernmost position (Charterhouse of Gaming).

5 The Enclosure and Its Demarcation

The spaces where the austere life of the Carthusian monks unfolds, in accordance with the liturgical times and harmonizing with the rhythm of the seasons, undoubtedly constitute a paradigmatic example of hidden geographies, since the regime of monastic enclosure is very rigid: in a Charterhouse you cannot enter.

To be sure, we can even go so far as to say that there are four progressive enclosure levels, so to speak concentric. The Carthusian monastery—although there were exceptions, such as the Charterhouse of Padova—is normally located in an area that is not easily accessible, far from inhabited centers, often with considerable environmental value, surrounded by extensive forests that guarantee in themselves a first form of protection of the solitary life of the Carthusians. In fact, we speak of Carthusian desert.

The hermitage—and its immediate appurtenances—is also clearly delimited by the enclosure wall, a wall which is undoubtedly impressive at times and which often also achieves the purpose of concealing the spaces enclosed within it from prying eyes. In turn, the cells where the fathers lead most of their existence represent a sort of real enclosure in the enclosure, even though the decisive one is the enclosure of the heart. The Carthusian monk is in fact separated from everything and everyone,

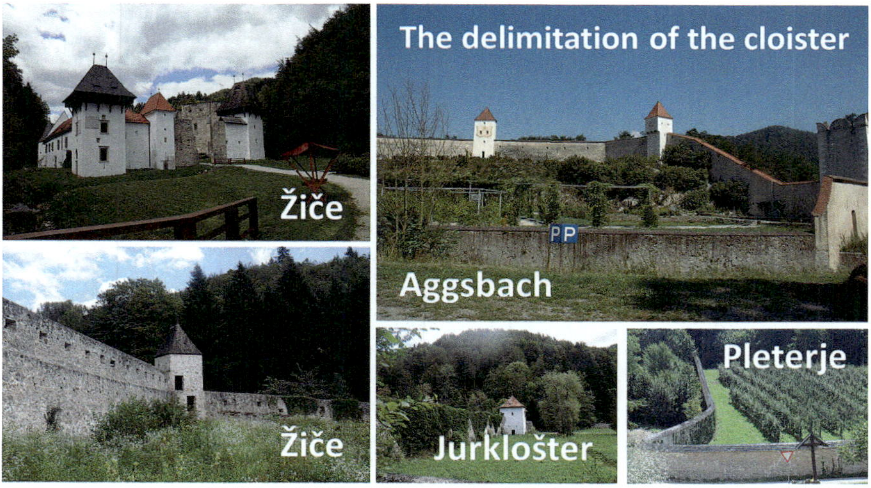

Fig. 7 The enclosure wall

he lives by and for God alone but precisely for this reason he is placed in the beating heart of the Church, carrying out an incessant essential work of intercession that makes fruitful the life of the Church.

In the case of some Carthusian monasteries of the eastern Alpine-Adriatic region, the enclosure wall is also interspersed with imposing towers designed in an eminently defensive manner (see Fig. 7), a characteristic present in Carthusian monasteries that were particularly subject to Turkish raids (e.g., the most spectacular fortified cases represented by the Charterhouse of Žiče and Aggsbach, but those of Jurklošter and Bistra are also recalled) (Fig. 8).

6 The Fathers' Spaces

The fathers' spaces are represented by the large cloister also called Galilee, with a square or rectangular base, where the topography of the site allows it. The large cloister is a sort of covered connection, closed in the case of the Carthusian monasteries located in places characterized by a rigid climate or open in places characterized by milder ones. It connects the heremitical cells with the monastic church and the other spaces used for the life of the Community.

At the center of the large cloister there is a lawn/garden that also houses the cemetery of the Carthusians and the annexed chapel, in some cases called the chapel of the prior. Note that the Carthusian cemetery is essential: at their death the bodies of the departed monks are buried directly in the bare earth, protected only by the monastic robe with the hood sewn to hide the face. Above the burials a simple wooden cross is placed without even mentioning the name of the deceased.

Fig. 8 Fathers' accommodations. Above: the large cloister and the cemetery church. Below: the Carthusian cell (a cloistered enclosure; Aggsbach Charterhouse: model in the cloister museum)

The fathers, also called cloister monks, lead mainly solitary life inside the cell. Each father has his own cell, a two-story dwelling with various rooms and a small garden, in which the monk leads most of the day (he sleeps, prays, devotes himself to *lectio divina* and meditation, studies, eats his meals, carries out manual and gardening work, according to extremely punctual schedules). The eremitical cells are arranged on the outer sides of the large cloister and are designed according to the needs of the solitary life.

Each cell communicates with the corridor of the cloister through a door and a little door used for delivering the meals to be consumed in solitude. When the monk enters the cell coming from the corridor of the cloister, he finds a first space called Hail Mary where, precisely, each time he elevates this prayer to the Virgin. The cubiculum is instead the only heated space (from a wood-burning stove). Inside there is a wooden structure including a kneeling-stool and a bed. In the cubiculum the monk also eats his meals. On the ground floor instead there is a laboratory—mostly carpentry—and a woodshed, in addition to the garden.

The cell does not have windows overlooking the garden of the adjoining cell, so as not to disturb the solitude. The same desk for the delivery/collection of the meal has an angled shape (see Charterhouse of Mauerbach) such as to avoid the interaction between the father and the brother who does this service. At precise hours of night and day, recalled by the tolling of the church bell, the fathers leave their cell, walk silently through the corridors of the cloister and reach the monastery church.

7 The Spaces Used for Community Life

The heart of the Charterhouse is represented by a sober main church, usually with a single nave without a transept and sometimes with a linear, sometimes polygonal, sometimes semicircular apse. The Church is normally divided into two parts by a wall that divides the space into an area reserved for the brothers and one—that of the presbytery—reserved for the choir of the fathers. In the Charterhouse of Mauerbach this separating septum consists of the walls of the corridor of the large cloister (see Fig. 9).

In the Church the Carthusian family gathers at various hours of the night and day for the official services (Mattutino, Lodi, Vespero) and for the Holy Conventual Mass, celebrated during the feriae by a weekly priest and usually concelebrated only on the occasion of the festivities, while each Carthusian monk also celebrates a further S. Mass in solitude every day in one of the many chapels scattered in the various buildings of the Charterhouse.

Other particularly important spaces for community life are the chapter room where the Carthusian monks come together under the guidance of the prior to make the decisions concerning the community and the refectory. Here instead they consume the meal in community but in strict silence, listening to the proclamation of edifying readings taken from the Holy Scriptures, from the writings of the church fathers or from Carthusians, from the lives of the saints or from the Statutes of the order, normally only on the occasion of the festivities. Other common areas are the library and the archive.

Also worth mentioning is the spacing, a sort of walk outside the enclosure, lasting four hours, which the fathers perform once a week (and the brothers once a month). This winds through the natural environments adjacent to the monastery—which occasionally become ephemeral spaces—with a recreational function but also aimed at consolidating the cohesion of the Carthusian family.

8 The Spaces of the Brothers and the Obediences

The brothers' spaces include the brothers' cells, usually a chapel and laboratories, as well as kitchen gardens, gardens and the farm, connected to the various *obediences*, where the work activities necessary for the economic and food subsistence of the Carthusian family are carried out.

Here the monks widely practice fruit growing, viticulture and production of liqueurs which are then sold to pilgrims. In this regard we should remember the *Pleterska Hruška*, a liqueur that is bottled using glass containers in which a pear had previously been grown (Kartuzijani in Kartuzija Pleterje; Pletér) (Fig. 9).

The brothers' cells are usually located inside a common building: each of them is limited to one, maximum two rooms, as unlike the fathers, the brothers spend most of their time outside the cell itself in the different obediences, where they work in solitude or in any case in strict silence (Fig. 10).

9 Conclusions

The Charterhouses are not made up of a single building, as it happens in the case of foundations belonging to other religious orders but are compact settlements. From a religious point of view, they can be considered as a sort of *regions*, identified on the basis of the spiritual needs connected to the austere kind of life of the Carthusians. As such they have not a mere architectural relevance but represent objects of an undoubted geographical interest. It is, in fact, a set of concentric spaces, characterized by a gradually greater degree of concealment proceeding from the outside to the inside.

The outermost concentric space, represented by the widespread presence of the forest, constitutes a form of gradual transition—a blurred border—between the surrounding environment and the Carthusian world, which contributes, preceding the impenetrable wall of the enclosure, to preserve the quiet of the monks and the great silence that reigns beyond it.

The cloistered wall, as a rule, cannot be crossed except in very exceptional cases by duly authorized persons, e.g., two days a year for the visit of relatives of the monks, or on the occasion of the funeral of a Carthusian or of a papal or cardinal visit. It identifies an internal space hidden *ad extra*, therefore completely unknown to those who live in the external world at least as long as a charterhouse hosts the concrete presence of these monks.

The large cloister and the Carthusian cell constitute, instead, concentric spaces hidden *ad intra* as, in the first case, they isolate the space of the fathers from that of the brothers and, respectively, in the second, the space where each father leads most of his life from that of the other fathers. In the final analysis, the most inaccessible space is represented in the Charterhouse by the enclosure of the heart, that is the interior space, where each father intertwines in contemplation, his constant silent conversation with the creator.

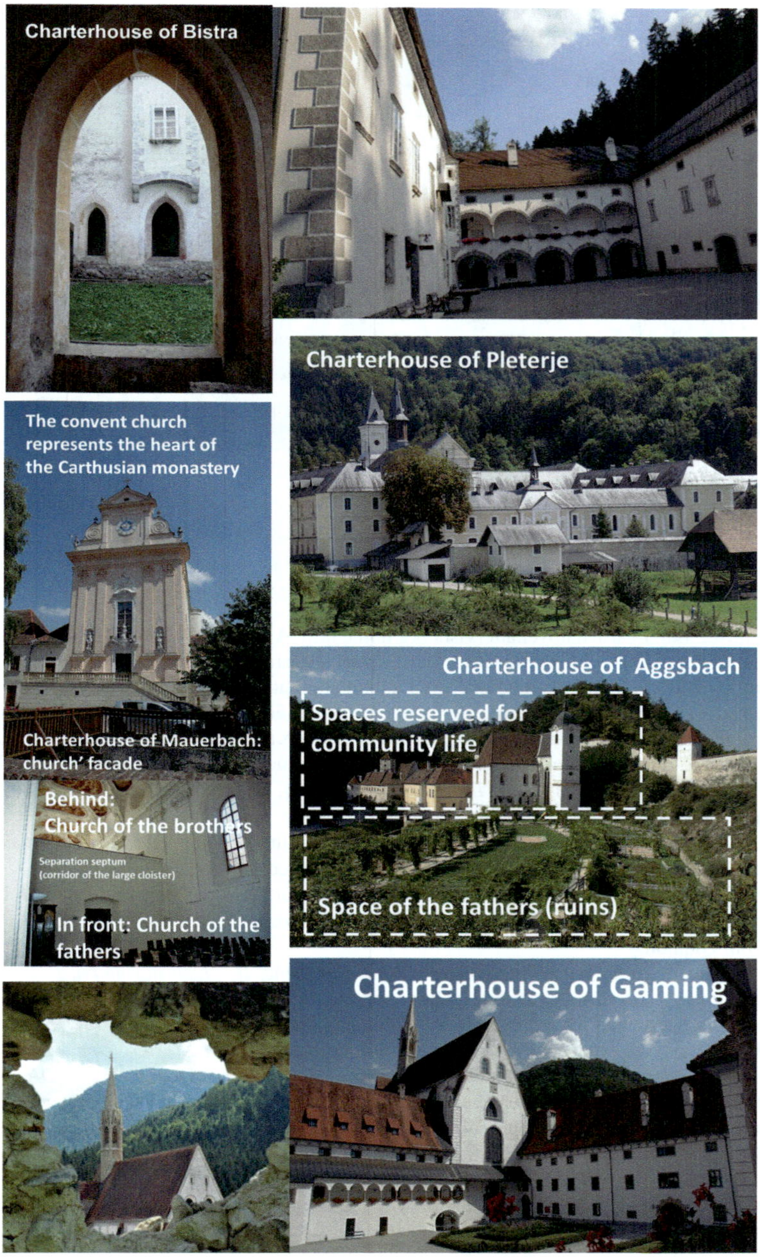

Fig. 9 The spaces designated for community life

Fig. 10 The obediences

References

Cartusialover. http://cartusialover.altervista.org/home.htm. Accessed 24 July 2019.

Galliano, G. (ed.) 2003. Orizzonti spirituali e itinerari terrestri. *Geotema* 21.

Galliano, G. 2002. Geografia e religione. Una lettura alternativa del territorio. *Geotema* 18.

Geografie del sacro: nuove prospettive per la ricerca geografica. https://www.ageiweb.it/gruppi-di-lavoro/geografie-del-sacro-nuove-prospettive-per-la-ricerca-geografica-proposta-di-istituzione-di-un-gruppo-di-lavoro-agei-coordinamento-prof-gianfranco-battisti/. Accessed 24 July 2019.

Hermann, M., and A. Spandl. 2007. *Die Kartause Gaming. Stifter, Mönche, Pioniere*, ed. Walter Hildebrand. Scheibbs: radinger.print.

Hogg, J. 2004. Geografia delle certose: espansione dell'Ordine Certosino. *Analecta Cartusiana*, dl. 220: 5–31. http://www.cartusiana.org/node/4715. Accessed 24 July 2019.

Huber, A. M. *Die Kartause Mauerbach.* Bundesdenkmalamt (Wien), Salzburg: Verlag St. Peter.

Padri certosini, I. 1997. *Una parola dal silenzio. Fonti certosine, I - Le lettere.* Comunità di Bose, Edizioni Qiqajon.

Padri certosini, I. 2000. *Fratelli nel deserto. Fonti Certosine. II - Testi normativi, testimonianze documentarie e letterarie.* Comunità di Bose, Edizioni Qiqajon,

Kartuzijani in Kartuzija Pleterje. http://www.kartuzija-pleterje.si/. Accessed 24 July 2019.

Knall-Brskovsky, U. (ed.) 1999. *Kartause Mauerbach. Vom Kloster zum Zentrum für Baudenkmalpflege.* Denkmal Forschung, Schutz, Pflege, Focus Denkmal 3, BDA Bundesdenkmalamt, Hofburg, Säulenstiege, Wien, Mauerbach: "agensketterl" Druckerei GmBH, ristampa tratta da Österreichische Zeitscrift für Kunst und Denkmalpflege, LIII, Heft 273/4.

L'Ordine Certosino. https://www.chartreux.org/it/. Accessed 24 July 2019.

Mlinaric, J. 2001. *Kartuzija Bistra,* Ljubljana, Družina.

Moleres, F. & Tassone, F. 2009, *Certosini a Serra San Bruno. Nel silenzio la comunione,* Serra San Bruno (VV): Edizioni Certosa.

San Bruno e i certosini. http://www.certosini.info/. Accessed 24 July 2019.

Thir, K. 2018. *Der hl. Bruno, die Kartäuser und die Kartause Aggsbach*, "Freunde der Kartause Aggsbach – Kartäusertexte 1, 8. Korrigierte Auflage, Aggsbach-Dorf: Hausedruckerei, Diözese St. Pölten.

Zadnikar, M. 1995. *Kartuzija Pleterje.* Novo mesto: Dolenjska Založba.

Hidden Urban Geographies

Hidden Urban Geographies: The Case of Barcelona

Carles Carreras

Abstract This chapter aims to epitomize some of the hidden dimensions of the cities, which in fact is a suggestion of the different processes required for their scientific analysis. The starting point is the various experiences in urban research, mainly in the case of Barcelona. From there, the path continues to the presentation of a set of different and complex variables, isolated due to the analysis' requirements but nevertheless considering relevant relationships between them. Firstly, the urban forms and planning are discussed. Secondly, all environmental aspects; important underground city; structures of land property; dynamics of the urban economy; domestic spaces and everyday life; urban segregation, both social and ethnical; urban marginal life; and political power are discussed. Finally, as a conclusive reflection, the debate on the limits of the cities is presented.

Keywords Barcelona · City · Social conflicts · Power · Urbanization

1 Introduction

The complexity of the urban in contemporary cities is positively proportional to the possibility of formulating many different hidden urban geographies. Even more so when the stimulating hypotheses of urban revolution and planetary urbanization (Lefébvre 1970; Brenner 2014, 2019) are assumed. This contemporary urban complexity itself is the result of diverse overlapping of various visible and non-visible dimensions, qualities and forms on all analysis' scales, both spatial and temporal.

In fact, all scientific efforts to search for explanations deal with the development of different reality's hidden dimensions. In order to clarify this urban complexity and to facilitate the explanation of different elements and factors in contemporary urban geography, it is necessary to isolate some variables, in spite of their always complex and dynamic interrelationships. There is no clear hierarchy among these variables because complexity itself implies a certain deregulation of any kind of hierarchy.

C. Carreras (✉)
Department of Geography, University of Barcelona, Barcelona, Spain
e-mail: ccarreras@ub.edu

M. Krevs (ed.), *Hidden Geographies*, Key Challenges in Geography,
https://doi.org/10.1007/978-3-030-74590-5_13

These variables are presented according to their level of apparent visibility, from the more to the less visible. It should be underlined that these variables are themselves dynamic and changing, thus increasing the complexity. Many of the examples presented come from the author's long experience in urban researches, mainly not only on the city of Barcelona but also on some other cities, such as Sarajevo in Bosnia or São Paulo in Brazil. The findings are partly a result of the author's many years of teaching urban geography at the university level. The final result is perhaps a sort of a brief abstract of urban geography in general, as the quoted bibliography reveals, in spite of its non-didactical presentation.

2 The Hidden Forms of the City Plan and Planning

One of the first elements important for a city's recognition and used in its definitional efforts is the visible compactness of urban landscapes, the so-called third dimension of cities. It comprises a more or less dense group of different buildings, streets, gardens and monuments that gather on the observer's horizon. Some techniques of contemporary popular urban marketing have succeeded to symbolize this complex third dimension and its volumetric impact with one simple, selective and artistic skyline.

The landscape impact of cities over the territory sustains the false image of a certain perpetuation of the country–city contradiction. High densities of several elements (population, activities, capital, knowledge or culture) and the material compactness currently offer the most popular definition of the urban, of cities in general, in opposition to the rural, the country: agglomeration.

Largely beyond all the different cityscapes, however, it remains very difficult to discern the formal plan of the city, which is only clearly represented in the city cartography.[1] The volume of the built environment, its massive character and verticalization, tends to hide the regular and to an even further extent the irregular plan of urban public spaces, its order and its structure. Only some baroque outlines and their postmodern copies allow the integration of certain comprehensive perspectives based on the model of the fifteenth to seventeenth centuries *Piazza del Popolo* and its famous *Tridente* in Rome, and this can also be found in a few modern federal capitals such as Washington or Brasília.

Thus, knowledge of city plans requires a serious research on urbanistic evolution and its social consequences. It should be noted here that there is not always a direct correlation between the conceiving of a plan and its effective construction and urbanization. Fragments of the former outline or buildings could remain, explicitly or not, in the present landscapes, generating the well-known urban palimpsest (Huyssen 2003). Another important hidden dimension remains and not only in historical cities—the overlapping remnants from different times.

[1] One of the consequences of this real invisibility is the great difficulty with which common people read and understand city plans, both in traditional, paper versions as well as in their digital form.

The study case of Barcelona gives many examples of the progressive concealment of its so self-apparent plan, with its three traditional historical sections: concentric medieval, gridded nineteenth century and irregular contemporary city (Fig. 1). The standardization of construction and the massive verticalization of buildings (Souza 1994) highly contribute to the homogenization of different neighbourhoods with a constant process of partial reconstruction. That is not the case for other European historical cities, e.g. the majority of French or Italian, whose historical centres have been relatively well-maintained.

Moreover, urban planning, officially established at the beginning of the twentieth century, is today present in almost every city in the world. Its diffusion through the International Style of rationalist architects after World War II was so rapid that it was applied in very similar forms as a response to explosive urban growth on the five continents. Since then, planning has become a unique technical discipline (Santos 1996: 151–170), decreasingly related to social sciences that take the responsibility for designing city plans, as already commented.

Planners and, in general, anyone involved in cartography production, tend to assume a demiurgic role over the territory and its societies through the map (Varanini and Ginevri 2012: 58–59). Inspired by gods from mythical times or by their specialized knowledge of modernity, they act as a sort of general landowner, deciding where to place housing, infrastructures, services and activities, with or without democratic control, depending on the political regime of each country and city. Ordinary people are generally not used to read in a proper manner and understand the cartographic

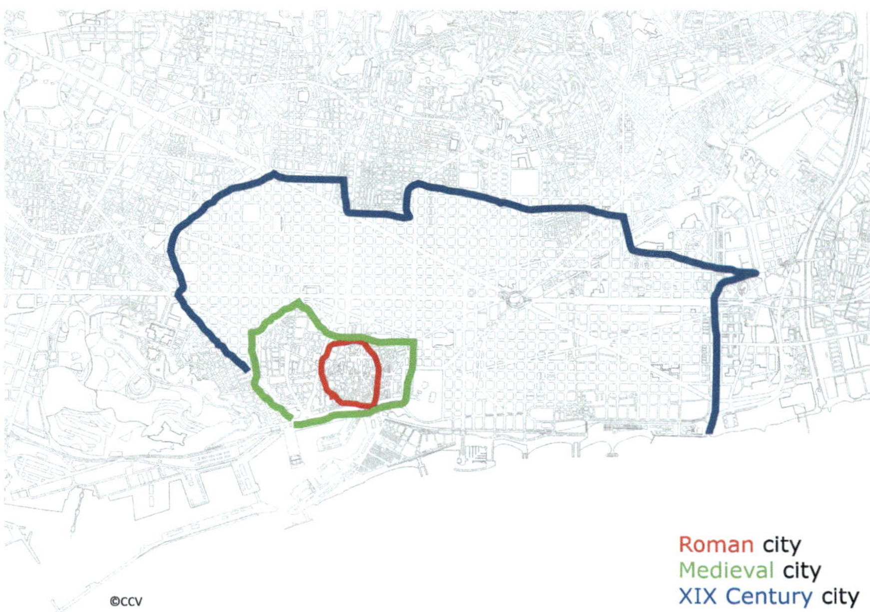

Fig. 1 Historic evolution of the morphology of Barcelona

Fig. 2 The automobile SEAT factory in Barcelona after the deindustrialization process (*Author* Carles Carreras)

language of planners, and that fact reinforces the power of planning. One of the most famous conflicts between planners and citizens was the fight of the North American journalist Jane Jacobs (1916–2006) with the New York official Robert Moses (1888–1981) during the '60 s (Flint 2009).

Even in the best-planned cities with sometimes highly creative forms, their planned character tends to be hidden, to dissolve under the vastness and compactness of its urban fabric. It happens in the Georgian Royal Crescent of Bath, as well as in the federal capital Washington, or in the original bird or plane form of Brasília. Only through tourist maps or satellite images, it is possible to restore the hidden planned character of such conceived spaces.

In more dynamic cities, the architects' general plan tends to melt under the changes that economic and social life imposes on any part of the city. The changes in the planned land use distort many initial goals and directions. For example, since 2019, Barcelona's general metropolitan plan, approved in 1976, has undergone 1,240 partial modifications (*Avanç del Pla...* 2019: 42). Even a large number of local regulations could contribute to the distortion of general planning, as in the Barcelona study case where, during the '60 s and due to several changes in its local regulations, it was allowed that four new floors be added to the nineteenth-century-planned buildings. The main central streets of this Mediterranean city appear today with two jagged edges of different architectures, and different periods and heights are completely breaking the uniformity of the city's famous nineteenth century plan. That once happened to the French Hausmannian streets, and it indicates a feature of rapidly growing cities, as has mainly been the case in many recent Third World agglomerations.

3 The Visible Environmental Invisibility

Urban landscapes currently appear as the biggest human artificial product. They are opposed to nature, and in popular thought urban is considered nature's negation. Even many contemporary environmentalist approaches consider the natural features of cities as something external to them. It is a great misunderstanding especially because the planetary urbanization since the last decades of the twentieth century has thoroughly revolutionized the traditional contradiction of artificial—natural and the capitalist contradiction country—city (Lefébvre 1970; Brenner 2014).

The more hidden natural urban feature are human beings, the citizens themselves. People tend to ignore their own natural essence trying in vain to forget or ignore illness and death[2] because of their confidence in technological progress. In the same way, all natural processes in their perpetual dynamics normally remain hidden in the majority of cities' inhabitants and urban researchers. It is only when external threats appear, such as serious earthquakes and floods, that they are considered natural disasters.

Increasingly less hidden is the atmospheric pollution. The consequence of industrialization and the constantly growing use of energy, especially in urban transportation, is the diffusion of respiratory illnesses and cancers. Intense smog visibly covers the majority of urban skies, especially in anticyclonic days, even after the temporary success of the symbolic Clean Air Act of London in 1956. However, many dangerous polluting particles could be hidden in more or less blue skies. The visibility of these environmental facts, in general, is a proportional function of their negative character. In this sense, bright blue or impressive cloudy skies, marine shores or snowy mountains are managed more as urban sceneries than as real and significant environmental elements.

Urban vegetation is much more visible. The historical precedent of the 1847 Birkenhead Park in Liverpool had influenced the development of a new landscape architecture, with the diffusion of gardens and parks in many cities around the world. Historically, only imperial and royal cities had big royal gardens and parks, but modernity expanded urban and public green spaces. The traditional planting of trees along the cities' streets has pioneer examples such as the *Unter den Linden* promenade in Berlin from 1647. However, the widespread planting of trees began with the construction of Parisian boulevards in the second half of the nineteenth century. Very symbolic are, for instance, the palms of Los Angeles, planted for the occasion of the 1932 Olympic Games. The massive use of cars and the urban explosion with

[2] The last corrections of this chapter were made during the terrible pandemic of Covid-19, which has dramatically ruptured our ordinary life.

its transport infrastructures after the Second World War has challenged many of these green spaces (Jacobs 1961). The civil protests and the hippie revolution in the 1960s vindicated green spaces for people, and a real process of urban greening started to appear around the world (Nicholson-Lord 1987). Numerous present-day cities probably have a greater number of trees than many natural forests. Surely, however, many of these trees are exotic to the city climate, their environmental effects remaining hidden under the aesthetic goals.

Relatively concealed among this urban vegetation lives a variegated fauna. The traditional nineteenth century zoos are today almost disappearing because of the environmentalist critics. Nevertheless, many urban animals inhabit the cities. Various bird species, not always very well adapted to the local climatic conditions, live in trees, as do squirrels and monkeys, depending on the latitude. The majority of hidden urban fauna consists of numerous rats that invade sewer systems, together with newer, exotic specimens, such as crocodiles. Even more numerous are all kinds of insidious insects, for example, flies, mosquitos, ants, or cockroaches, as well as every kind of more or less dangerous bacteria and viruses. They have nothing to do with the visible and important domestic animals, the enormous amount of all kinds of pets, from conventional cats and dogs to the increasing number of exotic species that generate segments of the public spaces and a variegated number of economic activities.

The artificial buildings' construction materials, with interior air conditioning and general lighting, are so apparent that the environment seems to be external to the city. However, nature is always there, hidden above and under the city, all around and inside the buildings.

4 The Evidently Hidden Undercity

The underground parts of the cities are at the same time notorious and hidden. The urban underground growth has started very early in mainly domestic buildings as oil or wine cellar, or as stores for a wide range of materials and items or as prisons and refuges[3] for people. Sometimes, these underground areas violate the public space of squares and streets because of their hidden character. In some cases, the expansion of these cellars transforms a city into a real ant's nest, as the champagne production in the Catalan village of Sant Sadurní d'Anoia did (Valls 2009).

The rise of urban infrastructures has expanded and diversified the underground of many public spaces. Sewer systems, drainage, energy, communication, subways, car parks, tunnels and so on make the hidden undercities each day more complex and complete (Pike 2005). Some parts are open to public access and many others are not, except for some kind of alternative tourism. From the early Rome's sewage system, the *cloacæ*,[4] to contemporary Paris, considered "*la cité des cataphiles*"

[3] The most famous are probably the Roman Christian catacombs.

[4] Many other Roman cities have their cloaca system, like Naples, where it is a tourist site; the same goes for Barcelona and its 19th century sewer system.

(Glowczewszky 1983), there is a long history of engineering. The underground transport systems, especially the fast ones, have been an object of general attention, gradually becoming the more public and non-hidden part of the big cities (Ovenden 2003, 2019). Because of their homonymy, especially in the Roman languages, many people consider metropolises as only those cities with underground transportation. Furthermore, many writers settled their novels into this underground world; for example, Argentinian Julio Cortázar (1914–1984) and his *Manuscrito hallado en un bolsillo*, dedicated to the metro of Paris' passengers (Cortázar 1974).

No less remarkable is the underground duplication of some central areas and streets, with their shops and offices, due to comfort reasons. Possibly one of the most spectacular examples is the case of the winter cold Canadian city of Montreal (Deglise 2008). Even central Barcelona had an underground commercial street between 1940 and 1990: the paradoxically called Light Avenue, in spite of the mild climate of the city (Carreras et al. 2016). Many of these underground premises are connected to the transportation networks. The famous Moscow metro, inaugurated in 1925, could be considered an urban expansion of the communication needs.

5 The Hidden Dimensions of the Urban Land Property

Because of its theoretically immaterial character, the land property, which is one of the most important elements in the structure of the cities, remains totally hidden. When almost the entire Earth's surface belongs to someone, the impact of urban land ownership is as strong as it is invisible. In fact, it is evident that the general outline of urban streets and squares mainly corresponds to the legal delimitation between public and private urban properties. Only a few founded cities, from Miletus to Brasília, and some urban renewal plans based on the Haussmannian Paris, have been able to partially avoid this private property weight. The legal instrument used to achieve it is the expropriation law, which of course, at the same time, always protects the rights of private property. Even during the short twentieth century experience of the socialist city, especially in the case of the Soviet Union, the private land ownership was attacked, but they never succeeded to entirely destroy it.[5] In addition to that, many urban renewal works have had an important impact on the revalorization of the affected lands, as was first demonstrated by the pioneer historical reform of Paris in the first half of the nineteenth century (Suttcliffe 1970). In this sense, it is possible to assume that many planning interventions finally only succeed in increasing the private land rent (Lefébvre 1972; Hall 1982).

During the city's history, the land ownership structure was maintained along the slow process of transformation from the big former rural plots of land to the more expensive urban little ones. The common urbanization process consisted of self-fragmentation of plots of land, mainly along the main ways of communication, and sometimes opening new streets—with or without a general plan. Landowners were

[5] In this regard, Sarajevo was one of the study case cities (Carreras and Moreno 2007).

usually local or regional, at the same time supported and encouraged by local credit banks. Urban planning and local financialization started almost simultaneously at the very beginning of the second half of the nineteenth century. Since the beginning of globalization processes, from the capitalist restructuration crisis of 1973, a global urban land market has developed, with huge consequences for the local inhabitants.

Today, international individuals and big foreign corporations are investing in many cities of the world, besides the little local owners, both with no visibility in the urban landscapes (Bernardos et al. 2014). In spite of this material invisibility, the urban property structures are today the most significant variable in the dynamics of contemporary cities, with profound consequences for all other economic activities and social inequalities.

6 The Hidden Dimensions of the Urban Economic Life

The traditional medieval commercial city was dramatically transformed by the sprawling of factories and workshops in the industrial city since the second half of the nineteenth century in some European regions. Big brick buildings with high smoking chimneys, surrounded by large, poor working-class neighbourhoods formed the Mumfordian Coketown and their Dickensian social landscapes (Mumford 1961: 446–480). However, the growth of the manufacturing processes both disseminated and concentrated new factories during the twentieth century. The first step was the peripheralization of manufacturing after World War II, and the second was the deindustrialization process at the end of the century. Production is mainly concentrated in a few Asian cities, and the entire global economy is involved in the so-called tertiarization. This concept implies the multiplication of service activities, directed either to consumers or producers and the fragmentation and specialization of all economic processes, progressively hiding the real base and structure of urban economies (Fig. 2).

The architectural pattern of productive buildings has also changed, progressively hiding their content and function. With the diffusion of rationalism, the pattern that has proliferated is a set of similar geometric containers, adaptable to different kinds of manufacturing productions and to various commercial functions or a wide range of services (museums, air or railway or marine terminals, hospitals or universities). The artistic creativity of some famous starchitects conceived different styles but similar and expensive buildings around the world (La Cecla 2004), in a sort of a hard, competitive race between cities for possessing their own famous and symbolic building. Apparently, beautiful and modern containers for disparate functions, in a process of sparing and making form independent of its content, hiding both, their economic use and their fabulous costs.

The economic dimensions are perhaps more evident in monofunctional urban buildings (petrochemical factories, cathedrals and temples, jails…). However, in compact cities, such as Barcelona or the majority of the Mediterranean (Carreras

Fig. 3 A present skyline of the city (*Author* Sergio Moreno)

2018), or in the downtown of disperse cities, with their predominance of plurifunctional buildings, the mix of uses increases the invisibility of their economic activities (Fig. 3). The complexity and heterogeneity of functions shades the economic specialization and the real land use of large urban areas. Furthermore, the economic relations among different cities that could explain urban hierarchies and the global life of entire regions have also become progressively hidden in the present territorial complexity. We are very far from the important Christallerian modelling that dominates geographical research between 1930 and 1970 (Berry 1967).

Cities are becoming more homogeneous everywhere, with the same typology of buildings, similar types of shops and commercial centres, as well as similar leisure districts. The case of many recent Chinese cities, with poor copies of some occidental cities' buildings, could be considered the peak of this homogenization process. As already remarked, the present-day urban landscapes symbolize the city of the capital, which is the form in the financial period of this late capitalism. The cities themselves appear as simple goods to be sold in a very competitive global market (Ashworth and Voogd 1990). Their landscapes and their skylines become marketing wrappers, while the real economic base[6] of every city remains hidden from both their inhabitants and visitors and even from political administrators and social researchers.

In the case of Barcelona, for instance, the organization of the 1992 Summer Olympic Games had great touristic success and complex results. It created several economic activities, especially in the service sector, but it also had a more relevant hidden consequence in the urban land market, with huge benefits for the owners of buildings in the city centre, many of them foreign corporations.

[6] The concept corresponds to the old methodological debate of the 1950s around the basic and non-basic urban economic functions (Alexander 1954).

7 The Paradox of the Eclipsed Domestic Spaces and the Everyday Life in General

The currently visible and generally studied city is above all its public spaces: squares and streets, bounded by architectural façades and with few, more or less accessible buildings only; public administration buildings as well as markets, stations or commercial centres. Even many public buildings lack open access to people, for example, jails and military or government premises. However, the main part of the urban space, the residential one, is the domestic space. It remains hidden in the majority of citizens and visitors, including social researchers. Architects and interior designers deal with prototypes of apartments with an almost uniform distribution of rooms and their functions, especially after the definition of Le Corbusier (1887–1965) of *La machine à habiter* (Le Corbusier 1923). Only the increasing use of glass façades and big windows and balconies allow some partial glimpses in the night time.[7]

Of course, there are many cultural differences in how this visibility of domestic spaces to outsiders is handled. In Mediterranean urban homes, the combination of culture and climate that uses the closure of domestic life prevails. The Islamic houses are organized around their interior, with no windows at all on their façades. Similar is the Catholic ones, where the windows normally remain closed behind blinds that only allow the vision out of the interior into the exterior. The Protestant houses, on the other hand, usually have open windows, with the lighted interior at night. In history, human society has evolved from tribal and communitarian values and habits to individualism and privacy rights. This process partially explains the progressive confinement of domestic spaces.

The special patterns in which citizens organize their own domestic spaces is a much-hidden dimension that few anthropological studies have fragmentarily analysed. A significant example of this is the relevance that the British writer Virginia Woolf (1882–1941) gave to a room of one's (women's) own (Woolf 1929).

Even more difficult to analyse are the different processes of appropriation of the space by people in their everyday life. The American urban planner Kevin Lynch (1918–1984) elaborated a pioneering method in order to know, understand and explain the different processes of individual appropriation of the cities (Lynch 1960). He conducted an empirical study of Boston, Jersey City and Los Angeles, designing the mental map techniques in order to understand the ways that citizens read the image of their cities and how they behave in consequence. The relevance of that method is directly proportional to the enormous amount of constantly changing information. The analysis is very useful on a large scale, but almost impossible for general and larger studies, where this important aspect always remains hidden (Brenner 2019).

[7] Alfred Hitchcock's (1899–1980) film *Rear Window* from 1954, based on the short novel of Cornell Woolrich (1903–1968) *It Had to Be Murder*, dramatically remarks this paradox between domestic and public life.

Citizens, tourists and visitors, even urban analysts have always had the possibility to capture some information about the domestic and quotidian urban life by reading the façades and the external display of buildings and houses. It is an imperfect method. The urban palimpsest in general and the complexity of plurifunctional buildings and condominiums and plurifamiliar residences make the task very difficult.

Urban geography has generally dealt with the city as a material object, an abstract place, mainly public, mainly static, mainly masculine and mainly at the working hours of the day. It is the majoritarian productivist point of view. The consumer's city's diversity, difference and dynamic remain always evasive if not hidden.

8 The Hidden Processes of Social and Cultural Segregation

City in general and each of the concrete cities with their unique toponyms tend to be a relatively homogenous object. For this reason, the city and each of its toponyms are used as a subject of various actions in order to impose group or class political options to the majority of their citizens. For example, during the Summer Olympic Games in 1992, the slogan *Barcelona turns its back to the sea* became very popular. Barcelona, like any city in the world, has neither back nor face; moreover, its harbour has been used for commerce or fishing at least for the last 2,000 years. Researchers and Barcelonans need to examine who faces the Mediterranean Sea and why. For this reason, urban social and cultural diversity often remain hidden under a supposed or imposed general interest.

Urban social and cultural structures are always very complex and dynamic, and the political projects often tend to unify the class, gender, cultural, ethnic or age differences. The nation state implemented the first big modern program trying to unify all of its citizens, who were not any more mere subjects, under a general concept of homeland. Wilhelm von Humboldt (1767–1835) created the modern German university as one of the strongest bases for reunifying Germany. The German as well the French first schools of geography were asked to scientifically enrich this unification process. Cities are also very effective tools for the unification of their inhabitants, who are citizens both in a political and locational sense. Cities already existed in premodern times, before the state, at least in the Mediterranean world (Braudel 1949). In postmodern times, cities have once again achieved a homogenization role in their competition in the international market.

However, during their history, cities have created several ghettos, even homogeneous preindustrial cities. Jewish ghettos in the Christian world, the Forbidden City in Beijing, Chinatowns in the Philippines or Hispanic America are just a few examples. The development of capitalism has multiplied this kind of ghettos due to the increasing international migration and the expansion of colonization overseas. Overall, social segregation has been the most significant feature of capitalist cities, often combined with the cultural and ethnic ones. The first examples analysed were Little Ireland in Manchester or Little Italy in Chicago (Engels 1845; Park and Burgess 1925). The combination of the laws of the urban land market and the zoning principles of urban planning has morphologically organized and consolidated the contemporary fragmented city. The culmination of this fragmentation process could be the desired segregation of rich people living in closed condominiums; this process is thoroughly diffused in Brazilian cities (*condominios fechados*) and in general, around the Latin American urban world (Bellet and Sposito 2009; Sposito and Goes 2013).

In spite of the general awareness of this widespread social and cultural segregation, the main part of its empirical materialization in the cities remains hidden to citizens and visitors, to politicians and even many social researchers. The best-known areas and neighbourhoods of the cities are always their centres and tourist places that hide the majority of other urban spaces. Only a few social ghettos or ethnic neighbourhoods could attract particular interest, as the favela tours in Rio de Janeiro, or many Chinatowns everywhere, or the Raval in Barcelona (Martínez-Rigol, Carreras and Frago 2015; Wong and Tan 2013).

9 The Not-So-Hidden Marginal Life

During the nineteenth century, the fact emerged progressively in the Western cities that the capitalist progress generated inequalities and that the cities' space was shared between rich and poor people, mixed even with misery, vice and crime. The first Marxist analysis categorized that social reality into three social classes: the bourgeoisie, the proletariat and the lumpenproletariat. The nineteenth century philanthropists, such as doctors, priests, engineers or writers and even some industrial entrepreneurs considered the health and social degradation in the industrial cities as an illness that could be treated with hygienic urban planning and social policies, reformists or revolutionaries. Examples are numerous in many countries: Charles Dickens (1812–1870), Charles Fourier (1772–1837), Leo XIII (1810–1903) or the Barcelona's planner engineer Ildefons Cerdà (1876).

Within the frame of the urban reformism stream, some writers were interested in the description of the underworld of great cities. Urban marginal world, normally hidden from everyday life, became popular through journals and books creating a new literary genre, the thriller. Starting as serial literature, the genre was first published in the most popular journals and has had a big success, multiplied by the cinema, throughout the twentieth century.

Probably the first populist novel of mystery and crime was *Les mystères de Paris*, published in 1842/43 by the French writer Eugène Sue (1804–1857). The scandalous success of the book resulted in national and international imitations; at least 14 mysteries happening in different European cities appeared between 1849 and 1970. Until today, every year, every city has had books written about its secret life, although finally, results are not that secret.

This literary genre has the advantage of its popularity because such novels are read not only by learned intellectuals but by everyone. It would be possible even to identify different national types of this genre based on the detective protagonists of the serials.

For example, the British Arthur Conan Doyle (1859–1930) or Agatha Christie (1890–1976), the Americans Dashiell Hammett (1894–1961) or Raymond Chandler (1888–1959), the Belgian Georges Simenon (1903–1989), the Spanish Manolo Vázquez Montalbán (1939–2003), the Italian Andrea Camillieri (1925–2019) or the Swedish Hennig Mankell (1948–2015) (Carreras 2013).

10 Urban Hiding of Political Power

Materialization of power, political as well as religious and economic, has an enormous significance for the birth and development of cities. Castles and palaces, temples and markets have organized the first human settlements that are known under the generic name of cities. During the long process of formation of the nation states since the seventeenth century in Western Europe, some cities became symbols of the new political power in the form of capitals. Paris, London, Saint Petersburg or Berlin compete in their representative role. The twentieth century multiplication of states also multiplied the number of capitals, with more or less brilliancy.

However, with the increasing process of economic capital accumulation, non-capital cities take primacy in many countries and regions, for example, Barcelona in Spain, Milan in Italy, Frankfurt in Germany, New York in the United States or São Paulo in Brazil. In the process of globalization of the economy, through the diffusion of new information and communication technologies, some cities, capitals or not, acquire the global category, especially New York, London and Tokyo (Sassen 2001). Their stocks exchange activities play an important role in the definition and conception of the global city.

The increasing influence of the city that is visible in the concentration of the headquarters of different corporations at all the scales, in parallel contributes to hiding the organization and the functions of traditional and real political power. In this sense, the great architectonic creation of the twentieth century has undoubtedly been the skyscraper, which symbolizes everywhere the real power of the economy in the capitalist cities (Gottmann 1966).

11 Progressive Hiding Process of city's Limits, a Kind of Conclusion

In geography, the delimitation of the city is usually also an exercise of definition. Many preindustrial cities were clearly delimited by different kinds of walls, inside which even the rural population took its shelter during war periods. The separation between the country and the city was clear, although many rural landowners and peasants could live in the city. Barcelona has three different historical circuits of walls, and in the first half of the nineteenth century, all the gates still closed at 9 p.m. except one that closed at 10 p.m. By night, the city remained closed and safe (Carreras 1993).

The diffusion of capitalist industrialization ended with the material limits of cities at the same time that urbanization started to occupy more and more territories. The traditional compact city was overflown by a diffuse new city composed of large suburbia everywhere. Huge communication infrastructures completed an impressive and progressive land occupation. Furthermore, agricultural and natural areas were totally subordinated to the city. The contradiction between the country and the city has been completely reverted in favour of the city.

However, during the last expansion of capitalism around the world, since the big 1973 crisis, and the hegemony of financial capital and all types of new technologies, many differential processes of urbanization have been developed throughout the world. The traditional contradiction between the country and the city has been overcome by the investments of capital and knowledge everywhere. The cities are not anymore a bounded object as hypothesized by the French philosopher Henry Lefébvre in 1970, with the combination of the process of implosions and explosions of the cities (Lefébvre 1970; Brenner 2014).

Urban geography today, therefore, covers all territory due to the process of planetary urbanization. This requires a new definition of the cities and a new point of view to be able to explain the wide and deep changes of contemporary human society.

These and other important contemporary trends that characterize our post-truth times claim for a new scientific approach to the urban that our research group includes in a vindication of a renewed classical political economy. The main goal must be to allow the explanation of the urban totality against the mainstream fragmentation of the anecdotic and fetishist discourses in much scientific literature.

References

Alexander, J. 1954. The basic-non basic concept of urban economic functions. *Economic Geography, Worcester* 30: 246–261.

Ashworth, G.J., and H. Voogd. 1990. *Selling the city: Marketing approaches in public sector urban planning*. London: J. Wiley & Sons.

Bellet, C., and M.E.B. Sposito. 2009. *Las ciudades medias e intermedias en un mundo globalizado.* Lleida: Universitat de Lleida.

Bernardos, G., Martínez-Rigol, S., Frago, L., and C. Carreras. 2014. Las apropiaciones de la ciudad a la hora de la globalización: las estrategias del capital ruso y chino en el mercado inmobiliario de Barcelona. *Scripta Nova* 18, 493(52): 1–17. Barcelona: University of Barcelona.

Berry, B.J.L. 1967. *Geography of market Centers and Retail distribution.* Englewood Cliffs (NJ): Prentice Hall.

Braudel, F. 1949. *La Méditerranée et le monde méditerranéen à l'époque de Philippe II*. Paris: Armand Colin [second revised edition 1966]

Brenner, N. 2014. *Implosions/Explosions: Towards a study of planetary urbanization.* Berlin: Jovis Verlag.

Brenner, N. 2019. *New urban spaces: Urban theory and the scale question*. London: Oxford University Press.
Carreras, C. 1993. *Barcelona espai català, temps europeu*. Vilassar de Mar: Oikos-tau
Carreras, C. 2013. *La ciudad en la Literatura. Un análisis geográfico de la Literatura urbana*. Lleida: editorial Milenio.
Carreras, C. 2018. *Città nel Mediterraneo, città del Mediterraneo? Civiltà del Mediterraneo XVII (XXII), 29: 139–150*. Napoli.
Carreras, C., Martínez-Rigol, S., Frago, L., and A. Morcuende. 2016. *Cities, Public Space and Citizenship. Some Contemporary Mediterrranean Urban Social Movements*. Athens: Atiner's Conference Paper Series.
Carreras, C., and S. Moreno. 2007. Los procesos de modernización en Sarajevo. La incierta dirección de la flecha del tiempo. *Anales de Geografía de la Universidad Complutense* vol. 27, n. 1: 29–44. Madrid: UCM.
Cortázar, J. 1974. *Manuscrito hallado en un bolsillo*. Madrid: Octaedro.
Deglise, F. 2008. *Montréal souterrain: sous le béton, le mythe*. Montréal: Heliotrope.
Engels, F. 1845. *Die lage der arbeitenderklasse in England*. Leipzig: Verlag Otto Wigand.
Flint, A. 2009. *Wrestling with moses: How Jane Jacobs took on New York's master builder and transformed the American city*. New York: Random House.
Glowczewszky, B. 1983. *La cite des cataphiles: Mission anthropologique dans les souterrains de Paris*. Paris: Le Livre des Méridiens.
Gottmann, J. 1966. Why the skyscraper? *Geographical Review* 2 (56): 190–212.
Hall, P. 1982. *Great planning disasters*. Berkeley: University of California Press.
Huyssen, A. 2003. *Present pasts: Urban palympsests and the politics of memory*. Standford: Standford University Press.
Jacobs, J. 1961. *Death and life of the great American cities*. New York: Random House.
La Cecla, F. 2004. *Contro l'architettura*. Bologna: Bollati Beringheri.
Corbusier, Le. 1923. *Vers une architecture*. Paris: Éditions Vincent, Freal.
Lefébvre, H. 1970. *La Révolution urbaine*. Paris: Anthropos.
Lefébvre, H. 1972. *Le droit a la ville II. Espace et politique*. Paris: Anthropos.
Lynch, K. 1960. *The image of the city*. Cambridge (MA): MIT Press.
Martínez-Rigol, S., Carreras, C., and L. Frago. 2015. El Raval de Barcelona un laboratori d'estudis urbans. *Treballs* 79: 125–150. Barcelona: Societat Catalana de Geografia.
Mumford, L. 1961. *The city in history: Its origins, its transformations and its prospects*. San Diego: Harcourt, Brace and World.
Nicholson-Lord, D. 1987. *The greening of the cities*. London: Routledge & Kegan Paul.
Ovenden, M. 2003. *Metro maps of the World*. London: Capital Transport.
Ovenden, M. 2019. *Underground cities: Mapping the tunnels, transits and networks of our cities*. London: White Lion Publishing.
Park, R.E., and E.W. Burgess. 1925. *The city: Suggestions for investigation of human behavior in the urban environment*. Chicago: The University of Chicago Press.
Pike, D. 2005. *Subterranean cities: The world beneath Paris and London 1800–1845*. London: Cornell University Press.
Santos, M. 1996. *A Natureza do espaço. Técnica e Tempo. Razâo e Emoçâo*. Sâo Paulo: Hucitec.
Sassen, S. 2001. *The global city: New York, London, Tokio*. Princeton: Princeton University Press.
Souza, M.-A. de. 1994. *A identidade da metrópole: Verticalização em Sâo Paulo*. São Paulo: EDUSP.
Sposito, M.E.B., and E. Goes. 2013. *Espaços fechados e cidades: Insegurança urbana e fragmentação socioespacial*. São Paulo: Editora da Unesp
Suttcliffe, A. 1970. *The autumn of central Paris*. London: Edward Arnold.
Valls, F. 2009. *The international competitiveness of cava: Success of a particular firm or the district*. Barcelona: Universitat de Barcelona-Fac. Economía I Empresa.
Varanini, F., and W. Ginevri. 2012. *Projects and complexity*. London: CRC Press.

Avanç del Pla Director Urbanístic Metropolità 2019. Barcelona: AMB https://urbanisme.amb.cat/divulgacio/publicacions/detall/-/publicacio/avanc-pla-director-urbanistic-metropolita/9303969/6724201. Accessed 10 April 2020.

Woolf, V. 1929. *A room of one's own.* London: Read Books Limited.

Wong, B.F., and Ch-B Tan. 2013. *Chinatowns around the world: Gilded Ghetto, Ethnopolis, and Cultural Diaspora.* Leiden: Brill.

Geography: A Hidden Antidote to Rescue Modern Architecture

Mostafa Norouzi and Somayeh Khademi

Abstract Critical regionalism can be seen as an approach to architecture that tries to stand up for places culture and identifies the identity of a place where modern architecture has failed to, by using the building's geographical context and reference of local characteristics. This idea was initiated by Lewis Mumford and then by Frampton, Tzonis, and Lefaivre. Frampton in his essay argues that it is critical to adopt universal values of modernism, taking into account the geographical context of the building. He refers directly to the climate, light, topography, and local tectonic form, which should be understood as historical and geographical conditions of the construction industry. This study discusses the critical regionalism theory and its geographical expressions in architecture. The paper opens with a theoretical review, presenting a criticism of modernism and the role of geographical factors as a hidden antidote to rescue modern architecture. Thus, geography gives us an incredible lens through which to see the architecture and their elements. These statements raise the question of what is architecture pertaining to geography. What can still be used to identify the originality of a place or region when globalization and cultural homogenizing are ever-growing and, in many cases, have ceased to exist? Perhaps the most powerful story here is the narrative of how geographical aspects make it possible to trigger critical regionalism as a powerful paradigm in contemporary architecture.

Keywords Geography · Critical regionalism · Modern architecture · Place · Identity

1 Introduction

> Geography has preceded, subsist and will endure, while our civilizations will pass away.
>
> Le Corbusier

M. Norouzi (✉)
Geography and Urban Planning, Tarbiat Modares University, Tehran, Iran

S. Khademi
Geography and Urban Planning, Kharazmi University, Tehran, Iran

M. Krevs (ed.), *Hidden Geographies*, Key Challenges in Geography,
https://doi.org/10.1007/978-3-030-74590-5_14

Modern architecture is a consequence of the universal evolution of architecture which occurred after the global phenomenon of industrialization; itself being a result of the proliferation of design and construction equipment. Even though the issue of regionalism is historically prior to modernism, through the advent of modernism and the dominance of separating object from subject, a concern was raised as in how to still maintain paying attention to the place identity in such an environment and from an outside standpoint. Since those times, one of the major orientations in architectural criticism has been the issue of regionalism. According to this approach, architecture should profoundly follow specific regional pragmatics based on climate, geography, local materials and cultural traditions. This argument stems from which that the compactness of time and place in relation to modernity has changed the shape of social life in which there is a kind of integration and shared values that includes all human societies at any time and place. It has been recognized that globalizing forces of international modernism had a homogenizing tendency, differences elimination, local identity obliteration and remove the geographical boundaries (Frampton 1983; Lehmann 2016). Aided and abetted by homogenizing impacts of the form of globalization, some changes have led many to argue that architecture as an outcome for identity and place have lost their importance.

In fact, the same discussion on the degree of influence and control on the process of globalization and the international style in regionalist architecture is considered as the main concern. The universalization of technology and design may ignore specific local culture and geographic requirements. The term International Style captured the mood of this universal architecture. In this regard, critical regionalism can be understood in terms of anti-globalization, as the main focus is to reintroduce the place and identity into buildings (Abu Hammad and Abu Hammad 2017). The multiplication of critical regionalisms in the world, which is certainly a consequence of globalization, represents an immense intellectual challenge because it confronts every citizen of the world with an increasingly larger range of regional cultural expressions (Botz-Bornstein 2015). In this line, since the final decades of the last century to the present day, there have been numerous researches on the subject of regionalism in architecture, particularly about understanding concepts, backgrounds and criteria, as well as reading them in various geographical contexts (Tzonis and Lefaivre 1981; Frampton 1983; Curtis 1986; Frampton 1987; Eggener 2002; Lefaivre and Tzonis 2003; Canizaro 2007; Colquhoun 2007; Shadar 2010; Carlson-Reddig 2011; Tzonis and Lefaivre 2012; Nolan 2014; Haggerty 2017; Bahga and Raheja 2018; Le 2018; Salman 2018; Zoghi Hoseini et al. 2018).

Critical regionalism can be seen as an approach to architecture that tries to stand up for places culture and identifies the identity of a place where modern architecture has failed to, by using the building's geographical context and reference of local characteristics. This idea was initiated by Lewis Mumford[1] and then by Tzonis and

[1] It is not quite plausible to fully define the regionalism of Mumford, for it did go through a lot of change and development throughout his career. Lefaivre and Tzonis argue that Mumford "did not make things easy for anyone wishing to get a clear overview of his regionalist paradigm" (Lefaivre and Tzonis 2003).

Lefaivre (1981, 2012), Lefaivre and Tzonis (2003) and Frampton (1983, 1987). In one of the early endeavors to the notion of regional characteristics, Lewis Mumford discussed this topic in his lectures during the first half of the twentieth century, in which he marked the regional aspects of architecture in confrontation with the universal and international style. Frampton in his essays argues that it is critical to adopt universal values of modernism, taking into account the geographical context of the building. He refers directly to the climate, light, topography, and local tectonic form, which should be understood as historical and geographical conditions of the construction industry (Frampton 1983). In fact, the term regionalism was not the innovation of these architects and did not offer a new concept. They chose this term, since this new movement was similar to the extensive efforts of architects who were looking for an alternative approach—in designing buildings, landscapes and cities—which could carry the geographical features of a unique environment along with the specific cultural traits of that region (Tzonis and Lefaivre 2012).

The aim of this chapter is to explore the geographical features that contributed to architecture in the globalization context. From a geographic viewpoint, it is obvious that any regional characteristic has to be supported. The premise that underlies the exploration is that the geographical characteristics such as climate, topography, spatiality, place identity, surrounded environment and sustainability as a hidden antidote to rescue modern architecture. Thus, geography gives us an incredible lens through which to see the architecture and their elements. These statements raise the question of what is architecture pertaining to geography. What can still be used to identify the originality of a place or region when globalization and cultural homogenizing are ever-growing and, in many cases, have ceased to exist?

This paper begins with a review of region and regionalism as a pivotal sub-discipline in geography and then, relevant concepts and theoretical underpinnings of critical regionalism: regionalism as it is currently utilized, as well as relevant studies that link regionalist architecture to geographical effectiveness. A framework for this study is developed. Context for this framework is provided with a discussion of critical regionalism and their characteristics and their specific geographical needs.

1.1 Region and Regionalism: A Geographic Viewpoint

Region is an interpretation of geography, identity, plus cultures and institutions. Region in its origins denotes line, direction, as well as area and section. A region's specific characteristics are due to having traits and features based on that same area, which make region a fundamentally geographical term. This concept has a long academic history, dating back to the geography of Strabo. However, "the first systematic definition of the notion of region was made by Herbertson in an article dated 1905. With regard to its more methodological aspects, it can be said that the purpose of this author is to create a "systematic geography" and seeks to find geographical divisions orders on the globe" (Betioli Contel 2015; Jones 2019). Pursuant to this, efforts conducted in developing regional studies in the 1980s led to a general belief

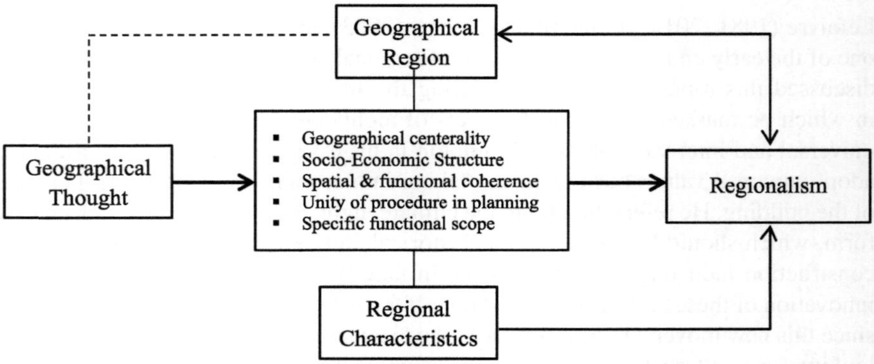

Fig. 1 Theoretical diagram of regionalism via geographic viewpoint

that geography could play a central role in social understanding (Bradshaw 1988). Nowadays regional studies have known as a vital part of geographical disciplines. The study of the essence of regio referred to the fundamental works of many geographers (Mumford 1938; Bradshaw 1988; Terlouw 2001; Paasi 2003; Howell 2013; Betioli Contel 2015; Jones 2019).

The concept of regionality depends on its being possible to correlate cultural codes with geographic regions (Mahgoub 2007). The significance would be for Alexander Von Humboldt in the study of geography of flora, the characteristics of a landscape, which is the most highlighted mark of a region (Silva et al. 2015). From Mumford's perspective (Mumford 1938), the meaning of region is the concept of geographical unit. Indeed, regionalism has a meaning beyond physical attention and toward spatial characteristics.

According to him, the region can be divided from bottom to top; from the smallest unit of human habitation, in terms of functions, activities and interests, or mainly from top to bottom; based on the features of the land, climate, plant and animal life which help distinguish between regions. By adding human beings to this image, the differences become more subtle and diverse (Fig. 1).

1.2 An Interdisciplinary Dialogue

Geography is often an analytical key factor to support the design of an architectural practice and can influence design choices such as color, materials, lighting, or structural elements of a building. A growing number of studies have discussed how geographers and other researchers might effectively analyze and study the built environment in a manner which not only acknowledges, but also surpasses representational readings of historical content and symbolism in various architectural forms. Accordingly, architects highlighted that not only design is influenced by the physical attributes of a location, like its climate, topography and site features, but is also influenced by the social attributes of a geographical area as well, like its culture, perception, design style or history (Rapaport 1977; Frampton 1983, 1987; Lees 2001; Jacobs 2006; Popescu 2006; Kraftl and Adey 2008; Jones 2019; Kraftl 2010). In this line, Boussora (1990) noted that the characteristics of both physical place (such as climate, topography, landscape, and local vegetation) and local society (as adapt to the current social communal needs of the residents, the local construction tradition, and the local resources) are the factors of local architectural contents.

The diversity of architecture, from Vitruvius's perspective, is the result of the physical, mental and behavioral features of the people, and is essentially defined by geography (Eggener 2002). It is also notable for Tzonis and Lefaivre (2012) that Vitruvius—in addition to architectural issues—discusses geopolitical and global concepts which are divided into regions lacking equal quality. He declares that just as natural and climatic conditions influence building design, human beings are affected consequently. Vitruvius, for example, argues that a moderate environment forms a moderate architecture and, as a result, breeds moderate people; and he believes this environment to be a superior one. Or could it be that harsh environmental conditions would make different types of people evolve, both in terms of lifestyle and physical and behavioral perspectives.

In contemporary architecture practices in the world, it is nearly impossible to see the implicit meaning without footprint of geography or as Gausa et al. (2003) stated: "Architecture really belongs to another meta-discipline: Geography." Mostly geographical features are hidden behind the designed images, signs, and structure. From Maudlin view, architecture is an inclusive social study. He emphasizes that architectural boundaries cannot be limited to a specific spatial or temporal realm: "Different places have a different geology and a different environment, so different materials are available; ordinary people with limited resources and limited cultural networks can act only within their local geographical limits. This thinking reveals the influence of post-war functional determinism" (Maudlin 2010). Similarly, Kingston Heath argues in *Vernacular Architecture and Regional Design: Cultural Process and Environmental Response* that architecture is not merely a technical or aesthetic experience, but is inextricably linked to similar environmental and social processes (Heath 2009). In the meantime, major cultural styles and principles, plus their chronological alternations have come into existence through human history, while climatic diversity, along with ecological manifestations, such as language, artworks, or architecture depend to some extent on geography (Popescu 2006, 2008).

Lees (2001) argues that architectural geography should transcend mere representation. She examined earlier architectural geographies—from the Berkley School too political semiotics—and asserted that geographers haven't had a great deal to say about the practical and influential or nonrepresentational importance of architecture. Therefore, she employed the dispute over Vancouver's new public library building as a jump-off point for portraying how geographers might assume a more critical and politically progressive architectural geography. The Colosseum design of the library denotes the origins of the western civilization, which to some Vancouver residents is an aloof portrayal of their beloved multicultural city. Her aim is to push geographers over this contemplative cultivation of architectural form toward a livelier interaction with the building.

To engender interdisciplinary dialogue, Craggs et al. (2013) put forward the concept of "architectural enthusiasm" and highlighted the relationship between architecture and geography with emphasis on relationship between people, buildings, and place. There are three ways they contribute to recent projects on the built environment and architectural geography: first, the importance of people's emotions is highlighted through their engagement with the buildings, which is of a shared and exercised nature; second, they emphasize the role of architectural enthusiasts as influential agents with the ability to reform and shape the built environment; and third, they help establish connections with constructions through maintaining the practice of integrating urban exploration, local history, architectural practice, education and training and finally, a vast range of architectural tourism.

1.3 The Spatiality of Architecture

Since the beginning of the twentieth century, architecture has always been known to fundamentally be engaged with space, which forms its nucleus (Popescu 2006; Dursun 2012). Space is also known as the nucleus of geography (Massey 2005). However, according to Hilde Heynen, architects and geographers do not necessarily need to offer similar definitions of space (Heynen 2013). It is not often known for the practitioners and scholars of either geography or architecture to formally link these two fields together. In this regard, Jacobs and Merriman define space from the mentioned scientists' point of view as follows: architecture lies at the core of a design-oriented discipline with a tendency to reform the space, while geography is at the core of an analytical order toward prescribing and recognizing existing spatial conditions (Jacobs and Merriman 2011, p. 219). Colquhoun (1989) identifies the discrepancies in attitudes toward space and identity among geographical and architectural sciences by attending to form and function. In his view, there are basically two approaches toward form and function: one that considers form as independent from function, and one that considers function to actually determine form, which advocates a direct interaction between them; the second view being closer to that of geographers. However, Kraftl believes the advent of cultural geography to be the study of form in architecture, and perceives the contemporary architectural geography within cultural geography (Kraftl and Adey 2008; Kraftl 2010). Yet, Madanipour (1996) points out that architecture is always eventually interested in form. Indeed, this mere passion for form-production is more likely observed in a type of architecture leaning toward modernism. This approach is clearly at odds with the Newtonian physics, and as Ando puts it: "a place is not an absolute space of Newtonian physics. It is a universal one; a space with meaningful orientation and heterogeneous density." According to Ando, the nature of architecture and its ultimate purpose is to create space. He continues: "Architecture is not only concerned with manipulating forms, but I also believe that architecture is the construction of a "space," and most importantly, the construction of a "place" which acts as the foundation for space." (Shirazi 2012).

Christian Norberg-Schulz in the book *Genius Loci: Towards a Phenomenology of Architecture* describes the recognition of spatial features as one of the fundamental pillars of architecture and insists on pay attention to and shape these factors while designing (Norberg-Schulz 1980). For McNeill these spatial features are appropriate for clarifying such architectural meaning, because of the rich geographical theory available pertaining to place-making and identity of place (McNeill 2005). Person concentrates on methodologies assembled by geographers whose analytical researches seek architectural definitions through various geographic viewpoints and reaches a conclusion by suggesting how these perspectives might be applied through a single-building study. Incorporating issues about *place* into architectural studies paves the way for a keenly phenomenological comprehension of the built environment (Person 2011).

This discussion is developed by Gissen (2010) under the title of "territorial architecture," which recognizes architecture as a broader concept within the patterns and frameworks of geographical and environmental perceptions, aspiring to emerge out of environment, nature, society and technology; while an independent architecture can merely be accountable to itself. He points out that the writings of critical geographers can be very constructive, since while taking advantage of many scientific theories and philosophical critiques, they tend to always look toward aesthetic and spatial ideas and environmental concerns.[2] However, most architects also accentuate the break of architecture from a one-dimensional architectural speculation in the form of its independence—or what James (2013) defines specifically as "building"—and its move toward a comprehensive and spatial design. Similarly, Kenneth Frampton (1991) uses the term "the least autonomous" for architecture as another cultural production, recognizing it not only by his own technical methods, but also by the forces generated outside and around it. In this regard, the book Indigenous Capitals of Africa comes to mind, where Adjaye considers most native buildings in Africa to be a direct response to climatic conditions. Referring to his book, he indicates just how African cities are responsive to different climates. He divides cities into six different geographical regions and depicts the interaction between a set of general conditions and their locations; such as temperature range and seasonal rainfalls, plus how they affect vegetation and landscape, and moreover, the manner of people's adaptation to these conditions in their own vernacular architectural style (Adjaye 2013).

[2] Interestingly, Warf and Arias (2009, p. 1) highlighted that Geography, "has transformed into one of the most dynamic, innovative and influential of the social sciences", and that it "has moved decisively from being an importer of ideas from other fields to an exporter."

1.4 *Regionalism and the End of "The End of Geography"*

Through time, globalization and regionalism have been in conflict with one another. This is due to the fact that globalization has always had the tendency to level the barriers for the purpose of interaction between places and transforming isolated areas in order to create a homogeneous world (Hettne and Söderbaum 2002; Lefaivre and Tzonis 2003, 2012). The contemporary understanding of the nature of regionalism has abandoned the geographic determinism of the late nineteenth and early 20th centuries (geographical region as a natural region), giving way to an intellectual domination, and as a result, globalization—relying on factors such as neoliberalism, rationality, post-industrial capitalism, worldwide communication, universal awareness, organizational developments, and financial markets—took on an extranational form. Indeed, regionalist theories have gone through change in several aspects. Non-territorial geography, semanticism, multidimensionality, and democratic pluralism are among the most important theoretical changes.

Paul Vidal de la Blache as a French geographer tried to find a measure of defining the identity of a region and especially the environmental factors influencing the region, as he says, "the geographic personality".[3] He added human approach to the geographical milieu that focuses just on the "natural determinations" of regions. "The Isolated State" (1826) Johann Heinrich von Thünen developed a model that is considered to be the first serious treatment of spatial economics and economic geography-connecting it with the theory of rent. With organizing regions in this book, he considered an ideal and isolated state in a completely homogeneous area. According to Tzonis and Lefaivre (2012), Von Thünen used geographical keywords such as "place", "center" and "region" and his model was a brief view of the world. After him, other scientists like Walter Christaller[4] presented their models. This generation of researchers was trying to present realistic models against what is called globalization.

Globalization is the compression of time and geographical space that has eliminated geographical distances due to new technologies (Harvey 1989). The constraints of geography are shrinking and the world is becoming a single place (Waters 2001). This quote of Waters is indeed that "the end of geography" (O'Brien 1992), "death of distance" (Cairncross 1997), "borderless world" (Ohmae 1995), and "the world is flat" (Friedman 2005), which have been mentioned in numerous studies. However, this hypothesis has been challenged by the historical trajectory as well as the economic, political, social, cultural and institutional characteristics of different regions and cities. Indeed, opponents of the idea of "the end of geography" claim

[3] For Vidal de la Blache, "geographic personality", a fundamental concept in the Tableau de la Géographie de la France (portrait of French geography) (1903), refers to the ingenuity shown by each human group—and, more specifically, each people or nation—in taking full advantage of the resources drawn from the milieu in which it lives (Mercier 2009).

[4] Walter Christaller was a German geographer whose principal contribution to the discipline is Central Place Theory, first published in 1933.

that globalization brought about poverty, anxiety, and a decline in the quality of the environment (Warwick 2005; Harvey 2006; Christopherson et al. 2008; Cox 2008).

Concepts such as cumulative economics, tacit knowledge, face-to-face communication, social capital, and organizational networks aligned with the development of innovative procedures have reached success. Therefore, the processes of "globalization" and "localization" are not separate concepts, rather intertwined ones (Morgan 1997). Or as Dicken puts it, globalization—in its very broad meaning—does not portray the end of geography, but in fact, is geography itself (Dicken 2009, p. 563). Geography's traditional authority reveals that not only is it sustainable, but globalization has even vastly promoted it; since due to its enlightening tendencies, it has a universal nature. Despite the debates on globalization, it must be acknowledged that localization and regionalism are gaining more power. As a result, we have to oppose the death of geography, since it is developing, not perishing. A closer look at this subject reveals the growth and proliferation of both localization and globalization in their cultural, historical and particularly geographical aspects. Regionalism expresses a tendency to oppose globalization, which creates a different understanding of space creation and landscaping. Globalization does not highlight the importance of land, but regionalism promotes its significance. Rise of local identity sentiments, reinforcement of economic activities, and the changing nature of political activities are at this level only a part of a larger-scale process of structural changes. Globalization in no way eliminates other geographical scales; but by intensifying them, regionalism and localization lead to growth and can balance the opposing changes.

2 The Failure of Thesis and Antithesis

2.1 Toward the Emergence of Regionalist Architecture

Through the critique of modern and postmodern architecture, the paradigm shift is best understood. In modernism, regionalism approach was mainly defined as a confrontation with a form of universalization. Frampton believes that culture is under attack by Modernism's optimizations by restricting it (Frampton 1983). Hartoonian (2014) states that Frampton was always aware of the necessity of the word identity in revealing the shaping of society's flourishing. From his perspective, regional architecture is able to induce this identity to society. Thus, Frampton was particularly interested in those architects whose works were on basis of identity rather than aesthetics, historic, and technological. Furthermore, all three writers, Mumford, Tzonis and Lefaivre, have a similar concern for the boldness of technology in Modern architecture and the consequences of the International Style (Panicker 2004).

After this time, a large number of architectural innovations which used to be considered beautiful and a reminiscent of progress and prosperity got demolished.[5]

---

[5] Occasionally, this has occurred due to depopulation and consequently, insecurity; such as what happened to the Pruitt–Igoe complex.

At that same time, Charles Jenckes—the historian and architectural critic—in his book *The Language of Post-Modern Architecture (1977)*, named and developed the new pattern which Robert Venturi had previously embarked upon in architecture. After countless failures of modernism, it was replaced by postmodernism since the end of World War II; a period in which a great deal of inefficient buildings and unpleasant urban projects were executed. In actuality, as a response to this identity crisis brought about by modernism, postmodernism was born in form of an antithesis to criticize the modern identity. Postmodernism believes that identity is not necessarily or consistently a fixed notion, but is changeable and dynamic. Therefore, most scientists and researchers agree on this interpretation of the postmodernist philosophy that in fact, postmodernism is an outgrowth of modernism and an attempt to find answers to the problems of the modern era and a break from the deadlock of modernism.[6] Initially, postmodernism was greeted with astronomical progress and popularity; however, it was not long before people realized that postmodernist projects—very much like modernist ones—had not earned due success, since similar to their preceding modernist generations, they maintained building and presenting fanatical global models, regardless of environmental characteristics, social individuality and the cultural unity of the regions they were performing construction projects in. In this regard, Harvey argues that postmodernism represents a crisis in modernism during which disintegration, political multiplicity, and transition are consolidated and established. This is while the possibility of any kind of consistency, stability of unity, community, unity, longevity and durability is accompanied by excessive skepticism and pessimism. He sees postmodernism as a complex set of reactions to the philosophy of modernism and its presuppositions, without the slightest agreement on the fundamental tenets of those who believe in it (Harvey 1989, 1992).

Postmodernism as an antithesis against the thesis of modernism produced a short-lived spectrum of critical studies that redefined many aspects in architectural fields and produced a superficial style that dominated the profession. With the postmodernist trend, these architects tried to reconcile their designs with the local context, but unfortunately this trend was very transient due to the rapid industrialization and internationalization of architecture. Thus, critical regionalism rejected the universalization and international style of modernism and also, ornamentation approach

[6] David Harvey begins the discussion on globalization with the analysis of space and time from pre-modernist periods and continues to compare the endurance of this process to two explosions. He attributes the first explosion to the crisis of over-accumulation in the capitalist system, which took place in the second half of the nineteenth century and was accompanied by a so-called "modernist" cultural movement. Another explosion occurred in 1970 (the symbolic end of modernism) with the density of time and space. This transformation began with the crisis of over-accumulation in the mass-production system (the advent of postmodernism). In fact, it is modernism and postmodernism which have provided the grounds for the manifestation of globalization. The subject and connections of modernism, postmodernism, and globalization have been sources for the theoretical thinking of various scholars, but the prominent figure who has actually established this connection is David Harvey. He believes that postmodernism is the unfinished project of modernism, since postmodernism has emerged after the crisis of modernism and globalization since has become a hallmark of finding a way out of the stalemate of capital accumulation in the capitalist system (Waters 2001).

of Postmodernism.[7] (Frampton 1987; Lefaivre and Tzonis 2003; Moore 2005/2007; Tzonis and Lefaivre 2012; Abu Hammad and Abu Hammad 2017).

2.2 The Question of Critical Regionalism

What is a critical regionalism? Is it a state of mind or attitude of architects? Is critical regionalism a dedicated approach for architecture to attract more attention and to better competition in awards? Or, alternatively, is critical regionalism focus on the question of how to challenge the effects of globalization through locality? Can critical regionalism be regarded as a guideline for a desirable style? Similar questions and arguments hold also for the concept of a critical regionalism, which has also prompted heated debates.

Critical regionalism remains a nascent field of study within the architectural field. It is not only regionalism but it is a progressive approach to seek answers from global and local language of architecture. Shadar (2010) discusses that "Critical Regionalism theory and its architectural expressions in literature have a prominent characteristic of local architectural shortage: the ability to change and adapt to the varying human and cultural conditions of the residents using them." Indeed, from his point of view, an architecture practice should interact with history, beliefs, and its own time and place. In architecture, the concept of Critical Regionalism gained popularity as a synthesis of universal, modern elements, and individualistic elements derived from local cultures (Botz-Bornstein 2015).

This term was initiated as an approach by virtue of the aspirations of freedom that is linked to the nationalism, liberalism, anti-authoritarianism, and rationalism. The name Critical Regionalism was first used by Alexis Tzonis and Liane Lefaivre in *The Grid and the Pathway* (1981), and it was later adopted by Kenneth Frampton in his essay *Towards a Critical Regionalism: Six Points for an Architecture of Resistance* (1983) and expansion and revision of these points in *Ten Points on Architecture of Regionalism: A Provisional Polemic* (1987). Frampton highlighted that the critical nature of his essays is confrontation with placelessness and uniformity. He draws attention to recognizing the region, native building in respect to light, wind and temperature conditions, all of which require an architectural response that deserves a special place.

The six points of Frampton were inspired from a passage from Paul Ricoeurs History and Truth that Frampton quoted as a starting point in his article (Frampton 1983). Indeed, critical regionalism is to seek answers to the question of Paul Ricoeur: "How to be modern and to continue the tradition, how to revive an old dormant civilization as part of universal civilization." Frampton argued against the pervasiveness and apparent inappropriateness of International Modernism in favor of an architecture that was distinct in its local feature and identity. The phenomenon of

[7] Critical regionalism (according to Eggener 2002), even though a reaction to modernity, is difficult to be distinguished from postmodernism; whether it is in itself antithetical to it or accompanies it.

globalization—while being one of the major advancements of mankind—constitutes a sort of subtle destruction, not only of traditional cultures—which may not be an irreparable harm—but also what shall for the time being called "the creative nucleus of great cultures;" the nucleus on the basis of which we interpret life and shall in advanced be called "the ethical and mythical nucleus of mankind." (Ricoeur 1965, pp. 276–277).

It is not modernism which is denied by many regionalist architects, but mainly an international and universal style as a result of globalization. During the final years of the 1960s, many people—especially some particular architects—initiated the revision in the international style (Steyn 2014). Because they believed that the implicit concepts of modern architecture—not the international style—were sometimes misinterpreted in terms of environment and surroundings. A prominent example of such architects would be the Indian architect, Charles Correa, who had studied in the United States and eventually used his knowledge and native background to create a new style in his own cultural and geographical context. His most outstanding works show the close connection he had managed to establish with the environment (Lefaivre and Tzonis 2012, p. 163). Although he used territorial architecture as an inspiration, it led him to very creative architectural experiences. His approach was most discernibly regional. In a project titled the Tube House in Ahmedabad, India—which was completed in 1964—he created a wide opening in the house in order to expel the hot air outside and direct the cold air inside, at the same time. Blending traditional aspects with new clever ways to promote the interior has made this project into a regional project in Correa's critical view. He holds the opinion that contemporary architecture mainly necessitates an understanding of principles and adapting them to local materials, habits, climates, and traditions, which can help modify architecture according to local conditions and identities. He considers the concept of climate as a fundamental determinant of structural forms, cultures, and traditions of various nations (Correa 1983).

In his writings, Kenneth Frampton mentions the work of Tadao Ando—the Japanese architect—and introduces him as a regionalist. He declares that Ando's work is conceptually "critical," since it culturally opposes the instrumentalism entailed the development of the megapolis of Tokyo, and resists the growing consumerism of the modern city (Frampton 1988). However, in this regard, Ando has neither spoken of the term critical regionalism for his work nor objected to the given label. Ando claims that his job is to create places that express regional, cultural and geographical features and establish a relationship between humans with nature as well as other people. Ando states that the universe is not an integrated space, but essentially consists of "topoi" in concrete spaces. These "topoi" are in fact heterogeneous, yet interconnected units, and their diversity relates profoundly to history, culture, climate, topography and urbanization. He also states that architecture has created a new perspective and is thus responsible for extracting the special features of the supposed region. One of the measures in Frampton's critical regionalism is a direct dialectical affiliation with nature; a conversation with the environment that Ando's architecture embodies in a structural relationship through changing the impacts of light and land side effects. This dialogue is well illustrated in Rokko Housing (1978–1983) in Kobe, Japan. Rokko Housing is marked by a 60-degree slope at the foot of Mount Rokko. Avoiding the modernist approach of tabula rasa in leveling the site for construction, Ando chose his building on a steep slope to create a "quiet structure; erected wholly in nature," which would maintain the tectonic quality of the uneven piece of land.

Critical regionalism necessarily involves a more direct discourse with nature, compared to the abstract and formal type of modern and avant-garde contemporary architecture (Frampton 1983). What is evident about topography, also applies to a similar degree of an existing urban fabric and this matter can also be claimed for climatic probabilities and local qualities of light. Frampton moreover emphasizes that incorporating these factors must fundamentally oppose the optimal use of global techniques. In fact, critical regionalism—in opposition to modernization—uses site topography as a crucial element of its regional geography. Destroying the site in order to level the land and consequently reducing costs and increasing efficiency can be considered as a gesture of sheer placelessness. This occurs when the geographical features of a site are not taken into account in the process of architectural rationalism.

2.3 *Coda-Geography: A Hidden Antidote to Rescue Architecture*

This study discusses the critical regionalism theory and its geographical expressions in architecture, presenting a criticism of modernism and the role of geographical factors as a hidden antidote to rescue modern architecture. Thus, geography gives us an incredible lens through which to see the architecture and their elements. It nevertheless seems that the subject of the relationship between geography and architecture has been abundantly explained and expanded in the studies of many geographers and architects. However, given the pervasion of principles such as change and multiplicity in structures governing the global way of thought, ambiguities and daily complexities will most probably overshadow the nature of the issue. Moreover, the method of attending to the geographical context faces diverse attitudes and there is no one standard in this regard. Therefore, in many cases, theorists have highlighted the role of architects in properly defining and implementing this process more prominently than other factors, and believe that by cultivating the architects' mind and genuinely encouraging their geographical and cultural sensitivities, we can hope to expect that the ultimate architectural product—while enjoying pure creativity—expresses manifestations of a careful attention for spatial–temporal features in the most appropriate form. This approach accentuates the architects' role in paying attention to global ideas and their ability to apply architectural grammar and principles (albeit in line with the school of modernism).

In the context of re-reading cultural globalization—given the role and nature of geography—it is easy to understand why in traditional societies and before the prevalence of the globalization process, the identity crisis in architecture was of significant importance. The localization of architecture as a cultural product in traditional societies could be well understood by examining the relationship between "place" and "space" in such societies. By transforming the realm of space into a modern subject, the proliferation of countless identical spaces representing Euclidean geometry and modern abstract thinking—without motivation and originality of experience—space is rushing toward place lessness over time. In reality, what led modernist architecture to this unification is a lack of concern for these spatial differences. Since modern architecture had negated the geographical and historiographical boundaries, it can be concluded that from the perspective of the modernist architecture movement, history and geography, just as the main conditions of human culture, do not have a significant impact on the development of architecture, compared to technological innovations and visual expressiveness. The abstraction and uniqueness of modern architecture is at odds with the ancient culture of architecture, which originates from the historical development of its elements in a specific geographical context.

Regionalism is a form of architecture occurring in the place, and is a way of constructing space, sensitive to specific climatic and geographical conditions, or in other words (Popescu 2006): an architectural response to geography, rewritten in different manifestations of regionalism. He considers the late nineteenth-century regionalist architecture to be very much influenced by human geography. Meanwhile, he believes that the innovative movement in geography has sizably contributed to the creation of critical regionalism. It can be argued that in order to present regionalist architecture, a geographical understanding and appreciation of the qualities of the supposed space and its connection with socio-cultural activities are essential. These approaches can be traced back to the history of architecture. Therefore, Christine Norberg-Schultz asserts that the modernist movement neglects "memories and symbols," and she therefore considers the need for regionalism to create a "place," inevitable (Norberg-Schulz 2000, p. 8).

There is no critical regionalism without regions. While valuing the uniqueness of a place it maintains a high level of self-criticism, ushering in a new form of regionalist writing. It looks for the uniqueness of the site and location. Indeed, the main idea of regionalism was that an architect should engage with the specificities of culture, place, tectonics, and tactile experience, otherwise the built environment will only consist of functional entities. Frampton Proposes that critical regionalism mediate between universally accepted practice and the particularities of place and locality. Although in a fact-finding manner, it can be claimed that the rescue of architecture is latent within the region and regionalism concepts in geography and it allows reflecting on some of the key contemporary debates of "critical regionalism." It would seem necessary to focus on the different cultural, environmental, and geographical characteristics that influence the theory and practice of regionalist architecture. It is about how these characteristics contributed and are contributing to generating a portfolio of critical regionalism. These geographical characteristics between the problems of global and local can certainly help in one of the main challenges for revival of architectural debate in the contemporary era, which is to seek new forms of "regionalist architecture" as defined by Lewis Mumford and then by Frampton, Tzonis, and Lefaivre.

Here, critical regionalism can be summarized not necessarily as a style, but as a process of attraction and an important discourse that determines historical and geographical features. Emphasizing the specific features of a site, its climatic conditions, geographical location, environmental context and its local cultural background—expressed through unique structural combinations along with the use of contemporary architectural language—can reinforce regional identity and location. What is witnessed is that the critical regionalist approach has a strong tendency to understand identity, local culture, and geography, a concept non-existent in the dominant and globalized style of the late 1970s and early 1980s. Regionalism seems to be the only way out of an international and global style of architecture. This concept has made architects prioritize the creation of place. Perhaps the most powerful story here is the narrative of how geographical aspects make it possible to trigger critical regionalism as a powerful paradigm in the contemporary architecture.

References

Abu Hammad, N., and M. Abu Hammad. 2017. Sustainable design thinking: Adaptability, resilience, and productivity at the core of regionally responsive architecture. *Architecture Research* 7 (3): 55–68.

Adjaye, D. 2013. Geography and architecture. In *Architecture is life*, ed. Mohsen Mostafavi, AGA Khan Award for Architecture, 42–45. Lars Müller Publishers: Zurich.

Bahga, S., and G. Raheja. 2018. An account of critical regionalism in diverse building types in postcolonial Indian architecture. *Frontiers of Architectural Research* 7: 473–496.

Betioli Contel, F. 2015. Concepts of region and regionalization: Aspects of its evolution and possible uses to health regionalization. *Saúde Soc. São Paulo* 24 (2): 447–459.

Botz-Bornstein, T. 2015. Is critical regionalist philosophy possible? *Comparative and Continental Philosophy* 2 (1): 11–25.

Boussora, K. 1990. Regionalism: lessons from Algeria and the Middle East. *Mimar* 36: 64–71.

Bradshaw, M. 1988. Conclusions: Regions and regionalism. In *Regions and regionalism in the United States. The contemporary United States*. London: Palgrave.

Cairncross, F. 1997. *The death of distance: How the communications revolution will change our lives*. London: Orion Publishing.

Canizaro, V., ed. 2007. *Architectural regionalism: Collected writings on place, identity, modernity, and tradition*. New York: Princeton Architectural Press.

Carlson-Reddig, K. 2011. Re-reading critical regionalism. In *ACSA fall meeting: Local identities global challenges*.

Cox, K. 2008. Globalization, uneven development and capital: Reflections on reading Thomas Friedman's the World is Flat. *Cambridge Journal of Regions, Economy and Society* 1: 389–410.

Colquhoun, A. 1989. *Modernity and the classical tradition, architectural essays 1980–1987*. MIT Press.

Colquhoun, A. 2007. Critique of regionalism: The concept of regionalism. In *Architectural regionalism: Collected writing*, ed. Vincent B. Cenizaro.

Correa, C. 1983. *Architecture and identity*. Singapore: Concept Media Pte Ltd.

Craggs, R., H. Geoghegan, and H. Neate. 2013. Architectural enthusiasm: Visiting buildings with the twentieth century society, environment and planning D. *Society and Space* 31 (5): 879–896.
Christopherson, S., H. Garretsen, R., and Martin. 2008. The world is not flat: Putting globalization in its place. *Cambridge Journal of Regions, Economy and Society* 1: 343–349.
Curtis, W.J.R. 1986. Towards an authentic regionalism. In *Mimar 19: Architecture in development*, ed. Hasan-Uddin Khan. Singapore: Concept Media Ltd.
Dicken, P. 2009. Globalization and transnational corporations. In *International Encyclopedia of Human Geography*, eds. Rob Kitchion, and Nigel Thrift. Amsterdam: Elsevier.
Dursun, P. 2012. Dialogue on space: Spacial codes and language of space. *ITU A|Z* 9(1): 104–115.
Eggener, K.L. 2002. Placing resistance: A critique of critical regionalism. *Journal of Architectural Education* 55 (4): 228–237.
Frampton, K. 1983. Towards a critical regionalism: Six points for an architecture of resistance. In *The anti-aesthetic: Essays on postmodern culture*, ed. Hal Foster. Port Townsen: Bay Press.
Frampton, K. 1988. *Tadao Ando: Buildings, projects, writings*. New York: Rizzoli International Publications.
Frampton, K. 1991. Reflections on the autonomy of architecture: A critique of contemporary production. In *Out of site: A social criticism of architecture*, ed. Diane Yvonne Ghirardo. Seattle: Bay Press.
Frampton, K. 1987. Ten points on architecture of regionalism: A provisional polemic. In *Center 3, New regionalism*, 20–27. New York: Rizzoli.
Friedman, T. 2005. *The world is flat: A brief history of the twenty-first century*. New York: Macmillan Publishers, Farrar, Straus and Giroux.
Gausa, M., W. Muller, and V. Guallart. 2003. *The metapolis dictionary of advanced architecture: City, technology and society in the information age*. Barcelona: Actar.
Gissen, D. 2010. Territory: Architecture beyond environment. *Architectural Design* 80 (3): 8–13.
Haggerty, M. 2017. A tropicalist turn: Architectural perspectives on regionalism from housing in Dhaka. *Fabrications* 27 (2): 231–258.
Hartoonian, G. 2014. Critical Regionalism Reloaded. *Fabrications* 16 (2): 122–139.
Harvey, D. 1989. *The condition of postmodernity*. Oxford: Basill Blackwell.
Harvey, D. 1992. Social justice, postmodernism and the city. *International Journal of Urban and Regional Research* 16 (4): 588–601.
Harvey, D. 2006. *Spaces of global capitalism: Towards a theory of uneven geographical development*. London: Verso.
Heath, K. 2009. *Vernacular architecture and regional design: Cultural process and environmental response*. Oxford: Architectural Press.
Hettne, B., and F. Söderbaum. 2002. Theorising the rise of regionness. In *New regionalism in the global political economy*, eds. Shaun Breslin et al. London: Routledge.
Heynen, H. 2013. Space as receptor, instrument or stage: Notes on the interaction between spatial and social constellations. *International Planning Studies* 18 (3–4): 342–357.
Howell, A. 2013. Is Geography 'dead' or 'destiny' in a globalizing world? A network analysis and latent space modeling approach of the world trade network. *Journal of Globalization Studies* 4 (2): 3–20.
Jacobs, J.M. 2006. A geography of big things. *Cultural Geographies* 13: 1–27.
Jacobs, J.M., and P. Merriman. 2011. Practising architectures. *Social & Cultural Geography* 12 (3): 211–222.
James, C. 2013. Architecture, the environment and spatial innovation, Conference: IBEA 2011 Conference Proceedings—Innovation and integration, science, technology and policy in the built environment, at London South Bank University.
Jones, M. 2019. Regionalism. In *The Wiley Blackwell encyclopedia of urban and regional studies*, ed. Anthony Orum. John, Wiley & Sons Ltd.
Kraftl, P. 2010. Geographies of architecture: The multiple lives of buildings. *Geography Compass* 4: 402–415.

Kraftl, P., and P. Adey. 2008. Architecture/affect/inhabitation: Geographies of being-in buildings. *Annals of the Association of American Geographers* 98: 213–231.
Le, D.V. 2018. Vietnamese folk architecture: Regional architecture based on climatic adaptation. *The Design Journal* 21 (6): 855–862.
Lefaivre, L., and A. Tzonis. 2003. *Critical regionalism, architecture and identity in a globalized world*. Munich, Berlin, London, New York: Prestel.
Madanipour, A. 1996. *Design of urban space: An inquiry into a socio-spatial process*. Wiley.
Mahgoub, Y. 2007. Architecture and the expression of cultural identity in Kuwait. *The Journal of Architecture* 12 (2): 165–182.
Massey, D. 2005. *For space*. London: Sage.
Maudlin, D. 2010. Crossing boundaries: Revisiting the thresholds of vernacular architecture. *Vernacular Architecture* 41: 10–14.
McNeill, D. 2005. Skyscraper geography. *Progress in Human Geography* 29: 41–55.
Mercier, G. 2009. Vidal de la Blache, P. In *International encyclopaedia of human geography*, eds. Rob Kitchin, and Nigel Thrift, 147–150. Elsevier. https://doi.org/10.1016/B978-008044910-4.00605-2.
Moore, S.A. 2005/2007. Technology, place and nonmodern regionalism. In *Architectural Regionalism. Collected Writings on Place, Identity, Modernity and Tradition*, ed. B. Canizaro, 103–109. Hudson: Princeton Architectural Press.
Morgan, K. 1997. The learning region: Institutions, innovation and regional renewal. *Regional Studies* 31: 491–503.
Mumford, L. 1938. *The culture of cities*. New York: Harcourt, Brace.
Norberg-Schulz, C. 1980. *Genius loci: Towards a phenomenology of architecture*. New York: Rizzoli.
Norberg-Schulz, C. 2000. Architecture: Presence, language, place (Skira library of architecture)
Nolan, D.B. 2014. Sustainable architecture is best achieved through the concept of regionalism. Original research proposal. https://usslide.net/document/sustainable-architecture-is-best-achieved-through-the-concept-of-regionalism.
O'Brien, R. 1992. *Global financial integration: The end of geography*. London: Royal Institute of International Affairs, Pinter Publishers.
Ohmae, K. 1995. *The borderless world: Power and strategy in an interdependent economy*. New York: Harper Business.
Paasi, A. 2003. Region and place: Regional identity in question. *Progress in Human Geography* 28 (4): 475–485.
Panicker, S. 2004. Criteria for critical regionalism? A case of contemporary Indian architecture in the age of globalization. INTA International Network for Tropical Architecture, at National University of Singapore, Singapore.
Person, A. 2011. Toward participatory interpretation: Cultural geographies of architecture. In *The peer-reviewed 99th association of collegiate schools of architecture annual meeting proceedings: Where do you stand*, eds. Annie Cormier, Annie Pedret, and Alberto Perez-Gomez, 397–404.
Popescu, C. 2006. Space, time: Identity. *National Identities* 8 (3): 189–206.
Popescu, C. 2008. Linking notions: A reconstructed perspective of identity and architecture. In *Architecture and identity*, eds. Peter Herrle, and Erik Wegerhoff, 289–298. Habitat International, Transaction Publishers.
Rapaport, A. 1977. Human aspects of urban form: Towards a man-environment approach to urban form and design. Pergamon Press.
Ricoeur, R. 1965. *"Universal civilization and national cultures", history and truth*. Evanston: Northwestern University Press.
Salman, M. 2018. Sustainability and vernacular architecture: Rethinking what identity is. In *Urban and architectural heritage conservation within sustainability*, ed. K. Hmood. IntechOpen.
Shadar, H. 2010. Evolution and critical regionalism. *Journal of Urban Design* 15 (2): 227–242.
Shirazi, M.R. 2012. An investigation on Tadao Ando's phenomenological reflections. *Armanshahr Architecture & Urban Development* 5 (8): 21–31.

Silva, S.N., R.S. Lima, and J.M. Palheta da Silva. 2015. Principles of human geography revisited, by Paul Vidal de La Blache. *Revista GeoAmazônia* 3 (5): 141–154.

Steyn, G. 2014. Globalisation, vernacularisasion and the invention of identities. *SAJAH* 29 (3): 49–63.

Lees, L. 2001. Towards a critical geography of architecture: The case of an Ersatz Colosseum. *Ecumene* 8 (1): 51–86.

Lehmann, S. 2016. An environmental and social approach in the modern architecture of Brazil: The work of Lina Bo Bardi. *City, Culture and Society* 7: 169–185.

Terlouw, K. 2001. Regions in geography and the regional geography of semiperipheral development. *Tijdschrift Voor Economische En Sociale Geografie* 92 (1): 76–87. https://doi.org/10.1111/1467-9663.00140.

Tzonis, A., and L. Lefaivre. 1981. The grid and the pathway. *Architecture in Greece* 15: 164–178.

Tzonis, A., and L. Lefaivre. 2012. *Architecture of regionalism in the age of globalization, peaks and valleys in the flat world*. London: Routledge.

Warf, B., and S. Arias. 2009. Introduction: the reinsertion of space into the social sciences and humanities. In *The Spatial Turn, Interdisciplinary Perspectives*, eds. Warf, B., and S. Arias. 1–10. Routledge.

Warwick, M. 2005. Geographies of globalization. London, New York: Routledge.

Waters, M. 2001. *Globalization*, 2nd ed. London: Routledge.

Zoghi Hoseini, E., D. Diba, H. Kamelnia, and M. Mokhtabad. 2018. New trends in critical regionalism through the lens of Tzonis and Lefaivre. *Creative City Design* 2 (1): 7–20.

Hidden City: The Footprint of Illegal Urbanisation in Spain

David Lopez-Casado

Abstract The spread of urban areas throughout the country, as well as the sprawl processes that it involves, have been a recurring theme in the growth of most Spanish cities over the last few decades. These phenomena have been studied extensively from various perspectives and disciplines, though this has generally been carried out without a detailed analysis of the origins of the settlements that comprise them. It is this aspect, in this case, based on its connection to the planned/unplanned origin of new urban developments connected to this city growth model that can become a key element for isolating the various related processes and helping us to understand their different dimensions. In certain cases, the surface area of the urban developments that arise without planning is so large that they manage to compete in significance with the established city. This phenomenon has been called the formation of the hidden city. The main aim of this study is to describe and analyse the problem of the illegal urban developments that have been carried out in Spain since the 1960s and 1970s. This will be carried out considering that it is connected to urban sprawl and suburbanisation processes that characterised the growth of cities in this spatial context during that time period. Secondly, we will show the effects of this illegal urban development on the territorial structures of the municipality of Cordoba, one of the most affected by this type of process.

Keywords Suburbanisation · Illegal urbanisation · Urban sprawl · Hidden city · Illegal parcelling

1 Introduction

The spatial extension of the city, which now appears dissociated from demographic growth and therefore obeys different rationales, has become a greatly significant territorial mutation, both from a quantitative and qualitative point of view. In the first case, this is due to the large amount of resources consumed (especially space) by the

D. Lopez-Casado (✉)
Department of Human Geography, University of Seville, Seville, Spain
e-mail: dlopez19@us.es

M. Krevs (ed.), *Hidden Geographies*, Key Challenges in Geography,
https://doi.org/10.1007/978-3-030-74590-5_15

shift from a compact Mediterranean city model towards more dispersed urban forms, forms that emulate, in a certain way, the urban systems of the Anglo-Saxon sphere. And, from the qualitative point of view, because these transformations have entailed the alteration of a great number of highly connoted landscapes, and their substitution by others that are described, in the best of cases, as banal and anodyne; to which we should add the loss of complexity of the urban structures associated with the new monofunctional urban residential settlements that give rise to them. The speed with which this has happened, especially during the last expansion cycle of the economy, together with the great superficial extension reached, has caused the development of a spatial organisation that is much tenser and more complicated than the pre-existing ones.

1.1 New and Not-so-New Approaches to Urban Growth

The coexistence of modern urban developments associated with new approaches to city growth and the processes of decentralisation of productive activities towards the periphery of the city has resulted in the emergence of more diverse and confusing spaces. In these spaces, a mixture of uses and activities of all kinds, together with the survival, in many cases, of agricultural activities and a cluttered natural environment, make up the everyday landscapes in which a large part of the population builds its living environment; it is its residential space, its recreational dimension, its routine, its uniqueness and its identity. In this context, it is particularly important to look at the origins of the urban settlements that have gradually shaped the periphery of the population centres and have accentuated the image of the dispersed city. This is because a significant percentage of these settlements have formed outside of urban and territorial planning.

The difficulties associated with identifying the origin of these urban transformations—planned/organic, legal/illegal—that have been colonising the territory in locations more or less distant from the city limits may be behind the lack of this distinction. This has led to the fact that studies aimed at characterising them, first in the context of the phenomenon of second homes from the 1950s onwards and, already in more recent stages, linked to the new dispersed and low-density residential models, have failed to discern their origin. However, the latter is a crucial issue, given that the consequences of all kinds that unplanned urban settlements can have on the urban systems and territorial structures in which they are inserted are radically different from those associated with regulated procedures. This element justifies the need to study how the former processes are related, not only to the modern processes of suburbanisation and urban sprawl but also to those of second homes and even to the access to housing of the classes with less economic resources.

The beginning of illegal land parcelling has been documented in Spain since the end of the sixties and, above all, throughout the seventies and eighties of the twentieth century. It is a process that emerged as a cheaper or more popular version of the existing second home developments, which targeted the more affluent classes and which, especially from the fifties and sixties onwards, began to take on special significance in the metropolitan environments of the country's main cities. Therefore, illegal parcelling would, at least in the initial stages, cover the demand for second homes and leisure areas of a growing working class that was beginning to have its basic needs covered, and that also had three important resources: free time, the ability to save money and its own means of travel. The reasons why displacement spread in such a notorious way and even led to specific legislative initiatives to counteract it in certain regions are varied and complex. Among them would be the fact that the promoters of these *pseudo-urbanisations* managed to put plots of land on the market at prices that were affordable for the economy of large sections of the lower-middle classes. However, there were also social reasons, especially in relation to what Gaviria Labarta (1971a, b) would call the *chlorophyll ideology*, alluding to the need to regain contact with nature coming from the same working class that, a few decades earlier, had been inhabiting the suffocating neighbourhoods that emerged during the developmentalist period. All this takes place in a socio-political context in which the new democratic local corporations created after the dictatorship lack the capacity to break with a dynamic inherited from the previous periods of certain disregard for urban planning regulations which, moreover, at that time offered few instruments to prevent the development of this type of territorial transformation. Over the years, some of these issues would gradually be resolved, while local councils and planners themselves would try to channel an increasingly important demand for a type of property product—the rural plot, but at affordable prices—which, moreover, did not easily fit into the successive regulatory frameworks put in place.

1.2 Objectives, Methodology and Sources

The main objective of this paper is to highlight the uniqueness of the phenomenon commonly known in Spain as illegal urbanisation. As will be shown in this study, the phenomenon is unique because it is a special type of urban settlement with the characteristics—especially in its origins—in that this type of zone arises *sui generis* (i.e. in the absence of urban planning). Moreover, this variety of urban area exhibits structural features, such as lack of infrastructure, basic services, public facilities and equipment, etc., which differentiate this type from the other settlements that are more typical of most Spanish cities' peripheries.

In terms of methodology, the research is proposed as a theoretical review that deals with the main elements involved in the process analysed: illegal urbanisations or plots, second home developments and more generally, the urban sprawl in which these can be found. To this end, a bibliographical review, fundamentally but not exclusively made up of scientific works, is proposed.

The sources used are, therefore, scientific works that have been published since the beginning of the phenomenon, around the 1950s, until the present day. There are also those other reports/inventories, etc. prepared by the public administrations concerned with the phenomenon and which have sought to analyse and diagnose it. Finally, the different documents that form part of the city's municipal urban development plans have been used as case studies to highlight some of the hypotheses and goals proposed.

2 The City Fragments

The evolution of the transformed surface area (*artificial surfaces* as referred to by Corine Land Cover) in Spain over the last few years has been one of the most important processes of the countries of the European Union (González et al. 2016). A large part of the transformed territory is associated with the development of new urban settlements or the growth of existing ones; a fact that, although it has not been exclusive to Spain, has become a unique case in the European context due to several factors. This is fundamentally due to two elements: (1) the speed with which it has occurred, especially coinciding with the last expansion cycle of the economy, and (2) due to the extraordinary extension of the land surface transformed by this type of process. In this sense, data show that, for the period between 1990 and 2018, Spain went from 6,384.13 km^2 of land occupied by artificial surfaces (type 1—level 1—Corine Land Cover) to the current 12,549.77 km^2; in other words, the surface area has practically doubled. Furthermore, it is worth noting that around 60% of this transformation relates to urban structures, and a significant proportion of these are dispersed, while simultaneously making up over 70% of all the urban structures identified in 2018. However, and logically, this process of intense land artificialisation has not been homogeneous throughout the country; on the contrary, it is strongly polarised, associated with the growth of the main urban agglomerations (Fig. 1).

The above data highlights a fact that is widely accepted by the scientific community: the progressive abandonment of the traditional model of a compact Mediterranean city towards others that are more sprawling. These new forms of city growth are relatively modern, especially in the context of southern European countries; new urban forms that produce new landscapes and, of course, a different way of living. On the other hand, in Spain, many of these new urban developments have been carried out outside urban and territorial planning, as well as the current regulatory framework, generally under the formula of so-called illegal urbanisations.

In this context of accelerated changes in all areas, a model of city expansion will begin to consolidate. However, as indicated in the previous section, this model is not new; what is new is both its pace and its intensity. This is what has been described as the process of urban sprawl.

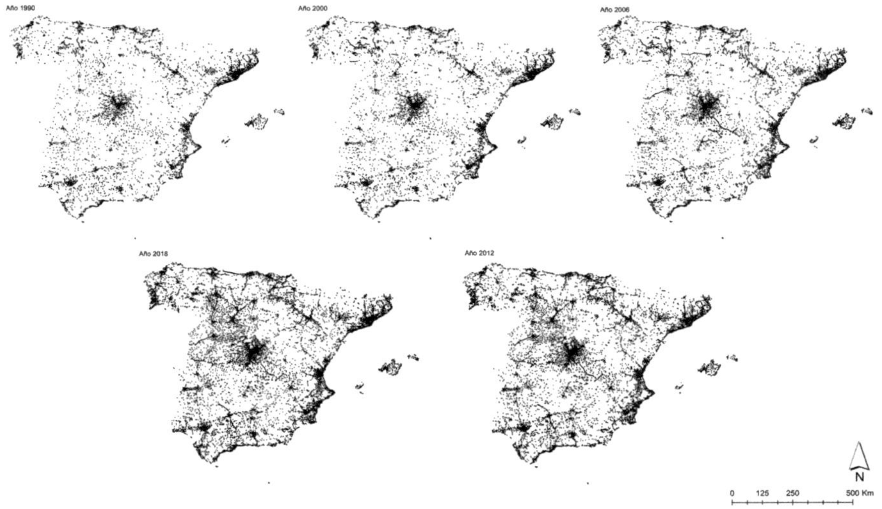

Fig. 1 Evolution of artificial areas in Spain between 1990 and 2018 (*Source* Corine Land Cover. Prepared by the author)

2.1 Forms of Dispersed Urban Growth

In 2006, the European Environment Agency published the report 'Urban Sprawl in Europe. The Ignored Challenge' where it draws attention to certain major changes that were taking place in the growth model of most European cities (European Environment Agency 2006). In this sense, it warned of the risks, both social and environmental, of this model of urban expansion in cities that had traditionally preserved a compact urban model until practically the 1950s (ibid.: 5). It also showed that the cities of the south of the continent, among others, were those especially suffering from this change of model.

For González et al. (2016), the data on the *artificialisation*[1] of land in Spain in the period 1987–2006 show not only that it was one of the countries in Europe—together with Portugal and Ireland—where the urbanised surface area grew most (ibid.: 11), but also show the unequal pace at which it did if the periods 1987–2000 and 2000–2006 are analysed. Beyond the striking nature of the data, it is important to highlight the way in which this expansion took place.[2] In that sense, the authors point out that, for the Spanish case, the following could be observed:

[1] According to the authors, this is understood as "… occupation of the same [territory] for residential, business or major public work purposes at the expense of natural and/or agricultural areas" (ibid.: 11).

[2] However, the work does not discriminate between planned/legal and unplanned/illegal urban growth, so it is impossible to know the incidence of one or the other.

> … a model that relied on urban sprawl, which separated and expanded different parts of the city across the territory and required powerful transport infrastructures to communicate them and ensure their functioning. (ibid.: 9)

Valdunciel (2013), among others, agrees with this diagnosis, clearly stating that "… the real estate bubble was based on an expansive urban development model" (ibid.: 709). The *urban sprawl* model is one that, as noted above, has been strongly criticised by the European Environment Agency because of the negative consequences it entails. Beyond these general questions, which are widely accepted by the scientific community, it is important to highlight in this work the proposal to characterise the different types of urban development in which these growths have emerged in the territory. The aim is to conceptually situate the phenomena of illegal parcelling and urbanisation within one of the proposed types. The establishment of types is an issue that has not been sufficiently well treated by the scientific literature or, at least, not in the sense of characterising the different types of urban settlements that have been created by the process of accelerated urban expansion developed during the last decades. This does not mean that the issue of urban sprawl has not been dealt with extensively by the geographical discipline from its most diverse perspectives, both theoretical and applied to specific case studies, and fundamentally in relation to large metropolitan areas, where its effects have been felt most acutely.

For some authors, continuing to use the idea of the city begins to pose serious problems of a conceptual and methodological nature (e.g. López de Lucio 1979; Nel·lo i Colóm 1998; Roca Cladera 2003). For his part, López de Lucio (1979) indicates that the use of this term to refer to the majority of those existing in Spain and the rest of Europe began to pose certain conceptual problems (ibid.: 25); and this, as the author goes on to say, is due—

> … to its radical transformation, to the rupture of the framework of collective and cohesive life that it always meant, to its dissolution or dispersion in a multitude of functionally homogeneous, socially "classified", spatially distant units, and whose only nexus of union is provided by a complex transport network… (ibid.)[3]

The author is establishing a definition of what cities would be like at the end of the 1970s, both from a morphological and a sociological point of view. In this context of change, López de Lucio considers that it is possible to clearly demarcate three types of city: the classical city, prior to the Industrial Revolution; the city "… of the *beginnings of industrialisation*, the 'liberal' city of the 'urban expansion' and the extension plans…" (ibid.); and, finally, the one he calls the *contemporary city*. The characteristics of the first two are well known, although, in terms of the second, it is worthwhile to consider some of the aspects mentioned, since they are causally related to the theoretical postulates of this work. In this sense, the author considers that "… the urban fabric [of the city at the *beginning of industrialisation*] is beginning to be defined in at least three categories: old town, planned urban expansions, spontaneous suburban or peripheral growth" (ibid.: 28). As it is easy to see, this is a much more synthetic classification than that offered by other authors (Capel Sáez

[3] Quotation marks are added by the author.

2002), although, on the other hand, it may help to offer a clearer vision of the city's expansion process at this stage. Therefore, it could be inferred that the phenomenon of second home developments would be included in what López de Lucio generically calls *spontaneous suburban or peripheral growth*. And, given that, as will be shown in the following sections, the phenomenon of illegal parcelling and urbanisation is a late temporal expression of second home developments, and an inferior one at the economic level and different at the social level, which would also be included in the aforementioned spontaneous suburban and peripheral growth. Other characteristics that stand out in the work of López de Lucio (1979) regarding the city at that time has to do with the beginning of the hierarchisation of the road system, among the consequences of which would be the physical separation of the three urban models considered. In this way, "… the old quarters are separated from the urban expansions by means of ring roads; the 'anarchic' developments on the 'outskirts' are separated from the urban expansions by means of outer ring roads or boundary roads, etc." (ibid.). This shows that, at the end of the 1980s, a clear distinction could be made between the ordered city, or the city with at least some origin in planning, and the *anarchic city on the outskirts*. Trying to find or establish order in that *territorial anarchy* will be one of the objectives, both of the successive reforms of the normative framework on urbanism and territorial planning, and of planning in its role as an element used for the implementation of the said framework.

However, López de Lucio (1979) also establishes the characteristics of the *contemporary city*, typical "… of advanced capitalism or of the so-called *State monopoly capitalism*" (ibid.: 29). Among them, the "… indefinite growth, no longer continuous, but fragmented and sprawling, superimposing itself and destroying the surrounding rural space…" stands out (ibid.), whose main consequence will be the gradual loss of the use of the term *city* to refer to this new urban reality or, at least, its subordinate use to other aspects that allude to its scale of influence. Some authors have analysed these phenomena from the point of view of "… urban deconcentration…" understood as "… the processes of urbanisation sprawl in territorial rings that surround the central nucleus…" (Mallarach Isern and Vilagrasa i Ibarz 2002: 57). On the other hand, López de Lucio (1979) indicates that these new realities will "… become what has been called, sometimes with clear pride, the *city-territory*, *conurbation*, *city-region, great metropolitan area,* etc." (ibid.: 29). Although, as Nel·lo i Colóm (1998) points out, these processes would already be, if not anticipated, at least perceived by some poets of the end of the nineteenth century, and it will not be until the last decade of the twentieth century when this phenomenon will definitively manifest itself at its fullest (ibid.: 35). For this author, the socio-economic and technological transformations that took place throughout the twentieth century have led to a physical and functional integration of space "… to such an extent that economic activities and urban lifestyles have spread over the whole territory" (ibid.), which has had the inevitable consequence that the terms *city* and *boundary* are now "irreconcilable concepts and the territory has become the *cittá sconfinata* of which some Italian authors speak" (ibid.).[4] In this context, Nel·lo i Colóm poses a paradoxical situation, in the sense that this *diffuse city* where limits are blurred in some way, which is why it is called *the city without borders*, nevertheless faces a reality that makes it, at the same time, "… also the city of borders. Borders and divides that are, in the first place, social and functional and, in the second place, political and administrative" (ibid.: 49).

Perhaps it is the latter, the political and administrative borders that most palpably interfere in the daily lives of the people who live or work in these areas. Issues as common as paying certain taxes, choosing a health centre, the schooling of children and a great many others, depend directly on the place of municipal registration, more than on the place of actual residence or work. On the other hand, the management of the uses and activities carried out on the territory also has a strong municipal component, given that the local authorities are almost exclusively responsible for town planning. Making the two realities converge is by no means an easy task, even though, at least in the area of land management and planning, there were certain experiences in the sixties and seventies, especially in some of the large urban agglomerations—Madrid, Barcelona, Valencia or Bilbao—to "… create political or management structures on a metropolitan scale that could tackle the various urban planning, infrastructure and service problems that arose on that scale" (Feria Toribio 1999: 312).

López de Lucio (1979) also points to the other characteristics of this *new city*. Perhaps the most significant features are those related to the tendency to homogenise the different urban networks or fragments that are sprawled throughout the territory, articulated through a highly hierarchical road system that allows connecting these fragments. Therefore, the author believes that the consequence of all this will be

> … total fragmentation—together with the dispersion and distancing of the different fragments—into multiple unifunctional units, with rhythms and periods of activity and vitality, users and residents, a shape and environment that are clearly differentiated and minimally related. (ibid.: 30)

Finally, this author shows that one of the invariants of most of these new *city fragments* that spread around the territory in what Pie i Ninot and Navarro (1988) call *territorial measles* (ibid.: 57) will be that of low density; an element understood as "… the brake on the evils of the city and a magic formula against land speculation…" (López de Lucio 1979: 32). The element of low density will be one of the issues that would be addressed in depth by scientific research in recent decades. Among them, perhaps we can highlight the work of Francesc Muñoz, who has been researching these issues for two decades (Muñoz Ramírez 2001, 2004, 2007, 2008, 2010; Nel·lo i Colóm and Muñoz Ramírez 2007). For him, low density, urban sprawl, monofunctionality or specialisation, fragmentation, etc. are a set of attributes that coexist in

[4] In the footnote accompanying this phrase, the author points to a group of works by Italian authors who have been addressing these topics since the 1990s.

the same area and will therefore give rise to what he calls vividly *urBANAL*isation. (Muñoz Ramírez 2004, 2008).[5]

For Indovina (2007), the really important issue, beyond the conceptualisation of the phenomenon itself or the related types that can be established, would be to assess

> … the impact on the city: what, traditionally, is still considered the city shape is the object of conflict inasmuch as those that for a long time have been considered "fragments", improper occupations of the territory (of the countryside), have ended up acquiring an urban physiognomy even though they lack all the physical and morphological characteristics of the city, that is to say, even though they lack intensity, density and lack of solution of continuity. (ibid.: 14)

The previous statement gives us a glimpse of the inherent attributes that the author considers a city must have, that is to say, those aspects that any settlement or those *fragments* that are used refer to the *improper occupations of the countryside (of the territory)*[6] must have to be considered properly urban, properly *city*: intensity, density and continuity (ibid.). But, in addition to these, Indovina identifies another important aspect to classify these settlements as cities: the existence of an adequate level of facilities and the use that the population makes of them (ibid: 18). From the above, we can see that it is possible for urban sprawl to exist without being considered a city and, on the other hand, that a truly *diffuse city* can exist. However, the author also draws attention to the inadequate use of the term *low-density city* when it is improperly associated with "… every settlement characterised by low density" (ibid.). However, in order to be able to speak of

> … low-density city [it] has to be both a city and characterised by low density; it is defined by the presence of urban complexity (functions and facilities) in a large area and by the urban use of that area by citizens. (ibid.: 18)

Therefore, the author again insists on the attributes of the city mentioned above, to which the issue of low density, and the need to occupy a sufficiently large area, are now added. If these were not the case, Indovina states, one would be faced with "… different levels of urbanisation" (ibid.), but not with examples of a diffuse city or a low-density city. Finally, the author states that the

> … low-density city and diffuse city are the formulas used in two different languages to name the same phenomenon, although one tends to highlight an element typical of architecture and construction, low density, and the other an element related to the organisation of space, diffusion. (ibid.)

The nuances expressed by Indovina in relation to the use of the terms mentioned are very significant, not only for the conceptual background they entail but, above all, for the theoretical framing of this work. This is so, firstly, because, as has been pointed out, and as we will try to show in what follows, the phenomena of urbanisation and

---

[5] The author uses this play of lower- and upper-case letters in the title of his 2006 work, *urBANAL*isation, which corresponds to the publication of his doctoral thesis, to emphasise the banal result of landscapes resulting from this type of urban development.

[6] The element that the author placed in brackets in the original definition has been consciously exchanged to emphasise the location of these fragments in the rural environment.

illegal parcelling are encompassed under the more generic phenomenon of second homes and, more specifically, in that of second homes developed in Spain from the 1950s and 1960s. However, since one of the characteristic features of illegal urbanisations is their lack of infrastructure and facilities, it seems logical to think that, given what Indovina has pointed out, the phenomenon that is the object of this research would not be a conceptual fit, at least not immediately, in the area of the diffuse city nor in that of the low-density city. This statement is fundamentally based on the grounds that, at least for this author, they cannot be considered cities, since they lack some of their main attributes.

2.2 *Urbanisation of Second Homes as a Precedent to Illegal Urbanisation in Spain*

The relationship between the traditional second home developments especially from the 1950s onwards (Ortega Valcárcel 1975: 9; Valenzuela Rubio 1976: 137) and illegal urbanisation or plots has been highlighted by various works (e.g. Comunidad de Madrid 1984; COPLACO 1981; Ezquiaga Domínguez 1983; García de Jalón Lastra et al. 1986; Requena Sánchez 1987; VV.AA. 1983). For its part, the work of Canto Fresno (1983) would be the first to attempt to contextualise this phenomenon on a national scale, highlighting the progress made by Spain with respect to the more developed countries in the European context (p. 83). Other, more limited works such as that of Guarnido Olmedo and Segura Gómez (1989) would address the same issues on the coast of Granada. For authors such as Bertolín Corbatón (1983) or Oliva Serrano and Rivera Escribano (2003), the reasons behind this trend would not be unique. Ortega Valcárcel (1975) would point out that the main reasons had to do with the strong economic development experienced by the country from the 1950s and 1960s onwards; but also with the consolidation of a shorter working day, which allowed to use part of the weekend for leisure, and the progressive widespread use of private cars which, together with the modernisation and extension of the road network, allowed for greater mobility and accessibility to spaces increasingly larger and more distant from the city. Along with all of these factors, the progressive increase of a feeling of dissatisfaction with life in the city, with ever greater problems of pollution, traffic saturation, noise, etc. brought with it a need to return to nature (Ortega Valcárcel 1975: 11). This last author draws attention to the appearance of new urban settlements in the rural space; however, at least from the functional point of view, these were not related to the space; a habitat, he adds, "… of an exclusively residential function, of temporary or secondary character, for urban occupants, promoted by urban people, and adapted to the formal demands of the population of the cities" (ibid.). What is relevant about these statements in the context of this work is that they are characterising the traditional second home developments that, especially from the 1960s onwards, will become a distinctive feature of many of the country's main cities.

The effects of this type of process on the territory would be understood differently by the researchers who focused on its analysis. Gaviria Labarta (1971b) finds them to be dangerous because "… the cultural and supra-structural conflicts derived from converting a rural area into an urban area are very important and compromise Society and the State" (ibid.: 205), and even questions whether this is really a necessity [that of the second home] or if, on the contrary, it is a process of manipulation by market agents through its various mechanisms (advertising, consumerism, credit, etc.) (ibid.). Opposing this position, Valenzuela Rubio (1976) highlights the positive effects, given that, among others, they are "… an effective agent of urban influence" (ibid.: 135), since the settlements that are born as second homes are soon susceptible to "… becoming residential suburbs" (ibid.). All this, she concludes, makes them one of the "… factors with the capacity to shape the region" (ibid.: 136). The area with greater consensus addresses the causes behind the emergence of the phenomenon. In this sense, Valenzuela Rubio (1976) considers that they are related to the progressive deterioration of the quality of life in the cities, "… consequence of the chaotic growth of the post-war period" and, secondly, by the also gradual improvement of the economic conditions that result in the increase of the savings capacity of increasingly more ample social sectors (ibid.: 137).

The emergence and rapid development of this type of settlement in the more or less remote peripheries of cities were caused by various factors, among which those associated with people's discontent with cities' living conditions can be highlighted (e.g. García-Bellido 1986; Gaviria Labarta 1971b; Valenzuela Rubio 1976). For Gaviria Labarta (1971a), this circumstance gave rise to the yearning of an ample sector of the city population to regain contact with nature, which led the author to coin the term *chlorophyll ideology*. Based on the analysis of advertisements for housing and urban developments on the outskirts of Madrid, the author notes how the texts refer to the benefits of life in contact with nature in contrast to the harmful effects of the city. Along with these, the work highlights those who seek to establish a more or less direct association between this way of life and the ideas of freedom and permanent leisure; all of this enclosed under a certain utopian halo (ibid.: 9). On the contrary, the potential users are not warned of the dangerous effects, mentioning only the supposed health- and even sex-benefits and selling a false sensation of embarking on a perennial holiday (ibid.: 119). But, along with the purely social aspects of the phenomenon, the author also mentions those of an economic nature, pointing out that the purchase of a plot in this type of development is synonymous with "… anti-inflationary guarantee—your savings on the mainland. The money has gone from the sock under the brick to the plot" (ibid.: 118). In other words, we take up again the arguments used by the authors quoted in previous paragraphs: free time, leisure and life in contact with nature, even if it is *domesticated*, as an antidote to the inconveniences of city life.

The return to democracy after the end of Franco's dictatorship would bring, among many other consequences, the progressive formulation of municipal urban plans in the main cities. These plans, for the most part, will be the first to be approved by the new democratic city councils. The territorial diagnosis made in the context of the formulation of these instruments will shed light on some of the problems arising from the urban sprawl processes associated with the phenomena described previously. Such is the case of the municipality of Cordoba, where its creators, with significant doses of irony, highlight some of the social consequences of this type of process. Thus, they point out that "… between the city of Vicenza and the anti-city of the Palladian Rotunda, there is the same relationship as between the non-peripheral city and the non-anti-city of the second home developments that our urban areas have generated in their last and destructive development period" (Ayuntamiento de Córdoba 1986: 32–33). That is to say, we can confirm both the dates of maximum development of the phenomenon and its negative effects. However, the writers of the general plan also strongly criticise "… the emergence of authentic rural suburbs without topography, in the case of the Sierra, being an obstacle to occupation—on the contrary, it is valued as a guarantee of escape from the urban world…", criticising, moreover, what they considered an ironic contradiction, since "… televisions and electrical appliances that can be connected to car batteries are massively brought in, which are the sign or the symptom of a denial of denial" (ibid.). In other words, they do seek to return to nature, but without giving up television.

The heterogeneity in the type of settlements that have traditionally been included under the umbrella of second home developments, especially in terms of their regulated/legal or non-regulated/illegal origin, has made it difficult to deepen their correct characterisation. This has been especially relevant, not so much in the former, which, as shown in the preceding paragraphs, do have a long history of scientific approach, but with respect to those that trace back to processes outside the legal system and urban and territorial planning. In this sense, authors such as Betrán Abadía and Franco Hernández (1994: 11) point out that the lack of consensus when choosing the most appropriate term to refer to this type of phenomenon is precisely one of the main problems for its study. According to Gaviria Labarta (1997: 133), the answer to the question of why illegal parcelling arises should be sought in the fact that it is the response of the less well-off classes, who want to have access to a type of property product traditionally reserved for classes with greater economic resources (Comunidad de Madrid 1984: 23). However, for this to happen, the users of this type of settlement had to accept a reduction in the facilities and provision of infrastructure, lacking, in most cases, basic infrastructures such as drinking water, electricity or sanitation. Similarly, Ezquiaga Domínguez (1983) considers that illegal plots involve a different way of using space, with the result that "… compared with the ornamental use of the plot in traditional [second home] housing estates, there is a tendency to grow vegetables or plant fruit trees…" (ibid.: 64). But, together with these, he points to the existence of complementary uses within certain areas, which highlight the development of economic activities "… typical of the *underground economy*, such as orchards, that are a means (whether complementary to other activities or not) of life at a time of economic crisis and unemployment" (ibid.).

In short, illegal housing developments or plots would have developed as an economical version of the traditional dispersed second home developments designed to meet the leisure demands of the more affluent classes. In this way, with the arrival of economic improvement in the country throughout the 1970s and 1980s, the growing demand would come from people who, although with fewer resources, had a certain degree of solvency to be able to afford a plot of land in *pseudo-urbanisations* which, despite lacking basic infrastructures and services, would be put on the market at prices that are affordable for increasingly large sectors of the population. This, together with their development on rural land and against urban and territorial planning, completes the main features that differentiate them from other types of urban sprawl.

3 Illegal Urbanisations in Cordoba: The Configuration of a Hidden City

The different inventories of illegal urbanisations drawn up by the Andalusian regional administration between 1998 and 2003 (Dirección General de Urbanismo 1989; Dirección General de Ordenación del Territorio y Urbanismo 2004) highlight, among many other aspects, the relevance of the phenomenon of illegal parcelling or urbanisation in municipalities such as Carmona in Seville, Chiclana de la Frontera in Cadiz, or the provincial capitals of Jaen and Cordoba. Similarly, the long historical trajectory that characterises this type of process in the Autonomous Community has seen an uneven evolution in the different municipalities affected. Thus, they are virtually absent in the cities of Granada, Malaga or Seville, which contrasts with the strong development that they have currently reached in other locations, such as Jaen or Cordoba; and it is Cordoba that remarkably suffers the greater impact of this type of phenomena (López-Casado 2020a, b).

Beyond trying to answer the question of why this situation has come about in the municipality of Cordoba, something to which some authors have already responded (Muñoz et al. 2013), what is relevant here is to quantify and characterise the process itself in this territorial area. In this context, the writers of the study on the housing market in Cordoba—carried out as a complementary work to the formulation of the General Land-Use Plan of 2001 (from now on PGOU.01)—pointed out, when referring to the phenomenon of illegal parcelling,[7] that the "... impressive extension of the phenomenon of abusive housing forces us to consider it as something more than a different housing sub-market" (Ayuntamiento de Córdoba 2001: 2–37). A phenomenon, they add, "... whose internal logic does not depend exclusively on its relationship with urban legality, but on a complex series of socio-economic factors...," which would mean that "... its effects on the local economy go far beyond what is usually considered" (ibid.). From the above, it can be seen that this is a complex process, with ramifications not only in the urban, territorial or legal sphere, but also in the social and economic sphere; to which should be added the associated environmental impact of the derived impacts.

The beginnings of the process of illegal parcelling should be placed in the area of the municipality's mountain range in the 1960s (López-Casado 2015). The selection

[7] In the municipality of Cordoba, the most common term used to refer to the phenomenon of illegal urbanisation is that of illegal parcelling; at other times, the term clandestine parcelling has also been used.

of this space is based on three different issues. First, by the good climatic conditions of that area versus those of the valley (Domínguez Bascón 1994), where the main nucleus of the city is located. The second reason had to do with the use of this space for some of the most popular traditions developed in the rural environment by the local population.[8] And finally, because it is the leisure area historically preferred by the people of Cordoba (Ayuntamiento de Córdoba 2001). Its rapid subsequent development and its extension to other sectors of the municipality—mainly the floodplain—would eventually lead it to be given the eloquent term of *abusive housing plots* (Ayuntamiento de Córdoba 2001). The document cited above points out that it is a completely unknown and little studied process, and this is in spite of the fact that, from the local scope, "… all series of explanations on the operation of the phenomenon had been produced, of feelings taken for certain… on a phenomenon of which paradoxically nothing significant was known except the geographic information offered by various censuses" (ibid.: 0–5). But what really is striking is that it was presented as "… something inherent to the local society, NATURAL [sic] and incapable of producing the astonishment that leads to the need to interpret the phenomena. On the other hand, seen from the outside… it emerged with a brutality that forced the need to explain [it]" (ibid.). In short, the emphasis is placed on the convenience of studying the explanation of the different aspects presented by these semi-urban centres that developed throughout the vast municipal territory, as a previous step to proposing specific actions to deal with them.

As a result of these investigations, the document under discussion offers some clues that helped not only to scale it quantitatively at that time but also to understand some of the explanatory rationales behind it. In this sense, together with the basic type of differentiation existing between the parcelling in the mountains and those in the floodplain (ibid.: 2–38), it should be noted that this is a very complex phenomenon that, therefore, transcends the excessively simplistic explanations that associate it with a different housing sub-market (ibid.: 0–5). Proof of this would be the coexistence, within the settlements, of a mix of uses ranging "… from more or less provisional facilities to authentic mansions of dubious aesthetic, including rural-type housing, all of which is interspersed with warehouses which sometimes serve as housing and which in other cases house productive activities, such as workshops or warehouses, corrals, stables and garden centres…" (ibid.: 2–37). However, it should be mentioned that the process of illegal parcelling that has been taking place in Cordoba since the 1960s and which still goes on today is not very different from what has been documented for other municipalities in the country. On the contrary, it has such similar characteristics that it might suggest that it is the response to the same socio-economic and territorial logic. An example of the above is seen in the photographs in Fig. 2, which show the incipient state of two illegal plots in two different territorial contexts. In addition to the clearest common elements, an in-depth analysis allows us to identify some of the characteristic features of the process. Thus, both images reflect the initial aspect of development that is tied to this type

[8] These are basically the traditional family meals, locally known as *peroles,* which are developed in different parts of the sierra.

Fig. 2 Images of incipient illegal parcelling. Left, Grazalema (Cádiz) in 2019; right, illegal parcelling of La Perla (Cordoba) in 2017 (*Source* Author's file (2019, 2017))

of settlement, beginning with the opening of the roads, which, at most, are provided with a layer of gravel to improve vehicle circulation. In this first stage, it is not usual for them to lack other elements of development; not even having electricity supply, which generally arrives when there is a critical mass of landowners to make their installation viable. Fencing work begins with the acquisition of the first plots. This starts with the placement of the access gate to the plot, as the images show (Fig. 2). The following steps of the process already present more variants, and will depend, to a great extent, both on the use that the plot owner is going to make of the property (housing, industrial activity, cultivation, leisure, etc.), and on the dynamics that the process itself is experiencing at that time.

On the contrary, there are also some differentiating elements with the processes that have been described for other municipalities. In this sense, one of those aspects that distinguish the illegal plots in Cordoba from those described in the work of Domínguez (1983) has to do with the type and quality of the buildings. For this author, these would be constructions of poor building quality, generally associated with self-construction processes (ibid., p. 61 and following). In contrast, the illegal plots in Cordoba feature predominately the coexistence in the same area of buildings of the most diverse types, qualities, surfaces and styles (see Fig. 3); all this while

Fig. 3 Examples of buildings of contrasting construction and architectural quality located in the illegal parcelling of Cordoba (*Source* Author's file (2017))

sharing space with other uses. Although, on the other hand, this fact may also be highlighting that we are witnessing two different stages of the same process.

The authors of the PGOU.01 propose a basic type of differentiation between illegal parcelling in the soils of the mountain range and those located in the valley. This approach being valid, recent research (López-Casado 2021) has shown that the complexity of the phenomenon makes it necessary to move towards a detailed analysis, which allows for the establishment of a typology that responds more precisely to the socio-demographic and territorial reality of the phenomenon. However, the illegal plots that have been developed in Cordoba for more than fifty years are located within two of the three large morpho-structural units present in the municipality: The sierra and the floodplain, with the third area, the fields, remaining practically unaffected (see Fig. 3). The map highlights, on the one hand, the unequal effect of the phenomenon on the municipal territory and, on the other, its special incidence on specific spaces, both on the soil of the mountain range and, above all, on the western sector of the floodplain.

As mentioned in the previous paragraphs, we can see, especially in the plots located in the floodplain, that at the beginning of the process, the first buildings may match the described pattern—that is to say, humble dwellings that are, on many occasions, associated with self-construction. However, at a more advanced stage, it is very common for buildings with higher levels of quality to coexist with them (see Fig. 3). This situation would reveal the existence of another type of user and, therefore, of other social approaches; thus, this new reality could be the consequence of the arrival of what the drafters of Annex II of the PGOU.1 call "... the group of the new rich or submerged rich (shopkeepers, jewellers, etc.) whose pressure is directed above all towards luxury housing in illegal plots due to a problem of social segregation" (Ayuntamiento de Córdoba 2001: 1–23). However, the arrival of people involved in activities linked to the cultivation of products destined for drug trafficking, although already in stages much closer to the present time and with a more limited scope, is also a factor.[9]

Regardless of the other nuances that can be added to the distinctive features of the plots that are developed in different areas of the municipality, what is intended here is to highlight the intensity that the process has reached throughout its long history. This is especially visible in certain sectors of the municipality, such as the western part of the floodplain, where the concentration of illegal urbanisations has come to form almost a conurbation with the neighbouring municipality (Fig. 4). The same happens in certain areas of the sierra, although in this case, the explanation for this concentration has more to do with issues of topography and accessibility. On the other hand, data show that between 1984, the year in which the first published data are available (Ayuntamiento de Córdoba 1984), and the present day, the municipality of Cordoba has undergone a profound process of suburbanisation (see Table 1) caused, to a large extent, by the intense development of the phenomenon of illegal urbanisation. This

[9] For some time now, acts of dismantling clandestine marijuana crops within illegal parcelling have been taking place, which would be another sign of the new uses given to these areas (Redacción 2017, 2018).

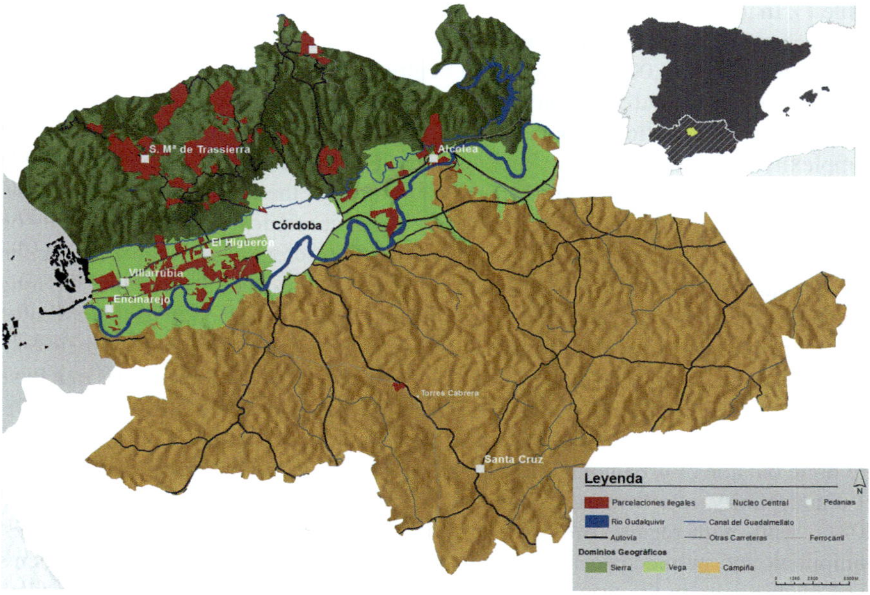

Fig. 4 Distribution of illegal parcelling in Cordoba and their relationship with the three large morphological-structural units of the municipality (*Source* Prepared by the author)

Table 1 Evolution of the population and urbanised area in the municipality between 1984 and 2019: Comparison between the consolidated city and periphery (*Source* Prepared by the authors based on INE and López Casado and Mulero Mendigorri 2015)

Year	Population			Area (ha)		
	Total (b)	Periphery (a)	% (a/b)	No. Main (b)	Periphery (a)	% (a/b)
1984	293,913	15,000	5.10%	2,705.61	2,777.97	102.67%
2019	325,701	35,788	9.76%	3,096.37	7,978.06	258.24%
Variation	9.76%	58.09%	–	14.44%	187.19%	–

is in a double sense: both because of the rate at which the population is leaving the consolidated city—the main centre—towards its more or less nearby periphery, and because of the superficial extension that the suburban settlements have reached in comparison with the city.

Although it is true that, according to the data shown, the amount of population that may be associated with the urban nuclei of the periphery is not very significant—just under 10%—it is so in absolute terms; especially if we consider that this figure is much higher than that of the resident population in most of the municipalities that belong to the Cordoba Urban Agglomeration (López-Casado 2016: 369). However, what may be more relevant is the degree of urbanisation and land consumption reached by the settlements dispersed throughout the vast municipal territory (see

Table 1). In this sense, we can see how the latter already represents more than double the space occupied by the main nucleus of the municipality. Thus, if the latter has slightly more than 3,000 ha, the urban centres of the periphery add up to almost 8,000, of which around 5,000 ha correspond to illegal urbanisations (López-Casado 2020a). The combination of both aspects, together with the intensity and degree of complexity that the phenomenon of illegal urbanisation has reached in certain sectors of the municipality—for example, the western area of the floodplain (see Fig. 3)—leads to the conclusion that we are in the process of configuring an unplanned city on the fringes of the consolidated city, which this study refers to with the term *hidden city*. The justification that supports such a decision is based on the fact that the long period of time associated with the development of the phenomenon has contributed to configuring a system of *semi-urban settlements* that are governed by their own rules, both in their physical configuration and in the code of social relations that have been established between the different parts of this system. In the first case, this is because both the appearance of the settlements and the construction of the buildings are done outside of urban planning and the legal system, which emphasises the idea of clandestinity, a term used in some of the stages of the process (López-Casado 2021). And, with regard to the second aspect, it is because the lack of facilities, equipment and public spaces has led to the appearance of new relational elements used exclusively by the users of this type of settlement and therefore, hidden from the rest of the city's population.

4 Conclusions

The process of urban sprawl that has characterised the growth of most Spanish cities, which was especially relevant during the last expansion cycle of the economy, has led to the configuration of urban peripheries with highly complex territorial structures. Together with this fact, the speed and intensity with which the process has occurred and the lack of a theoretical corpus of its own has led to the use of terms derived mainly from the Anglo-Saxon sphere, and these are not always correct. This situation has generated, on more than a few occasions, excessive generalisation or explanations that, somehow, have contributed to masking the heterogeneity of the settlements that configure these urban peripheries. This work has shown that the analysis and characterisation of these spaces have not been done correctly; this is because the root of urban nuclei has been overlooked, especially in terms of their legal urban situation. This is a relevant issue, given that the consequences of this circumstance for the territorial structures in which they are inserted are substantially different from those that have followed the guidelines and rules set out both by urban and territorial planning and by the current legal framework.

Secondly, the relationship between second home developments and the phenomenon of illegal urbanisation is maintained, and it is stated that the former are antecedents and conceptual and social references of the latter. In this sense, illegal urbanisations emerged in Spain in the mid-1960s and, above all, throughout the 1970s, as a response to a growing demand from the incipient working middle class. This sector of the population, having had its basic needs covered (mainly by having a regular home and a private vehicle), and with free time and the ability to save, felt the need to leave their everyday life in the city—generally in neighbourhoods with high population densities, with a lack of facilities and green areas for recreation—to go somewhere outside the city where they could spend their leisure time. The impossibility of accessing second home developments due to the high price of land—and of the final residential product—in these areas meant that *pseudo-urbanisations* gradually appeared on the market, where plots of land were made available to the demanding population at more affordable prices. This was done at the cost of reducing infrastructures and public services until they practically disappeared, as well as the urbanisation of the interior road, which is why the use of the term urbanisation (in its classic meaning) to refer to these new settlements is problematic since it is precisely the lack of the inherent urbanisation elements that best characterises the settlements.

In spatial terms, of all the Andalusian provincial capitals, Cordoba is the most affected by the phenomenon of illegal urbanisation, as almost 8% of its surface area is occupied, on a regional scale, by this type of unplanned urban process; it is followed by Jaén at 2.26% and Malaga and Seville with just over 1%. These figures, provided by the last inventory carried out by the regional administration in 2003, are not enough to show the scope and territorial significance that this has in the municipality of Cordoba. In this sense, research has been able to prove that the total surface area occupied by urban settlements identified as illegal urbanisations currently reaches just over 5,000 ha, a figure that is almost double the 2,860 ha of surface area of the main city centre. In other words, based on this data, it can be stated that the municipality has been forming an unplanned illegal (?) city over the last fifty years, formed by a significant number of semi-urban dispersed settlements that compete in importance with the planned city, at least in terms of surface area, and with certain significance as far as the population is concerned. But, additionally, the illegal character not only of the development process itself of this type of settlement, but also of the construction phase of the buildings or the implantation of diverse uses, makes the idea of clandestine space even more pronounced. Coupled with all this, the progressive implementation of an entire system of relations, both social and spatial, between the different settlements of this type, which runs parallel to the distancing of the planned city, makes the term *hidden city* an even stronger and appropriate option to define this process.

References

Ayuntamiento de Córdoba. 1984. *Plan General de Ordenación Urbana de Córdoba. Memoria de Información.*

Ayuntamiento de Córdoba. 1986. *Plan General de Ordenación Urbana de Córdoba.* Córdoba: Ayuntamiento de Córdoba.

Ayuntamiento de Córdoba. 2001. *Plan General de Ordenación Urbana de Córdoba: Anexo II*. Córdoba: Ayuntamiento de Córdoba.

Bertolín Corbatón, J. 1983. Motivaciones Ideológicas de La Segunda Residencia. In *Urbanismo y Medio Rural. Valencia: La Vivienda Ilegal de Segunda Residencia*, ed. Diputación Provincial de Valencia, 41–48. Valencia: Diputación Provincial de Valencia.

Betrán Abadía, R., and Y. Franco Hernández. 1994. *Parcelaciones Ilegales de Segunda Residencia: El Caso Aragonés*. Zaragoza: Diputación General de Aragón, Departamento de Ordenación Territorial, Obras Públicas y Transportes.

Canto Fresno, Consuelo del. 1983. Presente y Futuro de Las Residencias Secundarias En España. *Anales de Geografía de La Universidad Complutense* 3: 83–103.

Capel Sáez, H. 2002. *La Morfología de Las Ciudades. I, Sociedad, Cultura y Paisaje Urbano.* Barcelona: Ediciones del Serbal.

Comunidad de Madrid. 1984. *Urbanizaciones Ilegales. Programa de Actuación.* Madrid: Centro de Información y Documentación de la Consejería de Ordenación del Terriotorio, Medio Ambiente y Vivienda.

COPLACO. 1981. *Estudio Sobre Las Urbanizaciones de Segunda Residencia En La Provincia de Madrid*. Madrid: Ministerio de Obras Públicas y Urbanismo Comisión de Planeamineto y Coordinación del Área Metropolitana de Madrid.

Dirección General de Urbanismo. 1989. *Inventario de parcelaciones urbanísticas de la provincia de Córdoba (Diagnóstico)*. Sevilla: Dirección General de Urbanismo, Consejería de Obras Públicas y Transportes, Junta de Andalucía.

Dirección General de Ordenación del Territorio y Urbanismo. 2004. *Inventario de Parcelaciones Urbanísticas En Suelo No Urbanizable En Andalucía (Memoria)*. Sevilla: Dirección General de Ordenación del Territorio y Urbanismo. Consejería de Obras Públicas y Transportes, Junta de Andalucía.

Domínguez, J.M.E. 1983. Parcelaciones Ilegales En Suelo No Urbanizable: Nuevas Formas de Consumo Del Espacio En Los Márgenes de La Ley Del Suelo. *Ciudad y Territorio. Revista de Ciencia Urbana* 56: 59–72.

Domínguez Bascón, P. 1994. Rasgos Climáticos Del Término Municipal. In *Córdoba Capital, 3. Geografía*, ed. Antonio López Ontiveros and Bartolomé Valle Buenestado, 25–30. Córdoba: Caja Provincial de Ahorros de Córdoba.

European Environment Agency. 2006. *Urban Sprawl in Europe. The Ignored Challenge. EEA Report*. Vol. 10. https://doi.org/10.1080/02697451003740312.

Feria Toribio, J.M. 1999. Nuevas Periferias Urbanas y Planificación Pública. In *La Ciudad : Tamaño y Crecimiento : [Ponencias, Comunicaciones y Conclusiones Del III Coloquio de Geografía Urbana]*, ed. R. Domínguez Rodríguez, 309–316. Málaga.

García-Bellido, J. 1986. La Cuestión Rural. Indagaciones Sobre La Producción Del Espacio Rústico. *Ciudad y Territorio. Estudios Territoriales* 69: 9–52.

García de Jalón Lastra, Á., J.L.S. Guerra, E.J.M. Ezquiaga Domínguez, and L.M. González. 1986. Estudio de Las Parcelaciones Ilegales de La Provincia de Valladolid. In *Estudio de Las Parcelaciones Ilegales de La Provincia de Valladolid*: 199. Valladolid: Colegio Oficial de Arquitectos de Valladolid.

Gaviria Labarta, M. 1971a. La Ideología Clorofila. *Campo, Urbe y Espacio Del Ocio.*

Gaviria Labarta, M. 1971b. La Vida Cotidiana En Las Urbanizaciones Del Hinterland de Madrid. In *Campo, Urbe y Espacio Del Ocio*: 204–316. Madrid: Siglo XXI de España Editores.

Gaviria Labarta, M. 1997. La Caseta y El Huerto Como Espacio de Placer. *La Disciplina Urbanística: El Suelo No Urbanizable, Las Parcelas Ilegales y El Nuevo Código Penal: Actas, I y II Jornadas.*

González, R.L., Piñeira Mantiñan M.J., and Vives-Miró S. 2016. El Proceso Urbanizador En España (1990–2014): Una Interpretación Desde La Geografía y La Teoría de Los Circuitos de Capital. *Scripta Nova. Revista Electrónica de Geografía y Ciencias Sociales* XX (539).

Guarnido Olmedo, V., and I.M. Segura Gómez. 1989. Una Aproximación Al Estudio de La Segunda Residencia (La Costa Granadina). *Paralelo 37º*: 163–175.

Indovina, F. 2007. Introducción: Antes de La Ciudad Difusa. In *La Ciudad de Baja Densidad. Lógicas, Gestión y Contención*, ed. Francesco (Coord) Indovina, 13–23. Barcelona: Diputació Provincial de Barcelona.

López-Casado, D. 2015. *Los Procesos de Ocupación Irregular Del Suelo No Urbanizable En El Municipio de Córdoba (España): Primeros Resultados de Una Investigación En Curso.* Barcelona: Facultad de Arquitectura. Universidad de la República.

López-Casado, D. 2016. Consecuencias Sociodemográficas de Los Procesos de Suburbanización En Las Aglomeraciones Urbanas: El Caso de Las Parcelaciones Ilegales Del Municipio de Córdoba. In *Población y Territorio En La Encrucijada de Las Ciencias Sociales: Actas XV Congreso de La Población Española*, ed. Josefina Domínguez-Mujica and Ramón Díaz-Hernández, 365–379. Fuerteventura: Universidad de Las Palmas de Gran Canaria. Servicio de Publicaciones y Difusión.

López-Casado, D. 2020a. Illegal Parcelling in Cordoba (Spain): The Result of Illegal Urban Planning or Hidden City Development? *Dela* 54: 125–148. https://doi.org/10.4312/dela.54.125-148.

López-Casado, D. 2020b. La informalidad como nexo: Producción social del hábitat en ciudades latinoamericanas frente a parcelaciones ilegales en España. *ACME: An International Journal for Critical Geographies* 19(3): 706–724.

López-Casado, D. 2021. De urbanización ilegal de fin de semana a barrio precario: las parcelaciones ilegales en Córdoba. *Ciudades* 24: 225–246. https://doi.org/10.24197/CIUDADES.24.2021.225-246.

López-Casado, D., and A. Mulero Mendigorri. 2015. La Contribución de Los Datos Espaciales de Referencia de Andalucía (DERA) a La Investigación Geográfica : El Caso de Las Parcelaciones Ilegales Del Municipio de Córdoba'. In *XXIV Congreso de La AGE: Análisis Espacial y Representación Geográfica: Innovación y Aplicacion*, ed. J. de la Riva, P. Ibarra, R. Montorio, and M. Rodríguez, 195–204. Zaragoza.

López de Lucio, R. 1979. De La Ciudad Fragmentada y Compacta a La Disgregación Espacial Articulada. *Anales de Geografía de La Universidad Complutense* 115: 25–36.

Mallarach Isern, J., and J. Vilagrasa i Ibarz. 2002. Los Procesos de Desconcentración Urbana En Las Ciudades Medias Españolas. *Ería* 57: 57–70.

Muñoz Cuenca, J.M., J. Gómez de Hita, and A. Mulero Mendigorri. 2013. Procesos Metropolitanos Encubiertos: Córdoba Como Caso de Estudio. *Ciudad y Territorio. Estudios Territoriales* 177: 511–532.

Muñoz Ramírez, F.M. 2001. La Ciutat Multiplicada: Polítiques Urbanes i Condicions de Centralitat a La Metròpoli Postindustrial. *Revista Catalana de Sociología* 77–85.

Muñoz Ramírez, F.M. 2004. UrBANALització. La Producció Residencial de Baixa Densitat a La Província de Barcelona 1985–2001. *Tesis Doctoral*. Universitat Autònoma de Barcelona.

Muñoz Ramírez, F.M. 2007. La Producción Residencial de Baja Densidad En La Provincia de Barcelona (1985–2001). In *La Ciudad de Baja Densidad. Lógicas, Gestión y Contención*, ed. Francesco Indovina, 51–84. Barcelona: Diputació Provincial de Barcelona.

Muñoz Ramírez, F.M. 2008. *Urbanalización: Paisajes Comunes, Lugares Globales.* Barcelona: Gustavo Gili.

Muñoz Ramírez, F.M. 2010. La Densidad Urbana: De La Ciudad de Concentración Al Campo Urbanizado. In *La Agenda Cerdà. Construyendo La Barcelona Metropolitana*, ed. Joan Fuster Sobrepere, 75–115. Barcelona: Lunwerg.

Nel·lo i Colóm, O. 1998. Los Confines de La Ciudad Sin Confines. Estructura Urbana y Límites Administrativos En La Ciudad Difusa. In *La Ciudad Dispersa. Suburbanización y Nuevas Periferias*, ed. Francisco Javier Monclús Fraga, 35–58. Barcelona: Centre de Cultura Contemporània de Barcelona.

Nel·lo i Colóm, O., and F.M. Muñoz Ramírez. 2007. El Proceso de Urbanización. In *Geografía Humana: Procesos, Riesgos e Incertidumbres En Un Mundo Globalizado*, ed. Juan Romero González, 2nd ed., 275–354. Barcelona: Ariel.

Oliva Serrano, J., and M. Rivera Escribano. 2003. Utopías y Sentidos de Habitar La Ciudad Dispersa: Estrategias Residenciales En El Área Metropolitana de Pamplona/Iruñea. *Zainak. Cuadernos de Antropología-Etnografía* 23: 137–153.

Ortega Valcárcel, J. 1975. *Residencias Secundarias y Espacio de Ocio En España.* Valladolid: Departamento de Geografía, Universidad de Valladolid.

Pie i Ninot, R., and F. Navarro. 1988. De Los "Establiments" a Las Parcelaciones Ilegales. *Ciudad y Territorio. Estudios Territoriales* 75–1: 55–80.

Redacción. 2017. La Policía Local de Córdoba Decomisa Más de 5 Kilos de Marihuana En Majaneque. *Diario Córdoba.* Córdoba.

Redacción. 2018. Desmantelado Uno de Los Mayores Invernaderos de Marihuana Con Más de Mil Plantas. *Cordópolis* 25–26.

Requena Sánchez, M.D.. 1987. *Organización Espacial y Funcional de La Residencia Secundaria En La Provincia de Sevilla.* Sevilla.

Roca Cladera, J. 2003. La Delimitación de La Ciudad: ¿una Cuestión Imposible? *Ciudad y Territorio. Estudios Territoriales* 135: 17–36.

Valdunciel, J. 2013. Tipología de Las Formas de Crecimiento Del Urbanismo Expansivo. In *XXIII Congreso de Geógrafos Españoles. Espacios Insulares de Frontera, Una Visión Desde La Geografía*: 709–718. Palma de Mallorca.

Valenzuela Rubio, M. 1976. La Residencia Secundaria En La Provincia de Madrid: Génesis y Estructura Espacial. *Ciudad y Territorio. Revista de Ciencia Urbana* 2: 135–152.

VV.AA. 1983. *Urbanismo y Medio Rural. Valencia: La Vivienda Ilegal de Segunda Residencia.* Valencia: Diputación Provincial de Valencia.

The Residential Quality in the "Hidden" Roma Neighbourhoods: A Case Study of Harman Mahala, Plovdiv

Nadezhda Ilieva and Boris Kazakov

Abstract The Roma represent the second-largest ethnic minority in Bulgaria and the most dynamic population group in terms of population growth. Typically, they live in segregated neighbourhoods in cities or smaller settlements. The expansion of these Roma neighbourhoods both horizontally and vertically (in height) is difficult to trace, considering that the vast majority of the buildings are illegal and in most cases not depicted on cadastral maps and urban plans, or—due to the rapid illegal construction—the existing plans do not represent the real situation. These neighbourhoods remain invisible to the general public, as they are excluded from their minds as possible places of visiting, let alone habitation, and in their essence represent a type of unused spaces. This paper presents the result of a study aiming to make an assessment of the residential quality in the Roma neighbourhood of Harman Mahala, located in the second-largest Bulgarian city—Plovdiv. The residential quality has been assessed in both objective and subjective aspects, based on urban planning documents and legislation, field survey among the residents, and data collected with the use of an unmanned aerial vehicle (UAV). Spatial data has been combined with field studies, quantitative and qualitative methods of gathering information, which complement both the objective and subjective assessments of the residential environment (how the residents of the neighbourhood perceive the city, their neighbourhood, and their homes). The study proposes a methodology for analyzing the internal structure of Roma neighbourhoods by using data obtained from aerial photography, which compensates for the lack of information on the current state of the housing stock. The terrain has been captured from a height of 90–120 m using precise sensing instruments, recording images in the visible spectrum. Through the subsequent processing of the obtained image data and through the application of appropriate methods and algorithms, detailed maps of the actual status of the surveyed buildings—outline, height, density, surface area, etc.—have been created, which also depict the changes that took place over time. Basic spatial indicators, such as building density coefficient and green space percentage, as well as their dynamics over the period 2016–2019 have been calculated. The results of the study indicate

N. Ilieva (✉) · B. Kazakov
National Institute of Geophysics, Geodesy and Geography, Bulgarian Academy of Sciences, Sofia, Bulgaria
e-mail: nadeto.ilieva@abv.bg

M. Krevs (ed.), *Hidden Geographies*, Key Challenges in Geography,
https://doi.org/10.1007/978-3-030-74590-5_16

that while the objective assessment of the studied neighbourhood is relatively low, the subjective assessment of the residents is actually high, due to a strong sense of place and sense of security which the segregated community provides.

Keywords Roma neighbourhoods · Objective and subjective assessment · Residential quality · Harman mahala · Plovdiv

1 Introduction

One of the major challenges which modern European cities will face in the coming decades is the increasing social inequalities and spatial polarization, which are directly related to the integration processes (Cities of tomorrow etc. 2011).

Over the last decades, Bulgaria has seen a clear trend of growing number of Roma population living in cities (Ilieva et al. 2020). In most cases, the Roma inhabit existing Roma neighbourhoods or newly emerged ones. The intensive change of the Roma neighbourhoods both in horizontal and vertical (in height) aspects are difficult to trace, considering that most buildings are illegal and in most cases not depicted on cadastral maps and urban plans, or—due to the rapid expansion of illegal construction—the existing cadastral plans do not represent the real situation. These Roma neighbourhoods remain hidden to the ethnic majority, as they are excluded as possible places of habitation and represent a type of unused spaces, as far as the ethnic majority of Bulgarians is concerned.

The creation of sustainable neighbourhoods not only takes into account the quality of the environment but also increasingly focuses on creating stable, integrated communities where people live and work (Dempsey et al. 2011). The way these communities influence the space they inhabit largely determines the residents' perception of the living environment and the way it is assessed in terms of safety, health, aesthetics, comfort, and general welfare.

According to Bonnefoy (2007) "residential or housing quality refers to the physical condition of a person's home as well as the quality of the social and physical environment in which the home is located. Aspects of housing quality include air quality, home safety, space per individual, etc." Wellner et al. (2015) point out that the dimensions of residential quality are "differentiated into built features (location, quality of building, etc.), economic (rent level, tenant mix, etc.), social (neighbourhood atmosphere, security, fluctuation, etc.), ecological and health-related (noise, pollution, etc.) aspects, thus contributing to the conceptualization of residential quality."

The topic of evaluating the quality of the residential environment is discussed by Kain and Quigley (1970), while Ha and Weber (1994) developed a residential quality index and tested its relationship to residential satisfaction. The latter authors identified seven dimensions of residential quality by using the principal component factor analysis—environmental safety, planning/landscaping, housing policy, sociocultural environment, public services, housing economics, and physical quality of housing. Later these indexes were applied by Kesalkheh and Dadashpoor (2012).

Aragonés et al. (2016) make a conceptual reflection of the relation between terms such as well-being and the quality of life and speak about domains of the residential environment, such as house, neighbourhood, and neighbours, as well as their function as predictive variables to explain residential satisfaction. Once the form of understanding residential satisfaction is presented, the indexes of residential quality and their transformation to predictors of this variable are analyzed and furthermore explained in two dimensions: subjective versus objective and physical versus social. The above-mentioned authors also highlight the importance of attachment to the place and its close relation with the residential satisfaction, especially focused on the neighbourhood.

According to Bonaiuto et al. (1999), "residential satisfaction can be considered an overall evaluation of the residential environment from inhabitants' perspectives. It can be defined as the experience of pleasure or gratification deriving from living in a specific place, that is, the global assessment that inhabitants give of their housing that can be considered at various levels of scale (e.g., house, building, neighborhood). Conceptually, residential satisfaction includes the three main components of the psychological construct of attitude: cognition, affect, and behavior." On the other hand, so far as neighbourhood environment quality is concerned, consistent evidence in the literature confirms that a theoretical model that includes three or four main areas can account for inhabitants' perceived neighbourhood quality. Specifically, the first three areas are traditionally established in the literature, whereas the fourth area has been proposed in more recent contributions. Those features are spatial features, human and social features, functional features, and contextual features (lifestyle, pollution/health, maintenance, and care).

Bonaiuto and Fornara (2017) speak about the residential (urban) quality evaluation of the various specific attributes of the residential (urban) environment, which can be technical and expert-based (i.e., objective) or observer-based (i.e., perceived or subjective). The two authors define residential satisfaction as the experience of pleasure or gratification deriving from living in a specific place, that is, the global evaluations that inhabitants give of their housing that can be considered at various levels of scale (e.g., house, building, and neighbourhood).

Abidin et al. (2019), based on a literature review, summarize that "residential satisfaction is defined as the feeling of contentment when one has or achieved what one needs or desires in a house. Residential satisfaction can also be defined as indicator of homeowners' view of the general quality of their life and it can mean that an individual's expectation of housing is met. Residential satisfaction is the descriptions of the quality of life of the inhabitants of a determinate residential environment and act as trigger factor affecting residential mobility". Abidin et al. (ibid.) also discuss the theories which are related to residential satisfaction such as housing needs theories, referring to Rossi (1955); housing deficit theories, referring to Morris and Winter (1975); and psychological construct theory, referring to Galster (1985).

The main purpose of the study is to carry out an objective and subjective assessment of the residential quality in the hidden Roma neighbourhood of Harman Mahala in Plovdiv, Bulgaria. It is interesting to know how a marginalized community evaluates its neighbourhood, dwellings, and neighbours. The authors of this study adopt the perception that the objective assessment of residential quality is related to the physical and infrastructural environment and to what extent it corresponds to the basic urban planning indicators set in legislative documents, while the subjective assessment relates to the residents' idea about the neighbourhood they live in.

The objective approach focuses more on the objective standards and scientific criteria for assessment (Shieh et al. 2011). The subjective assessment, on the other hand, refers to the way in which the residents perceive the city, their neighbourhood, and their homes. The subjective evaluation sometimes does not correspond to the parameters which decision-makers can control. Sometimes a residential environment can be perceived as of high quality, regardless of how that environment is deemed by objective observers or measures (Shieh et al. 2011). The subjective assessment of the residential quality is closely related to the so-called sense of place. The sense of place is a multidimensional, complex construct used to characterize the relationship between people and spatial settings and can be established at different geographic scales: a neighbourhood, a city, or a country. The term unites two related meanings of sense: (1) understanding and order, as in "making sense" and (2) feeling and sensation, such as smell, taste, and sight (Tuan 1977). The sense of place is defined by the following:

- Affection to the place
- Identity formation
- Dependency
- Security
- Belonging
- Developed social networks and relationships
- Duration of residence and residence status.

Knox and Pinch (2000) state that population grouping is a function of the degree of social interaction that is partly due to social distance. The neighbourhood cannot be viewed from a physical point of view alone: as a place where representatives of the Roma group live. The connection between space and population is deeply rooted, with the community itself transforming that space which represents a "mirror" of the community's values, attitudes, and perceptions.

According to Shieh et al. (2011), "people are active agents in constructing perceptions of their environments and their perceptions are influenced by individual characteristics and circumstances. Second, people's perceptions of their surroundings are often closely correlated with independently-derived observations of the surroundings, and are often stronger predictors than measured conditions of individuals' place identity as expressed in neighbourhood attachment and satisfaction" (Shieh et al. 2011).

Quantifications of neighbourhood perceptions have focused either on positive social characteristics, such as control, collective efficacy, stability, informal support, and security, or on negative social characteristics, such as disorder or crime (Taylor 1996). Knox and Pinch (2000) note that the segregation of ethnic groups in space helps immigrants and vulnerable and marginalized groups to support one another, preserve their culture, provide security, and strengthen their political representation and influence. Van Kempen and Özüekren (1998) describe the advantages of maintaining close relationships within the community and the mutual benefits of receiving support in terms of finding a job and financial resources. Densely clustered ethnic neighbourhoods are also perceived as protective spaces, where members of the respective group reduce their isolation, allowing the development of protective mechanisms (Boal 1981). For these reasons, ethnic-based solidarity models are used as a "strategy for surviving urban inequalities and urban poverty" (Kogan 2004).

The Roma are rejected both at the institutional and at the personal levels by the surrounding ethnic majority. Enclosed within their ethno-territorial community, the social functions performed by segregated urban structures, such as Harman Mahala, can be summarized as follows: provision of security, preservation of cultural identity, provision of services, assistance and support for the community members, etc. The secure, enclosed environment of the Roma in the neighbourhood is perceived as a fortress protecting them from a hostile external environment. While individuals are distancing themselves from the general society, the opposite process occurs within their own group: due to the physical proximity of residents with similar ethnic, religious, and social backgrounds, social contacts and specific culture are maintained, which are not based on the norms and values of the ethnic majority (the macro society). The members of the Roma community are connected to social networks through which they benefit from each other and offer mutual support, more easily maintained at a higher level of spatial concentration. The Roma form a space with common ethno-cultural characteristics of the residents: common ethnic and social background, shared interests, worldview, beliefs, norms, values, similar expectations, and lifestyle. The subjective assessment of the neighbourhood is influenced by the time of settling and the sequence of the generations that inhabit this area: usually, the longer the stay, the stronger the space-population relationship and the deeper the imprint on the surrounding environment. Therefore, short-term transient residents rarely develop a sense of attachment and belonging to the place. Asenov's (2018) survey and quantitative studies show that the third to the fourth generation of Roma reside in the Harman Mahala quarter. This creates prerequisites for forming a community that is deeply rooted in the space it inhabits and intertwined through strong social relationships. Very often, the neighbour—close or distant—is a relative, and the division of space is a matter of internal arrangements, which are defined depending on the closeness between neighbours.

The Roma are true members of the neighbourhood in its physical aspect, given that they are linked to very strong social networks that provide support to the members of the community, reduce their solitude, provide a safe environment, and help them solve their everyday problems. On the other hand, the acute need for solidarity and social support at the collective level also serves as a shield against discrimination by the surrounding ethnic majority and even provides certain advantages of economic (creation of specific livelihood strategies and alternative economic structures), cultural (the ability to maintain and unite common cultural patterns), social (linking to social support networks), and political nature (creation of alternative political institutions) (Boal 1981; Marcuse 1997).

Another factor that influences the formation of an ethno-cultural community is the so-called spatial stigmatization, which is related to the process of attributing certain features and characteristics to a particular person or a group, which are derived from perceived stereotypes, rather than from their actual behaviour. Previous research (Ilieva et al. 2019) has established an increasing capsulation of the Roma community from the surrounding urban environment. The autonomous spatial unit, such as Harman Mahala, is becoming increasingly remote and decreasingly reliant on the state, while at the same time, the role of the internal social networks, loyalty to the place, long-standing coexistence (for more than a century), the internal rules, laws, and norms are becoming more and more important. These processes lead to the ever-greater isolation and growth of the invisible wall that exists between the Roma quarter and the surrounding urban environment. The segregated neighbourhood is perceived as a white spot on the city map by the ethnic majority, excluding it as a possible place of visiting, believing it is a high-crime and dangerous area. This spatial stigma extends not only within but also beyond the neighbourhood itself, as evidenced by the decline of real estate prices adjacent to the Harman Mahala quarter. The construction of this invisible wall between the ethnic majority and the Roma minority, together with the accompanying rejection of the Roma on institutional and personal levels, plays the role of a strong consolidating factor, which reinforces the main characteristics defining the sense of place: affection to the place, identity formation, security, belonging, developed social networks, and relationships.

2 Brief Historical Description of the Dynamics and Distribution of Roma in Bulgaria

The Roma people live dispersed among the general population, and, therefore, no typical areas of high Roma settling density have been formed in the country. One of the reasons why Roma live across the whole territory of a given country, not just in Bulgaria, is the specific character of their professions and crafting which they practice to make a living, not allowing them to support a large group of people. This forces them to split into relatively small units and maintain closer economic relations with other ethnicities, rather than with their own. It is those relations that contribute to the formation and transformation of the Roma ethnic self-consciousness and make the estimation of the exact number of Roma difficult since many tend to identify themselves as Turks or Bulgarians during the population census.

From The Liberation (1878) to shortly after the end of WWII (1946), the Roma population number in Bulgaria grew by 90%—from 89,000 to 170,000 people. During the so-called socialist period—between the end of WW II and 1989—the number of Roma nearly doubled and had reached 313,000 people by 1992. Fast growth of the Roma population in the country followed, and by 2011, their number had reached 325,000 people. According to the so-called expert assessment, however,

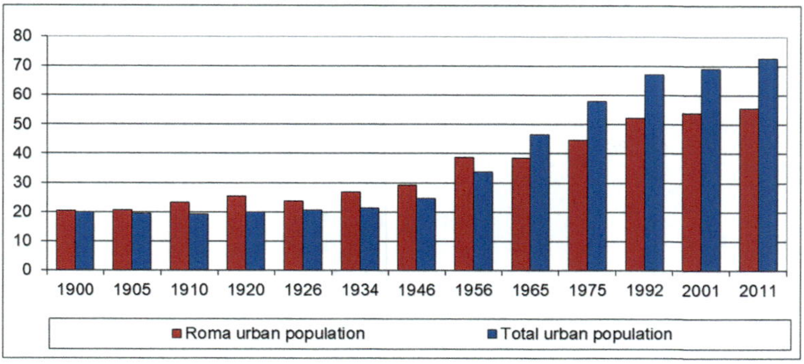

Fig. 1 Shares of Roma urban and total urban population in Bulgaria (%)

the number of Roma in Bulgaria nowadays is actually twice higher than the official numbers—around 750,000 people (Ilieva et al. 2019).

Until the mid-1950s, the proportion of Roma in cities was higher than the national average. The increase of that share was mainly due to the twice higher growth rates of Roma in cities than in villages, as well as the administrative transfer of rural residents into urban ones due to the change of status of some villages which were declared cities (Fig. 1).

After WWII, there has been a clear trend of increased concentration of Roma in cities. The Roma live mainly in medium-sized (30–100 thousand residents) and large (above 100 thousand residents) cities. Almost 50% of the Roma live in such cities, although in most cases their share is low (2–5% of the city's total population number). In recent years, increased migration to cities has been observed, as a result of the severe economic crisis and high unemployment rates in the rural areas of the country.

Since 1992, an increase in cities with a high share of Roma population has been observed in the Eastern Balkan Range Mountain, Bulgaria's north-western and north-eastern region, and some cities along the Maritsa River valley, including Plovdiv (Fig. 2).

Plovdiv is the second-largest city in Bulgaria and is located in the south-central part of the country. The Harman Mahala quarter itself is located in the north administrative region of Plovdiv, covering an area of approximately 5 ha (Fig. 3) and is home to around 1800 people according to field research (Asenov 2018). According to that same research, 75.8% of the population of Harman Mahala identify themselves as Turks or *Millet*, 19.8% as Roma, and 4.5% as Bulgarian, while for the surrounding macro society of ethnic Bulgarians are all regarded as Roma or gypsies.

Since the accession of Bulgaria to the EU (2007), there has been a gradual increase in emigration abroad—currently around one-third of the residents of Harman Mahala effectively live abroad. The socio-economic status of a household is a decisive factor for the location of its dwelling, the conditions, and the permanence/duration of its occupancy. Income levels, together with demographics, are also seen as key factors

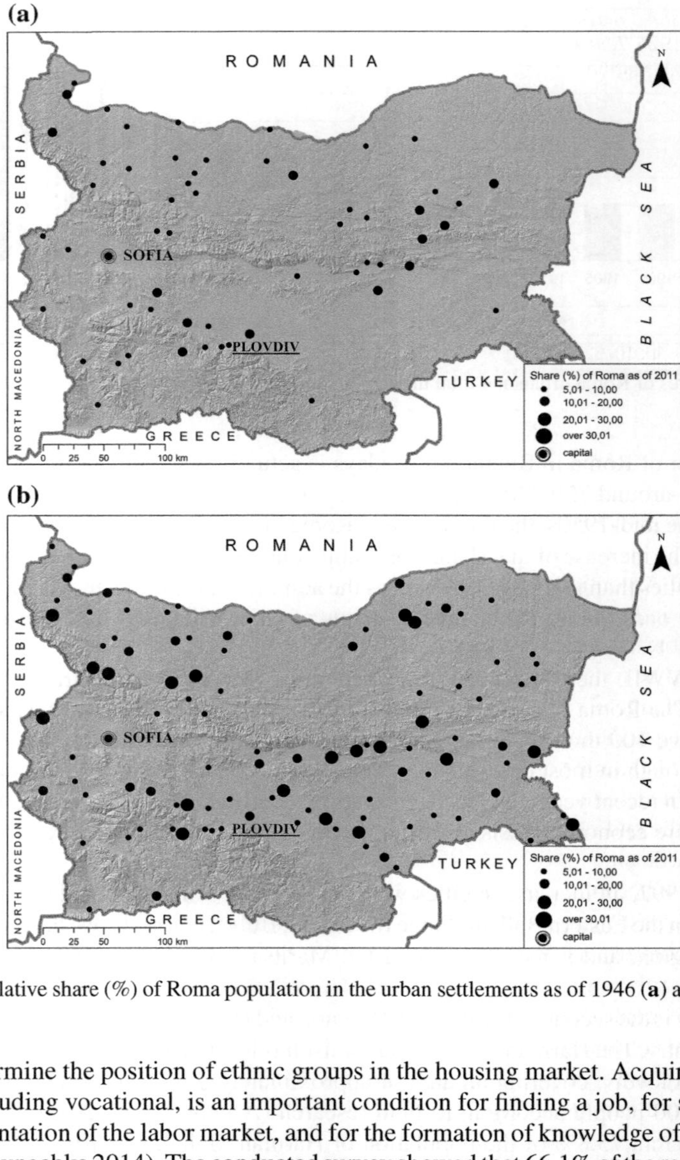

Fig. 2 Relative share (%) of Roma population in the urban settlements as of 1946 (**a**) and 2011 (**b**)

that determine the position of ethnic groups in the housing market. Acquired education, including vocational, is an important condition for finding a job, for successful implementation of the labor market, and for the formation of knowledge of the workforce (Ravnachka 2014). The conducted survey showed that 66.1% of the respondents were currently unemployed, 17.7% were working part-time, and only 16.2% were working full-time. The low educational attainment of the respondents (57% of them have primary and lower education, 34% have secondary education, and just 2% have undergraduate and higher education) determines to a great extent their position in the labor market: the majority of the employed respondents work in various industries, construction, landscaping, transport, and service sector. The budget of 35.5%

Fig. 3 Locations of the Harman Mahala quarter in the urban structure of Plovdiv

of the households is formed by one source of income only (which reflects the high degree of uncertainty and dependence on possible negative changes), followed by three sources (32.2%), two sources (25.8%), and four or more sources (6.5%). One of the main tools for analysing household wealth is to consider the structure of their total income. In only 39% of the households in the studied neighbourhood, wages are listed as an income source. In forming the structure of the total household income in Harman Mahala, 38% falls on the non-regulated income: suitcase trading, return migration to Western countries (mainly Germany), collecting herbs, mushrooms, etc. The share of households relying on financial support from relatives working abroad is significant (32.1%). Passive sources of income (those that are not the result of active employment) occupy an essential part in shaping the budget of the households surveyed. For example, child benefits are observed in 43.8% of the cases, pensions 25%, social benefits 17.1%, and unemployment benefits 6.3%.

3 Methodology of the Research

The objective assessment of the residential quality in the neighbourhood is made on the basis of the adopted thresholds set out in various legislative documents. Detailed maps of the actual state of the buildings were created on the basis of aerial photography with the use of the unmanned aerial vehicle (UAV), cadastral plans (2010), and the Integrated Plan for Urban Regeneration and Development of the city of Plovdiv (2013). Those images were georeferenced and digitized in a GIS environment. GIS technologies were used to trace, visualize, and analyse changes in the urban space's structure and to calculate the major spatial indicators which were later attached as attribute information to the corresponding ArcGIS layer. For the purposes of the current study, two UAV scannings of the studied area were conducted: one in 2016 and another one in 2019. The obtained images were later processed by Pix4D software. Through the application of appropriate methods and algorithms, detailed maps of the actual status of the surveyed buildings—outline, height, density, surface area, etc.—were created, which also depicted the changes that took place over time. Main spatial indicators such as building density coefficient, and green space percentage, as well as their dynamics over the period 2016–2019, have been calculated.

Objective indicators for evaluating the quality of residential environment include the following:

- General characteristics of the housing stock (number of dwellings, suitability, degree of degradation of the building stock, etc.) according to the set standards in The Spatial Planning Act, Ordinance No. 5 for rules and regulations for spatial planning, 2001, and Ordinance No. 8 for the volume and content of development schemes and plans, cadastre, and property register, 2001;
- Structure of the housing stock (number of rooms, number of inhabitants, etc.);
- Ownership structure (state, municipal, and private);
- Structural indicators (density of construction, intensity of construction, built-up and unfolded built-up area, gross and net residential areas, gross and net residential areas per inhabitant, average number of storeys, housing satisfaction, gross and net density of habitation, etc.);
- Proximity to industrial areas;
- Degree of improvement of the housing stock and the residential infrastructure;
- Effective and favourable distribution of basic human functions (habitation, labor, recreation, transportation, and services) for which regulatory values are determined by Regulation No. 7 of 22 December 2003, stating the norms for urban planning in the different urban areas;
- Degree of development of technical infrastructure (roads; electricity supply; water supply; sewerage; gas pipelines; telecommunication; and other types of networks, internet access, etc.); and
- Garbage collection service and frequency.

Regulation No. 7 of December 2003 on the rules and regulations for the development of the different types of areas and zones and the urban planning regulations for cities, such as Plovdiv (pop. over 100,000), and for low-rise residential areas (up to 10 m height), such as Harman Mahala, provides the following:

- The minimum gross built-up area per inhabitant is 35–50 m^2.
- The net habitation area (or the so-called regulated residential property) must be 50–55% of the gross habitation area.
- The density of construction must be 20–60%.
- The intensity of construction must be 0.5 to 1.2.
- Required landscaped area must be 40 to 60%.

The gross habitation area of the residential zones includes regulated land plots for construction; regulated estates for public and service buildings, streets, and parking lots; as well as regulated estates for public landscaped areas.

Spatial data has been combined with field studies, which complement both the objective assessment of the residential environment and its subjective assessment. The survey involved the distribution of a questionnaire among 500 randomly selected inhabitants of the studied neighbourhood, aged between 18 and 76 (63% male and 37% female), and was conducted in 2018. The answers provided by the respondents were marked by the interviewer, who gave additional explanations, since most respondents are not fluent in both written and spoken Bulgarian; therefore, educational and health mediators, who are part of the community of the Roma group, were hired as interviewers. The questionnaire consisted of both open-ended and yes/no questions, as well as graded-answer questions (the Likert scale).

The open-ended questions were used to clarify the reasons for the answers received to the yes/no questions. Respondents were asked to describe their current residential environment and indicate the main advantages and disadvantages of their current living conditions, what they would like to preserve as before, what they would like to change, and what they would organize differently in their home, neighbourhood, and city, based on their life experience. The respondents were also asked to state five specific things they like or dislike about their neighbourhood.

The following (objective) data regarding the housing stock was provided by the respondents: number of stories of the building they live in, construction material used, year of construction, number of rooms, electrification, sewerage, water supply source, built-up area of the dwelling, and provision with home appliances.

The theoretical model of the residential environment quality assessment proposed by Shieh et al. (2011) was used to design the questionnaire described above, while respondents were asked to evaluate their present residential situation with regard to each attribute in the model (Fig. 4).

This model starts with the top-level attribute environmental quality, represented by residential satisfaction. The top-level attribute branches out into more specific, lower level attributes, in this case, satisfaction with the dwelling, the neighbourhood, and the neighbours. In turn, some of these attributes branch out further into even lower level attributes. This continues until the end-level attributes are reached, that is, attributes on which the object may be validly measured. Respondents evaluate their residential situation on all of the attributes in the model (Shieh et al. 2011).

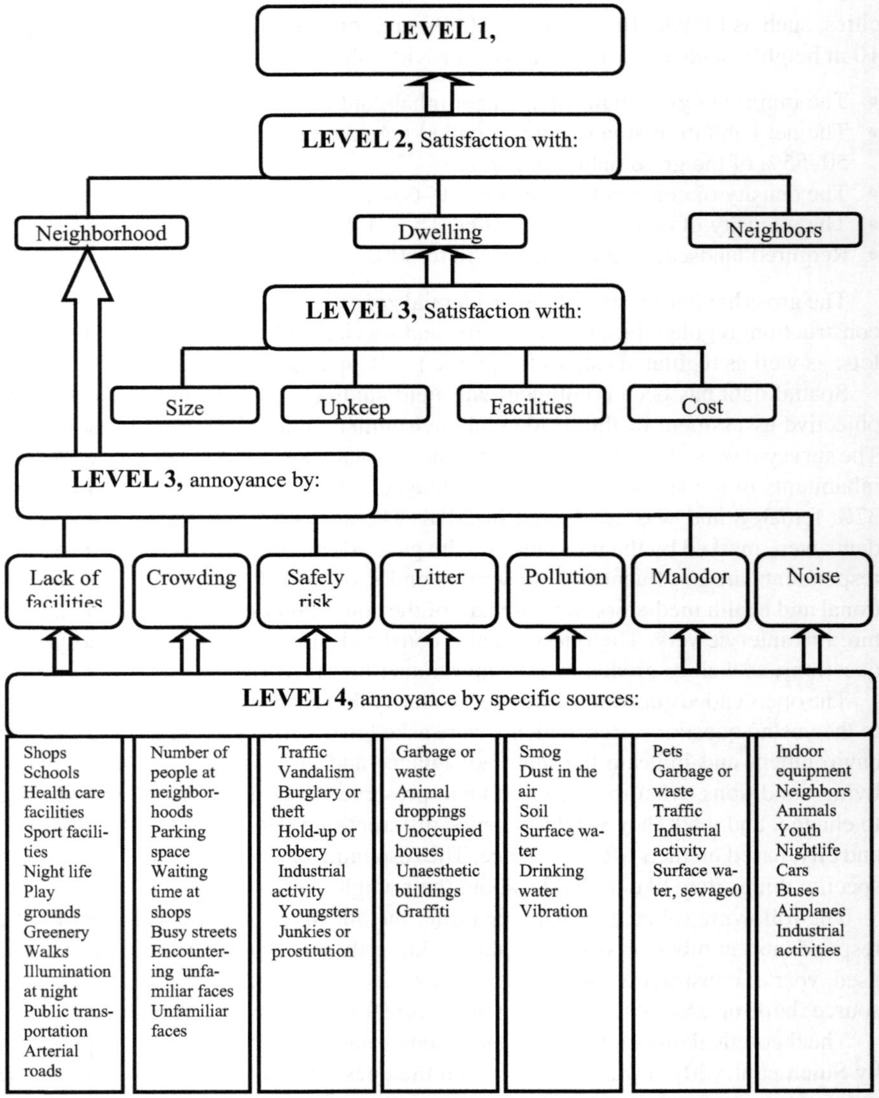

Fig. 4 Theoretical model of attributes contributing to residential satisfaction, Shieh et al. (2011)

In the questionnaire, the attributes were *residential satisfaction* (Level 1)—measured by asking the respondents to what extent they would regret leaving their present neighbourhood if they had to move (a 5-scale response format: from 'not at all' to 'very much'); *satisfaction with the neighbourhood, the dwelling, the neighbours* (Level 2), and the *dwelling features* (Level 3)—measured through questions, such as: 'How satisfied are you with …'; (response format: from 'not at all' to 'very much'); *(Dis)satisfaction with the neighbourhood attributes* (Level 3) and the specific *sources of annoyance* (Level 4), on the other hand, were measured by asking respondents to what extent they were annoyed by these attributes/specific sources. After finishing the relevant sections, respondents were asked to name and evaluate sources of annoyance that were not stated in the relevant sections of the questionnaire. Each section finished with the request to state the most annoying source. The respondents were also asked about their desire to move to another neighbourhood, attitude towards living in social housing, satisfaction with the amenities, the safety, the relation with neighbours, etc.

4 Results of the Objective Assessment of the Quality of Residential Environment

The major part (76%) of the housing stock in Harman Mahala was built after the socio-political changes of 1989 as a response to the natural growth of the population and due to the improvement of the residents' financial well-being as a result of the significant share of people currently working abroad (about 34%). In the second half of the 1990s, there was a trend of mass construction of new residential buildings and reconstruction of the existing ones which remained hidden for the institutions, researchers, and general public. More recently, in just three years (2016–2019), 19 new buildings were built with a total built-up area of 981 m^2 (an average built-up area of 51 m^2) and 16 buildings were reconstructed and upgraded, with a total built-up area of 1618 m^2.

In 2018, following a resolution of the Municipal Council, 23 illegal houses located away from the core of the neighbourhood were demolished (Fig. 5). Those were inhabited by members of the Burgudzhii Roma group, who, unlike the rest of the residents, are Christians and speak Romanes language (as opposed to Turkish in the case of the vast majority of the residents). After the houses were demolished, the Roma were provided with municipal housing and were moved to the largest Roma neighbourhood in the city of Plovdiv–Stolipinovo.

Only 3% of the houses in the neighbourhood actually have a legal ownership document, while 97% of the buildings were built on municipal (council) lots. A random and unregulated construction of homes has made the housing infrastructure in the quarter insecure since the houses lack any building permits and were built without meeting the required standards. All these problems pose dangers related to the occupation of the buildings by the residents. The housing stock itself is diverse, represented by different construction approaches which basically reflect the different construction periods, while solid buildings are the predominant type. Nearly 90% of all dwellings are built of bricks. Only 4% of the houses are built of adobe and represent the housing stock from the earliest construction period of the quarter (the early 1900s). Regardless of the worsened urban development characteristics, Harman

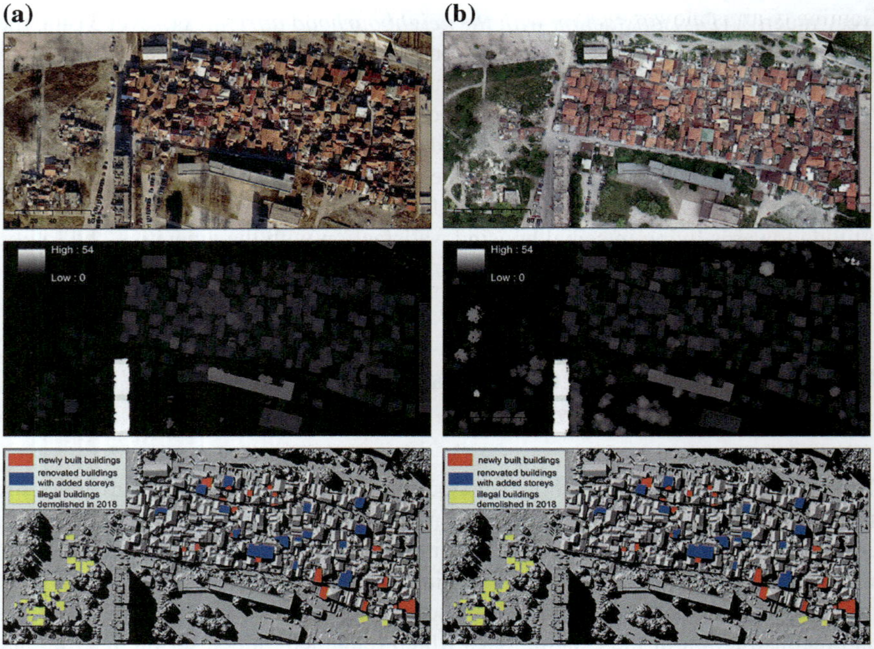

Fig. 5 Aerial and processed UAV images of the Harman Mahala quarter as of 2016 (**a**) and 2019 (**b**) depicting the general structure, the height, and the changes of the housing stock

Mahala stands out with better overall conditions compared to most other Roma quarters, given its relatively good household provision. Ninety percent of the dwellings have an internal water supply and 94% of the buildings are connected to public sewerage, while all buildings have electricity. Ninety-two percent of the inhabited dwellings have a bathroom and an internal toilet (above the national average of 82%). Heating systems mainly use mixed sources—electricity, firewood, and coal—43.5%, followed by heating only with electricity—41.9%. Dwellings with three (39%) and four (28%) rooms make up two-thirds of all dwellings in the neighbourhood. Sixty-three percent of the houses are three-storey buildings, 30% are two-storey buildings, and just 7% are one-storey or ground floor houses (the oldest ones).

No urban planning regulations apply in the neighbourhood: who is going to build and where are all matters solved by internal arrangements within the community. No green spaces exist in the quarter. The service infrastructure is only represented by primitive shops for basic everyday products. Space has become the most valuable asset in the neighbourhood. The gross building density rose from 13% in 1952 to 63% in 1982 and to over 95% in 2018. The street network is inadequate, and in an extremely poor condition, the streets are narrow, sometimes to a point where two people can barely pass through. In many places, the neighbourhood represents a construction site (Ilieva et al. 2020).

The spatial data enables the visualization of the continuous horizontal and, due to the existing limitations, vertical expansion of the Harman Mahala quarter in recent years. The average size of a household in the neighbourhood is 5.4 members. According to our survey, only 32% of all households are nuclear families where two or three generations share the same home. Whenever possible—if the housing conditions allow it—married sons normally remain with their parents. In cases where housing conditions don't allow that, outbuildings are built in the yard, reconstructions of old buildings are done, or collecting of funds comes into practice: all members of the household are involved, and the old buildings are torn down, while new houses are built in their place. The increase in population is the reason for the extremely high gross population density, which reached about 370 people/ha in 2018. Significant vertical expansion and continuing construction of homes on any lots available have been done, including through acquiring parts from the street network and the pavement areas. The desire to maximize every space available often creates severe conflicts between neighbours, although the vast majority of the existing homes, as well as the ones about to be built, are on municipal lots.

The degree of proximity between the individual families is crucial, as each family negotiates and decides whether a certain family is "one of ours" or not so much. The only laws applicable are the internal community laws. Any action by authorities to eradicate even illegal life-threatening constructions is perceived by the residents as a personal encroachment on their identity (Asenov 2018).

The prevailing dwelling size in the studied neighbourhood (51% of all homes) is 60–89 m^2, while for the city of Plovdiv as a whole, the prevailing dwelling size is actually smaller, 30–60 m^2 (62% of all homes). However, considering the number of household members and the structure of the households, a significant imbalance is observed between a dwelling in Harman Mahala and the city of Plovdiv (2.4 people per dwelling on average). In the case of Harman Mahala, there are only 12 m^2 of average useful floor area per person, while for the city as a whole that area is 24 m^2 or twice higher. Almost half of the population of the studied quarter (47%) gets between 10 and 20 m^2 of the floor area, and if we add the population with less than 10 m^2 of the living area, the percentage rises to 77% (Ilieva et al. 2020).

Green areas for wide public use exhibit two significant functional weaknesses: the only park in the North Administrative Region of Plovdiv—Ribnitsa Park—lacks the required by Regulation No. 7 15-min-walking accessibility from the studied Roma quarter. The supply of green areas of wide public use per inhabitant in the North Administrative Region is three times (2.8 m^2) lower than the city average (11.51 m^2) and seven times lower than the requirements of Regulation No. 7 (20–28 m^2 per inhabitant) for cities of the rank of Plovdiv. These extremely low values result from the fact that there is only one park that has no sufficient capacity.

Green areas with limited public use are also extremely insufficient in the area (Ilieva et al. 2019). Although not subject to specific requirements, such spaces are vital to the green system of the city and its ecology.

The *sports fields* are few and are mainly designed for professional use and not so much for public use.

The *healthcare system* is represented by the so-called Outpatient Centres (OC). As far as those are concerned, there are not equally distributed throughout the city of Plovdiv. A total of three OCs operate in the North Administrative Region (out of thirteen OCs in the entire city). The regulations state that the service perimeter of OCs should be 800 m, while Harman Mahala is located beyond that radius.

The number of existing *nurseries* (representing the pre-school educational system) is insufficient—out of the existing sixteen nurseries in the city, just three are located in the North Administrative Region of Plovdiv, while of the provided thirteen additional nurseries to be built, only one is to be located in the vicinity of Harman Mahala. The situation with *kinder gardens* is not much different: four new kinder gardens are planned in the North Administrative Region of Plovdiv, one of which is in Harman Mahala.

The pupils from the studied Roma quarter go to a typical *segregated school* where the vast majority of students are Roma. Accessibility standards for schools and kinder gardens provide a service radius of 500 and 300 m respectively—none of which are met in the case of Harman Mahala (Ilieva et al. 2019).

5 Results of the Subjective Assessment of the Residential Quality

The analysis of the questionnaire shows that a mere 0.8% of the respondents stated they were dissatisfied with their *neighbours*, while 14.3% said they were mostly dissatisfied. The assessment of *the neighbourhood itself* is highly positive—49.8% of the respondents stated they were satisfied and 45% were mostly satisfied with it. Despite the overcrowded homes and their condition, *housing* also received a very positive rating: 50% said they were satisfied and 23.1% that they were mostly satisfied with their dwelling. The positive assessment of the housing has been confirmed as well—74.4% are satisfied with the size of the dwelling, 76.6% with the amenities, and 49.7% with the price of the estate (Fig. 6). Just 12.8% find their jobs more important than the security of their surroundings and relationships with their neighbours.

The share of respondents who would not move to another neighbourhood is high—84% (Fig. 7), while half of those who say would move out, state that they would only do that if they settle in the area adjacent to the neighbourhood. Of those who would choose another neighbourhood if they could, 16% indicate they would move to the Hadzhi Hasan Mahala—a Roma neighbourhood located in the city centre, known for its higher social status among the Roma in Plovdiv. Therefore, relocation to that neighbourhood is not only a result of material well-being but also an expression of prestige.

The narrow streets and open spaces are common places for the community where people hang out and children play, while adults communicate with each other and share their daily work responsibilities. The streets of the neighbourhood are usually quite lively, to some point, due to the overcrowded homes. According to the survey,

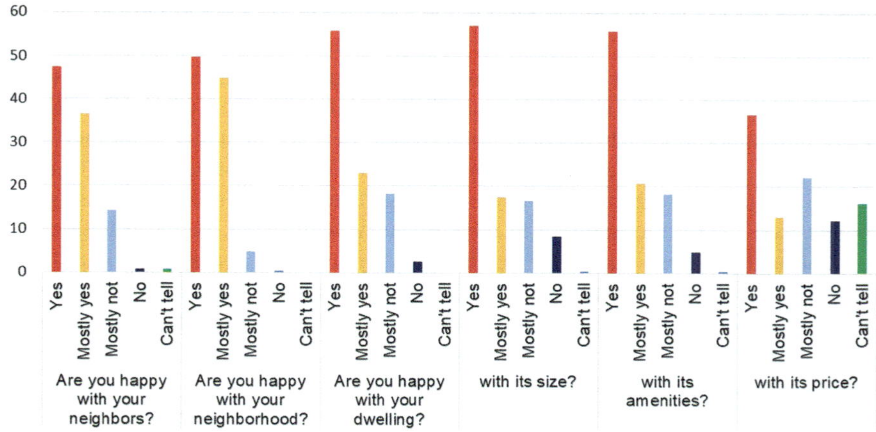

Fig. 6 Level of residents' satisfaction with the neighbours, the neighbourhood, and the dwellings

Fig. 7 Attitudes of the respondents toward moving to another neighbourhood

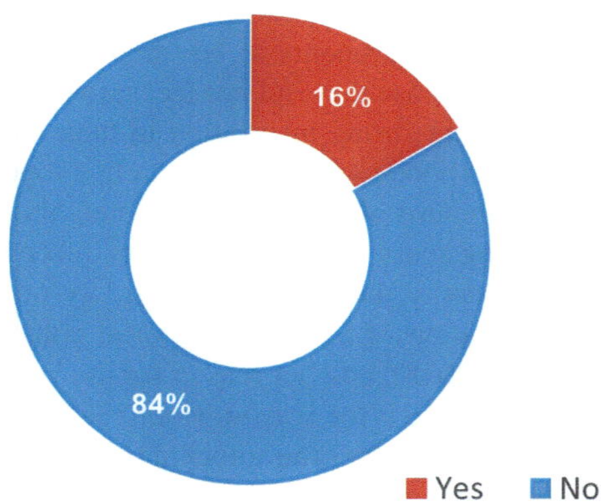

practically all respondents define their neighbourly relations as good (97.4%). The fact is that most of the residents know each other: 77.5% declare that they know everybody and spend their time with their neighbours (73.8% hang out together on a daily basis) for several reasons, such as sharing food, heating, talk and entertainment, and strengthening social networks. The most important topics the respondents discuss with their neighbours and friends are not only finding work (47%) and money (42.4%) but also illness or death (49%) and house repairing (31.1%). On the question "How did you find a job?", more than 70% of the respondents pointed out that it was through friends, relatives, or neighbours, which is indicative of the fact that the established social networks are particularly strong in terms of business. In this respect, it can be said that those networks go beyond the neighbourhood itself. Bearing in mind that about one-third of the residents of the neighbourhood effectively live abroad, chain migration is very characteristic. Usually, after first settling in a certain city (in this case most often in Dortmund, Germany), the immigrants help their relatives, friends, or neighbours to settle in the same city and get a job.

Upward social mobility, in some cases, leads to a change of residence and of the household, which moves into a more prestigious, infrastructurally better developed, richer neighbourhood, which corresponds to the new social status of an individual or a family. In other cases, the preference of ethnic communities to live in isolation within a given urban space is the leading factor, rather than the social status change itself. With the increase in their financial well-being, some residents who work abroad purchase real estate outside the "ghetto", but at the same time do not give up their estates located within the Harman Mahala neighbourhood (Asenov 2018). These circumstances prove the strong connection between the population and the neighbourhood, and even in cases residents move out of the "ghetto", they do not want to break their connection with the community.

During the so-called socialist period (1944–1989), attempts were made to reduce the concentration of the Roma population and gradually disintegrate the Roma neighbourhoods by settling residents of these neighbourhoods in the newly built at that time apartment complexes. Since the early 1990s, residents who had moved out have returned to the Roma neighbourhood. They decided to sell their expensive, well-maintained apartments to buy underdeveloped and smaller properties in Harman Mahala because they did not feel safe in a different ethnic environment. This encapsulation and strong integration in their own, closed group, applies not only to the neighbourhoods where the Roma live together with the ethnic majority but also to the inhabitants of the other Roma neighbourhoods. Such behaviour proves not only the strong sense of belonging to the place but also the formation and establishment of an identity that is not based solely on ethnicity—here we can even talk about the formation of a strong territorial community (Ilieva et al. 2020). According to Asenov (2018), another reason for the very prominent affiliation Roma people have with the place is the way in which the individual Roma neighbourhoods are formed. Asenov emphasizes the autochthonous nature of Harman Mahala, unlike other Roma neighbourhoods in Plovdiv, such as Stolipinovo, which was artificially formed by the displacement of Roma from the inner city, by decision of local authorities before World War II, when the territories which the Roma neighbourhoods occupy nowadays were outside the city limits. This artificially created neighbourhood has also grown as a result of the immigration of Roma who came from villages around Plovdiv. That is how a heterogeneous community has been formed by various Roma groups, differing in origin and ethno-cultural characteristics, which hinder the formation of such a strong attachment to the place as observed in the case of Harman Mahala.

Spatial segregation is associated with internal polarization, which leads to the rejection of outsiders. Individual Roma groups have a high degree of isolation. According to the conducted survey, 45.7% of the respondents think their Bulgarian friends/acquaintances/colleagues behave rather distantly, only 11.3% believe that partial or full integration of ethnic groups in Bulgaria is possible in the near future, 42% said they were victims of ethnic discrimination, and 34% said discrimination was higher than it used to be (yet the proportion of those who find it hard to answer the question is quite high—44.3%).

The purpose of self-isolation is to provide protection from outside influences, self-defense, and protection of one's own interests through the mechanism of exclusion of others. The rejection of others applies both to Bulgarians and members of other Roma communities. Social distance is identified by the attitude of one group as a whole towards another group. The differences that exist between groups are barriers to their potential interaction. To the question "Do you mind Bulgarians settling in Harman Mahala?", only 3.3% of the respondents said they didn't mind. As for the other Roma neighbourhoods in the city, the highest share of respondents doesn't mind settlers coming from Hadzhi Hasan Mahala—39.6% (given the higher social status of that neighbourhood's residents), 14% don't mind settlers from Sheker Mahala (which was formed by migrants from Harman Mahala after the great flood in the 1950s), while the least desired new settlers turned out to be those from Stolipinovo quarter, with only 8.3% of approvement.

When talking about sustainable communities, which are one of the preconditions for creating a sustainable urban environment and positive assessment of the living environment, the sense of security experienced by the residents must be taken into account. Ethnic grouping in urban space provides a safe habitat; to the question "Do you feel safe in the neighbourhood where you live?", 82% of the respondents gave a positive answer, whereas that percentage is smaller when asked about whether they feel safe in the city as a whole (72.9%) or the country as a whole (53.2%) (Fig. 8). Many of the Roma live in constant fear that they will be evicted, but on the other hand, the authorities themselves are restraining themselves from such actions simply because there is nowhere to move those people. Episodic actions toward tearing down illegal hosing have been observed in recent years. The partial demolition of homes provoked serious discontent, expressed through civil protests. Since 2018, there has

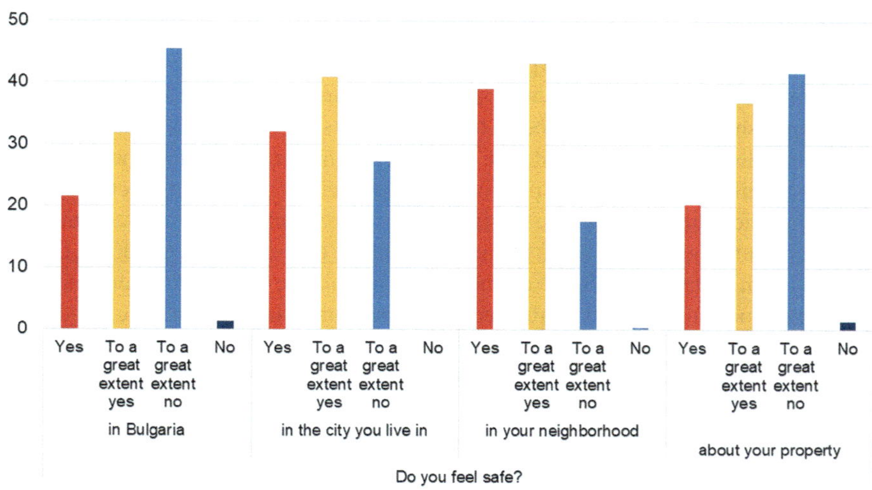

Fig. 8 Respondents' sense of safety concerning the neighbourhood, the city, the country, and their property

been an increased sense of fear and uncertainty about the future of the inhabitants' dwellings. Therefore, the proportion of residents who feel safe about the property they own is much smaller (57.1%).

Moving to a different neighbourhood with a heterogeneous ethnic environment is a very traumatic process for the Roma, as it leads to the loss of their traditional social structures, while at the same time they face the challenges of adapting to a new environment that is completely different from the one in the segregated urban structure. Here we find the answer to the question of what stops the Roma from moving to social housing (83% of the respondents) with better sanitary and living conditions and choose to live in the segregated neighbourhood together with "their own kind". Residents categorically refuse to leave their homes, even though most of them are overcrowded out of security reasons. The uncertainty comes not only from the fact that they have to pay rents, although at a relatively low cost, but also because they will not live in their current environment, close to "their own kind". Most of the offered social housing is located not in ethnically mixed areas, but in Stolipinovo—considered to be the largest Roma neighbourhood in Europe—where the residents of Harman Mahala would feel even more insecure. They prefer to live in crowded homes where they feel safe, rather than moving to another Roma neighbourhood, where the feeling of insecurity would be even greater than if they moved to an ethnically Bulgarian neighbourhood.

When asked how they would describe Harman Mahala, 67% of the respondents only emphasized the positive aspects of the neighbourhood, while the other 33% of the respondents emphasized the negative aspects. To the question "What are you not satisfied within the neighbourhood, you live in?", the most common sources of dissatisfaction turned out to be the street pollution, the lack of pavement, sidewalks, green spaces and playgrounds, insufficient street lighting, narrow streets, overcrowding, lack of personal space, etc. Some respondents pointed out their fear of eviction from the neighbourhood and of their illegal properties being demolished. It is noteworthy that none of the respondents expressed a negative opinion about their own home conditions or about their neighbours. The only downside mentioned was the lack of homeownership documents. None of the respondents expressed fear that the houses were unstable or insecure, despite the fact that a considerable part of them was not built by architects, and upon inspection of the properties under construction, it was found that they did not meet the engineering requirements. Positive evaluations of the neighbourhood mostly refer to the tranquillity, lack of thefts, security, friendship and family ties, childhood memories related to the neighbourhood, etc. The above information was also confirmed by the results obtained from the conducted survey, according to which the highest negative assessment concerned: (1) the pollution with household garbage, (2) the overcrowded streets and neighbourhood as a whole, and (3) the lack of green spaces. When asked what they would change in the neighbourhood if they could, the respondents' suggestions were mostly related to widening and paving the streets and sidewalks, construction of playgrounds, landscaping, installation of more garbage containers, frequent collection of garbage, laying of asphalt, putting benches, etc. Only 4.1% of the respondents stated they would take drastic measures such as demolishing illegal buildings and constructing new legitimate ones, while 5.1% proposed to solve the problem of illegal buildings by issuing a legal document of ownership by the authorities.

6 Conclusion

The study shows that as far as the Roma ethnic community is concerned, the objective quality of the residential conditions, in particular dwelling size and quality, come at a second place after the subjective factors for assessment. Regardless of the relatively poor living conditions, especially in terms of overcrowdedness, pollution, and lack of green areas, the Roma of Harman Mahala emphasize the fact that they actually feel secure in the community and in the quarter they live in. The physical (architectural) condition of their homes, although not the best, is something people prefer to overlook, at the expense of the favourable social environment which provides them with the so much needed sense of security—something ethno-cultural minorities generally seem to constantly look for. The respondents who took part in the survey clearly show their satisfaction with their community, despite the drawbacks of living in crowded homes, for which people do not even have legal ownership, and, therefore, living in constant fear of losing them. The Roma as a specific ethnic community differ greatly in the assessment of the quality of the living environment, in comparison within the case of the majority—the Bulgarian population. As indicated above, subjective factors are crucial in the overall assessment of the living environment. In future construction or reconstruction of the housing stock, local governments must take these features into account in order to maintain the existing high subjective assessment of the neighbourhood in which the studied Roma community inhabits. The established internal social networks have the greatest weight, and they should be preserved in future relocations of the Roma population. Any future action by the authorities must be preceded by a field study, keeping in mind that the genesis and formation of the various Roma neighbourhoods differ, thus influencing the assessment of the living environment. For this reason, the findings of this study cannot be mechanically transferred to other Roma neighbourhoods, and the study can be seen as an example that can be applied to any future interventions by local authorities.

Acknowledgements The study was carried out within the framework of a contract with The National Scientific Research Fund №. DN 15/7 2017. "Spatial segregation and its impact on the social integration of the Roma people in Harman Mahala, Plovdiv."

References

Abidin, N. Zainul, M. Isa Abdullah, N. Basrah, and M. Nazim Alias. 2019. Residential satisfaction: Literature review and a conceptual framework. In *IOP conference series: Earth and environmental science*. IOP Publishing.https://doi.org/10.1088/1755-1315/385/1/012040

Aragonés, J.I., M. Amérigo, and R. Pérez-López. 2016. Residential satisfaction and quality of life. In *Handbook of environmental psychology and quality of life research*, ed. Ghozlane Fleury-Bahi, Enric Pol, and Oscar Navarro, 311–328. Switzerland: Springer.

Asenov, K. 2018. *Antropologiya na getoto—prostranstvo i kultura* (*Anthropology of the ghetto—Space and culture*). Plovdiv: Studio 18

Boal, F.W. 1981. Ethnic residential segregation, ethnic mixing and resource conflict: A study in Belfast, Northern Ireland. In *Ethnic segregation in cities*, ed. C. Peach, V. Robinson, and S. Smith, 235–251. London: Croom Helm.

Bonaiuto, M., A. Aiello, M. Perugini, M. Bonnes, and A.P. Ercolani. 1999. Multidimensional perception of residential environment quality and neighbourhood attachment in the urban environment. *Journal of Environmental Psychology* 19 (4): 331–352.

Bonaiuto, M., and F. Fornara. 2017. Residential satisfaction and perceived urban quality. In *Reference module in neuroscience and biobehavioral psychology*, ed. J. Stein, 1–5. Oxford: Elsevier.

Bonnefoy, X. 2007. Inadequate housing and health: An overview. *International Journal of Environment and Pollution* 30 (3–4): 411–429.

Cities of tomorrow—Challenges, visions, ways forward. 2011. Luxembourg: Publications Office of the European Union.

Dempsey, N., G. Bramley, S. Power, and C. Brown. 2011. The social dimension of sustainable development: Defining urban social sustainability. *Sustainable Development* 19 (5): 289–300.

Galster, G.C. 1985. Evaluating indicators for housing policy: Residential satisfaction vs marginal improvement priorities. *Social Indicator Research* 16: 415–448.

Ha, M., and M.J. Weber. 1994. Residential quality and satisfaction: Toward developing residential quality indexes. *Home Economics Research Journal* 22 (3): 296–308.

Ilieva, N., K. Asenov, B. Kazakov, and T. Lyubenov. 2020. Dynamics of the internal structure and habitation conditions of the ghettoized urban structure of Harman Mahala quarter—Plovdiv, Bulgaria (through combined application of remote sensing and field research methods). In *Smart Geography*, ed. Stoyan Nedkov, 175–191. Switzerland: Springer.

Ilieva, N., B. Kazakov, K. Asenov, and O. Kouwatli. 2019. Location and interrelations of the roma quarter of Harman Mahala with the urban structure of the city of Plovdiv, Bulgaria. *European Journal of Geography* 10 (2): 118–133.

Kain, J.F., and J.M. Quigley. 1970. Evaluating the quality of the residential environment. *Environment and Planning A: Economy and Space* 2 (1): 23–32.

Kesalkheh, S., and H. Dadashpoor. 2012. Assessment residential environmental quality of traditional and new neighborhoods, in a rapid grown city, Tehran. In *New and Traditional Residential Environments, 48th ISOCARP Congress*, 1–11.

Knox, P., and S. Pinch. 2000. *Urban social geography: An introduction*, 4th ed. Harlow: Prentice Hall.

Kogan, I. 2004. Last hired, first fired? The unemployment dynamics of male immigrants in Germany. *European Sociological Review* 20 (5): 445–461.

Marcuse, P. 1997. The enclave, the citadel, and the ghetto: What has changed in the post-Fordist U.S. city. *Urban Affairs Review* 33: 228–264.

Morris, E., and M. Winter. 1975. A theory of family housing adjustment. *Journal of Marriage and Family* 37 (1): 79–88.

Ravnachka, A. 2014. Educational infrastructure of Blagoevgrad District—A factor of labour demand and supply in the labour market (2001–2011). *Europa XXI* 27: 47–61.

Rossi, P.H. 1955. *Why families move: A study in the social psychology of urban residential mobility*. Glencoe: The Free Press.

Shieh, E., A. Sharifi, and M. Rafieian. 2011. Identification of factors that assure quality of residential environments, using environmental assessment indices: A comparative study of two of Tehran's neighborhoods (Zafaranieh & Khaniabad). *International Journal of Architectural Engineering & Urban Planning* 21 (2): 119–132.

Taylor, R.B. 1996. Neighborhood responses to disorder and local attachments: The systemic model of attachment, social disorganization, and neighborhood use value. *Social Forum* 11: 41–74.
Tuan, Y.F. 1977. *Space and place: The perspective of experience*. Minneapolis: University of Minnesota Press.
Van Kempen, R., and A.S. Özüekren. 1998. Ethnic minority housing in the European Union: A case study of Turks. *Tijdschrift Voor Economische En Sociale Geografie* 89: 459–466.
Wellner, K., F. Landau, and C. Müller. 2015. Housing economics and changing residential quality in Berlin. In *22nd annual European real estate society conference*, ed. Kerem Yavuz Arslani. Istanbul: ERES Conference.

Hidden Population Geographies

The Hidden Geographies of Ageing in Rural Areas

Irma Potočnik Slavič

Abstract This paper underlines that ageing in rural areas is a very complex and multilevel interdependent process which is based on networks—the later is being created by interrelations among a mentally constructed (individual perception of person aged 65+), the formalized (involvement of different relevant stakeholders) and the material (rural areas) environments. We argue that these networks are crucial for the creation of age-friendly rural areas. However, the networks—their establishment, the functioning, the form—are less frequent in research on ageing in rural areas. Therefore, we focus on the hidden geographies of these networks: where are the networks placed, who is included, what kind of relationships is performed, how they are addressing the actual needs and expectations of ageing populations in rural areas. For the further analysis of the significance of these networks, a wide longitudinal research is needed. The hidden needs and expectations, hidden information, hidden knowledge, hidden skills, hidden networks of actors are all creating the contemporary landscape of ageing that is not a win–win situation for ageing in rural areas. We would like to underline that ageing in rural areas becomes friendlier when these networks become visible and operational—since this would open up the field for the creation of new job opportunities in rural areas, also for the fine-tuning of existing and the future development of essential social and economic infrastructure, which is relevant for age-friendly rural communities. Re-organization of public care for elderly on national as well as on the local level is expected, new forms of health and social services need to be developed aiming at more efficiency and financial sustainability.

Keywords Rural · Ageing in rural areas · Rural communities · Rural networks · Ageing in place · Slovenia

I. Potočnik Slavič (✉)
Department of Geography, Faculty of Arts, University of Ljubljana, Ljubljana, Slovenia
e-mail: irma.potocnik@ff.uni-lj.si

M. Krevs (ed.), *Hidden Geographies*, Key Challenges in Geography,
https://doi.org/10.1007/978-3-030-74590-5_17

1 Introduction

The future socioeconomic and spatial development of Slovenia will be shaped by the intensive absolute and relative growth of older populations (aged 65+) and general population ageing. Already in 2003, Slovenia encountered more elderly than the young population. In 2010, there was 16,5% of population aged 65+, ranking Slovenia in the 20th place in the world. It is likely that the share of 65+ will increase to 20% in 2020, and reach 30% in 2050 (Slovenia will be ranked as the 9th oldest country of the world; Davies and James 2011)—similar to several European countries (such as Andorra, Spain, Bulgaria, Bosnia and Herzegovina, Italy, Czech Republic, Montenegro, Estonia, Greece, Latvia, Poland, etc.; Davies and James 2011). Van de Kaa (1987) modelled what he termed the second demographic transition: it is marked by continued rates of fertility at below replacement levels. This results in structural ageing of the population, with death rates higher than birth rates for an extended period.

The present demographic dynamics in Slovenia indicate general demographic erosion: suburban zones are slightly more demographically vital in comparison to urban and rural areas (Kerbler 2015). In 2050, those who are currently aged between 25 and 60 or in other terms the majority of the current working age population, will be aged 65+. The fertility patterns of this population and their residential, employment, provision of services, education and cultural involvement and leisure time patterns will influence the formation of older populations' spatial structures.

The contemporary elderly people care system in Slovenia is mostly grounded on institutionalization principles. Taking into the consideration the predicted growth of older populations, the preservation of mentioned principles would triple the existing demand on capacities for senior homes, demanding huge investment capital and also relevant long-term public financial means. According to Rudel (2008), the re-organization of public care for elderly on national as well as on the local level is expected, new forms of health and social services need to be developed aiming at more efficiency and financial sustainability.

Quantitative data related to population ageing on national level is quite often elaborated without or with very limited interrelation to space and place, and especially surveys on ageing in rural areas are rare. Evident is the lack of the awareness of huge heterogeneity of the age group 65+ living in rural areas (regarding their mobility, level of self-dependence, level of income, provenience, social representation of rural and identity). Lately, several projects (within the state funding scheme or via EU projects- LEADER/CLLD or INTERREG) address some of the above-mentioned issues. Due to the positive impacts of the age group 65+ to several fields (local economy, transfer of knowledge and experiences, preserving local identity, etc.), the hidden potential of this age group should be addressed.

This paper underlines that ageing in rural areas is a very complex and multilevel interdependent process which is based on networks—the later is being created by interrelations among a mentally- constructed (individual perception of person aged 65+), the formalized (involvement of different relevant stakeholders) and the material (rural areas) environments. We argue that these networks are crucial for the creation of age-friendly rural areas. Therefore, we focus on the hidden geographies of these networks: where are the networks placed, who is included, what kind of relationships is performed, how they are addressing the actual needs and expectations of ageing populations in rural areas. The hidden needs and expectations, hidden information, hidden knowledge, hidden skills, hidden networks of actors are all creating the contemporary landscape of ageing that is not a win–win situation for ageing in Slovenian rural areas. We would like to underline that ageing in rural areas becomes friendlier when these networks become visible and operational—since this would open up the field for the creation of new job opportunities in rural areas, also for the fine-tuning of existing and the future development of essential social and economic infrastructure, which is relevant for age-friendly rural communities.

2 Methods

Since Slovenia is lacking data on individual's perception of ageing and especially on ageing in rural areas, we conducted a representative field survey (conducted in period 2010–2019; Fig. 1), which included 1223 persons aged 65+ living in Slovenian rural areas. According to the population density and level of urbanization, the majority of Slovenian territory is defined as rural—and we conducted our research in settlements,

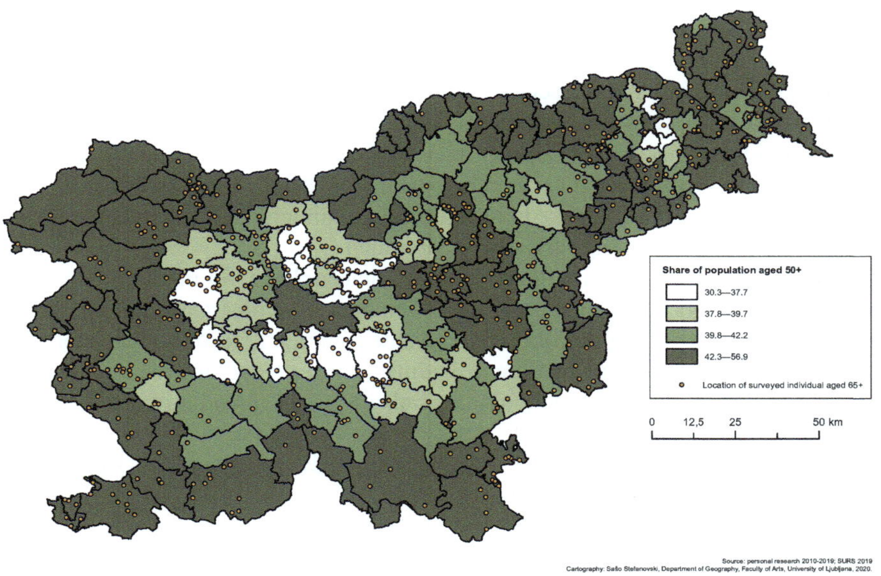

Fig. 1 The geographical distribution of surveyed ageing population in Slovenian rural areas

which are defined as rural according to the criteria of Slovenian Rural Development Programme (2019).

We tried to avoid generalizations and shortcomings of data on the elderly population through a field survey, using the survey questionnaires that addressed the de facto living elderly (65+) in rural areas with the following two topics: (1) how the elderly live in the rural areas and (2) how rural areas should be like for senior citizens.

Data on how the elderly population lives in rural areas were obtained through 18 questions (structured and semi-open) covering the following subject areas:

- identification data (gender, age, settlement and municipality of residence, marital status, type of employment during the working period);
- individual attitude to the surrounding living environment (how long they live in the existing settlement of residence, satisfaction with living in the settlement of residence, what they are most proud of in the domestic settlement and what they miss the most or where they plan to move in the future); and
- independence in daily life and the need for assistance (whether and what assistance/service they are already receiving or would need, financial resources to pay for the assistance, are they willing to receive help, knowledge on the various forms of assistance).

However, we have also obtained information about how rural areas should be like for elderly people by evaluating the factors that make up the model of so-called ideal environment for the elderly. In doing so, we proceeded from the findings of Goltz and Born (2005) which included several factors (attractive landscape, acceptable/low cost of living, existing network of central settlements, ability of residents to live with the elderly—intergenerational coexistence, stable price of real estate, proximity to children, relatives, friends, spending time in/or close to home), and we added two additional factors (living in a senior home, involvement in various activities for the elderly). The respondent evaluated each factor on a scale from zero (insignificant) to nine (very important). A set of similar relevant factors (housing, social life, etc.) was also recognized by Slovenian researchers (Ramovš 2013; Valenčak 2013), as well as by Davies and James (2011): housing, local community and social networks are among the key factors of age-friendly environment.

We conducted also several interviews with stakeholders: 54 in-depth interviews with the managers or other crucial staff members of senior's institutions, 35 in-depth interviews with responsible persons on the community level, and 10 interviews with firms and associations that are providing services (food and health, cleaning, shopping) for elderly. Besides qualitative, also a set of quantitative data (official statistics on ageing, monitoring of various projects related to ageing in rural areas, reports of clubs and associations, etc.) was investigated for the purpose of step-by-step construction of an analytical model.

We are aware of several shortcomings of our methodological approach, below we will list just few:

- There is a high level of subjectivity contained in the responses of more than 1200 respondents aged 65+ since the questionnaire is based on introspection.

- We were conducting research for nearly ten years, and we noticed that the surveyed population has been changing (for example, the recent younger respondents have more computer literacy, are more self-confident on rights of people aged 65+, have gained more positive media representation, etc.).
- Our research was mostly focused on mobile and healthy persons aged 65+ since we wanted to identify the existing networks they are being involved in, and we did not focus on elderly with severe health problems or persons with dementia.

3 Theoretical Approach

The recent proliferation of research on ageing in rural places can be attributed to at least three factors: the environmental context of ageing with issues relating to place and space (Iecovich 2014); the environmental experiences being reshaped by the processes associated with effects of globalization on rural areas; and the ways how demographic ageing intersects with other challenges in rural areas—herewith also building on the multifarious contributions that older people make in rural places (Keating et al. 2013).

Human geographers quite often look upon the spatial distribution of elderly people, identifying clusters with bigger or smaller concentrations of people aged 65+. Nevertheless, on the other side, the impact of ageing populations on social and economic infrastructure has largely been under-researched (Scharf et al. 2016). In some countries, the presence of local facilities (such as a shop, a post office or public transport) is regarded as a central feature of a vibrant local community and as an element of the physical environment that renders communities age friendly. Also, in rural areas older people may become even more vulnerable to the absence or gradual loss of what might be regarded as essential services—e.g. where gaps in service provision exist, the community and voluntary sector is often expected to intervene to meet the old people's care and support needs. Research in England (Scharf et al. 2016) underlined that without volunteers many rural services would be unsustainable. Being aware of shortcomings of service provision in rural areas, it is a widely held assumption that older people in the rural areas are securely embedded within supportive social networks and that demand for formal services is lower in rural than in urban places. The lower rates of loneliness in rural than in urban places were confirmed in Austria, Sweden and the UK, whereas in the Netherlands, Luxembourg and Italy no significantly relevant differences were reported (Burholt 2011).

In this paper, we followed an approach to age-friendly rural communities. It is based on the report on the International Rural Ageing Project (1999), which listed seven key topics concerning ageing in rural places which were identified as priority areas for the international research community (demography, health, intergenerational relationships, life course perspectives, participation of rural elders, impact of technology and evidence of successful rural policies). In a review paper, Burholt and Dobbs (2012) argued that 13 years after this project there was still a substantial way to go in these and other topic areas. They had noticed an increase in the number of studies in these and other topics, but generally, research lacked a critical and analytical focus. In many cases, »rural« had not been seen as an ever-changing context that can potentially shape experiences and outcomes for older people (Scharf et al. 2016).

In designing the theoretical and methodological frameworks of our research, we proceeded from a critical human ecology that states—

(1) interactions between humans and their environment are dynamic, and therefore we focused on the processes and factors that guide the development of human society in real geographical (rural) environments;
(2) the boundaries between different levels and environments are permeable and that cooperation is crucial; and
(3) older adults are not only passive receivers of the environment in which they live but have many capabilities to make, operate and adapt to their environments.

Based on the existing literature and our empirical research (Potočnik Slavič 2019), we developed an analytical framework model of age-friendly rural community (i.e. the rural space and place; Fig. 2) as a mental construct, being marked by interrelations and spatial dimensions. The model is illustrating the interrelations among three spatial dimensions (individual person, stakeholders and broader rural context), pointing out the most relevant elements for ageing population related to social and economic infrastructure of rural community.

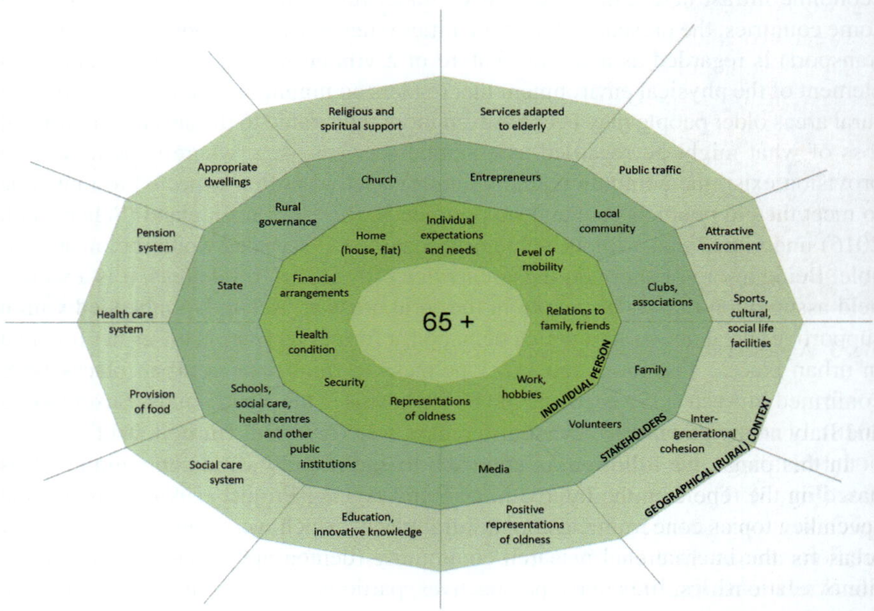

Fig. 2 Model of age-friendly rural community

In everyday life, older people as individuals formulate their own *mental construct* by which they imagine living and ageing in rural areas. This includes their living, health and material status, relationships with family and friends, work and hobbies, level of mobility, their individual expectations, needs and ideas, etc.

Ageing in rural areas is also constituted and significantly shaped by the work of *stakeholders* (family, institutions—social care, health centres, local community, volunteers, societies, school, church, entrepreneurs, media, etc.) which also have their own ideas or programs on ageing in rural areas.

Ageing is taking place in rural areas, so the residence of the elderly in rural areas also forms a *material space* in rural areas. It consists of the various services available to the elderly (i. e. social and economic infrastructure): food supply, various shops, health and social care services, public transport services, educational, sports, cultural and other activities and services tailored to the needs of the elderly, an attractive natural environment, retirement system, religious and spiritual support, intergenerational cooperation and an overall positive attitude towards ageing in society.

This simplified model allows us to create a notion that ageing in rural areas is a very complex and multilevel interdependent process which is based on networks—the later are being created by interrelations among the mentally constructed environment (individual perception of person aged 65+), formalized environment (involvement of different relevant stakeholders) and the material environment (rural areas).

In our research, we mostly focused on the identification of existing networks, but we did not investigate further either symmetric relations or asymmetric relations between elements of our model. But the identification of network itself enabled us to argue that there is a significant power in these networks related to the creation of an age-friendly rural community.

4 Results

4.1 The (Un)hidden Individual Expectations of Elderly in Rural Areas

We undertook research in all municipalities in Slovenia, which are defined as rural, and excluding urban parts (settlements with more than 5000 inhabitants). We were focusing on rural environments and trying to provide relevant number of surveyed people aged 65+ (in total 1223 individuals) according to the statistical representation of elderly population on NUTS3 level.

In evaluating the importance of the individual factors of the model (so-called ideal environment for the elderly; Goltz and Born 2005), we found that three internal factors related to the elderly's immediate environment were crucial for the interviewed elderly (Fig. 3):

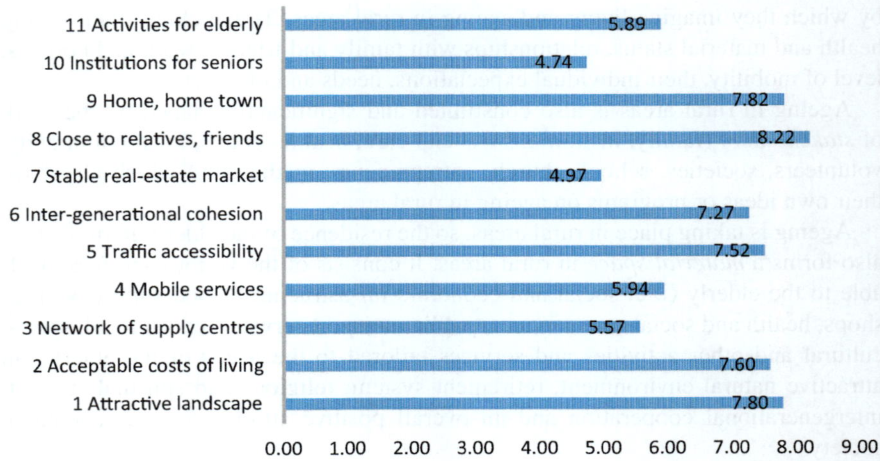

Fig. 3 The importance of factors creating an age-friendly rural community; evaluation provided by respondents (N = 1223) in Slovenia (*Source* Own survey, 2010–2019)

- the highest value was given to closeness of children, relatives and friends (8.22 on the scale 0–9),
- to perform ageing in or near home/residence (7.82) and
- to live in attractive rural areas (7.80).

One has to be aware that Fig. 3 provides the general analysis of surveyed population. Each person in 65+ might have different ideas on ageing in place. This is confirmed by the analysis of answers of four persons of the same age, living in the same rural settlement (Fig. 4). They evaluated certain factors quite similar (i.e. the importance of traffic accessibility—factor 5), three persons provided alike answers, but one (Person 1) usually had a different opinion (for example, regarding acceptable costs of living—factor 2, or being close to relatives and friends—factor 8).

The later might be confirmed also with the detailed analysis of respondents' answers to two questions (Fig. 5):

- the respondents had more uniformed view on where they would like to spend their retirement (ageing close to home, home town);
- however, their opinion whether they would like to go to senior homes is much more diverse.

We would like to underline that our modest, incomplete knowledge and understanding of seniors' needs and expectations create more and more hurdles in the creation of an age-friendly rural community. *A regular, systematic monitoring of elderly peoples' needs and expectations, performed on local (municipal) level is the first step towards the unhidden geographies of ageing in rural areas.*

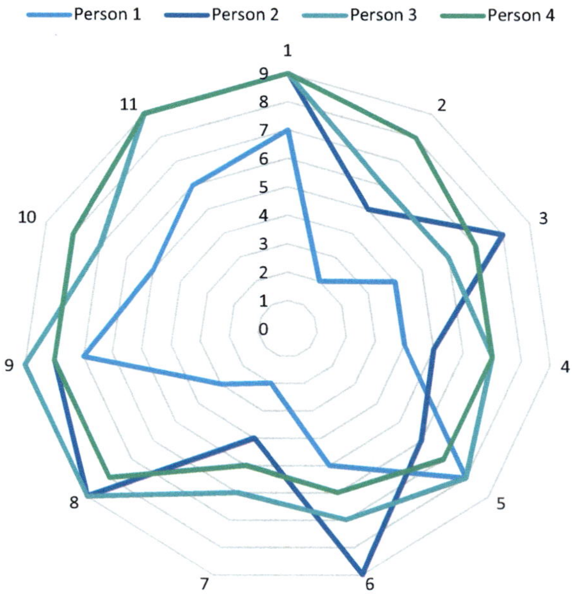

Fig. 4 A detailed breakdown of the evaluation of factors creating an age-friendly rural community provided by four respondents (same age, living in the same rural settlement) (*Source* Own survey, 2010–2019)

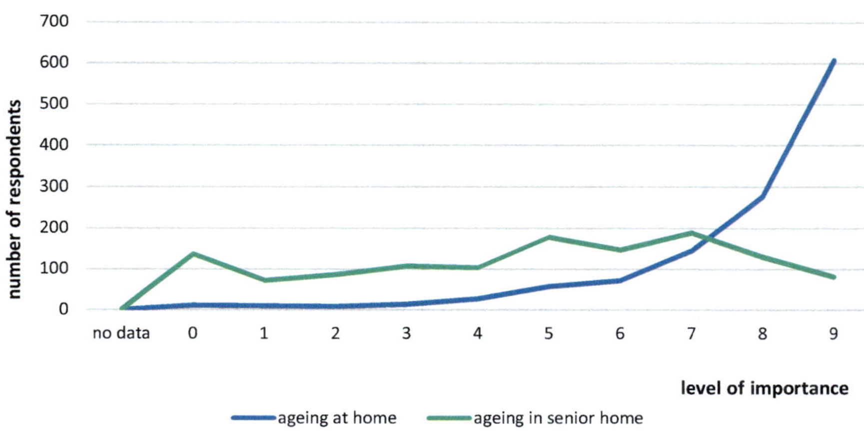

Fig. 5 The distribution of answers regarding the expectations of respondents on ageing at home or ageing in senior home (*Source* Own survey, 2010–2019)

4.2 The Hidden Geographies of (in)formal Structures Related to Elderly in Rural Areas

Since our surveyed population valued the importance of their immediate environment very high, it is of no surprise that our respondents would most often ask children and relatives for help (two-thirds of the answers), one-fifth would ask neighbours for help, 7% the health centre, the Red Cross and Caritas (each by a good 3%) and 1% the municipality. Similar findings are identified in Germany (Goltz and Born 2005); while surveys from South Australia and Northern Ireland also indicate that the elderly do not always want to live close to their families and relatives. According to senior care providers, the proportion of older people who do not want to burden their offspring with their age (because they have their own families and are active) is growing, as they believe that responsibility for their lives is primarily on their shoulders and on the side of the local communities/countries or are willing to hire private service providers for this kind of care (Davies and James 2011, 114–116).

Therefore, we looked upon the existing (in)formal structures related to elderly in rural areas. Existing formal networking of elderly could be observed through seniors' associations and clubs as non-profit and NGOs. There are 512 of them in Slovenia (2019), are organized as a network (Zveza društev upokojencev Slovenije, ZDUS) and officially represent 230,000 members (out of 457.496 pensioned residents aged 65+ in 2020; SORS 2020): one could obtain bigger clusters in areas with more urban characteristics and in bigger agglomerations (Ljubljana, Maribor, Celje, Koper, Nova Gorica, Novo mesto, etc.; Fig. 6). However, on the other side, every municipality has its own senior club; bigger municipalities (according to the size of the territory and number of inhabitants) registered several clubs (2–7). Occasionally, there are even more specialized seniors' clubs; for example, several are registered for the purpose of one big company (pharmaceutical company, bank, food-processing enterprise) or they follow professional orientation (pedagogical workers, retired journalists) or broader interests (for example, the literature club).

The geographical location of the senior clubs is mostly associated with localities that have a certain kind of centrality, i. e. providing services for local population. We have approximately 6000 settlements in Slovenia, and 10% of settlements have certain elements of centrality—addressing also senior clubs organization. This means that the basic structure of elderly people self-organization is quite well distributed over the entire county, and rural areas are not legging behind from this aspect. The seniors' organizations take care of several activities: information transfer, medical hints and measurements, sports and cultural activities, creating social bonds and mutual knowledge sharing. The number of people being actively involved in this clubs differs, but we did not collect data on this.

At the same time, it is of special relevance that elderly people in rural areas do not actively contribute their time, work, knowledge and experience to senior clubs only. Usually, if their mental and physical health allows them, they are quite often well embedded in several clubs and informal settings as well. We followed a day-course of our respondents.

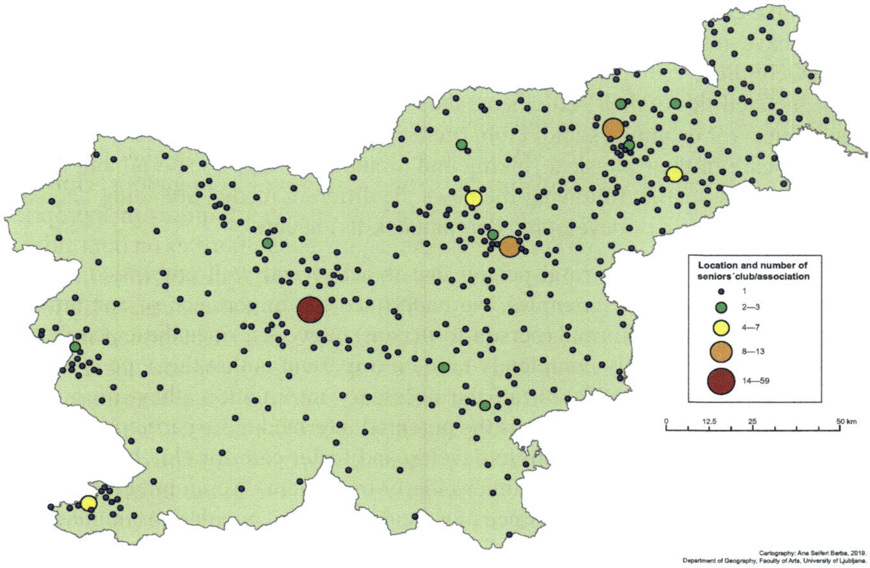

Fig. 6 The spatial distribution of senior clubs in Slovenia (*Source* of data: Zveza društev upokojencev Slovenije (http://www.zdus-zveza.si/, 2019))

> "Firstly, personal hygiene. Then I go and buy a newspaper and listen to daily news. Then I need to prepare some food, if the weather is fine I am occupied with gardening. *In the afternoon, I am visiting the club and taking care of club's administrative issues.* When I am asked (quite often) I am looking after a grandchild, or I am occasionally visiting relatives and *attending English course.* There is *a wish to develop my computer skills.* At week-ends: I enjoy in mountaineering, I attend Sunday mass and still *participating in fire brigade* – mostly in organizatorial issues."

(Male, married, previously employed in public institution, since the birth living in the rural locality, very pleased with the life in the village, not thinking to move, but if—he would prefer a senior home in rural area).

> "In the morning I prepare coffee and read the newspaper. Afterwards I do the house works, check the computer, do some shopping in the local store, and prepare lunch. In the early afternoon, I go to kindergarten to pick-up my grandchild and I look after him for several hours. In between, I practice gardening, take care of flowers, trees, *attend our choir* and do the sports once a week. I listen to the news and TV programme in the evenings. Still I practice hiking during week-ends, Sunday mass, other obligations with choir, visit relatives and help my brother on the farm."

(Female, married, previously employed in public institution, since the birth living in the rural locality, very pleased with the life in the village, not thinking to move, but if – she would prefer a secured dwelling).

These short quotations further on undercover the daily action space and place of active and healthy seniors living in rural environments:

- they take care of their homes (dwellings, gardens, bricolage);
- they have interrelations with their family members and friends (take care of grandchildren, visit and help relatives);
- use the facilities of their settlement (shopping, participating in several communal activities: for example, senior club, fire brigade, choir, church); and
- taking care of their physical (hiking and mountaineering, do sports) and mental (reading newspaper, following the news via different media, attending language course, practising or developing computer skills) health.

This is a type of elder rural person that is active and well embedded in local environment. With these examples, we underlined the importance of combination of formal (clubs) and informal (person-to-person) networks, since these people do a lot of work that is usually completely hidden. Important is that this type of seniors has a huge social network with all the necessary information (they are network brokers) and might be considered as the potential intermediators participating in the future de-institutionalized system to develop and foster care for elderly. They have networks, knowledge, skills, and other elderly trust them. We identified them as a hidden nodes of rural networks. Since more and more aged populations might expect more help from their local communities (Davies and James 2011) and due to the need for re-organization of public care for elderly on national as well as on the local level (Rudel 2008), *it is expected, that new forms of health and social services need to be developed aiming at more efficiency and financial sustainability capitalizing on active aged persons being visible network brokers.*

4.3 (Un)hidden Potentials of Rural Networks: Rural Development Program Address the Issues of Ageing in Rural Areas and Builds on Networks

Interrelations among individual dimension of ageing, actors and stakeholders of formal and wider geographical environment are of crucial importance for the development of social and economic infrastructure for elderly. Slovenian rural areas with dispersed settlement structure and small municipalities have a proper basic structure for creation of age-friendly rural communities. In every municipality, there is a primary school (one general and occasionally smaller sub-unit), a parish (due to their shortage in number, usually priests take care of one or more parishes) and there are more than 110 locations of senior homes (Fig. 7). This creates a very convenient basic structure for institutionalized cooperation of schools-churches-care institutions for elderly people in rural areas.

Slovenian Rural Development Programme addresses the issues of ageing and social exclusion of deprivileged populations (including elderly) especially within the LEADER/CLLD (Community-Led Local Development) program. LEADER/CLLD applies seven principles: to actively involve local population in the preparation of

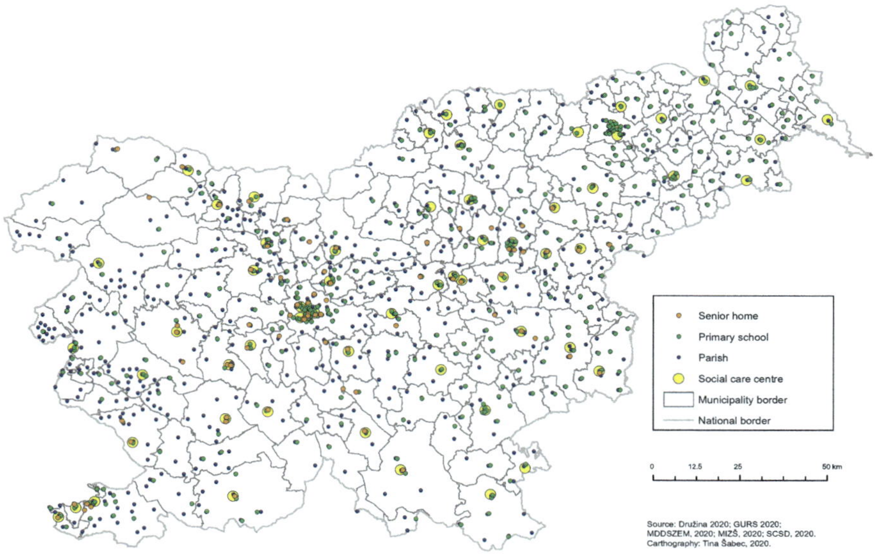

Fig. 7 The spatial distribution of institutions, which are relevant for creation of age-friendly rural communities

Local Development Strategy, follows the bottom-up approach, territorial identification of developmental priorities, followed by structural empowerment and final implementation via co-financed projects on site, the activation is based on networking and cooperation and especially tries to foster multisectorial approach. Within the LEADER/CLLD program, particular attention is given to the appropriate involvement of representatives of local persons: these are not formally elected or nominated members of the community but rather individuals, groups, businesses or organizations with an interest in the rural development agenda. Partners originate from public, private and NGO sectors, create a formal partnership, called local action group.

Several target institutions (municipalities, developmental agencies and centres mostly as lead partners, and social care and health care institutions, schools, clubs, associations, solo and micro-entrepreneurs, companies, social enterprises as project partners) have been involved in the programming period 2014–2020 since the intergenerational inclusion and services for elderly have been placed among top priorities of LEADER/CLLD program. A multi-funded financial scheme, including EAFRD (European Agricultural Fund for Rural Development Fund), EFRD (European Fund for Regional Development) and partners' co-financing is providing a framework for projects implementation. There is a positive correlation between the priorities of the LEADER/CLLD program in this programming period with the needs and expectations of elderly: according to our survey, respondents highly evaluated also the attractive landscape (7.80) and intergenerational cohesion (7.27 on the scale 0–9). Until the end of the year 2020, in total more than 350 projects were supported within 37 local action groups all over Slovenia. In every local action group, they invested in

2–3 (or even more; Program razvoja podeželja 2019) projects, related to the needs of the elderly. The following structures and programs have been delivered:

- intergenerational centres (new buildings, renovated old public buildings, rooms, libraries, parks, gardens, points for meetings; Ganc et al. 2019);
- programs and trainings for intergenerational cooperation (activities for skills and knowledge transfer, including local artisanal traditions, social events, creativity, etc.);
- programs for active and healthy ageing and quality of life (sports, reading, cognitive and motoric skills, arts, thematic groups on challenges of ageing, learning environments, etc.); and
- structures and topics for the maintenance and future development of cultural heritage (museums, exhibitions, thematic routes, research on traditions including the ethnographies of elderly).

It is important to invest and make the results visible via project implementation towards further strengthening of social and economic infrastructure that is relevant for ageing populations. In 2017, the Slovenian Ministry of Labour, Family, Social Affairs and Equal Opportunities established (with the co-financing of European Social Fund) a network of 15 multigenerational centres around the country, mostly allocated in urban centres—they work as nodes for facilities on sub-regional and local level (Ganc et al. 2019). *At the moment, we are able to quantitatively record some kind of positive developmental phase in the field of creating age-friendly rural communities (including the number of projects, allocated financial means, the number of people involved*, etc.*)*. But even more important developmental phase is to come after the projects are completed; then we will be able to evaluate whether the supporting program, allocated financial means and advertised activities, its implementers (actors and stakeholders) and target groups in certain rural localities have managed to *create sustainable and long-lasting structures and networks, willing to learn from each other and addressing the needs of elderly.*

4.4 The Hidden Opportunities of De-Institutionalization

There are approximately 110 state- or private-owned senior homes in Slovenia, providing 20,000 places, i. e. the existing facilities are sufficient solely for 5% of persons aged 65+. There is a constant need to increase the number of places in institutions (at the moment 50,000 seniors are on the waiting list to register in senior homes). The expected investments are enormous and the state is not willing to invest in further public (state-owned) institutionalization. Therefore, few private initiatives lead to the opening of smaller senior units also in rural areas, at the same time the establishment of these institutions ranks high among the development priorities of rural municipalities. This type of investment is usually too big for prevailing small municipalities with limited budgets; therefore, they usually practice public–private partnership and attract seniors with better financial circumstances. The existing pace of these initiatives (as institutionalized approach) is not able to meet the increasing demands; consequently, we should pay much more attention to the support of de-institutionalized approaches to ageing. Active and healthy seniors would like to stay at home as long as possible; therefore, active and healthy ageing in place with all necessary social and economic infrastructure is to be even more supported in the future.

The state provides social care and health care support from their centres (with social- and health-care professionals), but this is not enough. In the previous years, every municipality is trying to financially support social- and health-care of seniors that are offered by (smaller) private institutions and entrepreneurs. Seniors in our survey are willing to accept the help in their daily practices (shopping, cleaning, basic health services, socializing), but they are more open to the help from people they know—their friends, neighbours, relatives. Here, *a combination of public–private system and an assistance of small businesses (not just volunteers) might be a proper answer.*

We would like to point out an early-bird initiative. Farmers are able to register social care service for adult and elderly persons as a supplementary activity on the farm, thereby providing accommodation and day-care services to active and healthy elderly not being dependant on help with basic daily functions. Several farmers have indicated their interest, but the legislation has not provided relevant health and social care standards. Both, seniors and farmers would like to find a better and faster solution: there is an attempt of the so called ‘household community’, where a former farm tourism is adapted to the needs of active 6 seniors, all of them mobile and healthy, who would not like to go to senior home, and still find themselves capable enough for independent living in a small community. Every member is in charge of providing lunch once per week, they do shopping together (twice a week a farmer drives them to the local centre), they enjoy in joint socializing and now-or-then voluntarily participate in smaller on-farm works. Seniors and farmers signed a lease contract that allows them to use the individual room and common places (Horvatič 2020). For this type of initiative, the mobile units of social- and health-care workers should be developed.

4.5 Encouraging Solutions and Hidden Gaps of Projects Related to Ageing Issues in Rural Areas

In some more remote rural areas, the population is facing the decrease in supply of public services of general interest, which create a basic social and economic infrastructure in rural areas. Local schools, primary schools, pubs, local meeting places are closed down due to the lower level of demand. Further, the remaining populations (usually elderly) need to overcome longer distances to reach these services. At the same time, also the public transport is not available anymore. Therefore, seniors and local municipalities are forced to look upon the solutions—some are looking for mobile solutions (mobile shops, a very well and long-time working system of mobile libraries—*bibliobus*, the first steps towards telemedicine) and the potential of contemporary ICT solutions. The problem is that numerous seniors are quite often with or without modest experiences related to usage of available mobile and digital solutions.

There are very prosperous and innovative solutions provided by several projects being supported by EU funds (for example transnational project INTESI overarching five Alpine states; Marot et al. 2018). Solutions are based on integration governance model: integration of telecommunication services with transport, health care and social care services, while the basic goods provision should be supported with e-services and innovative spatial models (even regional models with a prudent re-use of formerly abandoned rural buildings; Kostanjšek and Marot 2019). Nevertheless, there are relevant issues after the project ends since small municipalities are usually not able to find additional financing and human resources to proceed with the activities. Actors and stakeholders of this supporting environment are very critical towards the state since we are lacking the fundamental law on long-term services provision.

5 Conclusion

The elderly in Slovenia are mostly interested in ageing-at-home. Currently, a different way of forming pensions is being established and, as a result, the financial strength of the older population derived from public social transfers is diminishing. Therefore, Slovenian rural areas should capitalize on existing structures and networks and upgrade them with publicly acceptable and innovative solutions. This would require the gradual preparation of an up-to-date database that would enable the development of modern ageing strategies and the development of measures, which would better address the characteristics of different types of rural areas as well as the real needs of the elderly.

Given the expectations and needs of the elderly and their prevailing spatial behaviour, both local communities and the state (with the adoption of relevant laws; Marot et al. 2018) should give a clear impetus to the working age population, and especially young people, that there are hidden working places—being interesting and innovative, offering public and private employment in the field of active ageing sector. It would also be necessary to attract innovative companies developing e-services tailored to the needs of the elderly (Kerbler 2018). However, by introducing novel standards, the state should give these new working places financial support and at the same time introduce a system of quality control for the emerging services of ageing sector.

The message is that the hidden needs of elderly should be continuously monitored and put in front of the strategic developmental documents and action plans, providing the proper financial and implementation mechanisms. Hidden potentials of elderly populations should be actively involved in the creation of support services for elderly. Hidden contribution of elderly people to social and economic infrastructure in rural areas should be measured and evaluated. Hidden power of existing clubs and associations, formal and informal organizations should be further developed. Hidden geographies of networks among all identified environments (individual, formalized, material) should be further monitored and used for the creation of age-friendly rural communities.

References

Burholt, V. 2011. Loneliness of older men and women in rural areas of the UK. In *Age concern Oxfordshire: Safeguarding the convoy: A call to action from the campaign to end loneliness.* Oxford, 35–39.

Burholt, V., and C. Dobbs. 2012. Research on rural ageing: Where have we got to and where are we going in Europe? *Journal of Rural Studies* 28 (4): 432–446.

Davies, A., and A. James. 2011. *Geographies of ageing: Social processes and the spatial unevenness of population ageing.* Surrey, England.

Ganc, D. et al. 2019. A network of multigenerational centres as an answer to the demographic changes in Slovenia: An example of good practice in Posavje. In *Demografske spremembe in regionalni razvoj*, eds. Nared, J., Polajnar Horvat, K., Razpotnik Visković, N., 287–296. Ljubljana: GIAM ZRC SAZU.

Goltz, E., and K.M. Born. 2005. Zuwanderung älterer Menschen in ländliche Räume–eine Studie aus Brandenburg. *Geographische Rundschau* 57 (3): 52–56.

Horvatič, M. 2020: Dejavnosti socialnega varstva kot dopolnilne dejavnosti na slovenskem podeželju. Filozofska fakulteta, Ljubljana, in press.

Iecovich, E. 2014. Aging in place: From theory to practice. *Anthropological Notebooks* 20 (1): 21–33.

International Rural Ageing Project. 1999. *Shepherdstown report on rural ageing: the result of the expert group meeting, May 22–25: 199.* Shepherdstown, WV: University of West Virginia.

Keating, et al. 2013. Age-friendly rural communities: Conceptualizing "best-fit." *Canadian Journal of Ageing* 32 (4): 319–332.

Kerbler, B. 2015. Population aging in Slovenia: a spatial perspective = Staranje prebivalstva v Sloveniji: prostorski vidik. *Acta Geographica Slovenica* 55 (2): 303–317.

Kerbler, B. 2018. Using information and communication technology in home care for the elderly. In *Caregiving and home care. InTechOpen, Web of science core collection*, ed. Mollaoglu, M., 203–228.

Kostanjšek, B., and N. Marot. 2019. Trends in the provision of basic goods in less accessible areas with proposals for alternative solutions in the case of Idrijsko-Cerkljansko. In *Demografske spremembe in regionalni razvoj*, eds. Nared, J., Polajnar Horvat, K., and N. Razpotnik Visković, 183–196. Ljubljana: GIAM ZRC SAZU.

Marot, N. et al. 2018. Izzivi zagotavljanja storitev splošnega pomena v 21. stoletju: celostnim pristopom naproti. Ljubljana: Biotehniška fakulteta, Univerza v Ljubljani.

Potočnik Slavič, I. 2019. Rural ageing: Facts, expectations and spatial dimensions. In *Demografske spremembe in regionalni razvoj*, eds. Nared, J., Polajnar Horvat, K., and N. Razpotnik Visković, 27–44. Ljubljana: GIAM ZRC SAZU.

Program razvoja podeželja 2014–2020, 2019. Ministrstvo za kmetijstvo, gozdarstvo in prehrano. Ljubljana.

Ramovš, J. 2013. Staranje v Sloveniji. Raziskava o potrebah, zmožnostih in staliŝčih nad 50 let starih prebivalcev v Sloveniji. Inštitut Antona Trstenjaka za gerontologijo in medgeneracijsko sožitje. Ljubljana.

Rudel, D. 2008. Zdravje na domu na daljavo za zdrave osebe. *Informatica Medica Slovenica* 13 (2): 19–29.

Scharf, T., Walsh, K., and E. O'shea. 2016. Ageing in rural places. In *Routledge international handbook of rural studies*, eds. Shucksmith, M., and D.L. Brown, 50–61. Routledge.
SORS 2020. Prebivalstvo. https://www.stat.si/StatWeb/Field/Index/17/104. Accessed 3 April 2020.
Valenčak, S. 2013. Starosti prijazno okolje v slovenskih mestih in na podeželju. In *Staranje v Sloveniji*, ed. Ramovš, J. Inštitut Antona Trstenjaka za gerontologijo in medgeneracijsko sožitje. Ljubljana.
van de Kaa, D.J. 1987. Europe's second demographic transition. *Population Bulletin* 42 (1): 1–59. Washington, D.C.

The Impact of Spatial-Demographic Disparities on Fertility Variations in Serbian Municipalities—An Example of Hidden Content Research

Suzana Lović Obradović and Gordana Vojković

Abstract Fertility rates in Serbia have been at a low level over several decades. They are also marked by spatial heterogeneity, determined by the effects of the geographical and socio-economic characteristics of the place. This means that at each particular location, a set of determinants are responsible for the volume of reproductive processes. This paper uses geographically weighted regression to study the spatial variation of the relationships between fertility rates, on the one hand, and demographic and socio-economic characteristic of the population (livelihood income, marital status, economic activity, and education), on the other hand, at the local and global levels, and to demonstrate the advantages of the local model in hidden contents' detection. Our results suggest that the relationships between the determinants and fertility rates are not constant but vary across space and that there is a difference in terms of direction and strength. These variations could not be detected using a global model, and in this sense, they remained hidden. On the other hand, the specific numerical and cartographically presented results of the local model have indicated, in a simple and sophisticated way, the locations of interest regarding fertility issues, creating a platform for numerous birth incitement strategies, as a leading population problem in many European countries.

Keywords Hidden contents · Spatial demography · Geographically weighted regression · Fertility variation · Serbia

S. Lović Obradović (✉)
Geographical Institute "Jovan Cvijić" SASA, Belgrade, Serbia
e-mail: s.lovic@gi.sanu.ac.rs

G. Vojković
Faculty of Geography, University of Belgrade, Belgrade, Serbia
e-mail: gvojkovic@gef.bg.ac.rs

M. Krevs (ed.), *Hidden Geographies*, Key Challenges in Geography,
https://doi.org/10.1007/978-3-030-74590-5_18

1 Introduction

Seemingly homogeneous content conceals numerous phenomena and relationships between them that can vary with space. Demographic phenomena and processes are not uniform, but their volume and intensity change in relation to the environment, creating a vicious circle of cause-and-effect relationships in the human–space relation. The development of geographical information systems and spatial analyses has brought novelty in demographic research, making it possible to detect hidden content and relationships. In that sense, researchers are using data to generate new insights into how the nature of places affects people and communities (González González 2017). The implementation of location information not only has enabled the study of variations in the relationships that exist between the determinants and demographic phenomena and processes but also has made it possible to measure the influence of the given determinants in a given place. In this sense, it is possible to identify for each location the determinants responsible for the demographic change, thereby creating a platform for population policy-making to prevent the deepening of the existing spatial-demographic disparities.

Fertility is one of the demographic phenomena to which the concept of spatial dependence can be applied with the aim of detecting its variations with respect to the place characteristics. As suggested by a number of recent contributions (Kamata et al. 2010; Mucciardi and Bertuccelli 2011; Wang and Wang 2015; Evans and Gray 2017; Wang and Chi 2017; Lovic Obradovic and Babovic 2018; Jung et al. 2019), this paper adopts a spatial perspective seeking to study the variation of the relationships between fertility rates, on the one hand, and demographic and socio-economic characteristic of the population, on the other hand. Fertility is chosen as an example to show the full strength of spatial demography in detecting the hidden content, which may have direct implications for population policy measures in a particular place. We have conducted geographically weighted regression (GWR), and the global and local model results are presented to demonstrate the advantages of the local model in hidden content detection.

2 Background and Literature Review

Hidden geography stresses that superficial phenomena may prevent access to the inner truth (Guohua 2019). Opposite to hidden geographies, there is a geography of visibility. The question is what should be hidden and what should be revealed, and, accordingly, how to choose appropriate methods. In the age marked by low fertility in most European countries and mass global migration, demographic research should be more than ever focused on finding details that may be hidden in seemingly homogeneous phenomena and processes, and that must by no means be neglected. The demographic development is taking place under the influence of general demographic rules and complex socio-economic factors (Madjevikj et al. 2016). The change in the level of fertility, more precisely its decline, is the focus of contemporary scientific research. The fact that in the 2000s most East European and South European countries had a total fertility rate equal to or smaller than 1.5 led to the adoption of the term "low fertility trap" (McDonald 2006; Lutz et al. 2006). According to McDonald (2006), TFR level 1.5 is supposedly the critical level, and in countries where it has fallen below 1.5, it will remain below the limit forever. Kohler et al. (2002) have discussed the emergence and persistence of low and particularly the lowest-low fertility in Europe, holding that if TFR levels equal to or below 1.3 persist over a long time in a contemporary low-mortality context, this would result in a 50-percent reduction in the annual number of births, halving the population size in less than 45 years (Kohler et al. 2006). On the other hand, there are countries with relatively high total fertility rates—about 1.9. As a consequence of spatial variations in fertility, a new wave of research has emerged seeking to improve the understanding of the relationships between fertility and economic, cultural, migratory, and institutional factors (Campisi et al. 2020).

Keeping in mind that there are marked differences both among and within countries, a systematic approach is needed to study the factors influencing fertility fluctuations, the impact of which varies depending on the place (Vojković 2007). The integration of place in demographic research has become a necessity, leading to the development of spatial demography, a research discipline that has flourished at the end of the 20th and the beginning of the 21st century thanks to the development of geographical information systems. Spatial demography is defined as a discipline that studies the impact of place on demographic phenomena and processes (Weeks 2004; Voss et al. 2006; de Castro 2007; Entwisle 2007; Devedžić 2013; Mathews and Parker 2013; Mathews et al. 2014; Balk and Montgomery 2015), presenting a reliable framework for detecting the hidden content. Specific methods, due to which, among other things, it established itself as a research discipline, were devised by refining classical statistical methods with location information. Why is the place (location) important in revealing the hidden content? And how can spatial demographics help this?

In contemporary research, spatial demographers use a wide range of spatial analyses focusing on the impact of place in explaining the studied phenomena and detection of spatial patterns that cannot be detected by visual analysis; they also seek to assess the significance of spatial patterns defined by visual analysis. Contemporary demographic studies have successfully used the method of spatial autocorrelation in exploring the spatial dependence of fertility levels (Weeks et al. 2000; Kamata et al. 2010; Goldstein and Klüsener 2014; Klüsener et al. 2016), identified predictors that influence variations in fertility levels (Ord and Getis 1995; Getis and Ord 1996; Mier-y-Terán and Martínez 2014), or measured the effect of population density on fertility (Loftin and Ward 1983). Spatial econometrics has also found application in fertility research (Goldstein and Klüsener 2014; Owoo et al. 2015; Kamata 2016), just like hierarchical modeling (Hirschman and Guest 1990; Khan 1997; Parr 1999; Amara 2015; Eyasu 2015). Geographically weighted regression can be applied in the study of phenomena in which the spatial dimension has a great influence on the interaction between variables. Therefore it is not surprising that the method is increasingly used in fertility studies, namely in detecting the factors that cause fertility fluctuations, as well as the variation of the relationships between the determinants and the location-specific fertility. In order to identify the main causes and the long-term effects of demographic change, it is necessary to "enter every settlement" (Vojković et al. 2012, 2018; Andjelković-Stoilković et al. 2018). In this respect, the use of a local GWR model is preferred as it has the potential to reveal local patterns in the spatial distribution of parameters that would be ignored by global models (Mucciardi and Bertuccelli 2011). This shows the value of geographically weighted regression in providing evidence-based local-population policy options in the pursuit of a fertility rebound (Jung et al. 2019). The results obtained using the local model make it possible to identify local variations in the relationship between location-specific fertility and determinants. This opens a new chapter, geared towards the volume of reproductive processes, which gives geography legitimacy to be included in the revision of the fertility transition theory (Lovic Obradovic 2019). The results obtained using local models show that the relationship between fertility and determinants varies spatially in terms of direction, strength, and magnitude, due to differences in local characteristics (Wang and Chi 2017). The greatest advantage of local models is the quantification of the factors' impact, i.e., it is possible to represent numerically the impact of each factor on fertility and indicate, in a simple but sophisticated way, locations of interest when it comes to fertility problems.

This research is motivated by the fact that in Serbia, as in most European countries, there is an evident decade-long decline in fertility, marked by spatial disparities. The current age model of fertility in Serbia is marked by a small share of women beginning their reproductive history before the age of 20, followed by a sudden increase and concentration in births during the optimal fertility period (20–34), and a gradual decline in the number of births by mothers over 35 (Magdalenić and Vojković 2015; Kokotović Kanazir et al. 2017). The complexity of the term *fertility* rests in the fact that it appears as a research subject in many disciplines: sociology, psychology, geography, demography, etc. This is perhaps one of the reasons why we still do not have a single comprehensive theory regarding low fertility, despite a series of complex considerations—supported by statistics, research findings, and high applicability (Rašević 2019). Also, there is no consensus on the dominant factors that cause fertility fluctuations.

One of the most significant factors impacting reduced fertility rates is the education level of the female population (Kamata et al. 2010; Aassve et al. 2012; Rašević 2015; Wang and Wang 2015; Wang and Chi 2017; Mirić 2019a). The direct impact of the high level of education attained is reflected in the delayed marriage, as a common and recognized framework for reproduction for the majority of the population, but also in the reduced duration of the fertile period. A higher education attainment among women is accompanied by different perspectives, and those related to fertility reduction include the desire for a smaller family, the use of contraceptives, economic independence, individual aspirations, etc. (Šobot 2014).

Numerous studies indicate that the impact of male unemployment on fertility rates is important (Kohler et al. 2002; Kamata et al. 2010). Male unemployment usually results in economic insecurity, primarily in traditional societies. Accordingly, a negative impact is expected. The values of fertility rates in Serbia, as well as trends, are in line with the changing status of women in the post-communist regime, which was accompanied by a transition from a state-oriented to a market-oriented economy and a greater share of women in the labor market (Blagojević 1997; Lović Obradović et al. 2016). Increased fertility rates driven by increased economic activity rates may be encouraged by part-time employment, increased women's educational attainment, an increased first-child birth rate, and an increased share of children enrolled in preschool programs (Engelhardt 2011), while decreased fertility rates are determined by the rigidity of the labor market and disadvantages regarding women's employment opportunities, the lack of development, poor availability of childcare services (Del Boca 1999), workload burden (Mills et al. 2008), etc. Women's economic activity is a positive framework if unequivocal social support and parenting gender equality are ensured (Šobot 2014).

Although a high level of correlation between marriage and fertility has been empirically documented in all historical stages in Serbia (Petrović 2011; Magdalenić 2016), there are conflicting views on the impact of marriage on fertility. Changing perceptions of marriage throughout history, associated with overall changes in social systems, may have been the most influential factor in fertility pattern changes. The shifting impact of marriage on fertility has been accompanied by increased nonmarital fertility (Stankovic and Penev 2010; Smith 2019), increased cohabitation (Bobić 2013), marriage age (Dommaraju 2013), changes in the family reproductive sphere, ideational changes, and increased individualism and independence (Magdalenić 2016).

Fertility rates also vary with the level of urbanization, resulting in the rural–urban dichotomy. Fertility is believed to be inversely proportional to the level of urbanization, i.e., higher in rural areas and small towns and lower in large cities (Campisi et al. 2020). This is possibly due to the fact that people inclined to having fewer children migrate to urban areas to enjoy opportunities for a greater income offered by cities, whereas people inclined to having more children move to rural areas where the costs of raising children are lower (Yung et al. 2020). However, peripherally located border and mountain villages in Serbia, which are affected by low fertility rates and migrations, vanish first (Babovic et al. 2016).

This paper is a pioneering endeavor in incorporating the spatial dimension in fertility research in Serbia with the aim of identifying the dominant factors of fertility fluctuations at the local level. Spatial-demographic analyses, which have recently become increasingly important in demographic science, enable a comprehensive investigation of each correlated factor of fertility in accordance with the geographic and socio-economic characteristics of space. The application of methods at the local level makes it possible to detect hidden relationships, which is another step forward in research. Regularities in these relationships are of great importance for the prediction of future demographic trends (Lovic Obradovic and Vojkovic 2019). Each lower administrative level gives an even more accurate picture and makes it possible to detect content that is not detected when data provided by higher administrative levels are used.

3 Data, Measurements, and Analytical Strategy

3.1 *Data and Measurement*

This study covers the territory of the Republic of Serbia[1] and the analysis relies on the 2011 Population Census (data are obtained by special processing of the Statistical Office of the Republic of Serbia) and vital statics data (The Statistical Office of the Republic of Serbia 2012). The research has been conducted at the municipality level (166 units).

3.2 *Dependent Variable*

The fertility rate[2] is used as the dependent variable, since the differences between the minimum and maximum values of the total fertility rates range from 0.9 to 2.3. Accordingly, the obtained results do not unambiguously indicate the variations in the relationships between the dependent and independent variables in the local model, and, therefore, they cannot reveal hidden content.

3.3 *Independent Variables*

The study includes eight independent variables, the choice of which has been determined by the availability of relevant data. They are classified into four groups: livelihood income, marital status, economic activity, and education. By taking into account the impact of socio-economic characteristics of the population on fertility, while incorporating the spatial dimension, a holistic understanding of fertility is achieved, which further provides significant new insights into how these aspects are related to fertility variation (Campisi et al. 2020).

3.3.1 Livelihood Income

In our analysis, we have observed the share of women aged 15–49 to whom social welfare (WSW) is the only source of income in the fertile contingent (%) and the share of households without income (HWI) in the total number of households (%). According to the 2011 census 51,056 women aged 15–49 in Serbia were social

[1] The data cover the population of Serbia, not including the AP Kosovo and Metohija. The municipalities of Preševo and Bujanovac are not included as the data are incomplete due to the census boycott by the majority of the Albanian ethnic community.

[2] The number of live births per 1,000 women aged 15 to 49 in a calendar year.

welfare recipients (material security, care allowance and assisted living benefits, one-off assistance, parental allowance, child benefit, disability allowance, etc.) as their principal (sole) source of income. The data indicate that 1.6% of women from the fertile contingent who were socially disadvantaged and had limited cash resources were in an unfavorable financial situation. As for households without income, according to the 2011 census, there were 68,042 registered households with no income (2.7%). The amount of household consumption depended on the scope of income sources. Social welfare recipients had a limited, basic budget needs. Therefore, it is expected that there are different fertility rate patterns depending on the source of livelihood. On the other hand, the unfavorable economic situation in households without income is an aggravating factor in family planning, which leads to the hypothesis that spatial fertility patterns are greatly influenced by the amount of household income.

3.3.2 Marital Status

Our analysis covers the share of unmarried women (UW) aged 30–39 in the fertile contingent (%) and the share of married women (MW) aged 15–30 years in the fertile contingent (%). According to the presented hypothesis, the marital status has a strong influence on the fertility rate. The explanation is that reproductive processes mostly take place within the marital framework. In this regard, it is expected that a larger number of live births will occur in the areas where the share of married women aged 15–30 is higher than in the areas with a higher proportion of unmarried women aged 30–39, precisely because of the length of the reproductive period in the marital framework.

3.3.3 Economic Activity

The following indicators of economic activity are selected: the share of economically active women (EAW) aged 15–49 in the fertile contingent (%) and the share of unemployed men (UM) in the total active male population (the unemployment rate of men) (%). The socio-economic transition has also led to increased employment rates among women in line with gender equality. While in some European countries higher fertility rates keep up with high rates of economically active women, the opposite trend is noticeable in Serbia (Kupiszewski et al. 2012; Šobot 2014). On the other hand, numerous studies highlight the impact of male unemployment rates on fertility rates (Kohler et al. 2002; Kamata et al. 2010). The latest research in Serbia has shown that men who will soon become parents are more intensively engaged in the labor market. As a result, fathers' employment rates are significantly higher than those of non-fathers, and, at the same time, fathers' unemployment and inactivity rates are significantly lower than those of young people who are not yet parents (Stanojevic 2018). Keeping this in mind, it is assumed that disparities in the economic activity of men and women influence the variability of the fertility rate.

3.3.4 Education

Our analysis includes the share of women with secondary education (SE) aged 15–49 in the fertile contingent (%) and the share of women with high and higher education (HHE) aged 15–49 in the fertile contingent (%). Abundant literature, as well as the 2011 census data, confirm the theory of a negative correlation between women's higher education and fertility rates in Serbia (Raševic and Vasic 2017; Miric 2019a, b). Women aged 40–49 who are either without education or who have started but not completed primary school are the only group where the rate of cumulative fertility is above the level required for generational replacement. Compared to women who have completed college, these women have a higher fertility rate per live birth (2.41 vs. 1.37). Women with completed primary education also have the cumulative fertility rate higher than two (2.08), while the cumulative fertility rate of women with the highest level of education is only 1.37 (Rašević 2015). Reaching higher and higher levels of education are directly related to delayed marriage and a shorter fertile marriage period. Besides, higher education is associated with increased aspirations for individual activities and further development, which is in line with the hypothesis that educational attainment affects the values of fertility rates.

3.4 Analytical Strategy

The modeling of the spatial variation of the relationships between independent variables (demographic and socio-economic), on the one hand, and fertility rates, as a dependent variable, on the other hand, which are correlated, has been done to demonstrate that local models are more relevant than global ones in detecting hidden content. We first calculated the derived demographic indicators (the determinants). To avoid multicollinearity and to reduce independent correlated factors, the variance inflation factor was calculated. The limit value of VIF was set to five. The values lower than five indicated that variables were not correlated, while values greater than five indicated the presence of multicollinearity. The locations of municipality centers (latitude and longitude), necessary for geocoding, were defined in the Google Earth application.

The next step involved a non-spatial descriptive statistical analysis, and then the global model was applied, the results of which indicate the values of coefficient estimated at the global level. To identify place-specific relationships between fertility rates and selected determinants at all of the 166 locations, we applied a local GWR model. The basic function of the GWR model is expressed as

$$Y_i = \beta_{0(i)+}\beta_{1(i)X1i} + \beta_{2(i)X2i\ldots} + \beta_{n(i)Xni} + \varepsilon_i \quad (1)$$

where Yi is the fertility rate in municipality *i*, (u_i, v_i) denotes the coordinates of the municipality center *i*, $\beta_{0(i)}$ is the local intercept for municipality *i*, and $\beta_{n(i)}$ is the local coefficient for predictor *n* for municipality *i* (Fotheringham et al. 2002).

In this case, regression coefficients were estimated for each location (municipality) based on location-specific weighting schemes. Accordingly, there were as many local regression models as there were observations (Wheeler and Tiefelsdorf 2005).

The regression coefficients and local R^2 were estimated for each location independently and mapped to detect hidden content and spatial heterogeneity. The GWR models (global and local) were conducted in *GWR4.0*, and the mapping of local coefficient and R^2 were conducted in *GeoMedia*. Only statistically significant values of the regression coefficients were mapped, i.e., those for which statistical significance (t = Est/SE) in the column was smaller than −1.96 or greater than 1.96 (95% confidence level).

4 Results

4.1 Descriptive Statistics

Table 1 shows the descriptive statistics of independent and dependent variables, their minimum (Min), maximum (Max), and mean values (Mean), as well as the standard deviation (SD). The mean value of the fertility rate is 36.6, but it varies from 8.9 as the minimum to 71.3 as the maximum value. The minimum value for women to whom social welfare is the only source of income is 3.99%, and for households without income is 2.55%. The descriptive statistics of the marital status shows that the mean value of unmarried women is 6.10%, whereas for married women it is 9.73%. On average, there are 41.4% of economically active women and slightly more than one-fifth (20.81%) of unemployed men. The results also show significant differences in the average share of women of different education levels—only 10.86% have high and higher education, whereas 39.6% have secondary education. Significant differences between the minimum and maximum values of the dependent variables indicate that

Table 1 Descriptive statistics

Variable	Mean	SD	Min	Max
Fertility rate	36.6	7	8.9	71.3
WSW	3.99	2.77	0.6	16.6
HWI	2.55	0.92	0.9	5.54
UW	6.1	2.21	2.38	16.64
MW	9.73	2.09	3.77	17.45
EAW	41.14	8.05	16.36	59.49
UM	20.81	7.54	4.8	44.37
SE	39.6	7.68	18.36	53.31
HHE	10.86	7.27	2.6	51.67

municipalities in Serbia are marked by different socio-demographic characteristics, determined by differential physical, ethnic, and cultural characteristics of a place.

4.2 GWR Global Regression Model

Table 2 presents the estimated regression coefficient (Est), standard error (Std. Err), and t-statistic (t = (Est/SE), and it also shows the values of the coefficient of determination (R^2), the adjusted coefficient of determination (adjusted R^2) and the Akaike information criterion (AIC). The global model results indicate a positive relationship between fertility rates and the following predictors: the share of women aged 15–49 for whom social welfare is the only source of income in the fertile contingent (WSW), the share of unmarried women aged 30–39 in the fertile contingent (UW), the share of married women aged 15–30 years in the fertile contingent (MW), the share of women aged 15–49 with secondary education in the fertile contingent (SE), and the share of women aged 15–49 with high and higher education in the fertile contingent (HHE). The relationship between the share of households without income in the total number of households (HWI), the share of economically active women aged 15–49 in the fertile contingent (EAW), and the share of unemployed men in the total active male population (UM) is negative. The t-test values show that the influence of six variables is statistically significant: WSW, UW, MW, UM, SE, and HHE.

The fertility rates are directly proportional to the determinants in municipalities with positive Est values. Increased determinant values lead to increased fertility rates. On the other hand, a negative sign indicates an inversely proportional relationship between the correlated variables, i.e., increased values of the determinants lead to decreased fertility rates and vice versa.

Table 2 Global regression model

Variable	Est	Std. Err	t(Est/SE)
Intercept	−103.019	78.999	−1.304
WSW	0.007	0.002	3.634
HWI	−0.002	0.006	−0.401
UW	0.025	0.004	6.926
MW	0.026	0.003	7.849
EAW	−0.001	0.001	−1.106
UM	−0.002	0.001	−1.961
SE	0.003	0.001	3.628
HHE	0.002	0.001	3.137
R^2	0.31		
Adjusted R^2	0.19		
AIC	2,126.812		

The Akaike Information Criterion (AIC) of the global model is 2,126.81. The value of the coefficient of determination, which indicates the model's fitness, is 0.31, which means that only 31% of the variation in fertility rates on average is explained by the chosen demographic and socio-economic determinants, which means that the global model does not have a satisfying explanatory power. The value of the adjusted coefficient of determination (calculated based on the number of determinants) is 0.19.

4.3 GWR Local Regression Model

The descriptive statistics results of the estimated local regression coefficient show their minimum and maximum values, the range, and the values of the first quartile (first 25% of the distribution), the median, and the fourth quartile (last 25% of the distribution) for each determinant. The DIFF-of-criterion values are also presented, where negative values suggest that all eight determinants are spatially non-stationary. The AIC of the local model is 1,793.484. The coefficient of determination is 0.79, which means that as much as 79% of the fertility variation is explained by the selected determinants, and the model fits in most of the Serbian municipalities. The explanatory power of the local model is satisfying in terms of explaining the fertility variation based on the chosen determinants. The value of the adjusted coefficient of the determination is slightly smaller (0.54) (Table 3).

Table 3 GWR Descriptive statistic and test for spatial variability of determinants

Variable	Min	Max	Range	Lwr quartile	Median	Upr quartile	DIFF-of-criterion
Intercept	−1230.54	815.154	2045.693	−208.325	−83.102	33.515	−1423.52
WSW	−0.021	0.031	0.052	0.001	0.005	0.011	−10.868
HWI	−0.094	0.132	0.227	−0.023	−0.008	0.011	−32.408
UW	−0.024	0.07	0.094	0.015	0.022	0.032	−113.866
MW	−0.002	0.077	0.079	0.02	0.027	0.034	−88.603
EAW	−0.015	0.011	0.026	−0.004	−0.002	0.001	−225.333
UM	0.011	0.011	0.021	−0.004	−0.002	0	−12.528
SE	-0.006	0.013	0.019	0.001	0.002	0.004	−135.572
HHE	0.017	0.026	0.427	0.002	0.004	0.007	−18.273
R^2	0.79						
Adjusted R^2	0.54						
AIC	1,793.484						

4.3.1 Fertility Rate

The main features of fertility in Serbia are a multi-decade decline and distinct spatial-demographic disparities. The fertility rate is the highest in the municipality of Tutin (71.3‰), and it is also high in the municipalities Novi Pazar (59.6‰) and Sjenica (51.9‰), located in southwest Serbia. The values of fertility rates above 50‰ can be observed in the municipalities of Stari Grad (51.5‰) and Bojnik (50.6‰). The reason for high fertility rates in the first three municipalities is the religious structure of the population. Namely, the Muslim population, which constitutes the majority in these municipalities, is marked by a highly reproductive model. On the other hand, in most municipalities (119 or 72%), fertility rates are below the national average (40.16‰) (Fig. 1).

4.3.2 Local R-Squared

The values of the coefficient of determination (R^2) for each municipality were mapped to identify the locations of interest in a faster and easier way. The cartographically presented coefficients of determination were classified by a five-quantile scale. Blue and red are used to represent the extreme values: blue indicates the minimum values, either positive or negative, while red indicates the maximum values.

The lower values of the coefficient of determination indicate that the model is less efficient in explaining the spatial heterogeneity of the relationship between the selected determinants and fertility in municipalities, while the better performance of the model is indicated by higher values. The total variance explained by the local model ranges between 0.582 and 0.891.

For example, the value of the coefficient of determination of 0.582 in the municipality of Opovo indicates that the model is less efficient, and the variation in the reproductive process cannot be fully explained based on the selected determinants. A lower value of the coefficient of determination was obtained for the neighboring municipality of Pančevo (0.583). The highest value of the coefficient of determination (0.891) has been recorded in Vršac. The selected determinants explain as much as 89% of fertility variations and have a major impact on the reproductive behavior in this municipality (Fig. 2).

4.3.3 Estimated Local Regression Coefficients

Figure 3 shows the estimated local regression coefficients, indicating the local variation in the relationship between fertility and the eight selected predictors. They were classified by a five-quantile scale.

Figure 3a presents the spatially varying relationship between the share of women aged 15–49 to whom social welfare is the only source of income in the fertile contingent (%) and fertility rates. The result indicates that the relationship is statistically significant in only 39 municipalities (23.5%). The highest negative value is −1.812.

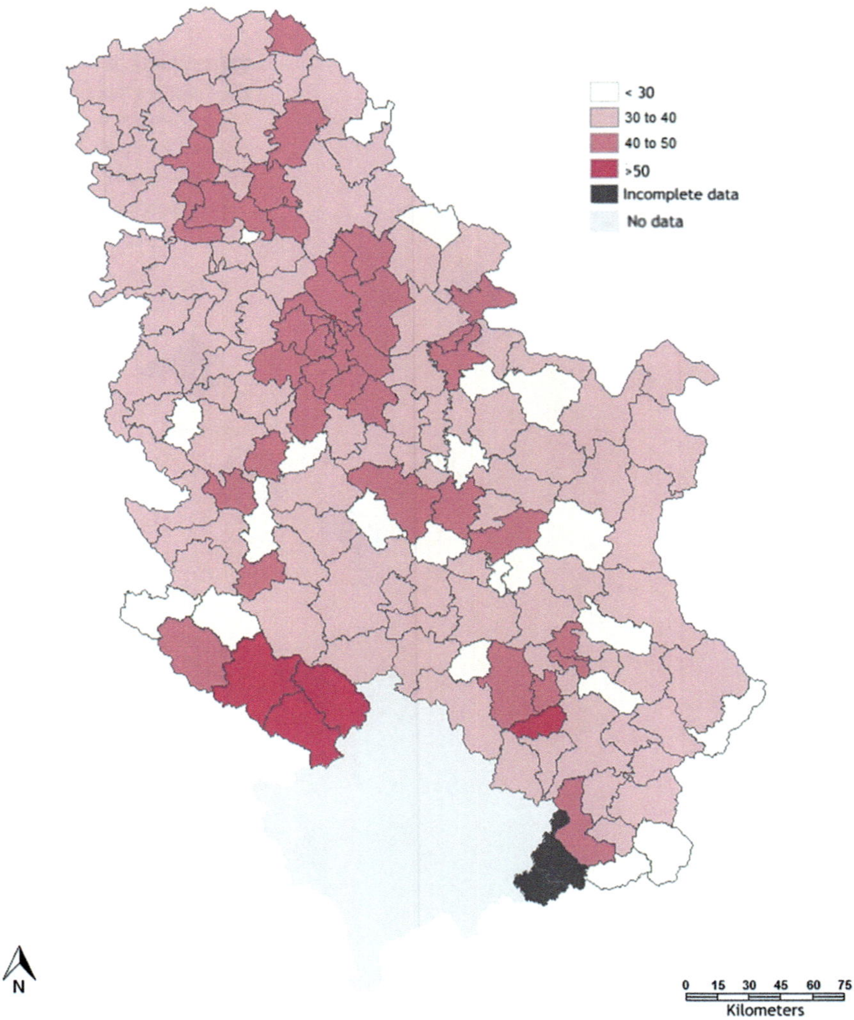

Fig. 1 Fertility rates, 2011 (Data are adapted from Opštine i regioni u Republici Srbiji, by Statistical Office of the Republic of Serbia, 2012 (https://publikacije.stat.gov.rs/G2012/Pdf/G20122008.pdf)

There are zero municipalities in the second, third, and fourth quantiles, whereas the fifth quantile includes 38 municipalities. Most of these municipalities are underdeveloped. The negative sign indicates that the relationship between the two variables is inversely proportional. Accordingly, if the share of women with the listed demographic characteristics to whom social welfare is the only source of income increased by 10%, fertility rates would decrease from 0.7 to 18.1%.

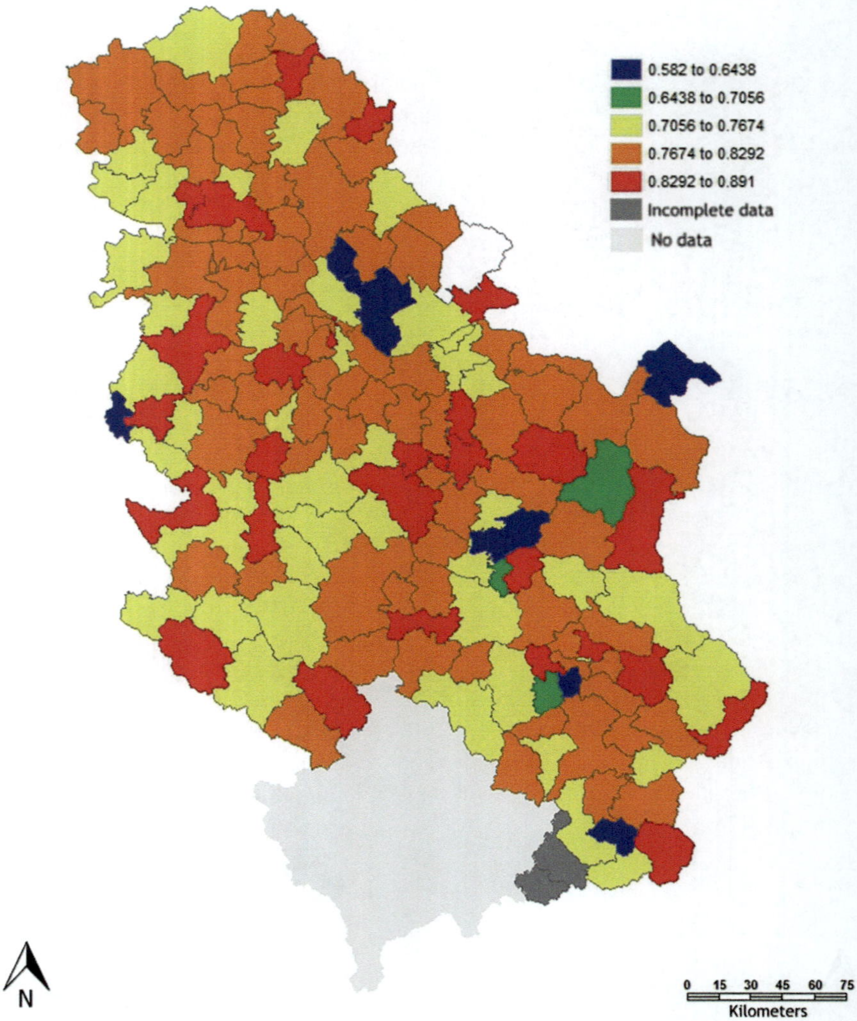

Fig. 2 Local R^2 values, cartographic visualisation of the results

As far as the relationship between fertility rates and the share of households without income in the total number of households (%) (Fig. 3b) is concerned, the results of the local model show that the relationship is statistically significant in 18 municipalities (10.8%), and the sign is negative. This means that there is an inversely proportional relationship between the two variables. Most municipalities are located in southeast Serbia, which is economically the least developed area.

Figure 3c shows the values of estimated regression coefficients of the share of married women aged 15–30 in the fertile contingent (%) and fertility rates. It is

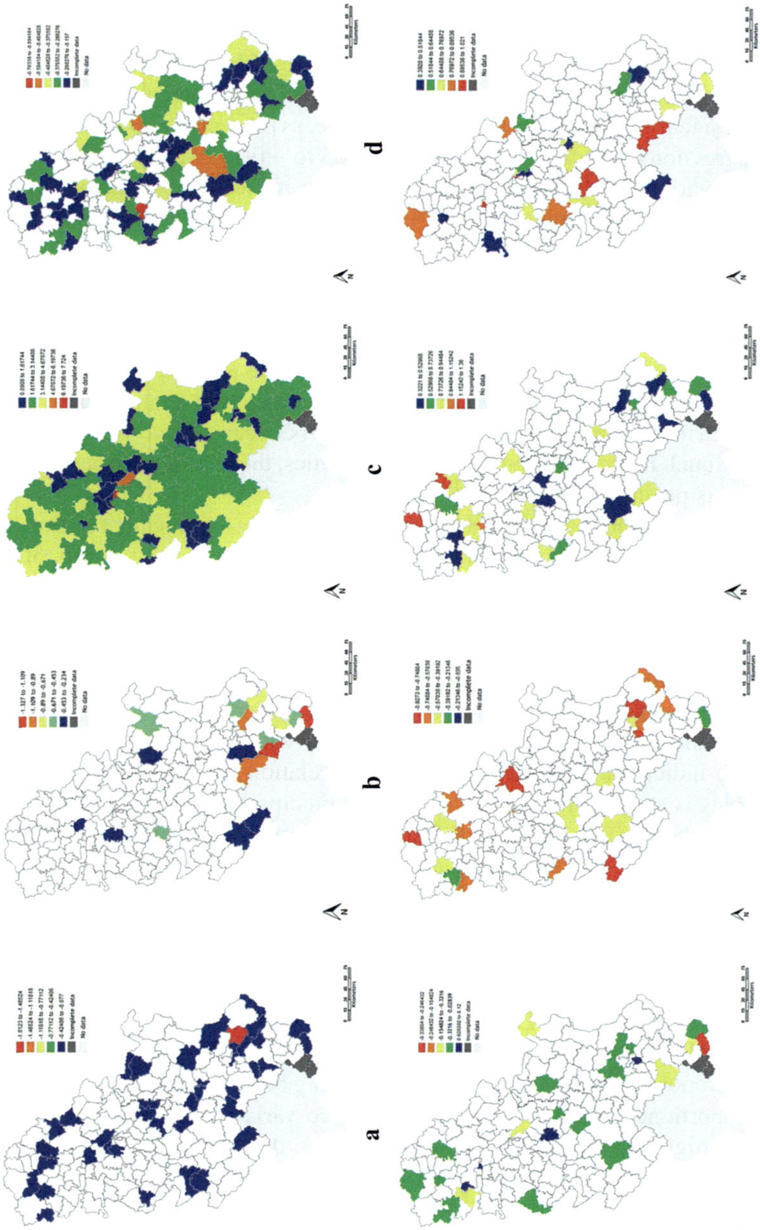

Fig. 3 Spatial distribution of estimated GWR local coefficients (significant areas only): **a** social welfare; **b** households without income; **c** married women; **d** unmarried women; **e** economically active women; **f** unemployed men; **g** women with secondary education; **h** women with high and higher education

interesting that only in this case the relationships are statistically significant in all 166 examined municipalities. The nature of the relationship is positive, i.e., if the share of married women aged 15–30 in the fertile contingent increased by 10%, fertility rates would increase from 0.9% in the municipalities of the first quantile to as much as 77.24% in the municipalities that belong to the fifth quantile.

As expected, the relationship between the share of unmarried women aged 30–39 in the fertile contingent (%) and fertility rates is negative, as presented in Fig. 3d. The values of the regression coefficients vary from −0.703 to −0.157. The relationship between the two variables is inversely proportional, so if the share of unmarried women with the listed demographic characteristics increased by 10%, fertility rates would decrease from 1.57 to 7.03%.

The relationship between the share of economically active women aged 15–49 in the fertile contingent (%) and fertility rates is the most interesting, and it is presented in Fig. 3e. The results of estimated regression coefficients show that the strength of the relationship varies and it even changes directions. The negative relationship between the two variables can be observed in 10 municipalities, and it ranges from −0.338 to −0.037. In these municipalities, the relationship between the two variables is inversely proportional. In the remaining four municipalities, the relationship between the two variables is positive.

The male unemployment rate shows a negative, statistically significant, relationship with fertility rates in 22 municipalities (13.2%), which can be seen in Fig. 3f. The maximum negative value of the coefficient is −0.927. If the share of unemployed men in the total active male population in this municipality increased by 10%, the values of the fertility rate would decrease by 9.27%. With the exception of two Belgrade municipalities (Savski Venac and Stari Grad), the others belong to the group of underdeveloped municipalities.

The values of the estimated coefficients of women with secondary education shown in Fig. 3g indicate that the expected positive relationship between the two observed variables is statistically significant in 33 municipalities (19.9%). A positive relationship indicates that an increased share of women who have completed secondary education in the fertile contingent may be expected to cause an increase in fertility rates. The lowest value of the estimated regression coefficient is 0.322, and the highest is 1.36.

The values of the estimated regression coefficients of women with high and higher education are presented in Fig. 3h, and the relationship is statistically significant in only 23 municipalities (13.9%). The coefficient values range from 0.393, as the minimum, to 0.989. In all municipalities, the relationship between the two variables is directly proportional. This is somewhat surprising because one would expect an inversely proportional relationship between the two variables keeping in mind that women with high and higher education marry later, due to which their fertile period is shorter. Besides, higher education is directly linked to higher aspirations for individual activities and personal improvement. However, recent research in Serbia indicates higher fertility among highly educated women (Mirić 2019b).

5 Discussion and Conclusion

This study investigates the spatial variation of fertility rates caused by variation in demographic and socio-economic determinants at the local and global levels, seeking to demonstrate the advantages of the local model in the detection of hidden content. The values of the coefficient in the global model, as a determinant of the influence of one or more independent variables on the dependent one, are estimated globally, indicating a constant and uniform influence of the selected variables on fertility rates. This means that the results obtained using the global model cannot detect spatial variations in the relationship between the determinants and fertility, directly influencing local population policies as a function of spatial disparities. The application of geographically weighted regression has demonstrated the ability of the local model to detect local variations as hidden content, which can be masked when global models are applied.

This is proved by the increased value of the coefficient of determination and the decreased value of the AIC. The high coefficient of determination in the local model, which shows that 79% of fertility variations on average are explained using the selected determinants, indicates its great importance in understanding spatially heterogeneous fertility patterns.

Our results suggest that the relationships between the determinants and fertility rates are not constant. They vary across space and there is a difference in terms of direction and strength. The results show that marriage is the most important determinant of fertility—it is statistically significant in most municipalities. It is interesting that, contrary to the expectations, the increased share of women with high and higher education cause increased fertility. The relationship between the economic characteristics of the population (WSW, HWI, and UM) and fertility is inversely proportional, indicating an inverse relationship. An exception is the determinant EAW, where the sign of the relationship changes—in 10 settlements it is negative, as expected, and in four it is positive. According to the results of the global model, the relationship between the two variables is negative in all statistically significant municipalities. This is the best example which demonstrates that some content can be hidden, i.e., the hidden content can be detected using adequate models, in this case—a local model. Specific numerical and cartographically presented results indicate, in a simple and sophisticated way, the locations of interest regarding fertility issues, creating a platform for numerous birth incitement strategies, as a leading population problem in many European countries. Based on the obtained and interpreted results, we may conclude that this study has fulfilled its primary objective, which is to demonstrate the advantage of using local models over global ones in detecting hidden content, which is in this particular case the varying relationship between factors and fertility, relative to the location.

The limitations observed during the research have inspired further fertility variation research. Namely, the administrative level lower than the municipality level for which demographic data are available in Serbia is the settlement level. The application of geographically weighted regression at the settlement level would lead to the detection of specific hidden content that may be masked in the data for higher administrative levels. Unfortunately, the degree of data availability at lower administrative levels, and especially geocoded data, is rather poor in Serbia, which limits more detailed research. Another limitation is the statistical significance of the data since the number of inhabitants in settlements varies from one (as many as 968 or 20.5% one-fifth of the settlements have less than 100 inhabitants, and these are mostly rural settlements) to 231,798 (Novi Sad). This limitation can be overcome if the research covers only municipality centers, which are demographic development axes of the belonging and, in some cases, neighboring municipalities.

Acknowledgements The study was carried out within the framework of Project 47007 and 47006, funded by the Ministry of Education, Science and Technological Development of the Republic of Serbia.

References

Aassve, A., T. Lappegard, and A. Vitali. 2012. A spatial analysis of gender equality and fertility. European Population Conference 2012, 13–16 June, Stockholm, Sweden. http://epc2012.princeton.edu/papers/121010. Accessed 13 Jan 2020.

Amara, M. 2015. Multilevel modelling of individual fertility decisions in Tunisia: household and regional contextual effects. *Social Indicators Research* 124 (2): 477–499. https://doi.org/10.1007/s11205-014-0793-5.

Andjelković-Stoilković, M., M. Devedžić, and G. Vojković. 2018. The border regions of Serbia: peripheral or marginal areas. *TRAMES Journal of the Humanities and Social Sciences* 22 (2): 211–227. https://doi.org/10.3176/tr.2018.2.07.

Babović, S., S. Lović Obradović, and I. Prigunova. 2016. Depopulation of Villages in Southeastern Serbia as Hindrance to Economic Development. *Journal of the Geographical institute "Jovan Cvijić" SASA* 66(1): 61–74. https://doi.org/10.2298/IJGI1601061B.

Balk, D., and M.R. Montgomery. 2015. Guest editorial: "Spatializing demography for the urban future". *Spatial Demography* 3 (2): 59–62. https://doi.org/10.1007/s40980-015-0017-x.

Blagojević, M. 1997. *Roditeljstvo i fertilitet – Srbija devedesetih.* Beograd: Institut za sociološka istraživanja Filozofskog fakulteta u Beogradu.

Bobić, M. 2013. *Postmoderne populacione studije.* Beograd: Filozofski fakultet – Institut za sociološka istraživanja.

Campisi, N., H. Kulu, J. Mikolai, S. Klüsener, and M. Myrskyla. 2020. Spatial variation in fertility across Europe: Patterns and determinants. *Population, Space and Place.* https://doi.org/10.1002/psp.2308.

de Castro, M.C. 2007. Spatial demography: an opportunity to improve policy making at diverse decision levels. *Population Research and Policy Review* 26 (5): 477–509. https://doi.org/10.1007/s11113-007-9041-x.

Del Boca, D. 1999. Participation and fertility behaviour of Italian women: The role of market rigidities. https://ideas.repec.org/p/wpc/wplist/wp10_00.html. Accessed 20 March 2020.

Devedžić, M. 2013. Raskršća demografije. *Stanovništvo* 2: 23–38. https://doi.org/10.2298/STNV1302023D.

Dommaraju, P. 2013. Marriage and fertility dynamics in India. *Asia-Pacific Population Journal* 26(2): 21–38. https://doi.org/10.18356/dcf9f3c9-en.

Engelhardt, H. 2011. On the changing correlation between fertility and female employment over space and time: A pooled time-series analysis on the impact of social indicators. *Zeitschrift fur Familienforschung* 21 (3): 245–263.

Entwisle, B. 2007. Putting people into place. *Demography* 44 (4): 687–703. https://doi.org/10.1353/dem.2007.0045.

Eyasu, A. M. 2015. Multilevel modelling of determinants of fertility status of married women in Ethiopia. *American Journal of Theoretical and Applied Statistics* 4(1): 19–25. https://doi.org/10.11648/j.ajtas.20150401.14.

Evans, A., and E. Gray. 2017. Modelling variation in fertility rates using geographically weighted regression. *Spatial Demography* 5: 1–20. https://doi.org/10.1007/s40980-017-0037-9.

Fotheringham, A., C. Brunsdon, and M. Charlton. 2002. *Geographically weighted regression: the analysis of spatially varying relationships*. Chicester: Wiley.

Getis, A., and J.K. Ord. 1996. Local spatial statistics: An overview. In *Spatial analysis: modelling in a GIS environment*, ed. Paul A. Longley and Michael Batty, 261–277. New York: Wiley & Sons.

Goldstein, J.R., and S. Klüsener. 2014. Spatial analysis of the causes of fertility decline in Prussia. *Population and Development Review* 40 (3): 497–525. https://doi.org/10.1111/j.1728-4457.2014.00695.x.

González González, M.J. 2017. Planing, urban sprawl and spatial thinking. *European Journal of Geography* 8 (1): 32–43.

Guohua, H. 2019. The hidden geography of the industry transfer: a case study of the Shenshan Special Cooperation Zone in China. In: *Abstract book of the international conference "Hidden Geographies"*, 29–31 August, Ljubljana, Slovenia, Department of Geography, Faculty of Arts, University of Ljubljana, Slovenian Association of Geography Teachers, EUROGEO – European Association of Geographers, 18–19. http://hiddengeographies.geografija.si/. Accessed 20 January 2020.

Hirschman, C., and P. Guest. 1990. Multilevel models of fertility determination in four Southeast Asian countries: 1970 and 1980. *Demography* 27 (3): 369–396. https://doi.org/10.2307/2061374.

Jung, M., W. Ko, Y. Choi, and Y. Cho. 2019. Spatial variations in fertility of South Korea: geographically weighted regression approach. *International Journal of Geo-Information* 8(6): 262. https://doi.org/10.3390/ijgi8060262.

Kamata, K., M. Iwasawa, and K. Tanaka. 2010. Spatial variations in fertility: geographically weighted regression analyses for town-and-village-level TFR in Japan. *Journal of Population Studies* 45: 1–20. https://doi.org/10.24454/jps.45.0_1.

Kamata, K. 2016. Diffusion process of fertility transition in Japan regional analysis using spatial panel econometric model. https://epc2016.princeton.edu/abstracts/160805. Accessed 14 Feb 2020.

Khan, H. T. A. 1997. A multilevel modelling approach to the determinants of urban and rural fertility in Bangladesh. *Asia Pacific Population Journal* 12(1): 55–76. https://doi.org/10.18356/2702b8d0-en.

Kohler, H.-P., F. Billari, and H.A. Ortega. 2002. The emergence of lowest-low fertility in Europe during the 1990s. *Population and Development Review* 28 (4): 599–639.

Kohler, H.-P., F. Billari, and H.A. Ortega. 2006. Low fertility in Europe: Cause, implications and policy options. In *The baby bust: who will do the work? Who will pay the taxes?*, ed. Fred Harris, 48–109. Maryland: Rowman & Littlefield Publishers.

Kokotović Kanazir, V., J. Stojilković Gnjatović, M. Filipović, S. Babović, M. Ivković, and S. Lović Obradović. 2017. Stanovništvo. In *Geografija Srbije*, ed. Milan Radovanović, 506–613. Beograd: Geografski institut „Jovan Cvijić" SANU.

Kupiszewski, M., D. Kupisyewska, and V. Nikitović. 2012. *The impact of demographic and migration flows in Serbia*. Belgrade: International Organization for Migration, Mission to Serbia.

Klüsener, S., M. Dribe, and F. Scalone. 2016. Spatial and social distance in the fertility transition: Sweden 1880–1900. http://www.demogr.mpg.de/papers/working/wp-2016-009.pdf. Accessed 20 Jan 2020.

Loftin, C., and S.K. Ward. 1983. A spatial autocorrelation model of the effects of population density on fertilty. *American Sociological Review* 48: 121–128. https://doi.org/10.2307/2095150.

Lović Obradović, S., S. Babović, and N. Shpak. 2016. Serbia and Russia on the demographic map of Europe two decades after the fall of communism. *TRAMES—Journal of the Humanities and Social Sciences* 20 (1): 59–73. https://doi.org/10.3176/tr.2016.1.04.

Lović Obradović, S., and S. Babović. 2018. Application of geographically weighted regression method in population distribution modelling. In *Abstract book of the international scientific conference "Information technologies, economics and law: state and development perspectives"*, 19–20 April, Chernivtsi, Ukraine, Bukovinian University, 65–66.

Lović Obradović, S., and G. Vojković. 2019. The role of spatial demography in visualization of population phenomena and processes. *Abstract book of the international conference "Hidden Geographies"*, 29–31 August, Ljubljana, Slovenia, Department of Geography, Faculty of Arts, University of Ljubljana, Slovenian Association of Geography Teachers, EUROGEO – European Association of Geographers, 55. http://hiddengeographies.geografija.si/. Accessed 20 Jan 2020.

Lović Obradović, S. 2019. *Model prostornog ispoljavanja demografskih procesa u Srbiji (doctoral dissertation)*. Faculty of Geography: University of Belgrade.

Lutz, W., V. Skirbekk, and M. Testa. 2006. The low fertility trap hypothesis: Forces that may lead to further postponement and fewer birth in Europe. *Vienna Yearbook of Population Research* 2006: 167–192. https://doi.org/10.1553/populationyearbook2006s167.

Magdalenić, I., and G. Vojković. 2015. Promene u starosnom modelu rađanja u Srbiji i zemljama EU – komparativna analiza. *Stanovništvo* 53 (2): 43–66. https://doi.org/10.2298/STNV1502043M.

Magdalenić, I. 2016. The influence of marital status on fertility in Serbia and the European Union. *Demografija* 13: 175–190.

Madjevikj, M., B. Apostolovska Toshevska, and M. Ljakoska. 2016. Regional differences in the population natural increase in the Republic of Macedonia. *Journal of the Geographical Institute Cvijic* 66 (3): 417–431. https://doi.org/10.2298/IJGI1603417M.

Mathews, S.A., and D.M. Parker. 2013. Progress in spatial demography. *Demographic Research* 28: 271–312. https://doi.org/10.4054/DemRes.2013.28.10.

Mathews, S., D. Janelle, and M. Goodchild. 2014. Future directions in spatial demography. https://www.researchgate.net/publication/259971624_Future_Directions_in_Spatial_Demography_Report. Accessed 4 Feb 2020.

McDonald, P. 2006. Low fertility and the state: The efficacy of policy. *Population and Development Review* 32 (3): 485–510. https://doi.org/10.1111/j.1728-4457.2006.00134.x.

Mier-y-Terán, M., and M. Martínez. 2014. Socioeconomic and spatial determinants of fertility amongst young women. https://iussp.org/sites/default/files/event_call_for_papers/MieryTeran_Mart%C3%ADnez.pdf. Accessed 14 Feb 2020.

Mirić, N. 2019a. Difference in fertility between women with and without tertiary education in Serbia. *Zbornik Matice srpske za drustvene nauke* 170: 245–257. https://doi.org/10.2298/ZMSDN1970245M.

Mirić, N. 2019b. *Fertilitet visokoobrazovanih žena u Srbiji (doctoral dissertation)*. Faculty of Geography: University of Belgrade.

Mucciardi, M., and P. Bertuccelli. 2011. A GWR model for local analysis of demographic relationships. *China-USA Business Review* 10(3): 236–244. https://doi.org/10.17265/1537-1514/2011.03.009.

Parr, N. 1999. *Application of multilevel models in demography: What we have learned?*. Sydney: Macquaire University.

Ord, J.K., and A. Getis. 1995. Local spatial autocorrelation statistics: Distributional issues and an application. *Geographical Analysis* 27 (4): 286–306. https://doi.org/10.1111/j.1538-4632.1995.tb00912.x.

Owoo, N.S., S. Agyei-Mensah, and E. Onuoha. 2015. The effect of neighbourhood mortality shocks on fertility preferences: A spatial econometric approach. *The European Journal of Health Economics* 6: 629–645. https://doi.org/10.1007/s10198-014-0615-3.

Petrović, M. 2011. Changes of marital behaviour and family patterns in post-socialist countries: Delayed, incomplete or specific second demographic transition? *Stanovništvo* 1: 53–78. https://doi.org/10.2298/STNV1101053P.

Rašević, M. 2015. Low fertility in Serbia: New insights. In *Fifth international conference of Balkans demography—the population of the Balkans at the Dawn of the 21st Century: book of abstracts. DemoBalk*, 21–24 October, Macedonia. http://iriss.idn.org.rs/187/1/MRasevic_Low_fertility_in_Serbia.pdf. Accessed 18 Jan 2020.

Rašević, M., and P. Vasić. 2017. Obrazovanje kao faktor fertiliteta i populacione politike u Srbiji. *ANNALES · Ser. hist. Sociol* 27(3): 599–610. https://doi.org/10.19233/ASHS.2017.42.

Rašević, M. 2019. *Religions and low fertility today. Stanovništvo* 57 (2): 1–11. https://doi.org/10.2298/STNV190814005R.

Smith, R.D. 2019. Marital fertility patterns and nonmarital birth ratios: An integrated approach. *Genus* 75: 9. https://doi.org/10.1186/s41118-019-0056-z.

Stanković, B., and G. Penev. 2010. Trends of extramarital births in Serbia in the second half of the 20th and at the beginning of the 21st century in the wider European context. *Stanovništvo* 1: 1–24. https://doi.org/10.2298/STNV1001001S.

Stanojević, D. 2018. Young parents on the labour market in Serbia: balancing professional and private domain. *Towards Better Demographic Future of Serbia*, eds. Vladimir S. Kostić, Slavica Đukić Dejanović and Mirjana Rašević. Beograd: SANU.

Statistical Office of the Republic of Serbia. (2012). Opštine i regioni u Republici Srbiji [Municipalities and regions in the Republic of Serbia]. Retrieved from https://www.stat.gov.rs/publikacije/?a=18&s=1807&d=&r=.

Šobot, A. 2014. On low fertility from the aspect of the economic activity of female population: Possibilities and restrictions in encouragement of childbearing. *Stanovništvo* 2: 43–66. https://doi.org/10.2298/STNV1402043S.

Vojković, G. 2007. *Stanovništvo kao element regionalizacije Srbije*. Beograd: Srpsko geografsko društvo.

Vojković, G., V. Kokotović, and A. Spalević. 2012. Demographic sustainability of the settlement system of South-eastern Serbia. In *The Population of South-eastern Serbia: The Impact of Demographic Changes in South-eastern Serbia on Social Development and Security*, ed. Mitrović Ljubiša, 411–423. Niš: UNIGRAF-X-COPY.

Vojković, G., Z. Živanović, and I. Magdalenić. 2018. Spatial-demographic misbalances as a challenge to the population policy. *Zbornik Matice srpske za drustvene nauke* 167: 663–672. https://doi.org/10.2298/ZMSDN1867663V.

Voss, P.R., K.J. Curtis, and R.B. Hammer. 2006. Exploration in spatial demography. *Population Change and Rural Society* 16: 407–429. https://doi.org/10.1007/1-4020-3902-6_19.

Wang, D., and J. Wang. 2015. A study of county-level fertility rate in China using geographically weighted regression method. *Population Association of America Annual Meeting 2015*. https://sheps.princeton.edu/papers/151543. Accessed 21 Jan 2020.

Wang, D., and G. Chi. 2017. Different place, different stories. A study of the spatial heterogeneity of county-level fertility in China. *Demographic Research* 37: 493–526. https://doi.org/10.4054/DemRes.2017.37.16.

Weeks, J.R., M. Gadalla Saad, T. Rashed, and J. Stanforth. 2000. Spatial variability in fertility in Menoufia, Egypt, assessed through the application of remote-sensing and GIS technologies. *Environment and Planning A* 32: 695–714. https://doi.org/10.1068/a3286.

Weeks, J. 2004. The role of spatial analysis in demographic research. In *Spatially integrated social science: examples in best practice*, ed. Michael Goodchild and Donald Janelle, 381–399. New York: Oxford University Press.

Wheeler, A., and M. Tiefelsdorf. 2005. Multicollinearity and correlation among local regression coefficients in geographically weighted regression. *Journal of Geographical Systems* 7 (2): 161–187. https://doi.org/10.1007/s10109-005-0155-6.

Hidden Social and Economic Geographies

Social Changes and Conflicts in Vora Municipality After the 1990s

Ornela Hasrama, Albana Kosovrasti, and Edlira Kola

Abstract Social developments in Albania have been closely linked to natural, economic, political, and historical factors due to the impact they have on the free movement of the population. Migratory movements in Albania, mainly after the 1990s, have played an important role in the transformation of socio-geographical spaces, having a multidimensional impact in the context of transformation and development of the territory and shaping new social characteristics of the population. Specifically, these transformations are evident in the territorial space of Vora Municipality, which can be identified as a typical case of territorial transformations and social changes of the population, as a result of the geographical position of this municipality. Vora Municipality is located between Tirana and Durrës, the two largest Albanian cities and has always served as a prime location for the most important businesses and people from different regions of our country, thus creating the *sponge effect*. The main aims of the paper are to identify and analyze the factors that have influenced the social transformation of this space, analyze the social effects in the population, identify new territorial transformations, and undertake the study of the geographical distribution of social conflicts. This study reveals the new social features of the population and the trends of these territories.

Keywords Territorial transformation · Population characteristics · Social development · Social conflicts · Albania

O. Hasrama (✉) · A. Kosovrasti · E. Kola
Department of Geography, Faculty of History & Philology, University of Tirana, Tirana, Albania
e-mail: ornela.hasrama@fhf.edu.al

A. Kosovrasti
e-mail: albana.kosovrasti@fhf.edu.al

E. Kola
e-mail: edlira.kola@fhf.edu.al

M. Krevs (ed.), *Hidden Geographies*, Key Challenges in Geography,
https://doi.org/10.1007/978-3-030-74590-5_19

1 Introduction

Albania is geographically located in the southwestern part of the Balkan Peninsula and features a long coastline facing the eastern shores of the Adriatic and the Ionian Sea. The Republic of Albania is bordered by Montenegro to the northwest, Kosovo to the northeast, Greece to the south, and the Former Yugoslav Republic of Northern Macedonia to the west, while the Strait of Otranto separates Albania from Italy (Qiriazi 2006). This geographic position has favored Albania to create links not only with other Balkan countries but also with Europe. Through the valleys, formed by the three major rivers of Shkumbin, Drin, and Vjosa were passing very important routes interlinking the Adriatic coast with the inland of the Balkans, Central Europe, Eastern countries, and vice versa (Academic of Sciences 1990). Because of these geographical conditions, Albania is an integral part of the economy of southeastern Europe and a key country in the development of the Adriatic-Ionian macro-region (NAPT 2015).

Defined in an area of 28,748 square kilometers, with an estimated population of 2.9 million, and an average population density of 98 inhabitants/km^2, compared to 116.7 inhabitants/km^2 which is the EU-28 average, Albania has been a NATO member since 2007 and a candidate country for EU membership since 2013 (NAPT 2015). Currently, Albania has been in transition since 1990 after the overthrow of the authoritarian communist regime. After the 1990s, the political, social, and economic transition brought about major changes in Albanian demographic transition. Massive immigration began in 1990, roughly 15% (Çabiri et al. 1998) of the population, but is still going on today.

First, the political factor has a major role in Albania, so it was thought that problems could be solved by reforms and these reforms should create new social relations, as well as the promise of politics in social justice, solutions to property claims, and building the rule of law. As a result of the fact that politics was thought highly important, most of the social conflicts were politicized and related to the political affiliations (Çabiri et al. 1998). During the years 1945–1990, the ideology was based on achieving the development of the country in a single country, with a centrally planned economic system, in which the workforce was key to the state. Internal and external migration of the population was nearly banned, thus affecting the sharp identification of the customs, traditions, and cultures of the population, where their geographical distribution was clearly distinguished. From the post-1990s period to the present day, the market economy and pluralistic political system are prevalent. Our country was no longer based on dictatorial ideologies but on capitalist ones, which has influenced to open the country and to promote international cooperation, especially in terms of trade. Also, the policies that allowed free movement of people have resulted in the rapid urbanization of central areas by expanding towards their peripheral areas and thus increasing the concentration of population in the major cities, mainly towards Tirana (the capital) and Durrës; only these two cities attracted about ¾ (Karaguni 2009) of the population involved in this migratory movement.

Second, the economic factor has played an important role in the country's socio-spatial transformations due to the large impact of private sector developments. Due to better opportunities of increasing the quality of life people were prone to move to large cities with better physical and social infrastructure and higher investment concentration, which contributed to the concentration of population in the main centers in the country, thus affecting the increase in population density. Also, the way of economic development, realized according to the "laissez-faire" model, has created a high pressure on housing demands in urban areas, insufficient infrastructure for social, educational, and health services, as well as an increased demand for transport. Meanwhile, in the peripheral areas, the gap between social and economic inequalities has been deepened (NAPT 2016); however, in 2014, the value of Albania's Gini Index was 29, while that of EU-28 was 31 (NAPT 2015).

Third, the social factor has played a determining role in spatial transformations due to the importance of social elements such as quality and standard of living, education, labor market, and lifestyle. After the 1990s our country has undergone a tremendous social transformation. At the beginning of the 1990s, the young population was dominant, regarding the gender structure the male element was identified, and regarding the settlement structure the rural population was dominant by 65% (World Bank Document 2007).

According to the Institute of Statistics (INSTAT) data in 2011, the Albanian population recognizes new characteristics related to the average age of the population which had increased from 30.6 years in 2001 to 35.5 years in 2011, which is caused by the massive migration of people and declining birth rates associated with family planning. In terms of the gender structure, the male element continues to dominate, but in certain age groups (35–39 years) due to migrations the female gender stands out in relation to the male gender. In terms of the settlement structure, however, it reversed for the first time, in 2011. According to INSTAT, the urban population was 53.5% and the rural, 46.5% (United Nation 2018) (Fig. 1).

According to World Urbanization Prospect, the settlement structure of the Albanian population will continue with the same trends, towards the dominance of the urban population, whereas by 2035 the urban population is projected to be over 70%. This demonstrates the important role played by the main centers in Albania by creating features of the urban-dominated center-periphery model, with high population density in the center, diminishing of the intensity of located economic and social activities from the center to the periphery, etc. However, rapid economic and political transformations have also been reflected in the social aspect, which has led to changes in population behaviors and gaining new features, thus often creating social conflicts.

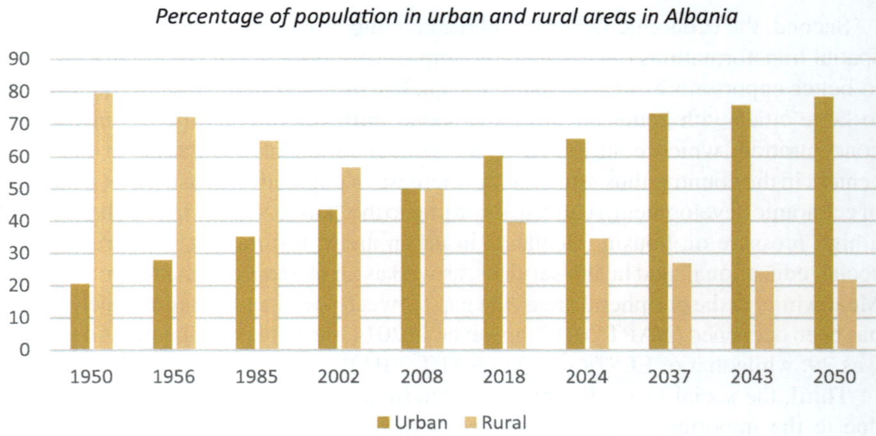

Fig. 1 Urban and rural population in Albania from 1950 to 2050 based on United Nation 2018

2 New Social Features of the Population After the 1990s in the Municipality of Vora: Comparative Approach

2.1 Geographical Extent of the Municipality of Vora and Its Importance in the New Features of the Territory

The Municipality of Vora is located in the central part of Albania, bordered by the Municipality of Kruja to the north, the Municipality of Kamza to the west, the Municipality of Tirana to the south, and the Municipality of Durrës to the east (Fig. 2).

The new municipality[1] consists of three administrative units: Vora, Preza, and Bërxullë. The municipality consists of a town and 18 villages (NAPT 2015). The territory of the Municipality of Vora has served as an important destination in attracting population from different parts of the country. For this reason, the geographical position of the territory has played a key role in bringing the settlers to this area due to its proximity to the capital, Tirana (Hasrama 2016), and the second-largest city in Albania, Durrës. Tirana-Durrës area is the main destination, including more than 61% of the population's migration, as a metropolitan area between the counties, thus creating the opportunity for human capital not only in the town of Vora but also in its villages. This area is also considered Albania's economic engine, comprising 35% of local businesses and 60% (A.RR.SH 2015) of foreign businesses; it is a preferred area for the expansion of shopping malls and other social places by offering an environment far away from the pollution and noise of the capital. Due to its geographical position, Vora was first populated around 1916–1917, when Albania was occupied by the Austro-Hungarian army which was set up at this crossroad several warehouses

[1] According to the last territorial administrative division in Albania approved in 2014.

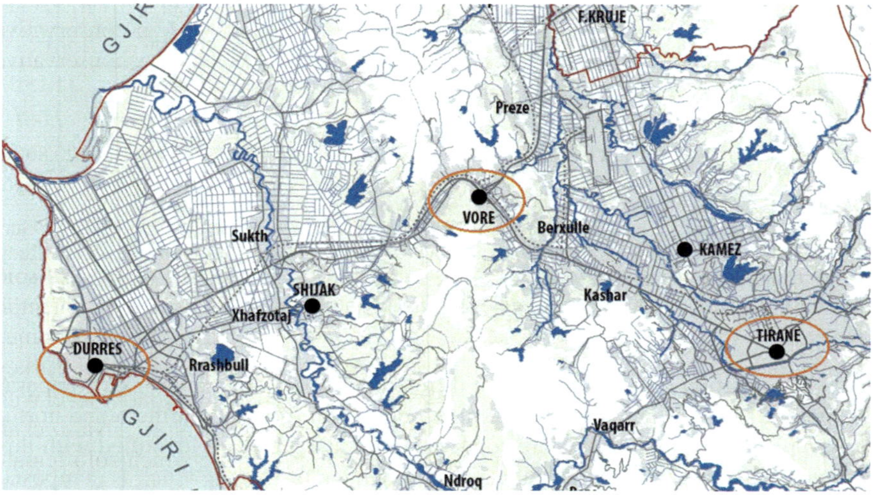

Fig. 2 Geographical position of the Municipality of Vora (NAPT 2015)

for army supply, and then the workers' families settled there during the construction of Tirana–Durrës railway as well as the construction of brick and tile factory (Subashi 2014). The employment of the population in these activities, which was followed by the settlement of other builder families, the construction of highways of national importance, and the establishment of some economic activities were the main socio-economic factors that have affected the population of the Municipality of Vora.

2.2 *General Features of the Population of the Municipality of Vora*

The population has been increasing since 1917. In certain periods, a gradual increase was influenced by factors such as natural population growth, immigration, and internal migration. In the centuries of the Ottoman Empire, especially in the 15th century, the documents prove the abandonment of many villages as a result of ruthless Ottoman exploitation (Subashi 2014). Part of this population fled to the mountainous areas and the other part migrated abroad. In subsequent years, the number of population increased, many of them immigrating to this area. Migratory movements are one of the most important elements in dealing with populations. On the basis of these movements, it was possible to distribute the population and settle some previously uninhabited areas, which had a direct impact on the socio-economic development of the geographical area. A distinguished feature of the Municipality of Vora has

Table 1 Population trends of the Municipality of Vora (several sources)

Municipality/commune	1917	1941	1979	2003	2011	2015	2018
Bërxullë*	–	–	–	7379	9883	10.341	11.151
Prezë**	1996	3974	3878	5838	6500***	6896****	7.067
Vorë	1147	1300	7000	16.949	19.608	20.434	20.867
Total	3.143	5.274	10.878	30.166	35.991	37.671	39.085

Sources *Vital Records, Bërxullë; **Subashi 2014, Preza in the Flow of History, 236–237; ***Guri 2012, "Strategic Environmental Assessment (SEA) Report. Environmental Impact Baseline Study for the Conceptual Urban Development Project. (Municipality of Vora, Commune of Prezë, Commune of Bërxullë); ****Vital Records, Municipality of Vora

been the attraction of population from different parts of Albania. From a comparative point of view, there is more incoming population from the northern and northeastern part of the country, where some geographical areas are identified with this particular characteristic. A typical case is the village of Domje which is composed of 90% of the incoming population and almost all of this population originates from Kukes, Tropoja (Hasrama 2016), etc., but in this area, there is also population from the southern or southeastern part of the country originating from Skrapar, Gramshi, etc. Migratory movements, decreasing infant mortality rate, and natural population growth have influenced the population trends over the years. The comparison between the administrative units is made with reference to Table 1.

In the data presented in the table, we note that prior to 1945 the population had a gradual increase. Factors such as the difficult economic and social situation of the population had an influence that is reflected in the high mortality rate, especially in the younger age groups. On the other hand, during the period between 1945–2000, the population increased rapidly as a result of a number of factors: the policy of that time to promote population growth, mortality reduction (especially infant mortality), health care increase, external migration ban (1945–1990), which contributed to this rapid and significant increase in the total population of the Municipality of Vora. In the 2000s, although factors such as external migration and declining of fertility rates were influential, the total population increased as a result of normal natural growth and especially the influx of people from all counties of Albania.

Initially, from 1990 to 1995, arrivals were scarce because there were no job opportunities available to support their living, but after 1995 immigration was at a high level and the net migration has always been positive. The villages that have undergone the most transformations are Marikaj, Domje, and Preza-castle but there were considerable spatial changes in other villages as well. Over the years, the positive ratio of incoming people with those leaving this area is always noticeable. In some data showing the net migration of the population of the country, the positive net ratio between the outgoing population and the incoming population is presented only for the territorial area of Vora (Table 2).

In terms of migration, the population of the Municipality of Vora is not significantly affected. Statistics of the Vora unit in 2011 show that the number of residents

Table 2 Incoming/outgoing ratio during 1998–2015 (Municipality of Vora)

Year	Incoming	Outgoing	Net migration
1998	190	140	+0.3
2000	388	251	+0.9
2003	543	266	+1.6
2006	144	116	+0.1
2015	1160	855	+1.3

who emigrated abroad is 1235 individuals, respectively: 41% to Greece, 49% to Italy, 6% to Germany, 2% to the UK, 1.5% to Switzerland, and 0.5% to the USA, where they work mainly in construction, transportation, trade and services, maintenance, and other occupations. The income they bring to families ranges ALL 15,000–50,000 per month (Municipality of Vora 2016). While there are about 1100 emigrants in 2014 for the same area, this was affected by the return of emigrants due to the economic crisis in neighboring countries, Greece and Italy, but currently there is an increase in the number of emigrants moving to Germany due to the facilities provided by the German government for the qualified population.

In terms of population structures, the aforementioned factors have had a significant impact on developing new features of the population of the Municipality of Vora today. The average age of the population has been increasing, which implies that population trends are aging. When comparing the 2005 and 2015 data, it is noted that the average age of the population has increased from 26 to 32 years, which is the consequence of family planning and migratory movements affecting about 70% of the population aged 20–35 (Guri 2012). Differences in the gender structure are small, namely 50.3% of the population are males and 49.7% females. With the new administrative division, in 2014, the ratios of the population structure by settlement are relatively 26.3% of the population living in the city and 73.7% of the population living in the rural areas.

According to INSTAT (2014), 67% of the population belongs to the age group 15–65 years, 22% to the age group 0–14 years, and 11% to the age group 65 years and over. The educational structure of the population for the Vora unit is expressed as a percentage: 10% illiterate, 10% with primary education, 20% with 8-year education, 30% with general secondary education, 10% with vocational education, and 20% with higher education (Municipality of Vora 2016). The current religious structure is the following: 90% Muslims, 5% Orthodox, and 5% Catholics. The Muslim population remains dominant, although, before the 17th century, the villages of Gjokaj and Kuç were two former capital centers of the Christian faith.

3 Changes and Social Conflicts

Quantitative systematic research on conflicting activity emerged in the 1960s. The study of urban conflicts, which may take the form of protest activities, is often conducted within the conceptual framework of urban social movements. However, it is already rooted that these conflicts are not always caused by social movements and that social changes do not come only from social movements (Trudelle 2003). Indeed, social conflicts, as they are reaching the crisis point, may be seen as one of the challenges that the social formalization, analysis, and modeling sciences should explore (Neuman et al. 2011) because not only do different disciplines view the social conflict from different perspectives but the meaning of the term is often interpreted differently by researchers within the same field of research (Neuman et al. 2011). Conflicting events (including protest events) are complex phenomena: they are highly conditioned in terms of content, origin, target, size, shape, intensity, time period, social position of actors, etc. (Trudelle 2003). Lying in different geographical areas, we can point to the fact that societies differ from each other in terms of customs, traditions and culture, and as a result, we can say that social conflicts in the population are expressed in different forms. Due to economic, social, and political conditions, societies have different stages of development, and our limitations lie in the fact that, according to some authors, the main cause of conflicts of interest is the inadequacy of material products that can satisfy all the needs of citizens (Zdravko 1972). Under these conditions, new features of the population of a country arise conditioned by the actions and activities that different social groups perform in the new geographical space for them.

Historically social conflicts and human conflicts, in general, have been considered as a field of reasoning by handling their solutions in the context of searching for the best explanations in terms of assuming a real objective (Morales-López and Floyd 2017). In modern societies, social conflicts find full embodiment carrying their specific characteristics in accordance with the features of each country of the world in which they appear. The history of social conflicts in contemporary Spain is rich in a variety of works and approaches (Godicheau 2013). Public libraries in the outlying suburbs of French cities, which are called peripheries, face many conflicts, occasionally violent. These conflicts take place within the context of important social, cultural, and political transformations that have accelerated during the last 15 years (Merklen 2016). Through somebody's eyes, social cohesion is concurrent with identifying some common problems (difficulty in integrating certain social groups, unemployment, crime, collapse of education, peripheral problems, population movement, economic insecurity migration, and fear of socio-economic degradation) (Lafaye 2011). All these factors are expressed in crises created by different social groups in society, where the geographical distribution is not the same all over the place and expressed in the same intensity.

3.1 Social Conflicts in the Municipality of Vora

Population movement and economic development of the area under study are the main driving factors that have influenced the socio-spatial change of this territory and as a result, it is concluded that the territory is expanded and transformed in the context of the social places in this area. One of the most prominent social places in changing their numbers is the places of housing, which, despite the intensity associated with them, have been increased in each period. The following table is a comparison of the evolution of settlements in a comparative plan for all three administrative units of the Municipality of Vora (Table 3).

Referring to the table above, for all three units, a considerable evolution of social housing is observed. The increase in the number of settlements is attributable to the substantial immigration in the area, the reasons for the population settling in this territory are economic, natural, and social factors. The economic factor has been primary due to employment and opportunities for better living, generally flat terrain and favorable climate are the two major natural factors, but also lifestyle or social conditions stand out as distinguished social factors. Regardless of the number of dwellings always showing positive indicators, in the Municipality of Vora the demand for territorial space is higher, not only for residential purposes but also for facilities for economic activity, however, the social crisis is presented in the legal aspect, property ownership issues in Albania have been present since 1995, when property transitioned from state-owned to private-owned. After 45 years of increasingly constrictive agricultural collectivization, the fall of the communist régime in Albania has, since 1990, resulted in the almost total fragmentation of co-operatives and state farms into 379,000 family farms, with an average holding of 1.85 hectares—a phenomenon unique in Eastern Europe (Deslondes and Sivignon 1995). Similarly, in Albania, the Municipality of Vora has had an evident ownership problem, which is considered one of the social conflicts with very negative effects. By law 7501, 19/7/1991, "On Land", the criteria for the division of land would not be according to the old boundaries. The Land law of 1991 defines new principles, allotting the estates to those who cultivate them (Sivignon 1992). Therefore, under such conditions, some territories were unclear in their possession in the context of documentation.

A significant part of the land of the Municipality of Vora prior to 1951 belonged to the Beys, the highest class of society at that time, who had acquired this land by purchasing (at very low prices) from the poorest strata of the population who exchanged it for little money to survive to the difficult living conditions. Then in the time of the communist system, that upper class of society no longer existed and the land was state-owned with no owners. By the 1991 law, with the fall of the communist regime, the land was allocated to persons living in the village, where each family in the Vora area owned 2000 m^2/capita. The persons who left that village had no right to own their former land because according to the ideology of that time the land belonged to those who worked it, including in this case persons coming from other parts of Albania. Under these conditions, it turns out that one land has two owners, so in the territory of the Municipality of Vora, the settlement of some of the social sites

Table 3 Buildings for residential purposes according to the municipality/communes and construction period (INSTAT 2011)

Municipality/commune	Total	Up to 1960	1961–1980	1981–1990	1991–2000	2001–2005	2006–2011
Bërxullë	2087	42	87	209	644	422	434
Prezë	1447	93	224	183	317	165	290
Vorë	2770	48	227	303	480	277	460

Table 4 Number of businesses in the Municipality of Vora during 2012–2018 (INSTAT, Active enterprises by municipalities and economic activity, 2018)

Year	2012	2013	2014	2015	2016	2017	2018
No. of businesses	749	835	907	968	915	972	1350

has been associated with land ownership problems creating social conflicts through the owners, including cases where these disputes have ended in physical conflicts up to life-threatening of the pretending persons as owners. The ownership conflict has impeded not only the further development of the territory but also the deterrence of international investment.

Another defining element in this territory is the establishment of a high number of economic activities. If we compare the business objects over the years they represent an increasing curve since the period of the population settled in this area. Specifically, for the Preza unit since 1922 where 13 businesses were listed, in 2014 the Preza unit numbered 176 businesses. In the Vora unit, the change has been more dynamic, wherein 1992 from no business reached to 765 businesses in 2012, so for 10 years this territorial area has had a very rapid development. While in the Bërxull unit for 2012 there were estimated about 76 businesses (Table 4).

What is highlighted in this case is the evolution of the number of businesses, 90% of which are starting up by incoming population or businesses located in this prime location, and only 10% of them are set up by the population of this area, making a significant difference between the local people and the non-local people who have an opportunity to invest. The local people have previously been oriented to the primary sector of the economy, where they had developed themselves, whereas today the established businesses created an area with a profile of the secondary and tertiary sectors of the economy, transforming this territory from an agricultural area to an industrial area. The movements pass from a status of exclusion, outside of the labor conflicts, to a progressive reintegration into the more traditional framework of the labor conflicts and the old labor movement, there are the same examples as in Germany or France (Mazières-Vaysse 2011).

The transformation of the geographical space is very noticeable, especially for the area under study. A typical example is found along the national road that belongs to the villages of Vora and Preza which are connected with other national roads. One problem identified is the lack of space of residential areas from the industrial areas. This is shown on the figure of land use which belongs to a part of the Tirana-Vora axis, where the change of land use is on a large scale (Fig. 3).

Changes in such proportions are a contributing factor to large immigrant population movements. Traditionally the mobilization of social movements has been analyzed according to rational choice, in other words, by a calculation of (individual) interests (Neuman et al. 2011). Due to better life opportunities, migratory movements have been impetuous and at times chaotic (referring to the way of housing and other constructions built for economic and social purposes), in addition to these spatial transformations, migration of the population has also affected the cultural diversity

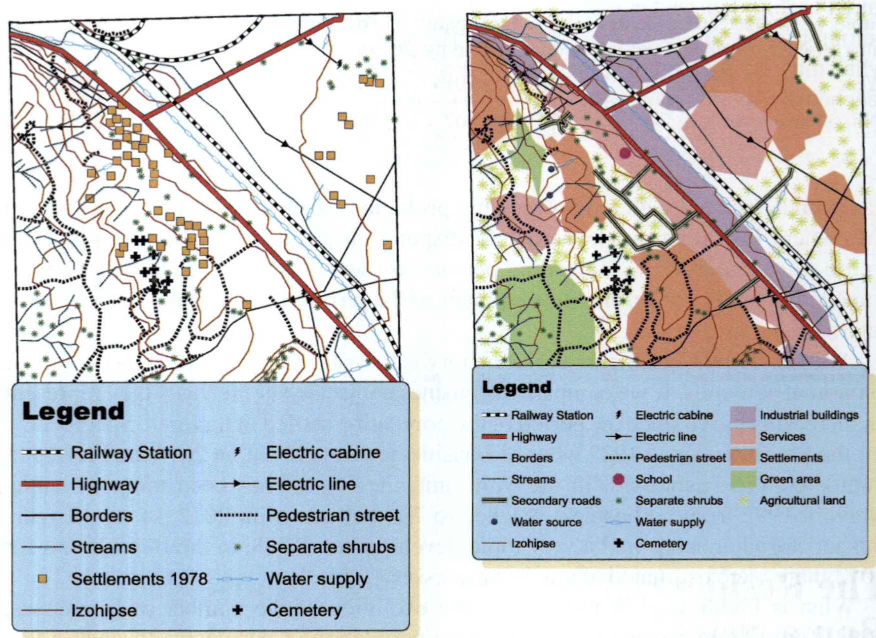

Fig. 3 Land use in 1978 and 2015

of the population of this territory. Currently the largest percentage of the Municipality of Vora is inhabited by the local population, but at the same time, the population coming from all over Albania is constantly growing. A typical case remains one of the villages of the Municipality of Vora, the village of Domje, which accounts for 90% of the incoming population, mainly from the northern part of Albania. These migratory movements also bring about the exchange of customs or traditions in the destination country, and in this incoming population we often find their traditions adapted to the traditions of the area, but there are also occasions when the incoming population fanatically preserves the customs and norms of their country of origin mentioning here: the dialect of the language, dress-up, lifestyle, etc.

However, moving in this area has set them free in the context of gender equality, building new relationships with children, living a new way of lifestyle, or respecting the rights and personal freedom of each member of the community, which previously caused cultural clashes with local people. If families were formerly patriarchal and composed of several members, nowadays families are small and each family member is free to express his or her own views openly. Living in the area has also made them more open-minded in the context of social life and the inclusion of women in many events. Also, it is noted that in recent years, much importance is being given to women who are no longer restricted from pursuing their studies or their professional development.

3.2 *Geographical Distribution of Social Conflicts and Social Effects*

Referring to the approaches covered in this paper from the geographical distribution point of view, it is evident that social conflicts are more present in those areas with the presence of local and immigrant populations. In the Municipality of Vora the geographical distribution is not uniform, in the comparative view among the three administrative units it is found that due to the geographical position, these three units are attractive for the incoming population; however, we can say that this population has preferred the surrounding villages, rather than the town itself, due to cheaper prices of land and larger area of agricultural land that they use to produce crops for self-consumption. These population movements, which have led to the social and spatial transformations of the territory, have created visible social effects in society, some of which are presented below.

Revenue Diversification The Municipality of Vora in the early 1990s was an agricultural country, in which the largest percentage of the population worked in this sector. Meanwhile today this area with a completely changed economic profile, in industrial-agricultural, has enabled about 60% of the population to work in the processing industry, a part in agriculture and the lowest percentage in the tourism sector, which takes positive trends in its development for the last five years. This implies that the population of this municipality is involved in all three sectors of the economy compared to the 1990s which were focused on the primary sector.

Skilled Labor As a result of the migratory movements of the population it is found that the benefits have been not only economical but also social, mainly the qualifications of the labor force are evident in the case of the migrant population of this area. This is evident in the private businesses that returnees have developed in their country of origin. These qualifications are more prevalent in the field of construction and handicrafts such as wood, marble, or decorative stone processing in this sector, mainly the inclusion of women in furniture and manufacturing jobs.

Social Welfare of the Population The social welfare and quality of life of the population in this area is increased, this is shown by the main elements that make up these two components. These include monthly income of the population, healthcare, education, economic and physical security, respect of human rights, etc. As a result of the economic growth, there is also an increase in the consumption of products, which is found a variety of products consumed and the demand for their quality. The majority of young people of the population want to pursue higher education and postgraduate education, regardless of the conditions their families are in, however, to mention the fact that this population is being educated at a higher rate compared to previous years.

Better Cohesion in Relationships Among Generations Families are now made up of a smaller number of members making that cohesion among members being higher, decisions being made in the family and young people being more independent and freer to determine their future or integration into the different social groups. Unlike years ago when families were made up of a large number of members, because of sharing the same house of many crowns in one family, decisions were made only by the men of the house, and the future of the younger ones was determined by the eldest in the family.

Social Integration of the Incoming Population and Adjustment to the New Environment The incoming population is well-adjusted to the new lifestyle in this area, some of them have borrowed many rituals that they were handed down to the local population or others have preferred to adapt them into their habits of origin. The integration of the incoming population is also evident in their employment in different sectors of the economy, their involvement in community decision-making, their integration into education, and their integration into different social groups. All of this has helped to ensure that there are no differences between different social groups in the context of human rights.

To get closer to reality, in a survey conducted in July 2019 with 15 immigrant residents of the Municipality of Vora, they were asked what better opportunities the city of Vora has offered, concluding that moving in this town enabled them employment, better social and physical infrastructure compared to their country of origin, more active social life, and better services. Also, when asked if they would like to go back to their country of origin, 77.8% of the respondents would not want to go back to their home country and 22.8% responded that they would return only to visit their family members or to a cleaner environment.

4 Future Trends: Advantages and Challenges

Conflicts of interest are solved through contracts, voluntary trade, and government policy (Huettel and Kranton 2012). In developing countries, conflicts are always linked to social inequalities and almost occur on natural resource use policy. It is because such countries possess some factors like low level of literacy, less media access, population pressure, incompatible planned projects, injustice, and political interference (Habibullah 2013). In such a territory, as in the present case study, migratory movements have always played a very important role in the features of the territory where the population is settled because the population holds its characteristics enabling the diversity of cultures. The territory of the Municipality of Vora can be considered as a transit area, due to its geographical position, because this position will always remain a desirable territory for the population in the future, which is also indicated by the values of the land which has continued to go up about 6 times over the last two decades.

Despite the advantages we can mention for this area, such as attracting skilled labor, economic growth of the area due to infrastructure upgrade, and investment increase in the area, this geographical area faces some challenges in the future, first, for due to the continuous increase in the number of population which has not been accompanied by the appropriate level of investments in social and physical infrastructure, in order to cover the service of the entire population of the municipality. Another important point is the urgent need to intervene in territorial re-composition in order to stop construction without specific planning but also to distinguish industrial areas from residential areas. In the territory of the Municipality of Vora, which is distinguished for its natural potentials that favor the development of agriculture, special attention should be paid to the development plan for agricultural areas in order to produce crops aimed at the national and international market. These measures could significantly reduce social conflicts related to the employment of residents, social services, or integration of different social groups. Finally, the main challenge remaining for this area is to mitigate social conflicts through policies developed by local governments and authorities.

5 Conclusions

Albania has experienced a major transformation in the social and spatial aspects, due to political, economic, and social factors. Compared to two different periods, before the 1990s and after the 1990s, the political factor has played a significant role in the development of Albania because of opposing regimes in leadership. During 1945–1990, the social classes, social status, and political status of the population were separate due to developments dictated by the dictatorial regime of the state, meanwhile after the 1990s, with the fall of the communist regime, everything changed and with a pluralistic society, these differences were no longer so evident. In the second period, after the 1990s, with the free movement of the population, the socio-spatial features of the different places in Albania were no longer the same. In this paper, regarding major transformations, we conducted a case study about the Town of Vora, in which it has been transformed from a territory with a dominant agricultural economic profile into a territory with an industrial economic profile, where these changes are totally related to migratory movements of population and properties are transferred from state-owned to privately own. According to the documents, the Municipality of Vora has been inhabited since the 5th century BC, but the first facilities in this area started around 1917, since then this territory has been transformed into an urbanized area for many reasons. Due to migratory movements, the territory now needs a new management and re-composition of the territory to create facilities for further development of the area.

Because of these changes in the society of the Municipality of Vora we meet some new features of the population that make social conflicts between them. Among them are the social conflicts caused by ownership issues based on the amendment in the law of 1991 on land-sharing; social conflicts arising because of cultural diversity due to immigration and social conflicts related to the labor market of local residents.

The geographical distribution of social conflicts is not uniform, but we can say that they are more present in the villages of this municipality due to the preference of the residents to move to these areas for cheaper prices of land and opportunities for subsistence agriculture to grow self-consumption food crops. Some of the social effects related to population movements that bring into social transformations and conflicts are income diversification, skilled labor, social welfare of the population, better cohesion in relationships among the generations, social integration of the incoming population, and adjustment to the new environment. Despite the positive trends in the development of this socio-geographical space, it is found that various problems are present which can be solved through more detailed planning of the territory due to the need for territorial re-composition, favorable social policies that mitigate social conflicts, and control of migratory movements aiming to prevent the abandonment of places by the outgoing population.

References

Academic of Sciences. 1990. *Physical geography of Albania*, vol. I, 39. Tirana: Poligraphic Combine-New Printing House.

A.RR.SH. 2015. Study, design and expansion of Tirana-Durrës Highway 10. Tirana

Çabiri, Y., M. Ekonomi, P. Misha, R. Lani, and Y. Pango. 1998. Human development report Albania 1998, 35. Tirana: Albdesing.

Deslondes, O., and M. Sivignon. 1995. L'agriculture albanaise: De la coopérative à l'exploitation de survie, 123. *Revue d'études Comparatives Est-Ouest*, 26.

Godicheau, F. 2013. Entre histoire sociale des conflits et histoire des concepts: Protestation et ordre public en Espagne dans le dernier tiers du XIXe siècle,1. Toulouse: Framespa.

Guri, S. 2012. Strategic Environmental Assessment (SEA) Report. Environmental impact baseline study for the conceptual urban development project. (Municipality of Vora, Commune of Prezë, Commune of Bërxullë), 135. Tirana: G&g Group

Habibullah, Maksi. 2013. Land use conflicts in developing countries: Framing conflict resolution and prevention strategies to ensure economic growth and human welfare. The case of Chotiari water reservoir from Pakistan. *Economics and Finance*, 46. AgroParisTech.

Hasrama, O. 2016. Population and its social elements in the sociogeographical area of Municipality of Vorë. Master's thesis. 35. Tirana

Huettel, S., and R. Kranton. 2012. Identity economics and the brain: Uncovering the mechanisms of social conflict. *Philosophical Transactions of Royal Society*, 680.

INSTAT. 2018. *Active enterprises by municipalities and economic activity*, 14. Tirana: Gent Graphic.

INSTAT. 2011. *Census of population and housing 2011*, 10. Tirana: Gent Graphic.

INSTAT. 2012. *Census of population and housing 2011*, 10. Tirana.

INSTAT. 2014. *Commune Typology and Infra-Urban Statistical Cartography*, 54. Tirana.

Karaguni, M. 2009. Sustainable urban development and urban Albania, 503. *Albanological Studies* IV. Tirana: Arbëria.

Lafaye, C.G. 2011. *Cohésion sociale et lien social*, 2. France: Centre Maurice Halbwachs.

Mazières-Vaysse, A. 2011. *Precarious workers and new forms of social conflicts at the turn of the century*, 5. Portugal.

Merklen, D. 2016. *Is the library a political institution? French libraries today and the social conflict between démocratie and république*, vol. 65 (2), 143. Johns Hopkins University Press.

Morales-López, E., and A. Floyd. 2017. *Developing new identities in social conflicts: Constructivist perspectives*, 67. John Benjamins Publishing Company.

Municipality of Vora. 2016. *Information office*. Albania.

NAPT. 2015. *Integrated cross-sectoral plan for Tirana-Durrës economic area 2015-2030*, 245. General National Plan (GNP).

NAPT. 2016. *The first national document for territorial planning 2015–2030*. General National Plan (GNP)
Neuman, M., A. Braun, E.-M. Heinke, and M. Sqalli. 2011. Challenges in modelling social conflicts: Grappling with polysemy. 4, *Journal of Artificial Societies and Social Simulation*, SimSoc Consortium 14 (9).
Qiriazi, P. 2006. *Physical Geography of Albania*, 9. Tirana: Ideart Publications.
Sivignon, M. 1992. La nouvelle agriculture en Albanie et l'exemple de Korça, 289. France: *Revue Géographique Des Pyrénées et Du Sud- Ouest*, 2.
Subashi, X. 2014. *Preza in the flow of history*, 47. Tirana: Edlora.
Trudelle, C. 2003. Au-delà des mouvements sociaux: Une typologie relationnelle des conflits urbains, 223. *Cahiers de Géographie Du Québec*, 47.
World Bank Document. 2007. *Evaluation of the urban sector in Albania*, 9. Department for Sustainable Development, Europe and Central Asia Region.
Zdravko, M. 1972. Les conflits sociaux et le développement social en Yougoslavie, 36. Revue de l'Est, 3.

The Fourth Industrial Revolution's Implementation—The Slovenian Case Study

Lucija Lapuh

Abstract Digitalisation and automatisation have been changing the economy since the digital revolution from 1969. Characteristic of the ongoing fourth industrial revolution are cyber-physical systems which communicate with one another. Every step of the manufacturing process is aimed to be interconnected with the 'Factories of the Future'. Theoretical characteristics and initiatives will be presented in the first part of this paper, considering the working force and sustainability as well. In the second part of the paper, the Slovenian companies' characteristics regarding digitalisation and (potential) use of high-performance computing in the automotive and electronic sector will be presented, based on the data gathered by questionnaires. The results show that the Slovenian economy is in the initial phase of new hidden technological development with some successful implementations.

Keywords Factory of the future · Industry 4.0 · Digitalisation · High performance computing · Smart manufacturing · Automotive sector · Electronic sector · Industrial revolution · Economic geography · Slovenia

1 Introduction

Throughout history, inventions have modernised the production and organisation of a society. Energy-related innovations have led to industrial revolutions. Mechanical production facilities were powered by water and steam during the first industrial revolution (late eighteenth century), while the second used electric energy to create mass production (late nineteenth century), and the third automated production by using electronics and information technology (since 1969) (Abdi et al. 2018).

Today's fourth industrial revolution is built on the third digital revolution. New technology is not a new type of energy, but information technology with digitalisation enables us to build a new virtual world. Physical and digital spheres are interrelated. "The major challenges that manufacturing companies face today are: the growing

L. Lapuh (✉)
Faculty of Information Studies, Institute for Innovation and Technology in Regional Development, Ljubljanska cesta 31a, 8000 Novo mesto, Slovenia
e-mail: lucija_lapuh@yahoo.com

M. Krevs (ed.), *Hidden Geographies*, Key Challenges in Geography,
https://doi.org/10.1007/978-3-030-74590-5_20

complexity of their processes and supply networks, cost pressures, growing user and customer expectations for quality, speed, customized products, worker's safety and assistance" (Ghimire et al. 2016: 2). The European Union (EU) fosters an economic growth to encourage companies' digital transformation.

This study contributes to the research on the fourth industrial revolution by demonstrating the recent development of the automotive and electronic industry in Slovenia by focusing on the use of high performance computing (HPC) within factories. The research seeks to address the following question: how successful digitalisation is within Slovenian enterprises in the automotive and electronic sector measured by the use of HPC. This paper aims to establish enterprises' needs, expectations and competencies in the field of HPC.

2 Theoretical Background of the Fourth Industrial Revolution and the HPC Usage in Industry

Technologies are changing the way manufacturers work, how they collaborate with partners, suppliers and customers. In the contemporary economy, there are factories with three different kinds of industrial systems. The first, Fordist production regime with routinised services (e.g. the electronic component industries); the second, flexibly specialised craft-based production using flexible machinery to produce small volumes; and the third, flexible production with a combination of information technologies emanating from Japan, presented in Table 1. New information technologies are driving transformations in the flexible production process, which can be thought of as post-Fordist (Coe et al. 2013). Industry becomes more efficient with improved processes and developed innovative products and services.

The fourth industrial revolution has been named differently around the world: Industry 4.0 in Germany (Neugebauer et al. 2016), Factories of the Future (FoF) in the EU (Factories of the future 2013), Smart manufacturing in the United States of America (Liao et al. 2017), etc. The fourth industrial revolution is changing traditional

Table 1 Characteristics of the Japanese flexible production (Adapted from Dichen 2011; Coe et al. 2013: 282)

Technology	Highly flexible (modular) methods of production, relatively easy to switch products
Labour force	Multi-skilled flexible workers, with some responsibilities, operate in teams and switch between tasks
Supplier relationship	Very close supplier relationship in a tired system; just-in-time delivery of stocks requires close supplier network
Production volume and variety	Very high volume, total partially attained through the production of a range of differentiated products
Geographical tendencies	Clustering of production plants, key suppliers, and logistical providers

business models with information and communication technology (Neugebauer et al. 2016). Computer progress has led to automation in manufacturing (Kusiak 2018). Communication via the internet allows a continuous interaction and exchange of information and communication between the machines themselves (Roblek et al. 2016) as well as human–machine interaction in the production line (Neugebauer et al. 2016). Machines are streaming data via wireless sensors and sending these data to smart provider's centres, where they are analysed (Roblek et al. 2016). Every step of the manufacturing process is aimed to be interconnected and able to communicate with and learn from each other (Neugebauer et al. 2016). New information technologies have also transformed the way products and services are made. The use of information technologies in machines and their operation allow more sophisticated control over the production process (Coe et al. 2013). Computers are becoming more powerful and faster, machines and sensors smarter, data storage cheaper and data transmissions faster and more secure (Neugebauer et al. 2016). The implementation of digitalisation and Industry 4.0 leads to resource efficiency and higher production (Jäger et al. 2016). The internet is increasingly taken for a widespread and placeless phenomenon (Coe et al. 2013). Factory-of-things is very much aligned with the internet of things (Shariatzadeha et al. 2016), which represents the network for massive digitalisation. Their technical solutions can be identified in the manufacturing area as cyber-physical production systems (Neugebauer et al. 2016). Information technologies enable virtually real-time feedback of processes and their adjustment in the whole production cycle (Coe et al. 2013). Information technologies of the new generation are also cloud computing, 3D printing, visualisation as well as machine learning (Neugebauer et al. 2016).

In the factory of the future, operational processes have been digitised and connected with the internet. "The digital factory is a model of a planned or real factory used for design, planning and operations", (Shariatzadeha et al. 2016: 512) integrated within the smart factory where digital things are linked with physical things. Virtual manufacturing is an approach of the factory of the future which involves the integration of the processes from engineering to manufacturing, to improve them, introduce new products more quickly in the market in a cost-effective way (Souza et al. 2006).

Essential for companies' digitalisation is the quality of data and information. The characteristics of the so-called Big Data according to Jäger et al. (2016) are as follows: very large amounts of data (volume), different data sources and formats (variety), data analysis in real time (velocity), accuracy and reliability of data (veracity) and data usability (value). With technologies such as Big Data Analytics, maintenance can be predicted, decision-making in real time improved and jobs coordinated (Amplexor 2020).

Human–robot interaction enables humans to instruct robots to perform complex human tasks such as using human gestures (Nieto et al. 2017) (Fig. 1). Lights-out manufacturing or unmanned factories refers to factories that operate autonomously and require no human presence. These robot-run factories often do not even require lighting as machines are functioning in the dark (Future factory 2019).

Fig. 1 Robotisation is an important element of the fourth industrial revolution (Author Lucija Lapuh)

The complementarity between automation and humans requires new methods and tools for the design and operation of workplaces in terms of complexity, management, efficiency, safety and work satisfaction. Factories of the future are labour- or human-centred with a greater emphasis on social aspects including the employees' education and training, skill-based production work, customers and suppliers' innovations (Nieto et al. 2017) as well as the need for improved integration of the factories to their local urban or domestic areas.

FoF are highly flexible, rapidly adaptable to external changes and aim for a high degree of sustainability, which has to address three dimensions: economy, ecology and society (Herrmann et al. 2014). The fourth industrial revolution is eco-effective with environmentally conscious and economically feasible sustainable production, on the path towards future eco-factories. It aims for reducing the environmental footprint of manufacturing (Nieto et al. 2017) by reducing the energy and resources consumption within manufacturing processes, considering each phase of a product's life cycle, from acquiring the materials, production and usage, to disposing off the waste (Eco factory 2020). FoF aim for greener, more customised products of higher quality through the transition to a demand-driven industry. The purpose of automation is the adoption of services to individuals (Roblek et al. 2016), with the central objective of fulfilling individual customer's needs. Computerisation enables efficient and rapid switching from one part of the manufacturing process to another so that firms can develop products to the requirements of an individual customer. Companies are now able to make a wide variety of products without compromising on the cost savings traditionally associated with large-volume production. Developing new products and services is central to maintaining profitability. They have to be able to respond rapidly to changing market demand. Factories have to improve production fast and constantly (Coe et al. 2013).

The fundament for new opportunities is the availability of relevant information everywhere and at any time. "The production of the future will be characterized by shorter product life cycles, an increasing number of variants and short-term production program changes triggered by the customer" (Jäger et al. 2016: 116).

How the fourth industrial revolution could have been implemented does not have a generally accepted solution for all companies. Most companies, especially small and medium-sized enterprises (SMEs), see the greatest risk from the fourth industrial revolution in rising investment costs (Jäger et al. 2016).

The role that geography plays is being challenged by new geo-opportunities of cloud-based, open data environments (EUROGEO 2019) and virtual space. New technologies such as the internet of things and service-oriented solutions, cloud computing, as well as HPC (Fig. 2) and cloud manufacturing are emerging. The development and applications of hidden HPC technology provide the possibility for solving more complex problems and carrying out large-scale collaborative manufacturing (Zhang et al. 2014), also in automotive product development, which is beside electronic industry the main field of research in this paper.

Fig. 2 High performance computers provide the possibility for solving complex problems in the industry (Author Jani Pogačar)

HPC influences almost every aspect of our daily life when we are trying to solve problems in science, engineering or business dealing with large amounts of data, e.g. weather prediction, climate modelling, aerospace, communications, transportation, finance, medicine, and the manufacture of both new and traditional consumer products. HPC is useful in all computation-intensive research areas like engineering, physics, earth sciences, national security (Ezell and Atkinson 2016). HPC involves the use of supercomputers and massively parallel processing techniques to solve complex computational problems through data analysis, simulation and computer modelling (Ezell and Atkinson 2016). HPC allows researchers and engineers to understand and study complex phenomena, to solve compute-intensive problems, reduce time-to-results, increase the speed of research and making more parallel tasks. HPC applications require large amounts of memory, high network performance, very high computing capabilities and faster storage.

HPC machine is more complex than a simple desktop computer. HPC aggregates computing power that delivers much higher performance than one could get out of a typical desktop computer (Insidehpc 2018). In comparison with a desktop computer which generally contains a single central processing unit, an HPC system represents a network of central processing units. The computers' speed is measured by their ability to calculate floating-point operations per second named flops (Ezell and Atkinson 2016). The HPC individual nodes can work together to solve a problem larger than any one computer can easily solve. A common cluster size in businesses is between 16 and 64 nodes or from 64 to 256 cores (Insidehpc 2018). Computer clusters can be owned by public entities or by private companies (Bohar and Povh 2019). Software vendors own a cluster infrastructure of HPC and provide their software as services via the internet (Suklan 2019).

Industry can benefit from HPC mainly in the fields of discovery and innovation, to enhance competitiveness, increase the effectiveness of innovations and the speed of the experimentation, as well as reduce product development time and cost. By using HPC, companies can design new products, improve the existing ones and decrease the time to market products and services (High-Performance Computing 2020). Companies can use HPC for advanced simulation, modelling, data analytics, for predicting performance and failure, improving quality, optimising processes and design (Suklan 2019). HPC technology can be used to solve problems particularly in the development and testing of new products or production methods. Employees must have competencies for working with HPC infrastructure (Bohar and Povh 2019). Only a few industrial companies have used simulation tools in their design processes at the turn of this millennium, however, nowadays they are an essential tool for design and optimisation within factories of the future. "HPC, although being expensive, is a cost-effective tool for accelerating the research and development process" (Suklan 2019: 49).

HPC has a major impact also on the automotive industry. "Testing prototypes of new cars under different physical conditions in virtual environments using HPC makes the car-development process both quicker and cheaper" (Bohar and Povh 2019: 231). "HPC has transformed how vehicles and their components are designed, modelled and simulated, safety tested, and manufactured, playing a key role in reducing vehicle design costs, introducing innovative new features, and improving the fuel efficiency and safety of vehicles" (Ezell and Atkinson 2016: 16).

"HPC also drives business: The European Commission sees HPC as a way in which the European electronics industry can catch up with the world's leading computer manufacturers by developing and producing European microprocessors and new structures for processors, memory and communication units" (Bohar and Povh 2019: 231).

The Slovenian industry is mostly component-oriented. In the Slovenian economy, factories that use optimisation and automation of production are essential for economic development (Perko 2016). The Slovenian economy's digital transformation has been slow, consequently the digital maturity of Slovenian companies is weak (Bohar and Povh 2019). In Slovenia, no factory can be called the factory of the future, but there are some early stages (Perko 2016). The goal of the Strategic Research and Innovation Partnerships Factories of the Future (SRIP 2020) is to create technologies, products and processes for the FoF and intensify the net of Slovene toolmakers and prepare them for the challenges of the fourth industrial revolution.

Many Slovenian companies have large amounts of data but are often unaware of their business value (Bohar and Povh 2019). Only in the automotive sector the robotisation rate is relatively high and will increase automation (Perko 2016). "Virtual prototyping and testing are used rarely and only for small-scale products. Simulations that need HPC are out of reach for the Slovenian automotive and electronics industry, since they do not have access to the necessary hardware and software and are often without the necessary knowledge from these domains. Some organizations, do not have their own HPC infrastructure, but they are nevertheless active in the HPC field and have qualified people for this job" (Bohar and Povh 2019: 237, 240).

The Slovenian government has recognised the importance of HPC and listed it as a key enabling technology in the Slovenian Smart Specialisation Strategy to optimise production. Slovenia signed the European declaration on HPC in 2017 (Slovenia … 2017). Slovenian research groups dealing with HPC established a national consortium for supercomputing called Slovenian National Supercomputing Network SLING in 2018, which has become the Slovenian member of the Partnership for Advanced Computing in Europe PRACE (Bohar and Povh 2019). In the same year, Slovenia became a founding member of the European High-Performance Computing Joint Undertaking—EuroHPC (2020).

The company Arctur in Nova Gorica is the biggest private HPC provider in Slovenia. The private company Kolektor Turboinsitute also uses HPC for its business needs (Bohar and Povh 2019). Some successful SMEs using HPC infrastructure in Slovenia are, according to Fortissimo (2019)—Pipistrel (HPC Cloud-based simulation of light-aircraft aerodynamics), Magneti (HPC high-resolution modelling of magnets), EMO (Cloud-based HPC optimisation of manufacturing processes), Elaphe (HPC Cloud-based simulation of coupled electromagnetic and structural-acoustics in in-wheel electronic motors).

3 Methods

This study has expanded the InnoHPC (2019) database on the use of HPC. Enterprises from the automotive and electronic sector were chosen by the convenience which was based on the literature review, the author expert analysis on the potential usage of HPC in the industry and the online search of enterprises, as HPC technology is far from being widespread but still quite rare and hidden. The snowball sampling was used in parallel: research participants were asked to suggest partners. Every enterprise had a chance of being selected for the sample, as the invitation with a link to the questionnaire was published online and was also sent to the Strategic Research and Innovation Partnership of the Factories of the Future (SRIP ToP). The research is based on an online survey, but some questionnaires were conducted by phone due to the preference of enterprises' representatives. Despite searching for enterprises using HPC, the answers of those that are not using this technology are of great value as well, because the survey also tries to find out about their needs, competencies and interests in potential HPC use.

The sample consists of 29 valid surveys within Slovenian enterprises, 11 of which were conducted by the InnoHPC project in 2017 and 18 by the author's postdoctoral project in 2018. Slovenia was chosen as a key study due to its position between economically developed Western countries, where most HPC infrastructure is located and the Eastern countries—thus having economic partners on both sides. Simple statistical analyses were used to present the results descriptively and graphically.

4 The Use of High Performance Computing Within Slovenian Enterprises in Automotive and Electronic Industry

In order to find out how successful the digitalisation of Slovenian enterprises in the automotive and electronic sector is, the use of HPC was analysed. Firstly, enterprises' characteristics regarding the use of HPC were presented; secondly, the companies' needs in the field of HPC; thirdly, their expectations regarding the new technology; and finally, the employees' competencies in the field of HPC.

Starting with the general characteristics of surveyed enterprises, the results showed that 64% of enterprises were active in the automotive industry, while 32% were from electronic and the rest from other sectors. 48% of enterprises were large-sized, 28% medium, 16% small and 8% micro. Almost half of the companies (48%) had less than 10 people employed in the information technology (IT) department, 29% between 11 and 30 people and the rest a higher number.

81% of enterprises taking part in the survey had a research and development department. More than two-thirds of enterprises (71%) thought that cooperation with science or industry could foster HPC usage and organisation development.

Firstly, 33% of the enterprises of the study population used HPC to meet their business requirements: 63% of them used their own bought HPC infrastructure, 50% commercial infrastructure and 38% the one developed by a company (multiple answers were possible). Half of the enterprises used HPC for less than 5 years. Those who responded that they use HPC are all from western Slovenia: four from the Central Slovenia Statistical Region, two from the Gorizia Statistical Region and one from the Upper Carniola Statistical Region.

Respondents were asked to suggest reasons for using HPC solutions in their daily work. Companies used HPC mostly not only to address problems more efficiently (e.g. lower cost; faster time to solution) and to develop new products/services, but also to solve problems that could not be addressed by other means, for improving business innovation processes and other reasons. All surveyed companies used HPC in the Research and Development (R&D), 44% also in engineering. Direct benefits that HPC has for the company are accelerated innovation (89%), increased competitiveness (67%), increased productivity (44%), faster time to work (44%) and reduced cost (22%) (multiple choices were possible).

Software application and infrastructure used in companies are predominantly bought (64%), followed by commercial infrastructure (50%) and the one developed in-house (38%). Companies' representatives were also asked about specific facilities to access HPC. They mainly use Linux OS (56%), followed by Windows XP (33%) and Windows Vista (11%). HPC resources were mainly used over a GRID network to access HPC resources from other organisations (33%), over the GRID network to access HPC resources from other departments of the organisation (22%) or within the organisation (22%) (multiple answers were possible).

On the other hand, 67% of the companies taking part in the research did not use HPC solutions because they were satisfied with the actual level of technologies (71%), had no financial resources to integrate HPC in the current work (29%), had lack of knowledge about HPC (29%), or because of other reasons (12%). Only 6% of the companies envisaged integrating HPC to meet business requirements in the next 12 months, while the majority of the companies is either unsure (22%) or against (67%) integrating HPC to meet business requirements of the company, and 6% of respondents were undecided/un-opinionated regarding the issue. The majority of companies' customers did not require HPC usage from them (94%), furthermore, companies did not require their suppliers to use HPC (82%).

Secondly, in the following paragraphs, the survey results of the companies' needs in the field of HPC are presented. Figure 3 shows how companies rated the HPC development in Slovenia.

According to the enterprises in the sample, the need of the companies in integrating/using HPC was the highest for (shown in Fig. 4)—finding well-trained human resources (3.7); awareness and knowledge about possible applications and the potential of HPC technologies (3.4); getting help with modelling for developing a product or service that requires HPC (3.4); finding partners from academia and research centres to collaborate (3.3); and access to infrastructure (3.3).

Figure 5 shows that the companies used HPC to develop new products or to redesign products (86%), to conduct large-scale research projects (57%), to carry

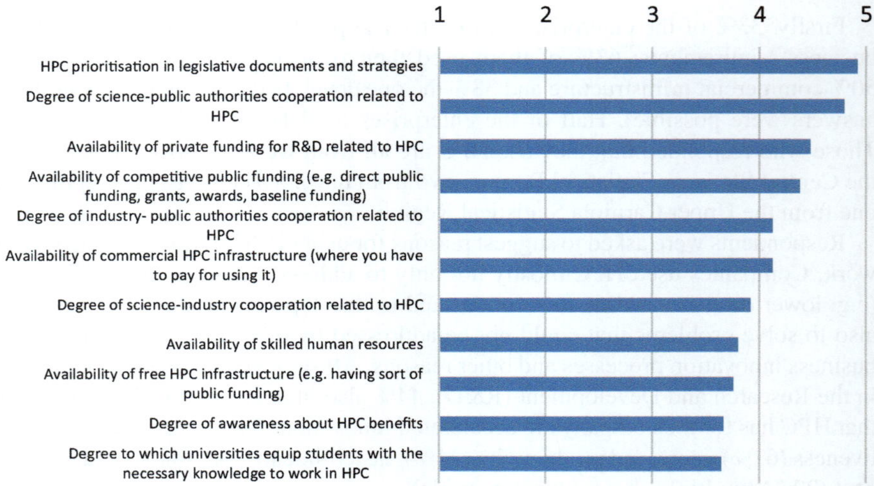

Fig. 3 HPC development in Slovenia (1 means poor and 5 means excellent) (n = 25)

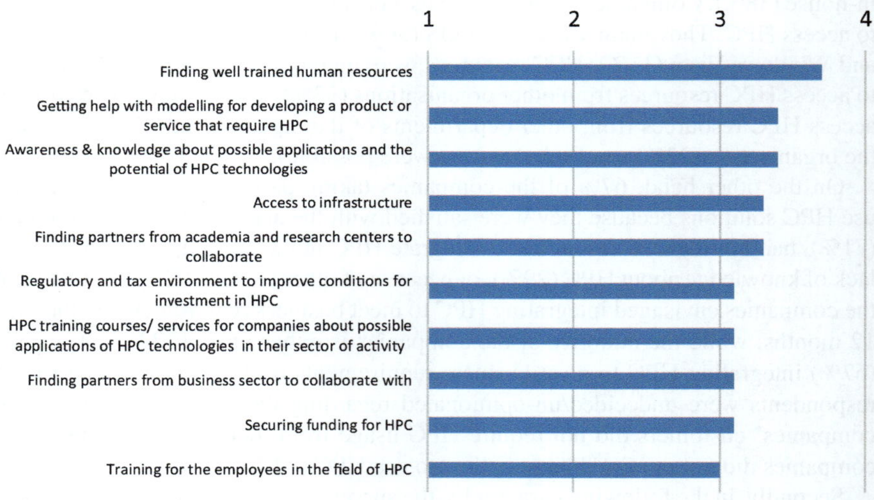

Fig. 4 The level of needs in integrating/using HPC in companies (1 means No need, 4 High need) (n = 15)

out simulations and modelling (50%), to improve their activity (36%), to optimise production and delivery processes (36%), to analyse or develop large datasets (36%) and to create visualisations (21%).

Thirdly, their expectations regarding HPC are presented in Fig. 6. The majority of enterprises saw the potential reasons to adopt HPC or to expand the use of HPC in cost reduction (M = 4.5), improvement in the quality of products/services (M =

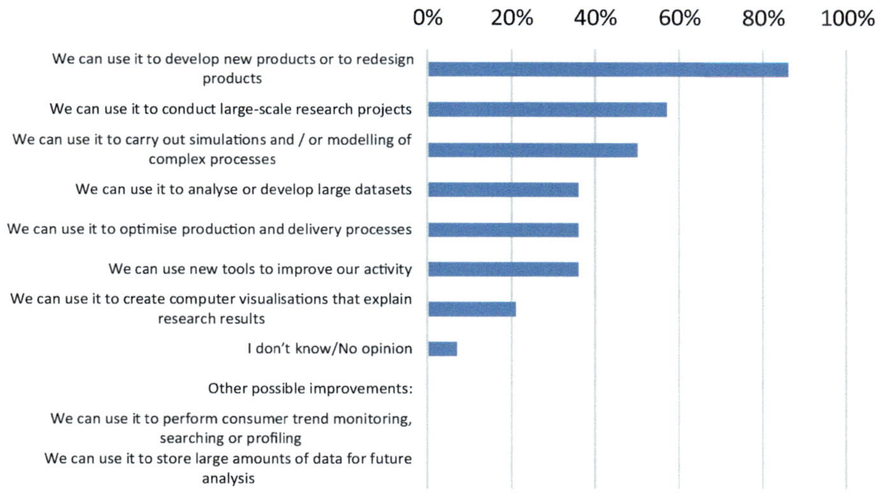

Fig. 5 HPC's use to meet business needs (n = 14) (Multiple answers were possible)

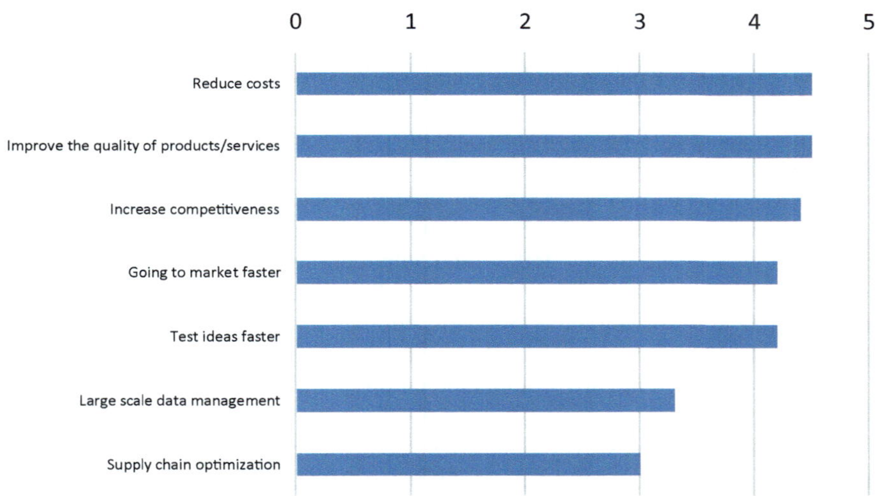

Fig. 6 Potential reasons to adopt HPC or to expand the use of HPC considering the companies' importance (1 unimportant, 5 very important) (n = 14)

4.5), followed by the increased competitiveness of the company (M = 4.4), testing ideas faster (4.2), going to market faster (4.2) but less so in the large-scale data management (M = 3.3) or the supply chain optimisation (M = 3).

The top three business problems related to the use of HPC were—high costs of using HPC (57%), lack of funds to support development based on HPC (50%) and finding well-prepared human resources (43%); they are shown in Fig. 7.

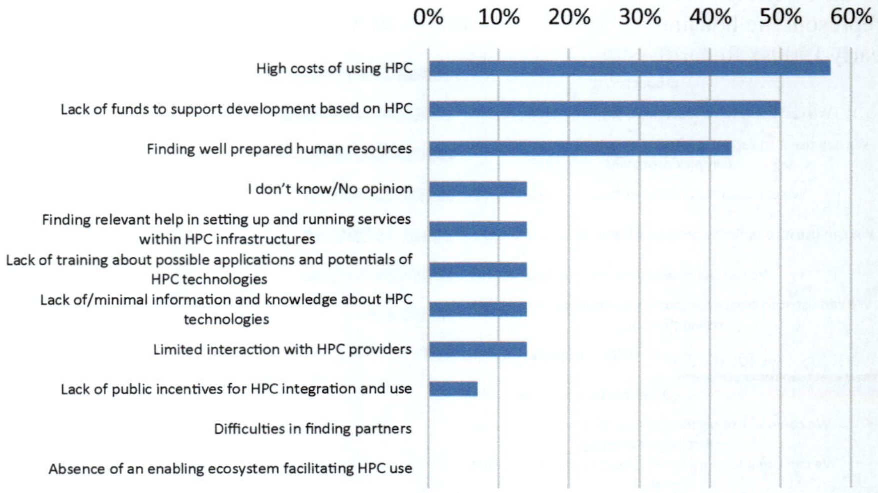

Fig. 7 Top three business problems related to the use of HPC (n = 14) (Multiple answers are possible)

29% of the enterprises reported that if they had access to free HPC infrastructure (e.g. having a sort of public funding) they would work with larger/more complex data or models, 13% of the companies would not, 54% would maybe, work with larger/more complex data or models, and 4% of the companies were undecided. 54% of the enterprises would consider HPC training if it was available in the sector of operation of the company, 29% of the companies would maybe, consider it.

43% of the enterprises would acquire HPC resources to solve computational problems, 29% were aware of how HPC could help them in solving problems, but they do not intend to use it because it is expensive, 14% of the companies knew how HPC could help them in solving these problems, but they do not intend to use it because they do not have the necessary knowledge yet, 7% of the companies did not intend to use HPC resources to solve computational problems as they did not know them yet, 14% did not know yet or had no opinion.

In the final part of the survey, companies' competencies in the field of HPC were studied within Slovenian companies. Within companies that worked with HPC report, their personnel working with HPC were equipped mainly with—HPC Code Development: Scripting languages (e.g. Python, Perl, etc.) (36%), HPC Code Development: Programming Languages (e.g. C, C++, Fortran, etc.) (36%), HPC Code Development: Linux Shell Scripting (e.g. BASH, CSH, ZSH, etc.) (14%), HPC System Usage: Basic Linux skills (i.e. Linux shell (e.g. BASH), SSH, etc.) (14%), HPC System Administration: Parallel File Systems (e.g. Ceph, Lustre, Hadoop FS) (%), HPC Code Development: Linux Shell Scripting (e.g. BASH, CSH, ZSH, etc.) (7.4%). 50% of the enterprises fall into the category of "no opinion" regarding the issue of demand for training in one of the following skill categories and 25% did not need it. Other companies reported that their personnel had a demand for training, with the higher percent HPC System Administration: Resource Managers/Job Schedulers (e.g. SLURM, TORQUE, etc.) (15%). 78% of the enterprises believed that HPC infrastructure and/or equipment could not be used by other companies, while 11% shared their equipment with others.

71% of the enterprises believed that cooperation with science/industry could foster the HPC usage and their organisation's development. The majority of the companies in this sample had not been involved in international projects related to HPC (71%). One company was involved in an EU project (Fortissimo) and one in a non-EU project. The majority of enterprises appreciated the importance of cooperation in HPC-related aspects among national enterprises (M = 4.4) and foreign enterprises (M = 4.3), followed by foreign research centres (M = 4.2) and national research centres (M = 4.1).

46% of the enterprises reported that they were aware that also other companies in the field use HPC, 13% believed that this is not the case, while 42% did not have relevant information to believe so. If enterprises had access to free HPC infrastructure, 29% of the enterprises would have worked and 54% of the enterprises would, maybe, have worked with more complex data. These results indicate that they are in general interested in HPC training, if they were available and that they need intellectual and funding support.

The overall response to the questions regarding their customers and partners was poor—probably because the companies were not willing to share their business secrets. The next chapter, therefore, moves on to discuss the theoretical part as well as the survey results.

5 Discussion

The manufacturing industry faces a growing pressure to constantly innovate and deliver new products at a faster pace, increase production and reduce costs, without compromising product quality, as the digital transformation is raising customer expectations (Amplexor 2020). Based on the review of literature on the fourth industrial revolution, the so-called FoF can become a factory in a good shape, ready to invest energy, capital and knowledge, raise quality, improve competitiveness and make better use of the potential of an educated workforce (Perko 2016). The role of HPC in the transformation of the industry to attain the FoF is nevertheless indirectly connected with the digitalisation of industry. Companies can benefit from the big data, which has been produced within them, economically (by reducing cost) and by decreasing time to market products and services. They can use this data by analysing them with HPC and consequently improving their processes from the development, through production to logistics.

This research, which was trying to focus in depth on Slovenian companies in the automotive and electronic sector, was a follow-up of the research done within Danube countries (InnoHPC 2019). The companies that could probably be using HPC technology were not that easy to find, as they are still quite rare in Slovenia, especially between SMEs. That is the reason why the use of HPC was examined as hidden (also virtual technology) within still quite hidden factories which can be proud of naming themselves FoF and being an active player on the global market of the fourth industrial revolution's development. Agreeing with Bohar and Povh (2019), HPC technology penetrates the Slovenian industry very slowly. The majority of Slovenian companies are still far from becoming factories of the future with the need for HPC analyses.

In order to compare this study's findings of Slovenian companies with the findings of previous work done in the entire Danube area, East–West division of the development regarding the usage of HPC infrastructure can be observed—Slovenia lags behind Austria and Germany but is better positioned than for example Bulgaria, Romania, Ukraine and Moldova (Besednjak Valič 2019).

Slovenian companies in the automotive and electronic sector, which are mainly at the level of tier 2 suppliers and are, according to Bohar and Povh (2019), very important in Slovenia, could have proven benefits from using HPC by developing, testing and producing small components. They have shown an extremely low rate of interest in cooperating and a lack of interest in the HPC topic—which is also the consequence of a small or even negligible interest in the use of HPC. Low response rate, the lack of knowledge about HPC and the negligibly small quoted recommendations of their partners were noticed among surveyed enterprises. A possible explanation regarding the results on HPC usage may be the lack of adequate knowledge, skills and money for investments. Due to the high cost of using HPC, firms can use it in connection with research organisations or in a cloud, not necessarily having their own technology and experts, which can be hired externally. Agreeing with Bohar and Povh (2019), it is important for the enterprises to be aware of HPC technology and the business value of their data as well as to follow the trends of Industry 4.0. The goal of the enterprises should be the ability to recognise the opportunities brought by new technologies and to take advantage of them. Enterprises that refuse to address these new developments cannot remain competitive in the long term (Jäger et al. 2016). Slovenia has to follow the example of many developed companies which are usually aware of the existence of HPC technology because they follow trends of the fourth industrial revolution and monitor the competitors on the market (Bohar and Povh 2019). New technologies have facilitated the development of new uneven geographies of economic activity (Coe et al. 2013).

This study has shown that the findings cannot be extrapolated to all Slovenian companies, as only the most developed ones, those who were foreseen using HPC were contacted and asked to take part in the survey. However, with relatively small sample size, caution must be applied, as the findings might not be transferable to the average current situation in Slovenia regarding the development of the factories of the future and the usage of HPC.

6 Conclusion

The fourth industrial revolution is based on the digitalisation of the economy with the so-called FoF. The initial objective of this study was to identify how successful digitalisation is within Slovenian enterprises in the automotive and electronic sector measured by the use of HPC. Returning to the question posed at the beginning of this study, it is now possible to state that the Slovenian economy is in the initial phase of a new technological development of the fourth industrial revolution characteristic of FoF with some successful implementations of digitalisation. 67% of the enterprises examined in the survey did not use HPC because they were satisfied with the actual level of technologies that they used; they considered HPC expensive or had a lack of knowledge about it. Besides the high costs of using HPC, one of the business-related problems was also finding well-trained human resources.

The key strength of this study is an overview of the initial phase of integrating HPC into the Slovenian economy. The research illustrates the needs and opportunities regarding HPC inside Slovenia's electronic and automotive sector with the potential of implementation in politics, as HPC is already mentioned in the Slovenian Smart Specialisation Strategy (2017) which encourages the development of the fourth industrial revolution in the country. The state's financial incentives can improve the situation within the Slovenian economy. The findings of this study suggest that companies should be better informed about the benefits they could gain by using HPC.

To sum up, the factories that are actively taking part in the fourth industrial revolution by using HPC are quite rare in Slovenia and consequently hidden not only behind the factory walls and invisible due to digital technology but also generally hidden within the Slovenian economy.

Acknowledgements This article is based on the research funded by the Republic of Slovenia and the European Union under the European Regional Development Fund (Grant reference C3330-17-529007).

References

Abdi, M.R., A. Labib, F. Delavari Edalat, and A. Abdi. 2018. *Integrated reconfigurable manufacturing systems and smart value chain*. Heidelberg: Springer.

Amplexor. 2020. https://www.amplexor.com/en/industries/automotive-and-manufacturing.html. Accessed 31 Jan 2020.

Besednjak Valič, T. 2019. Innovation, digitalisation, and the HPC in the Danube region. In *Go with the flow: High performance computing and innovations in the Danube region*, ed. Borut Roncevic, Raluca Coscodaru, and Urska Fric, 22–46. London, Budapest, Ljubljana: Vega Press Ltd.

Bohar, M., and J. Povh. 2019. National case study of Slovenia. In *Go with the flow: High performance computing and innovations in the Danube region*, ed. Borut Roncevic, Raluca Coscodaru, and Urska Fric, 228–244. London, Budapest, Ljubljana: Vega Press Ltd.

Coe, N.M., P.F. Kelly, and H.W.C. Yeung. 2013. *Economic geography, a contemporary introduction.* Hoboken: Wiley.
Dichen, P. 2011. *Global Shift: Mapping the changing contours of the world economy*. London: Sage.
Eco factory. 2020. http://www.madedifferent.be/en/projects/project/transformation-6-eco-factory?c=660. Accessed 13 Feb 2020.
EuroHPC. https://ec.europa.eu/digital-single-market/en/eurohpc-joint-undertaking. Accessed 3 March 2020.
EUROGEO. (2019). http://www.eurogeography.eu.gridhosted.co.uk/about-2/. Accessed 8 Oct 2019.
Ezell, S.J., and R.D. Atkinson. 2016. The vital importance of high-performance computing to US competitiveness. Information Technology and Innovation Foundation. http://www2.itif.org/2016-high-performance-computing.pdf?_ga=2.70059347.218455825.1581320442-155308587.1581320442. Accessed 10 Feb 2020.
Factories of the future, Multi-annual roadmap for the contractual PPP under Horizon 2020. 2013. https://doi.org/10.2777/29815. https://www.effra.eu/sites/default/files/factories_of_the_future_2020_roadmap.pdf. Accessed 11 Feb 2020.
Fortissimo Success Stories, Fortissimo's Small & Medium, Manufacturing Enterprises. 2019. Edinburgh and Nova gorica: EPCC and Arctur.
Future factory. 2019. https://www.cbinsights.com/research/future-factory-manufacturing-tech-trends/. Accessed 31 Jan 2020.
Ghimire, S., F. Luis-Ferreira, T. Nodehi, and R. Jardim-Goncalves. 2016. IoT based situational awareness framework for real-time project management. *International Journal of Computer Integrated Manufacturing*: 1–10. https://doi.org/10.1080/0951192X.2015.1130242.
Herrmann, C., C. Schmidt, D. Kurle, S. Blume, and S. Thiede. 2014. Sustainability in manufacturing and factories of the future. *International Journal of Precision Engineering and Manufacturing-Green Technology* 1 (4): 283–292.
High-Performance Computing. 2020. https://ec.europa.eu/programmes/horizon2020/en/h2020-section/high-performance-computing-hpc#Article. Accessed 31 Jan 2020.
InnoHPC project. 2019. http://www.interreg-danube.eu/approved-projects/innohpc. Accessed 26 July 2019.
Insidehpc. 2018. https://insidehpc.com/hpc-basic-training/what-is-hpc/. Accessed 20 Mar 2018.
Jäger, J., O. Schöllhammera, M. Lickefetta, and T. Bauernhansla. 2016. Advanced complexity management strategic recommendations of handling the "Industrie 4.0" complexity for small and medium enterprises. *Procedia CIRP* 57: 116–121. https://doi.org/10.1016/j.procir.2016.11.021.
Kusiak, A. 2018. Smart manufacturing. *International Journal of Production Research* 56 (1–2): 508–517. https://doi.org/10.1080/00207543.2017.1351644.
Liao, Y., F. Deschamps, Eduardo de Freitas Rocha Loures, and L.F. Pierin Ramos. 2017. Past, present and future of Industry 4.0—A systematic literature review and research agenda proposal. *International Journal of Production Research* 55 (12): 3609–3629.
Neugebauer, R., S. Hippmann, M. Leis, and M. Landherr. 2016. Industrie 4.0—From the perspective of applied research. *Procedia CIRP* 57: 2–7. https://doi.org/10.1016/j.procir.2016.11.002.
Nieto, F.A.M., V. Goepp, and E. Caillaud. 2017. From factory of the future to future of the factory: Integration approaches. *IFAC PapersOnLine* 50–1: 11695–11700. https://doi.org/10.1016/j.ifacol.2017.08.1689.
Perko, B. 2016. Tovarne na ključ za globalni trg. *Glas gospodarstva*: 73–74. https://www.gzs.si/Portals/SN-informacije-Pomoc/Vsebine/GG/2016-september/Avtomatizacija%20in%20robotizacija.pdf. Accessed 17 Feb 2020.
Roblek, V., M. Meško, and A. Krapež. 2016. A complex view of Industry 4.0. *SAGE Open* 6 (2): 1–11.
Shariatzadeha, N., T. Lundholma, L. Lindberga, and G. Sivard. 2016. Integration of digital factory with smart factory based on Internet of Things. *Procedia CIRP* 50: 512–517. https://doi.org/10.1016/j.procir.2016.05.050.

Slovenia signs the European declaration on high-performance computing. 2017. https://ec.europa.eu/digital-single-market/en/news/slovenia-signs-european-declaration-high-performance-computing. Accessed 31 Jan 2020.

Slovenian Smart Specialization Strategy. 2017. Slovenska strategija pametne specializacije S4. https://www.eu-skladi.si/sl/dokumenti/kljucni-dokumenti/s4_strategija_v_dec17.pdf. Accessed 26 July 2019.

Souza, M.C.F., M. Sacco, and A.J.V. Porto. 2006. Virtual manufacturing as a way for the factory of the future. *Journal of Intelligent Manufacturing* 17 (6): 725–735. https://doi.org/10.1007/s10845-006-0041-1.

SRIP Factories of the Future. 2020. https://www.tecos.si/index.php/en/clanstvo/srip-tovarne-prihodnosti. Accessed 3 Mar 2020.

Suklan, J. 2019. Gap analysis: HPC supply and demand. In *Go with the flow: High performance computing and innovations in the Danube region*, ed. Borut Roncevic, Raluca Coscodaru, and Urska Fric, 47–59. London, Budapest, Ljubljana: Vega Press Ltd.

Zhang, L., Y.-L. Luo, F. Tao, B.-H. Li, L. Ren, and X. Zhang. 2014. Cloud manufacturing: A new manufacturing paradigm. *Enterprise Information Systems* 8 (2): 167–187.

Borders as a Hidden Obstacle to the Organization of Public Transport

Matej Gabrovec, Primož Pipan, and Peter Zajc

Abstract The borders between the member states of the European Union do not represent an obstacle when travelling with personal vehicles, especially within the Schengen Area. Sadly, this does not apply to public passenger transport. The chapter presents cross-border public transport in the case of Slovenia, as it represents an ideal laboratory for the studying of changes in cross-border traffic flows. In the last three decades, due to political changes, the status and permeability of borders have changed several times. We analysed (1) the development of cross-border public transport in Slovenia before it gained independence in 1991, (2) the period of its accession to the European Union in 2004, and (3) the period when it was already a member of the European Union 2004–2019. The case of cross-border mobility between Austria and Slovenia in Carinthia was examined in more detail. Our results show that for a long-term operation of cross-border public transport an appropriate legal basis is crucial. This only opens the possibility of making the quality of cross-border public transport comparable to the quality of public transport within the country.

Keywords Border regions · Daily mobility · Cross-border transport · Public transit · Slovenia · Geography · Carinthia · Austria

1 Internal EU Borders

The borders between the member states of the European Union do not represent an obstacle when travelling with personal vehicles, especially within the Schengen Area. However, borders discourage spatial interaction. A lower demand for cross-border travel is linked to the linguistic and cultural differences, as well as legislative

M. Gabrovec (✉) · P. Pipan
ZRC SAZU, Novi trg 2, 1000 Ljubljana, Slovenia
e-mail: matej.gabrovec@zrc-sazu.si

P. Pipan
e-mail: primoz.pipan@zrc-sazu.si

P. Zajc
Regional Development Agency for Koroška, Meža 10, 2370 Dravograd, Slovenia
e-mail: peter.zajc@rra-koroska.si

M. Krevs (ed.), *Hidden Geographies*, Key Challenges in Geography,
https://doi.org/10.1007/978-3-030-74590-5_21

differences between the neighbouring countries, which can hamper cross-border economic activities. Border impacts tend to be bigger for rail and public transport in comparison to road transport (Rietveld 2012; Christodoulou and Christidis 2020).

In the European context the term borderline has been transformed from representing an obstacle to cooperation to acting as generators of development (Bufon 1995). The role of borders today is no longer that of division but increasingly that of integration of two countries, and borders are one of the factors of regional development (Klemenčič 1993). An integrated development of a border region has to be based on a sustainable and intensive cross-border cooperation and integration of the population living in that border region. Weak cross-border connections cause economic regression in border regions and emigration from them, as is the case of the Slovenian-Hungarian border and several Croatian border municipalities (Zupanc 2018). Cross-border mobility for work enables a higher income for employees, but on the other hand it can enable savings for employers due to cheaper workforce. Continuous cross-border integration and liberalization of cross-border currents consequently lead to the formation of integrated cross-border economic systems which require and enable joint management and planning. The planning of cross-border public transport is one of them. It facilitates sustainable mobility of the workforce and reduces external cost of traffic caused by pollution and congestion (Ratti 1991; Barth 2014).

The article presents cross-border public transport in the case of Slovenia, as it represents an ideal laboratory for the studying of changes in cross-border traffic flows. In the last three decades, due to political changes, the status and permeability of borders have changed several times. These changes have had a significant impact on cross-border traffic, in particular on public transport, which is more sensitive to changes in legislation and various international agreements compared to personal transport. The aim of the paper is to find out how political changes have influenced the supply of cross-border public transport. The emphasis is on the analysis after Slovenia's accession to the European Union. The question is why, despite increased cross-border daily mobility, public transport has not improved significantly. Why is it difficult for public transport to cross invisible internal borders between the individual member states of the European Union?

2 Development of Cross-Border Public Transport in Slovenia Between 1991 and 2019

We analysed the public transport links between Slovenian border regions of the neighbouring countries, i.e. Croatia, Italy, Austria and Hungary. We were not interested in long-distance international connections, such as direct lines between Ljubljana and Zagreb without intermediate stops. We were only interested in those lines that have a stop also (or only) in a border region, in the case of the connection with Zagreb in the Posavska statistical region in Slovenia. We examined connections across all road and rail border crossings and calculated the number of weekly connections. For

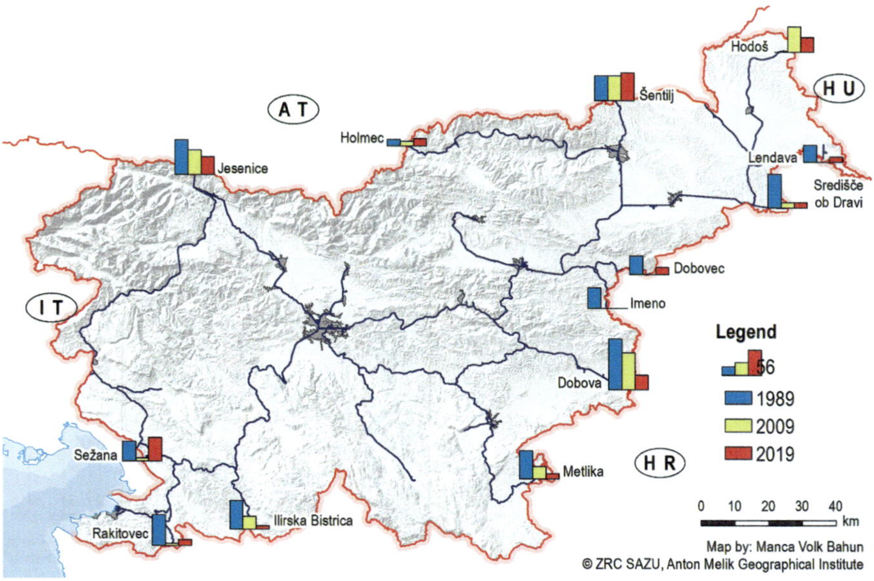

Fig. 1 Number of weekly cross-border railway connections (one way) in 1989, 2009, and 2019

1989, the source of the data was the Yugoslav Railways schedule, the bus schedule of the Socialist Republic of Slovenia, issued by the Transport and Communications Section of the Chamber of Commerce and Industry of Slovenia, and the schedules of individual bus carriers. For the years 2009 and 2019 we used the schedules of the Slovenian Railways and the register of international bus routes with the Ministry of Infrastructure of the Republic of Slovenia, and partly the published schedules of individual carriers (Fig. 1).

2.1 The Situation Before Slovenia Gained Independence in 1991

In 1991, before the Slovenian independence, cross-border daily mobility for work was negligible; there were only 404 residents travelling daily for work abroad. The key reason for public transport into the border regions of the neighbouring countries were occasional shopping journeys. The schedules of cross-border bus routes, which in many cases were operated only once or twice a week, were adapted to such journeys (see Fig. 2). In the case of Italy, a large percentage of passengers travelled to do part-time jobs in Trieste or Gorizia. The majority of them were women who worked daily or several times a week as housekeepers and considered it a "regular"

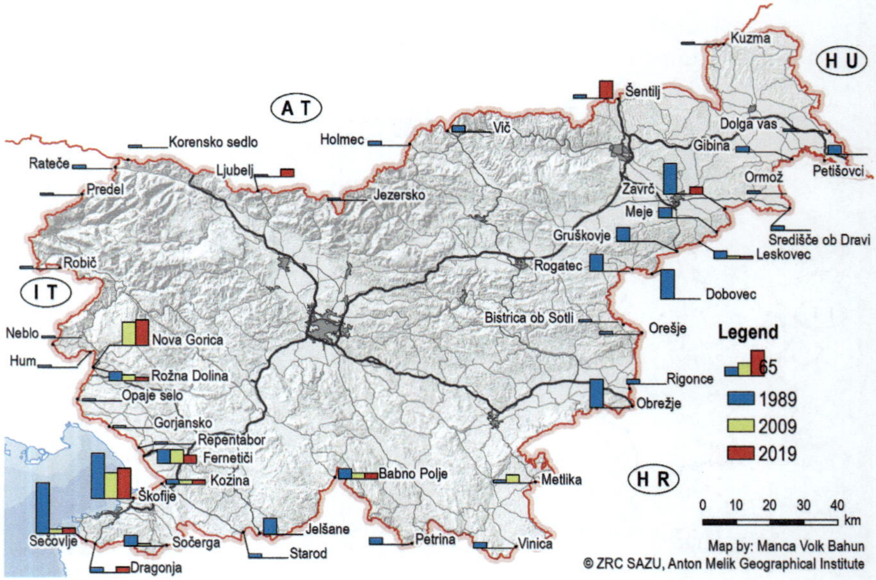

Fig. 2 Number of weekly cross-border bus connections (one way) in 1989, 2009, and 2019

job (Sedmak 2005). Adapted to these passengers, both cities were well connected to the cross-border hinterland by public transport.

Considerably more—about 2,000 Slovenian residents—commuted to work in Croatia. This number can also be rated as low. According to the census, there were 736 thousand employees in Slovenia this year, of which 441 thousand were daily commuters who worked outside their place of residence. An interesting example is the case of the Slovenian border municipality of Brežice, which lies only 40 km from the Croatian capital Zagreb. Despite the proximity to Croatia, of the 8,142 employed residents, only 354 commuted to work in Croatia. Significantly more workers travelled to work in the opposite direction, to Slovenia. According to the 1991 census data provided by the Croatian Bureau of Statistics, 8,700 Croatian residents commuted to Slovenia daily for work, mainly to the industrial centres of the Slovenian border municipalities. Public transport has also been adapted to these daily commuters. Since Slovenia and Croatia were then part of the same state, the Socialist Federal Republic of Yugoslavia, the organization of public transport across the then internal border was not an issue. Within the Yugoslav Railways, railway companies operated in individual republics, but they were harmonized with each other as the common schedules for the whole country were issued by the Yugoslav Railways. The Slovenian railway company also operated the railway network in the Croatian Istria. Consequently, the border between the Socialist Republic of Slovenia and the Socialist Republic of Croatia in the area of Istria did not present an obstacle to the provision of rail services. Bus services were provided by carriers from both republics. The bus companies at that time were socially owned. Both international as well as inter-republic and long-distance lines within Slovenia were implemented without any direct state subsidies on a commercial basis within the then self-governing socialist system. The schedule was coordinated within the Chamber of Commerce and Industry of Slovenia, which also provided for a single tariff scale. Slovenian carriers were also allowed to carry passengers on inter-republic routes within Croatia, and vice versa, Croatian carriers in Slovenia.

2.2 The Period Between Slovenia's Independence and Accession to the European Union (1991–2004)

This period was characterized by an accelerated motorization of Slovenia, related to the accelerated construction of the motorway network while neglecting the railway network and public transport. In the meantime, the number of passenger cars in the country increased by 50%, from 300 to 467 passenger cars per 1,000 inhabitants. Consequently, the use of public transport has decreased, the share of public transport users for commuting decreased from 47% in 1991 to 10% in 2002 (Bole 2004; Gabrovec and Bole 2009; Bole and Gabrovec 2012). In the 1990s, the state did not intensively interfere with the organization of public transport which ran on a commercial basis. The schedule arrangement designed during the time of Yugoslavia was maintained, with carriers gradually discontinuing the less profitable lines. The supply was getting worse. The decline in cross-border public transport during this period is not only linked to legislative obstacles in its organization, but above all to a general decrease in the supply of and demand for public transport in general.

After the breakup of Yugoslavia and the independence of Slovenia and Croatia in 1991, changes in both cross-border daily mobility and the organization of public transport were gradual. The previous border between the federal republics Slovenia and Croatia has become a prominent international border. Border crossings were built on the main roads, which contributed to a slowdown in cross-border public transport between the two countries. The border population was still able to cross the border freely in the first decade after independence on local roads. However, bureaucratic obstacles gradually increased. Croatian workers in Slovenia had to obtain work permits. Stiperski and Kochi Pavlakovich (2001) find that already the internal border between the republics in the time of Yugoslavia was an obstacle to economic cooperation, and after the independence of the two countries, this obstacle gradually increased. Bus carriers had to register the former inter-republic routes as international with the competent authorities of both countries. They lost the right to carry domestic passengers in now a foreign country, which made them less profitable. Both demand and supply of public transport gradually decreased. In northeastern Slovenia cross-border daily commuting from Croatia has almost completely stopped, as well as cross-border public transport, due to the reduction of labour force in the industry and the collapse of industrial plants. In contrast, cross-border daily commuting from

Croatia to southern Slovenia continued. Cross-border bus transport links were also maintained (see Fig. 2).

In contrast to the Croatian border, cross-border daily mobility was increasing at the Austrian and Italian borders. Zupančič (2002a, 2002b, 2003) estimated the number on the basis of field research—interviews with people employed at the border crossings and daily commuters. He estimates that around 8,000 Slovenes travelled daily to Italy in 2001 and 6,000 to Austria. The figure included full-time workers as well as part-time and seasonal workers. Cross-border commuting for work took place almost exclusively by passenger cars. A special group is women travelling from Slovenia to Italy via the Škofije border crossing between Koper and Trieste to provide domestic assistance. Relatively frequent public transport has been, and continues to be, adjusted to this group's needs today. Organized transport with passenger cars started to flourish in the beginning of the 1990s (Sedmak 2005). In this Mediterranean border region between Croatia, Slovenia and Italy the majority of cross-border travelling is related to shopping and trips (Bufon 2001). In the aforementioned field survey Zupančič (2002a) also noted that approximately 2,500 Croatian citizens travelled daily through Slovenia to Italy and Austria. There were direct bus services arranged for them from Croatia, for example, from Rijeka and Umag to Trieste. Due to bureaucratic obstacles, carriers on these routes, after the independence of Slovenia and Croatia, abolished stops on the territory of Slovenia. Thus, Rijeka was connected to Trieste several times a day through Slovenia, but at the same time there were no bus connections between Rijeka and Slovenia.

2.3 The Period After Slovenia's Accession to the European Union (2004–2019)

In 2004 Slovenia joined the European Union, and at the end of 2007 Slovenia also joined the Schengen area. This allowed for a free movement of labour without border formalities. The borders with Italy, Austria and Hungary have in fact become invisible, but they are clearly still a hidden barrier to public transport.

The Slovene state intensified the regulation of public transport as late as at the turn of the twenty-first century. It started to regulate the intercity bus and rail transport as public service obligation, which made public transport services eligible for financial compensation. However, only domestic lines, but not international ones, were included in public service obligation. While long-distance international lines may be successful in the market, this does not apply to local cross-border lines. Similar to domestic public transport lines, cross-border local lines on the market cannot compete successfully with personal transport without financial compensation or without restrictive measures in the field of personal transport. Despite the open borders between Slovenia, Italy, Austria and Hungary, cross-border public transport has not come to life. The situation is similar at other European internal borders, both for road and rail public transport (Gabrovec 2013; Sippel et al. 2018; Oszter 2019).

Barsch (2014) also notes for areas of major cross-border urban agglomerations in Switzerland, France, and Germany that cross-border public transport links are of lesser quality than comparable lines that do not cross national borders. In this period, the number of passenger cars continued to increase in Slovenia. According to the Statistical Office of the Republic of Slovenia, the number of passenger cars per 1,000 citizens increased from 467 to 556 in the same period, while the share of commuters on public transport, according to the survey Daily Mobility of Passengers conducted by the Statistical Office of Slovenia in 2017, declined to a good 4%.

Gradually, also the relevant European and Slovenian authorities started to become increasingly aware of the lack of cross-border public transport. On the basis of the Railway Transport Act of 2007, in 2008 the Slovenian Government adopted a Decree on the mode of providing public service obligations in inland and cross-border regional railway passenger transport. Following the issuance of this Regulation, the downward trend in the scope of cross-border rail passenger services has been stopped. At some border crossings, where previously passenger cross-border traffic had already been discontinued, it was reintroduced (see Fig. 1). At the European level, a legislative move was made only in 2016 when Regulation (EU) 2016/2338 of the European Parliament and of the Council of 14 December 2016 amending Regulation (EC) No 1370/2007 was adopted regarding the opening of the market for domestic transport services by rail. The new provision in the regulation reads: "Subject to agreement of the competent authorities of the Member States on whose territory the services are provided, public service obligations may concern public transport services at the cross-border level, including those covering local and regional transport needs." This is the legal basis that enables the EU Member States, with appropriate cooperation, to organize cross-border connections with public transport more effectively. In the field of bus transport, Slovenia only allowed the inclusion of bus cross-border lines in the public service with the amendment of the Road Transport Act at the end of 2019. In many areas in Europe cross-border bus or rail links have been re-established through transregional cooperation projects Interreg in the start-up phase (Gabrovec 2013; Sippel et al. 2018; Oszter 2019). A similar example in Carinthia on the Slovenian-Austrian border is described below.

The period under review is characterized by a steady increase in cross-border daily commuting. However, this increase is typical only in two directions, from Croatia to Slovenia on the one hand and from Slovenia to Austria on the other. The number of daily commuters from Croatia to Slovenia has been increasing since Croatia's accession to the European Union in 2013. From then until 2019, according to the Statistical Register of Employment kept by the Statistical Office of Slovenia, it increased from 1,400 to 3,000. Despite its rapid increase, the figure is still significantly lower than in 1991. Contrary to the Yugoslav period, today's daily commuters do not have public transport available for their route. The only exception and example of good practice is the bus line from Prezid in Croatia to Lož in Slovenia, which has a schedule adapted to the working hours of the industrial plant at Lož and provides public transport for the workers from the Gorski Kotar area in Croatia.

The flow of daily commuters from Slovenia to Austria is several times higher. According to the data from the statistical survey Active and Inactive Population, conducted by the Statistical Office of Slovenia by means of a questionnaire, the number of active persons working in Austria increased from 4 to 16 thousand between 2010 and 2018. It should be noted that this data is not comparable to the number of daily commuters from Croatia to Slovenia as it involves all those who did any paid work in Austria in the last week before the survey. In addition to full-time workers, others who performed various contract or occasional jobs were also considered. Since daily commuting from Slovenia to Austria is the most intense one among all cross-border regions, we discuss them in more detail in the next chapter, using the case of Carinthia. The total number of daily commuters from Slovenia to Croatia, Italy and Hungary is significantly lower, amounting to between 5,000 and 6,000 over the period considered.

Cross-border trips to school, especially high schools and universities, are also being on the increase. There have recently been daily cross-border journeys of primary school children, although less frequently (Gabrovec 2013).

3 Cross-Border Mobility Between Austria and Slovenia in Carinthia

The inhabitants of the Carinthia region in Slovenia are historically tied to the territory of southern Austria. In order to evaluate the purposes, frequencies and ways of their travel to Austria, we conducted a survey in 2018. In doing so, we restricted ourselves to a narrower research area, which included residents over the age of 15 in the settlements Prevalje, Ravne na Koroškem, Dravograd, and Vuzenica. These settlements are located along the Maribor-Dravograd-Prevalje-Bleiburg/Pliberk railway line. At the time of the survey, rail was the only option for cross-border public transport in this border region. The questionnaire included questions about the frequency of trips to each federal province in Austria in the last year, the modes and purposes of travel, and the satisfaction with the provision of cross-border public transport. With the mail survey method, we wanted to reach about 15% responsiveness without additional reminders. Respondents' responsiveness varies greatly by mail surveys (Tiran et al. 2019; Polajnar Horvat and Smrekar 2015). A representative random sample with the names, surnames, and addresses of the respondents was prepared by the Statistical Office of the Republic of Slovenia. The final number of surveys returned was 209, with a response rate of 19%.

The majority (85%) of all respondents were to Austria at least once in the last year. It is estimated that a slightly larger proportion of the respondents, compared to the proportion in the general population, were retired or were more likely to respond to the survey.

The respondents most often travelled to the Austrian province of Carinthia, which is geographically closest to the majority in terms of residence. 49% of respondents made a trip to Carinthia at least once a month, 48% of respondents less than 10 times a year, and 3% never travelled to the Austrian Carinthia.

As expected, the predominant mode of travelling with the respondents was by passenger cars. On the other hand, the vast majority of respondents never travelled by train (93%) or bus (89%). Almost half of the respondents were not able to assess their satisfaction with the existing cross-border public transport service, probably because they did not use it. 20% of respondents have made a bicycle trip to Austria at least once. The reason for this is probably cycling as a recreation and the popularity of organized cycling links in Austria.

The most common purposes of travel to Austria were regular and occasional shopping and visits of relatives, acquaintances, and friends (Table 1). 23% of the respondents made regular shopping at least once a month. The share of those who

Table 1 Purpose and frequency of respondents' travel to Austria

Purpose	Daily (more than 200/year)	Weekly (more than 40/year)	Monthly (more than 10/year)	Rarely (less than 10/year)	Never	No. of respondents
Regular shopping (food, cleaning products, …)	0 (0.0%)	9 (5.1%)	32 (18.1%)	65 (36.9%)	70 (39.8%)	176
Regular shopping (fuel)	0 (0.0%)	9 (5.1%)	29 (16.4%)	42 (23.7%)	97 (54.8%)	177
Occasional shopping (clothes, furniture, …)	0 (0.0%)	3 (1.7%)	28 (15.8%)	77 (43.5%)	69 (40.0%)	177
Journey to school or university	3 (1.7%)	1 (0.6%)	1 (0.6%)	3 (1.7%)	167 (95.4%)	175
Visiting relatives, friends, acquaintances	2 (1.1%)	6 (3.4%)	21 (11.9%)	45 (25.4%)	103 (58.2%)	177
Commuting to work	10 (5.7%)	4 (2.3%)	1 (0.6%)	3 (1.7%)	157 (89.7%)	175
Business trip	0 (0.0%)	1 (0.6%)	6 (3.5%)	11 (6.3%)	156 (89.7%)	174
Crossing Austria to go to another country	0 (0.0%)	5 (2.8%)	10 (5.7%)	80 (45.2%)	82 (46.3%)	177
Going to a restaurant, bar	1 (0.6%)	1 (0.6%)	10 (5.6%)	43 (24.3%)	122 (68.9%)	177
Tourism	0 (0.0%)	2 (1.2%)	21 (12.1%)	71 (40.8%)	80 (46.0%)	174
Recreation—winter sports	0 (0.0%)	7 (4.0%)	11 (6.2%)	45 (25.4%)	114 (64.4%)	177
Recreation—summer sports	0 (0.0%)	6 (3.4%)	15 (8.5%)	44 (24.9%)	112 (63.3%)	177
Entertainment	0 (0.0%)	1 (0.6%)	5 (2.9%)	27 (15.4%)	142 (81.1%)	175
Personal errands	1 (0.6%)	2 (1.1%)	18 (10.2%)	68 (38.6%)	87 (49.3%)	176

travelled to Austria at least monthly for the purchase of motor fuels was unexpectedly high. Such was 21% of the respondents. 54% of the respondents travelled from Slovenia to Austria at least once for tourism, 36% for winter and 37% for summer recreation. 50% of the respondents travelled to Austria at least once a year for personal errands. In winter, many well-tended winter resorts for skiing and other sports are attractive for tourism and recreation. Attractive in the summertime are the well-tended natural baths, cycling trails, and traditional events.

The number of daily and weekly trips was relatively small. 14 respondents (8%) travelled to Austria at least weekly for work, and 4 respondents (2%) travelled to school or university. It is estimated that around 2,000 people commute from the Carinthia region in Slovenia to Austria. This is comparable to the two regions in Slovenia that are most attractive for work. About 2,000 people travel to work to the Savinja region, and about 1,500 people travel to the Central Slovenia region (Zajc et al. 2019). Since the journeys to work, as opposed to the journeys used to do the shopping and other purposes, are daily, they represent a large share of the daily cross-border journeys of the inhabitants of Carinthia. This is evident in the daily distribution of journeys on weekdays. At the Holmec border crossing, there is a pronounced peak during the hours when employees commute to and from work at the Mahle factory near Bleiburg/Pliberk in Austria, which employs around 500 Slovenian citizens (Zajc et al. 2019). The consequence of the increase in the number of commuters to work since Slovenia joined the EU until today is also reflected in the constant increase in the average daily cross-border flow of vehicles—at the Holmec border crossing this number increased from 1,383 in 2004 to 2,289 in 2018, according to the Slovenian Infrastructure Agency, and a slight decline was recorded only during the economic crisis in 2009. Interestingly, between 1997 and 2004, a decline in traffic was recorded, which can be explained by a decrease in cross-border shopping trips. The increase in the share of commuting to work represents an opportunity for public transport, since with a greater concentration of passengers at peak times it is easier to organize competitive public transport at this time.

The cross-border region is attractive for cycling. In 2019 a cross-border bus Velenje-Dravograd-Lavamünd/Labot was promoted under the TRANS-BORDERS Interreg project with the promotional name Štrekna Bus. It was the first ever cross-border bus connection between the Carinthia region in Slovenia and the Austrian province of Carinthia after several decades. It connected three popular cycling routes on both sides of the state border and enabled the transport of passengers and bicycles. The thirty-seat bus is equipped with a 20-wheel trailer. The bus operated twice daily in July and August, and in May, June, and September on weekends and Slovenian and Austrian holidays. In the first season 1,182 passengers and 967 bicycles were carried. A common model of implementation and promotion was developed.

The travel time and frequency of runs on the Maribor-Dravograd-Prevalje-Bleiburg/Pliberk railway line do not provide a competitive alternative to car transport. Travel time, however, is less of a restriction when travelling for leisure than it is for everyday tasks, such as work or education. In 2018, under the Interreg TRANS-BORDERS project, Slovenian Railways also added a Saturday summer train between Maribor and Bleiburg/Pliberk on which one wagon is adapted to carry 23 bicycles.

Table 2 Number of pupils and students from Slovenia at selected educational institutions in Klagenfurt/Celovec and St. Peter im Rosental/Št. Peter v Rožu in the 2017/2018 school year

Educational institution	From Slovenia	Of which from the Carinthia region in Slovenia
Federal Slovenian Secondary School in Klagenfurt/Celovec	49	16
Bilingual Federal Trade Academy, Klagenfurt/Celovec	104	46
Secondary College for Occupation in the Service Industries Management, Št. Peter v Rožu	92	0
Alps-Adria University, Klagenfurt/Celovec	32	8
Carinthian State Conservatory	113	9
Educational institutions together	390	79

In 2019, the train also operated on three summer holidays. The number of passengers carried increased slightly from 746 passengers in 2018 to 811 in 2019. The growth in the number of bicycles transported was more pronounced, with 202 bikes in 2018 and 431 bikes in 2019.

Due to the Slovenian national minority, many bilingual educational institutions operate in the Austrian Carinthia region, where German and Slovene are equivalent teaching languages. The advantage of bilingual education in German and Slovene is the reason why the Austrian Carinthian educational institutions are traditionally enrolled also by residents from Slovenia. They stay in student residences during the week in Austria and return home to Slovenia at weekends. For this reason, we further explored the weekly commuting between Slovenia and Austrian Carinthia in terms of education. The data on the number of pupils and students (Table 2) and their mode of transport between Slovenia and Austria in the 2017/2018 school year were obtained directly from educational institutions in the Austrian Carinthia. In order to explore in more detail the modes of transport pupils and students use between Slovenia and Austria, we conducted in-depth interviews with the management of high schools. Furthermore, in-depth interviews were also conducted with 14 students. With their consent, the pupils were selected by the management of individual schools. From the data obtained, interviews conducted and our own knowledge, it is estimated that the number of Slovene schoolchildren attending other educational institutions in the Austrian province of Carinthia is probably insignificant.

In the 2017/2018 school year, approximately 390 students from Slovenia were educated in Carinthia, Austria. Of these, 79 were from the Carinthia region in Slovenia, with 62 pupils and 17 students (Table 2). Pupils attending schools in Klagenfurt/Celovec, Austrian Carinthia, make up a good 2% of all students with the permanent residence in the Carinthia region of Slovenia. The pupils and students from the Carinthia region in Slovenia often travel to Klagenfurt/Celovec by train from Bleiburg/Pliberk in Austria. They make the journey to Bleiburg/Pliberk as drivers or passengers in passenger cars. One of the reasons for this is the lack of frequency and the inconvenient train schedule on the Maribor-Dravograd-Prevalje-Bleiburg/Pliberk route. The pupils and students coming from central and western Slovenia mostly organize their own transportation due to the lack of cross-border connections, their frequency and the inconvenient schedule for weekly commuting. This takes the form of passenger car sharing and private bus transport on Sundays from Ljubljana to Klagenfurt/Celovec.

4 Closing Thoughts and a View to the Future

The analysis of cross-border daily mobility in Slovenia from 1991 to 2019 revealed frequent changes that were the result of political and economic changes in Slovenia and in the neighbouring countries. Public transport supply responded very poorly to changes in daily cross-border mobility. In 1991, it still managed to meet the needs of cross-border mobility. The schedules were adapted for both commuters and cross-border shoppers. Due to the small number of passenger cars and the consequent high number of captive riders, cross-border public transport lines were also economically efficient at the time. On the routes where the demand was declining, the carriers discontinued the lines. In cases where cross-border daily mobility was increasing, no new public transport lines were introduced. This is understandable since local public transport lines, both within the country and across borders, cannot compete with personal transport without any financial compensation from the state or the local community, or without the involvement in the public service obligations. However, the inclusion of cross-border public lines in the public service requires an appropriate legal basis and close cooperation between the competent authorities on both sides of the border. Due to these problems, many cross-border public transport lines in Europe have been launched as part of the Interreg projects. That was also the case in Carinthia, as described above. The TRANS-BORDERS Interreg project identified that for a long-term operation of cross-border public transport an appropriate legal basis is crucial. As part of the public debate on the changes to the Road Transport Act in 2019 we successfully proposed an amendment to the law that reads: "*Public service is provided in the territory of the Republic of Slovenia, and because of local and regional needs for transport, also at cross-border level, with the consent of the competent authorities of the Member States in whose territory such services are provided. Public passenger transport authority designs the interchanges of public service in cross-border public scheduled passenger transport.*" This legal provision opens the possibility of making the quality of cross-border public transport comparable to the quality of public transport within the country. Given the removal of one of the hidden bureaucratic obstacles, we can expect an improvement in the supply of cross-border public transport in the coming years. When supply improves, the challenge for the future research will be the question whether increased supply will be followed by increased demand, or what are the key measures for the transition of passengers from passenger cars to public transport.

Acknowledgements The authors acknowledge the financial support from the European Regional Development Fund for the funding of the project TRANS-BORDERS (TEN-T Passenger Transport Connections to Border Regions) under the Interreg Central Europe Programme 2014–2020, and the Slovenian Research Agency research core funding Geography of Slovenia (P6-0101).

References

Barth, E. 2014. How international borders affect local public transport: analyses and evaluations of cross-border agglomerations in Switzerland, France and Germany. Zürich: Institut für Verkehrsplanung und Transportsysteme, ETH Zürich. https://doi.org/10.3929/ethz-a-010336271.

Bole, D. 2004. Daily mobility of workers in Slovenia. *Acta Geographica Slovenica* 44 (1): 25–45. https://doi.org/10.3986/AGS44102.

Bole, D., and M. Gabrovec. 2012. Daily commuters in Slovenia. *Geografski Vestnik* 84 (1): 171–185.

Bufon, M. 1995. *Prostor, meje, ljudje: razvoj prekomejnih odnosov, struktura obmejnega območja in vrednotenje obmejnosti na Goriškem [Territory, borders and people. Cross-border relations, the structure of the border landscape and the evaluation of »borderness« in the Gorizia bordeland].* Trst: Slovenski raziskovalni inštitut.

Bufon, M. 2001. Oblikovanje čezmejnih vezi na tromeji med Slovenijo, Hrvaško in Italijo v Istri [Development of cross-border relations in the Istrian three-border area between Slovenia, Croatia, and Italy]. *Dela* 16: 39–60. https://doi.org/10.4312/dela.16.39-60.

Christodoulou, A., and P. Christidis. 2020. Bridges across borders: A clustering approach to support EU regional policy. *Journal of Transport Geography* 83. https://doi.org/10.1016/j.jtrangeo.2020.102666.

Gabrovec, M. 2013. Open borders with uncoordinated public transport: The case of the Slovenian-Italian border. *European Journal of Geography* 4 (4): 78–87. http://www.eurogeographyjournal.eu/articles/6.OPEN%20BORDERS%20WITH%20UNCOORDINATED%20PUBLIC%20TRANSPORT%20THE%20CASE%20OF%20THE%20SLOVENIAN-ITALIAN%20BORDERl.pdf. Accessed 10 Apr 2020.

Gabrovec, M., and D. Bole. 2009. *Dnevna mobilnost v Sloveniji [Daily Mobility in Slovenia].* Ljubljana: Založba ZRC.

Klemenčič, V. 1993. Geopolitični položaj ter teoretski in metodološki poizkus opredelitve tipov obmejnih območij na primeru Slovenije [A geopolitical discourse on the location of Slovenia and an attempt towards a sociogeographic typology of Slovenian border regions]. *Dela* 10: 9–20. https://doi.org/10.4312/dela.10.9-20.

Oszter, V. 2019. How to establish and operate cross-border public transport in a peripheral rural area? The example of the Central and Southern section of the border between Austria and Hungary. *Prace Komisji Geografii Komunikacji PTG* 22 (1): 52–65. https://doi.org/10.4467/2543859XPKG.19.006.10926.

Polajnar Horvat, K., and A. Smrekar. 2015. Veljavnost osebnega in spletnega anketiranja v geografskem raziskovanju [Validity of face-to-face and web surveys in geographical research]. *Geografski Vestnik* 87 (2): 115–126. https://doi.org/10.3986/GV87208.

Ratti, R. 1991. *Théorie du développement des régions-frontières*, Fribourg: Imprimerie St-Paul.

Rietveld, P. 2012. Barrier effects of borders: Implications for border-crossing infrastructures. *European Journal of Transport and Infrastructure Research* 12 (2): 150–166. https://doi.org/10.18757/ejtir.2012.12.2.2959.

Sedmak, M. 2005. Življenje z mejo [Living with a border]. In *Meje in konfini, Rakitovec, vas kulturnih, družbenih in naravnih prepletanj [Borders and Barriers. Rakitovec, a Village of Cultural, Social and Natural Intertwining]*, ed. Vida Rožac Darovec, 115–136. Koper: Založba Annales.

Sippel, L., J. Nolte, S. Maarfield, D. Wolff, and L. Roux. 2018. Comprehensive analysis of the existing cross-border rail transport connections and missing links on the internal EU borders. Brussels: European Commission, Directorate-General for Regional and Urban Policy. https://ec.europa.eu/regional_policy/sources/docgener/studies/pdf/cb_rail_connections_en.pdf.

Stiperski, Z., and V. Kochi Pavlakovich. 2001. Odnosi i veze u pograničnim područjima na primjeru hrvatsko-slovenske granice [Relations and connections in borderlands on the example of the Croatian-Slovene border]. *Dela* 16: 119–134. https://doi.org/10.4312/dela.16.119-134.

Tiran, J., M. Geršič, J. Kozina, and P. Kumer. 2019. Množično samoanketiranje v praksi: primerjalna analiza štirih geografskih raziskav [Mass self-administered survey in practice: Comparative analysis of four geographical studies]. *Dela* 52: 161–185. https://doi.org/10.4312/dela.52.161-185.

Zajc, P., M. Gabrovec, P. Pipan, and A. Rupreht. 2019. Čezmejna dnevna mobilnost v Koroški regiji [Cross border daily mobility in Koroška region]. In *Demografske spremembe in regionalni razvoj*, eds. Janez Nared, Katarina Polajnar Horvat, and Nika Razpotnik Visković, 223–235. Ljubljana: Založba ZRC.

Zupanc, I. 2018. Demogeografski razvoj hrvatskog pograničja 2001–2011 [Demogeographic Development of the Croatian Borderland 2001–2011]. *Migracijske i etničke teme* 34 (2): 113–142. https://doi.org/10.11567/met.34.2.1.

Zupančič, J. 2002a. Prekogranične dnevne radne migracije iz Slovenije u Austriju i Italiju [Daily crossborder commuting from Slovenia to Austria and Italy]. *Migracijske i Etničke Teme* 18 (2–3): 151–166.

Zupančič, J. 2002b. Grenzüberschreitende Pendelwanderung aus Slowenien nach Österreich und Italien. *Mitteilungen Der Österreichischen Geographischen Gesellschaft* 144: 145–157.

Zupančič, J. 2003. Čezmejne dnevne delovne migracije v slovenskem obmejnem prostoru [Crossborder daily labour migrations in the Slovene border regions]. *Razprave in gradivo: revija za narodnostna vprašanja/Treatises and Documents: Journal of Ethnic Studies* 43: 68–111. http://hdl.handle.net/11686/4921. Accessed 10 Apr 2020.

Hidden Geographies and Perceptions

Hidden Geography of Marginalization: A Case Study in Timişoara

Roxana-Diana Ilisei

Abstract In this chapter, we study the impact that the visible degradation of a public square has had on the social behaviour and responses of the residents and local actors in terms of place attachment and participation. The main objective is to identify the factors that generated the negative image associated by residents themselves with the neighbourhood, the elements that determined the degradation and marginalization of the neighbourhood, and the consequences of these processes for the people's behaviour. For this purpose, we collected data through surveys from residents and interviews with professionals on Traian Square, a public space, and the core element of the Fabric neighbourhood in Timișoara, Romania. This is a square with a long tradition and history in Timișoara, one of the symbolic squares of the city, which was marked by a process of transformation that led to a degraded place, associated in the collective imagery with a place of fear, mysterious and enigmatic. Without clear stages of the evolution of the square and its buildings, it is proven that the hidden elements are interconnected and can be revealed by a detailed analysis. The results indicate that the main factor that contributed to the marginalization of the square was represented by the local authorities, which allowed its physical and social degradation by not investing in improvements and allowing people with social problems to settle there. Facing this situation, many residents considered the option of moving to other neighbourhoods or reduced the degree of participation in the neighbourhood's life. These consequences can, in a bad scenario, lead to segregation, which should be avoided with adequate strategies.

Keywords Marginalization · Public square · Discrimination · Segregation · Place identity · Social cohesion · Hidden strategies

R.-D. Ilisei (✉)
West University of Timisoara, Bulevardul Vasile Pârvan 4, 300223 Timişoara, Romania
e-mail: roxana_ilisei@yahoo.com

M. Krevs (ed.), *Hidden Geographies*, Key Challenges in Geography,
https://doi.org/10.1007/978-3-030-74590-5_22

1 Introduction

Social communities are constructions made of people that have different cultures and ideals and live a different lifestyle (Hirschle and Kleiner 2014), but who interact in the social environment, even though they live a distinct reality of the same territory, according to their perception of it (Cottone 2001; Pan Ke Shon 2007). This represents the constructivist interpretation of reality, which focuses on the fact that each person constructs its own reality (Raskin 2002) based on its personal and subjective way of seeing it (Tobin and Tippins 1993 in Gail Jones and Brader-Araje 2002), which, in turn, is based on the lived experience (Au 1998). This very last element, the lived experience, refers to the fact that the world is felt and understood differently by social actors (Schwandt 1994), according to their experiences and interactions with their environment (Murphy 1997). Mainly, the constructivist interpretation of reality can be explained by the fact that an object, in this case, the reality, is perceived, interpreted, and explained differently by individuals (Ültanır 2012). The individual realities, however, are strongly related to the territory where they occur and, of course, have common points, even though these common points are differently interpreted. And here, in the territorial reality, is where many anomalies can occur, where many hidden elements "appear", and that is the focus of the present research. It is integrated with the constructivist perspective since it relies on investigating perceptions, attitudes, and behaviours of inhabitants and focuses on the identification of the hidden aspects of reality that determine its interpretation in distinct ways.

Historic neighbourhoods have always been core elements of the cities not only due to their proximity to the business centres, their central position, or their touristic potential but also because of their high potential to attract higher-income groups (Ergun 2004). All these reasons transform them into targets of different strategies adopted at the cultural, economic, social, or political level. These strategies aim to transform the neighbourhoods according to the general perception of a group of people regarding the right way to build a city and the right evolution that should be ensured, either by transforming them into gentrified neighbourhoods, residential neighbourhoods, business centres, tourist destinations, or degraded neighbourhoods. However, when investigating the particular situation of a neighbourhood, it is difficult to identify the exact purpose of the strategies or the exact phase of evolution, since the territory and the social community built on it are very sensitive topics that everyone tries to avoid being held responsible for.

Here is where hidden geography intervenes, with all those elements that are not visible or are kept out of sight, concealed, or masked on purpose. There might be also visible elements in the landscape, but whose causes or generating factors are unknown or, again, hidden on purpose. Hidden geography is always present in every aspect of the social life of historic places, but more so when certain strategies did not get to the initial purpose when some mistakes occurred and changed completely the evolutionary trajectory. Historic neighbourhoods tend to be more affected by different types of strategies, due to their importance and reputation in the long term. The situation of the buildings in historic neighbourhoods may be the result of some hidden agendas (Akkar Ercan 2011). The social regeneration of disadvantaged neighbourhoods (Blanc 2002) is related to hidden geographies, given the different focus of this process according to the distinct evolution phase. More than these types of actions, some factors are kept out of sight in a more or less intentional manner, such as poverty (Fahmy et al. 2002) or immigrants' experiences (Robinson and Reeve 2006). All these elements: buildings' situation, social regeneration, and poverty are interconnected in the present study, and, similar to other territories, the evolution was not natural, but it was allowed to happen (Slater 2013).

Nevertheless, whenever a strategy is adopted and related to a limited area, unintentionally some borders are traced and processes can be easily classified as inner and outer processes. When the concept of inner and outer is applied to socio-territorial reality, processes of marginalization, discrimination, or even segregation may occur.

Marginalization and discrimination are two interconnected processes considered by many authors as self-explanatory concepts. Thus, they are usually used as easily understandable concepts and are not clearly defined. However, a deep analysis of the two concepts reveals their complexity in terms of significance, causes, and consequences. Marginalization is considered a political process that appears as a result of particular configurations of power (Harris and Weiner 1998) and is linked to lack of material resources, representation and freedom, poverty, and powerlessness. It is often used in the literature on poverty and social policy, referring to "the disadvantages of individuals, households, social groupings, or spatial areas, in terms of some social, economic, cultural, or political activities or processes" (Ballard et al. 2005: 7).

A strongly related term is discrimination, which refers to "an ingroup's subjectively justified unequal, usually disadvantageous, evaluation or treatment of an outgroup, that the latter would deem unjustified" (Mummendey and Wenzel 1999: 159). It is a social concept, part of the broader concept of stigmatization, together with ignorance and prejudice (Thornicroft et al. 2007) which focuses on the unequal treatment of persons or groups (Pager and Shepherd 2008). To be considered social discrimination, the unequal treatment needs to fulfill two criteria: the target group should perceive it as illegitimate and the expression of that treatment should be based on the target's group membership (Hansen and Sassenberg 2006). The focus on discrimination in study related to hidden geography is justified by its obvious consequences in terms of access of the discriminated groups to goods and services (Ahmed et al. 2007) and the barriers that it generates in social integration (Lundberg et al. 2008). Even though the consequences can be easily noticed, causes of discrimination are unknown: it is known why the discrimination is perceived, but not the factors that generated the situation. These consequences, if not dealt seriously, can lead to segregation and, thus, a differentiated behaviour at the neighbourhood level (Broman et al. 2000; Major et al. 2002).

Segregation is a wide concept introduced by Chicago School in the 1920s (Arapoglou 2006; Bolt et al. 2008; Maloutas 2004), which has a double dimension: a social one, focused on the social criteria that determine the separation (Pasztor 2013), and a territorial or geographical one, focused on the influence of the space on the social processes. It represents the unequal distribution (Van Ham and Tammaru

2016) of social groups based on different criteria, but it also refers to the situation when a certain social group cannot benefit from all the conditions and services of the social system to which it belongs (Cândea et al. 1999). In North America, it was associated with the emergence of urban ghettoes (Lavía Martínez 2009) as an expression of extreme discrimination (Maloutas 2004). One of its consequences is reflected by a specific response of social groups in terms of social behaviour, degree of involvement in the community's life, which influences social cohesion and neighbourhood identity.

A cohesive society is characterized by persons who share common values and a civic culture and also by social order, solidarity, networks, capital, together with place attachment, and identity (Forrest and Kearns 2001). Social cohesion does not suppose only mutual help and friendship but also the share of common values at the community level, not individual or group level. Place identity represents a psychological investment of a person in the territory (Vaske and Kobrin 2001), whose presence can contribute to a higher degree of social cohesion by increasing the sense of belonging to the community (Williams and Vaske 2003), the participation (Hays and Kogl 2007), and involvement in local activities (Lewicka 2005).

The present study is focused on the situation of a historic public square from Timisoara, Romania, which was transformed from an important cultural area into a degraded space, a space of fear, generally avoided by the population. In this space, all the above concepts seem to be related, but a detailed analysis is needed to discover the story behind it. The visible elements of the landscape are represented by the social and physical degradation of the square, which determines people to feel discriminated and the whole square to be marginalized. The actual situation could lead to segregation (if not existent already), but what is known for sure is that it already affected people's responses and behaviour in their day-to-day life and also in their relationship with the square. What remains still unknown and seems to hover above the square are the factors and actors that contributed to this situation, the strategies adopted during this period and the future.

2 Location of the Studied Area

Traian Square is the third representative public square of the city of Timisoara, Romania (see Fig. 1), which still keeps the architectural charm of the old city (Ilieșu 2004) although at the moment it seems to be insufficiently renovated and restored. This is the centre of Fabric neighbourhood (Cuțara 1998), the commercial (Buruleanu and Medeleț 2004) and industrial neighbourhood (Cucu 2011; Cuțara 1998; Gyula 2011; Ilieșu 2004) of Timisoara in the XVIII and XIX century and a multicultural one, where big communities of Germans, Hungarians, Serbs, Roma people, Hebrews, and other nationalities used to live (Buruleanu and Medeleț 2004; Hațegan 2009; Ilieșu 2004; Munteanu and Munteanu 2002). Traian Square developed as the main centre of this flourishing neighbourhood, from the economic, commercial, transport, and social point of view and in there can be found some of the old buildings

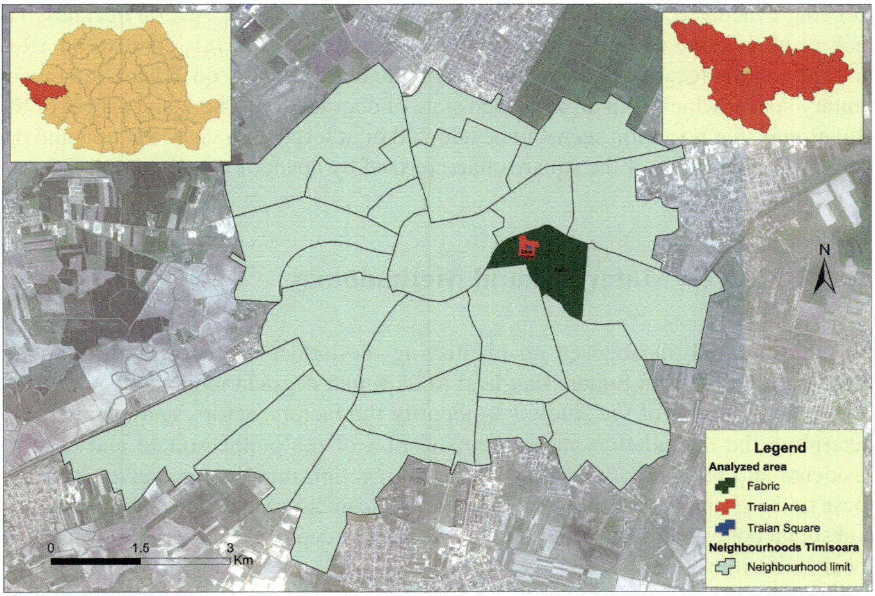

Fig. 1 Location of the studied area

from Timisoara, like the obelisk, Mercur Palace, Mirbach Palace, Stefania Palace (Cuţara 1998; Gyula 2011; Haţegan 2009). The special situation of these historical and architectural buildings corresponds to the hidden element of this study and the main reason for the research.

The particular evolution of the buildings in Traian Square, the legal state of property ownership, and the categories of owners have established challenging problems in recent years. Initially, the rich people of the neighbourhood used to live in the buildings from Traian Square. However, during the communist period, many of the representatives of the ethnic minority groups that lived there left the country, and, according to Law 119/1948, the houses were nationalized (Portal Legislativ 1948). Thus, most of the palaces were transformed into social housing, allowing people with limited financial resources to enter and live there. After the revolution from 1989, some of those houses were sold to the tenants who were able to prove that they had a lease agreement and lived in the respective buildings at the time of the promulgation of Law 112/1995 (Camera Deputaţilor 1995). Others were returned to the former owners or their descendants (Popescu 2017).

The fact that these houses were owned by the state for a long time or were inhabited by tenants without financial resources led to the deterioration of the buildings, especially in the case of historic buildings, which required a common action of the neighbours for their preservation. This situation of uncertainty continues until the present moment: many tenants are threatened by the return of the previous owners who claim their houses back. Thus, the actual tenants are scared about the possibility of being evacuated from the houses where they have lived for several decades. The lack of knowledge and interest from the local actors has, in this context, an effect not only on the local population but also on the buildings and on Traian Square as a unitary space, which is in an advanced state of degradation. The worst aspect of this situation is that no action seems to be adopted or at least not a visible action and this determines an image of the square characterized by abandonment and oblivion.

3 Objectives, Materials, and Methodology

The present study is focused on identifying the hidden elements and aspects of the situation of Traian Square that lead to its actual degradation and abandonment. The main objective of the study is to identify the factors, actors, and elements that determine the degradation and marginalization of the public square and how the residents and local actors respond to the segregation that they perceive. Based on these findings, another purpose was to identify the measure in which place attachment and social cohesion are affected by the negative perception and by the actions of the local authorities.

The approach was based on three hypotheses, tested during the research based on questionnaires and interviews in the study area: marginalization and segregation are associated, at the perceptual level, with the degradation of the buildings (H1); a negative perception of the residents regarding the neighbourhood and, consequently, regarding the square determines the diminution of place identity and social cohesion (H2); a particular response of the residents to the perceived segregation is the mistrust in the future of the neighbourhood (H3).

To fulfill the objectives, we used both quantitative and qualitative data collected through questionnaires and interviews done in July–September 2018 in Fabric neighbourhood, mainly in Traian Square and the side streets. It was proven in the scientific literature that segregation indices, which are often used in the segregation studies (Duncan and Duncan 1955; Massey and Denton 1988), are not able to build the general image of the phenomenon, as they cannot quantify the social situation and the social rejection that stays behind (Iglesias Pascual 2017). Thus, based on the focus of the present research on qualitative aspects of life in the neighbourhood and on neighbours' perception, together with the insufficiency of statistical data at the local level, the qualitative (interviews) and quantitative (questionnaires) methodology was chosen, both types of research methodology being strongly related to people's perception.

There were validated, in total, 106 questionnaires, with questions related to people's lives in the neighbourhood and included aspects related to people's perception regarding the square (positive and/or negative), marginalization perceived, actions of the local authorities, degree of involvement, and the square's future. The majority of respondents were represented by women (56%), while most respondents were 46–65 years old. The education level showed a predominance of the population with high school, followed by those with university studies. Eventually, the

Table 1 Sample size of the questionnaires in Traian Square

Identification criterion		Number	Percentage (%)
Gender:	Man	42	39.62
	Woman	60	56.60
	I prefer not to say	4	3.77
Age:	14–18 years old	3	2.83
	19–25 years old	7	6.60
	26–35 years old	19	17.92
	36–45 years old	17	16.04
	46–65 years old	49	46.23
	>66 years old	11	10.38
Education level:	No studies	0	0.00
	Primary or secondary school	16	15.09
	High school	36	33.96
	Vocational school	12	11.32
	University studies	28	26.42
	Postgraduate studies	14	13.21
Income level/month/family member	No regular income	4	3.77
	<1000 lei	15	14.15
	1000–2000 lei	40	37.74
	2000–3000 lei	42	39.62
	3000–5000 lei	5	4.72
Total		106	100.00

income level highlights the majority representation of the population with 1000–2000 lei/month/person (~200–400 €), followed by those with 2000–3000 lei (~400–650 €) (Table 1).

On the other hand, 14 interviews were conducted with respondents living mainly in the neighbourhood. These were classified into five main categories: representatives of public institutions, representatives of ONG's and private associations, representatives of the advisory council, teachers, and religious representatives. The purpose of this qualitative method was to get to know the perspective of both professionals that developed a certain activity in the neighbourhood during the last years and locals that have the function of leaders. Compared to the questionnaires, the interviews were applied based on the snowball sampling technique (Heckathorn 2011), they covered the same topics but were formed by open questions.

4 Results

The evolution of Traian Square over time generates a feeling of sadness and melancholy to the local population, who lives with the nostalgia of the stories of their parents and grandparents. Pictures from the past confirm the role and importance that the square had in the past and its flourishing character. Currently, buildings of the former square remain and could contribute to the revival of the square, if they were renovated and properly maintained. However, this is not yet the case, both the buildings and the plateau, and the square as a whole have a desolated image, which points to the lack of involvement and interest and abandonment. These aspects are reflected also by the image that people have in mind when they think about Traian Square. The majority of them consider it as disadvantaged or marginalized (almost 75%), while only 5% consider it as favoured. An important aspect that should be mentioned is that no one considers it as privileged (see Fig. 2), even though it is a historic square, the centre of a historic quarter, which has many beautiful buildings. Actually, the words that residents associate with the square are mainly negative, such as dirtiness (66%), insecurity (54%), noises (49%), and poverty (26%). The first positive word is on the 5th position in people's hierarchy, with only 13% of the respondents, and corresponds to the beauty of the square (see Fig. 3a). Despite the negative image, when asked if they heard about or noticed marginalization in the neighbourhood, the proportions are relatively similar: 46% negatives and 54% positives. Even though many people don't face marginalization at a declarative level, they live it in their day-to-day lives, given the fact that 48% of the respondents declared they feel discriminatory attitude coming from the people from neighbouring areas.

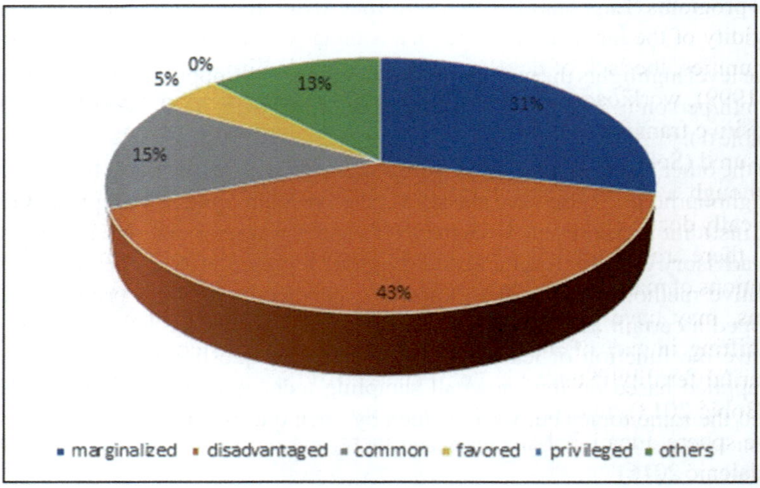

Fig. 2 Dominant feature of Traian Square

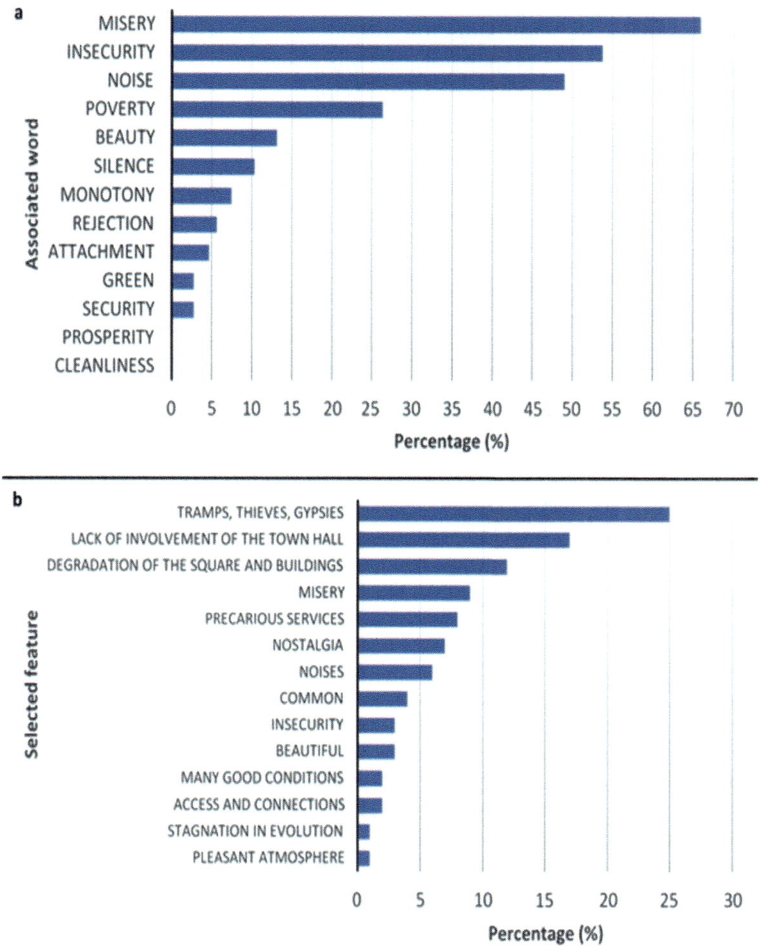

Fig. 3 Defining words (**a**) and features (**b**) of Traian Square

> Traian Square is not marginalized! It's just neglected! I think the town hall has a plan for the square, but probably it has not yet been reached... (Interview 14, church representative)

However, there is a consciousness of the importance of the square, but in the past, when it was a commercial and industrial centre.

> [It is] the most bohemian area of the city, because in the Citadel area there were not allowed night activities. There [in Traian Square] were the best restaurants, hotels. People were standing there and they were called Fabrukler. They were proud to be there! It was a saying, 100 years ago, that everything good comes from Traian! (Interview 13, ONG representative)

It is easily understandable that nowadays Traian Square is not the central square from the past, but totally the opposite and these are visible elements. What is hidden are the actions that lead to this situation, the factors that determined the degradation. So the question is: who is responsible for this situation? When people think about facts, they usually relate the square's image with the presence of tramps, thieves, and Roma population (25%), the lack of involvement of the administration (17%), and the degradation of the square and buildings (12%) (see Fig. 3b). The second option provided represents, actually, a partial answer to the question elaborated above, a key for this research, and a pole of the uncertainty and the hidden facts: the local authorities.

> We even organized the Neighbourhood Days every year, here in Traian! But, unfortunately, the existence of these gypsies, who each time, by their behaviour, somehow ruined our events, determined us to give up. To get out of this area. And we went to more outlying areas of the neighbourhood to hold our cultural events. (Interview 5, representative of the advisory council)

The administration is considered responsible for this situation because it allowed the square's degradation. It is considered guilty because it allowed the entrance to the historical buildings and the concentration of a category of the population with obvious socio-economic problems. Thus, it is accused that other people failed to obtain a contract for renting a home in the area, while the Roma population either got a contract easier or lived without having a legal agreement. After all, corruption is accused, which allowed the transformation of the historical area into a ghetto.

> Before, Serbs, Hungarians, Germans, Jews lived here. Many Jews! But during the communist period, they left and then the houses were rented. They were nationalized. Because the owners left, some died and then the houses were given to the state. After the revolution, some tenants bought the houses and became owners. Others have procured false documents as heirs. Gypsies also entered other houses. Because the municipality allowed that! There has been and there is a great corruption here! (Interview 14, church representative)

> Traian Square already… how to say it? It has a very serious problem: the fact that a historical area has been transformed into a small ghetto. (Interview 5, representative of the advisory council)

As the main actor, the administration is accused not only for its actions in the past but also for not being currently involved, for not being interested, and even for being indifferent to the situation of the square. Currently, more than half of the respondents consider its activity as being very bad or bad (58%), while only 6% considers it good, and none as very good (see Fig. 4), based on the fact that the main words associated to this activity are carelessness (69%), mistrust (29%), and chaos (25%). It is the one considered responsible for the degradation of the buildings, because it does not take drastic measures, and the rental agreements do not provide measures against the tenants who do not have a proper behaviour or even accept bribes to allow these behaviours to keep going.

> There are some pretty light lease agreements. Without too many obligations from the tenant. Poor control of the good use of space. If he is a tenant who engages in criminal acts, he creates a problem for the citizens and, implicitly, for the police, they think that the police must solve the problem! This means that they can still host them in those houses. (Interview 5, representative of the advisory council)

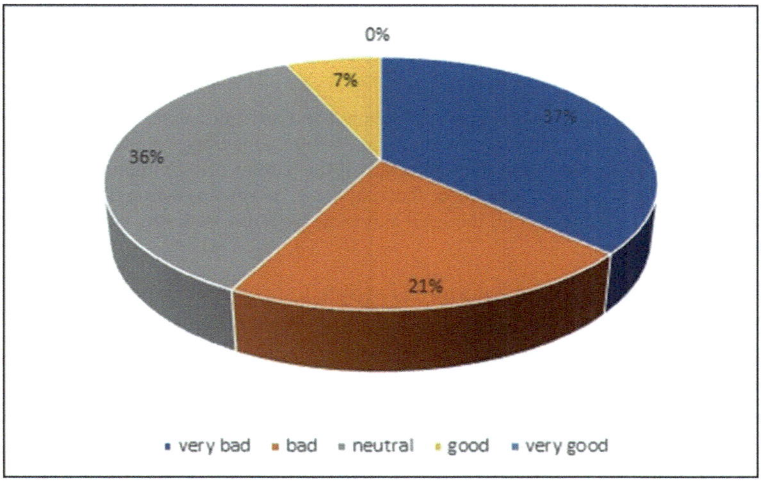

Fig. 4 Perception about the activity of the local authorities

Furthermore, there is a continuous struggle because each actor possibly involved in the evolution of the square accuses each other of being less involved. In this context, there does not exist a collaboration of all the actors for improving the situation, but more a fight to find a guilty and make it responsible for all the bad actions or the lack of actions.

Based on the information provided in the interviews, the conflict between the different actors can be easily noticed: the administration states that, regardless of the problem, it cannot intervene on private property, but in several similar situations it was found that it did not want to get involved, one reason being the lack of funds. Moreover, the lack of funds was also the one that prevented the development of the local celebration of the neighbourhood in 2018. Also, it is widely thought that the administration is not willing to support the initiatives coming from other actors, for example, those proposed by the neighbourhood advisory council, many remaining in the project phase, which is not implemented later. Last but not least, it is accused of lack of initiative, in the sense that, although it owns a building, it does not use it to improve the situation in the neighbourhood, but even prefers to leave certain apartments empty, which indicates carelessness. Also, the fact that it does not have a clear inventory of the tenants living in the buildings from the square, denotes a passivity concerning how the square evolves, confirmed by the delay in solving problems. Thus, the perception is that if any of the people with a certain influence had the respective difficulties or they were directly affected by those problems, these would be solved with priority.

Other actors accused of lack of interest are the Directorate for Culture and National Heritage and the private companies, such as mobile operators. The first actor is considered responsible for the deterioration of the buildings, not being involved in the improvement of the situation. Last but not least, the mobile operators are accused of not complying with the legal norms when placing the mobile phone cables, which contributes to the deterioration of the image of the buildings, but also of the entire square.

> The town hall allowed the degradation! Here should be drastic measures, fines! But the police do not support us at all! They say they are minors! But something could still be done, for example, community work. But they are afraid of the parents of the children! Instead, they fine honest people, because they know they pay them: if you cross irregularly, if you put a scaffold... But until you get all the legal approval, the house is destroyed! (Interview 14, church representative)

This lack of collaboration determines people to lose their hope and their trust in a possible amelioration or a possible better future. 49% of the respondents believe that the situation of Traian Square will not change, which is not a positive tendency, taking into account its current situation, while 35% consider it will be even more degraded than it is presently. Only 15% of the respondents consider that the square will be modernized (see Fig. 5a), following the trend of other historical neighbourhoods of different countries, such as Ruzafa in Valencia (Torres-Pérez 2007) or Lavapiés in Madrid (García Pérez and Sequera Fernández 2014), which passed through the process of gentrification. The negative perception regarding the future is also shown by the scepticism related to the new status of Timisoara of European Capital of Culture 2021 (ECoC 2021). The majority of respondents (61%) believe that this new role will have a very low or low impact on Traian Square since only the city centre will attract investments. However, a quarter of the respondents (28%) considers that it will have a high or very high impact (see Fig. 5b), but 14% of them stated the impact would be negative because the prices in the area would arise if it was included in the cultural events, or it would contribute to an even higher degradation or at least stagnation because it would remain without funds (all being directed to the city centre). In the possibility of the modernization of the square, the actors' opinions are again different, according to the interviews: some see it as a positive aspect, due to the improvement of the square's quality, while others are worried about the situation of the poor people and Roma people that live in the square.

> Well, it should have an impact! Because, finally, the Cultural Capital means a proliferation of cultural activities. The application file includes projects aimed at the inclusion of Roma and activities in disadvantaged areas. So, it should [have an impact]! (Interview 8, NGO representative)
>
> For Traian area, what I think will be a danger as we get closer to the title of Cultural Capital, is that the prices of rents and buildings there will increase. And then people will be somehow forced... they can't afford it anymore... you know? To stay. And they will be forced to... I don't even know where to go anymore... outside the city, in huts, in off-handed houses. (Interview 1, NGO representative)

Facing this situation, the attitude of the local population is one of negligence, of acceptance, without doing anything to change something. Thus, 37% of the respondents declare that they do not participate in actions (such as organizing cultural or social events or environmental activities) or by any means in the life of the community, while a percentage of 34% is involved, but passively, by respecting the general organization rules of the neighbourhood. Actually, at the participation level, there

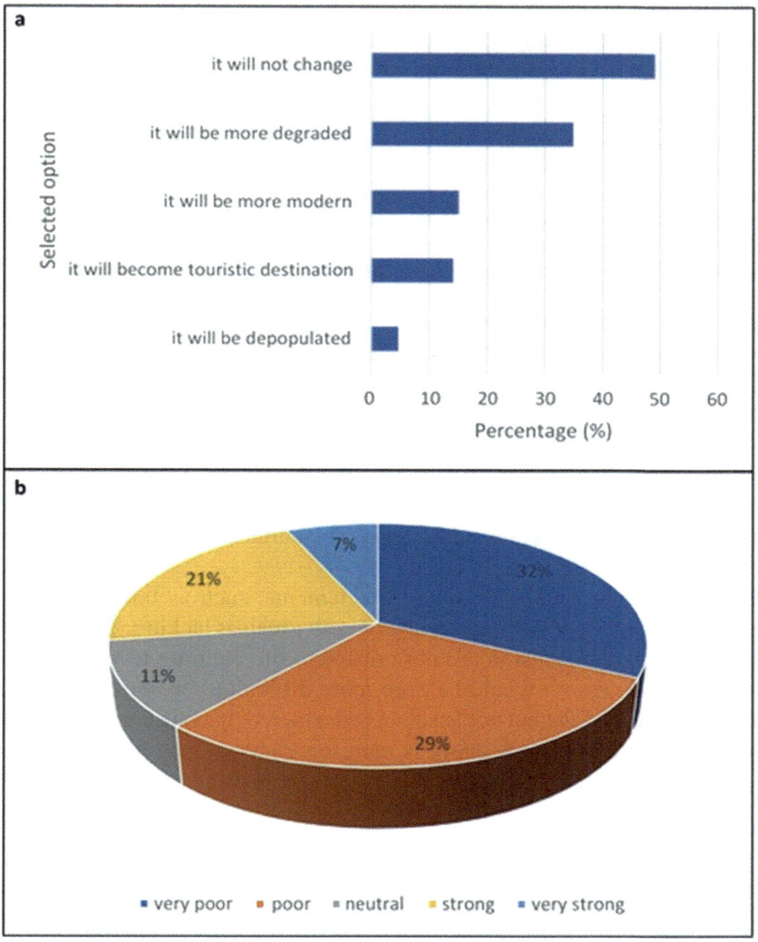

Fig. 5 Perception about the future of Traian Square (**a**) and about the impact of Timisoara ECoC 2021 on the square (**b**)

is a contradiction between the different actors, some of them stating that there is a high interest for participation, while others consider there is no participation, apart from the one for solving punctual, personal problems. Thus, the representatives of the neighbourhood advisory council are under the opinion that the local population is interested in participating in events and activities and in working to improve the existing situation. NGOs, however, state that people are not accustomed to working together for a common purpose due to the lack of activities, but also because Traian Square is perceived as a public space, where no community can be formed. However, in the projects they undertook, they were surprised by the openness of people to activities, albeit in small percentages. Actually, although all local neighbours say they want to get involved, just a few are the ones who actually participate actively in the end.

> Traian Square does not have... From our studies we have found that you cannot say that it has a community formed around it. Because it's a very public space. And then you don't feel attachment to it. So, you wouldn't want to intervene very directly. To make a more pleasant place... or more attractive... or more functional… […] At the declaratory level, everyone responded! I mean, everybody, I don't know the percentage, but the vast majority said yes. But from this, to actually doing something is a long way... Yet, at the same time you have to create the environment so that people can act. (Interview 8, NGO representative)

Based on these statements, it is easily understandable that social cohesion is at a very low level; actually, it is difficult to speak about social cohesion, when the majority of people do not even have friends in the neighbourhood (67%), but mostly outside. There are three factors evidenced in the interviews, considered to have deteriorated the cohesion that once existed in the square, namely, the loss of the hope of people, who saw that nothing was resolved in a long time; the absence of positive leaders, who could have an impact and influence the others to take action, and the incompetence of the authorities in managing the cultural diversity and the historical and architectural features.

Regarding the place identity, there are elements, such as buildings, that could create identity, but the sense of belonging of individuals is lacking. Thus, it is considered that identity and cohesion are not created at the building level because of the limited participation, so it would be even more difficult to create it at the square or neighbourhood level. Last but not least, it is considered that one cause of this absence of social cohesion and place identity is represented by the successive demographic changes, which have led to social alienation and, consequently, diminished interest in the community per se.

> And if the alienation exists, obviously there can be no interest for the heritage. For the beautiful. For the community. And obviously, no money will be allocated by the inhabitants for the neighbourhood. (Interview 11, teacher)

Facing this situation, many people consider also the possibility of changing residence and moving to another neighbourhood (51%). The main factors that determine their desire are the perspectives of evolution of the neighbourhood (40%), the feeling of unsafety (31%), and the bad relationships with the neighbours (24%), with a special focus on the noises and the conflicts with the Roma population. Actually, it was observed that there is a proportional relationship between the desire of changing residence and the time lived in the neighbourhood, with little exceptions: the higher desire of moving to another neighbourhood, village, city, or even abroad corresponds to people living longer in the square, more than 30 years, due to the fact that they have been in contact with the evolution of the situation and their level of disappointment is higher.

As a result of associating a negative image to the square, it is generally avoided, not only for transit but also for housing.

> Basically, this square has become… you know what? A kind of path shortening the road from one street to another! To go diagonally so as not to bypass the square, let´s say! (Interview 5, representative of the advisory council)

Thus, people who have the opportunity to choose another area considered better in the collective imagery will do so; only those who have no other option or better possibility will remain or choose to live in the square. As a general rule, they will be people with limited financial opportunities or who are excluded from other groups. Actually, the main element that stops people that want to move to other areas from changing their residence is represented by the insufficient income.

> The more negative the perception, the more only people who will be excluded from other parts of the city will move to those places. And then we will have an increase in segregation! This is the negative scenario... (Interview 11, teacher)

5 Conclusions

Traian Square is an emblematic public square of Timisoara which was affected by the different strategies adopted both in the communist period and in the actual moments. Its inhabitants are facing a multitude of problems and difficulties that are reflected at territorial, urbanistic, economic, social, educational, and labour levels. Even though in the present study only one public space was analyzed, the results can be extrapolated to other historical neighbourhoods with similar features to this one—heritage buildings transformed into social housing, degradation, presence of people with low income; or to other degraded or disadvantaged neighbourhoods. Based on certain patterns of evolution identified at the European level, the actual situation that characterizes nowadays Traian Square seems to be just a phase in the evolution of historic and central areas, meaning that the degradation may constitute a step before the actual gentrification, as in the case of Ruzafa in Valencia (Torres-Pérez 2007) or Lavapiés in Madrid (García Pérez and Sequera Fernández 2014).

The architectural design of the square proves the importance that it had in the past at the city level. However, a closer and deeper look reveals many difficulties and the contrast between the past landscape and the current situation makes every observer ask themselves: why? The current physical and social degradation of the square is visible, but, actually, the marginalization and segregation that the local population perceives are not related only to the degradation of the buildings. It is seen as a consequence of the political factor that did not succeed in managing the heritage buildings and allowed the continuous changes of the population together with the settlement of conflictual population. In this context, the first hypothesis is not confirmed, even though buildings degradation is visible, but it is not the main cause of the perceived marginalization.

People's responses to this situation are materialized in the desire to change residence, but also in a diminution of their interest to be involved in the community's life. Even though the heritage buildings create an identity of the square, they cannot create a place identity in people's consciousness due to their degradation and lack of preservation. Furthermore, the conflicts between the neighbours reduce the social cohesion, the square being mostly considered a public space, where no community can be formed. However, the general reputation of the square makes it to be avoided, so the specific role of a public space is also lost. Thus, the second hypothesis is confirmed, namely, that the negative perception and image of the neighbourhood and, consequently, of the square, determine the diminution of place identity and social cohesion.

As noticed through the people's responses during the field research, they lose their confidence in a possible future improvement, which means that the third hypothesis is also confirmed. It is not only that they believe the situation will be as bad as it is but also that, furthermore, they do not even trust in the future projects that are developed at the city level. In this case, of the few people that consider that the future status of Timisoara as European Capital of Culture 2021 will have a direct impact on the square, some believe that the impact would be a negative one. Thus, not even the projects intended to ensure a future development can give people hope.

Based on the findings stated above and expanding these to a wider context, it is demonstrated that elements of hidden geography are found in marginalized and disadvantaged neighbourhoods as an aftermath of the degradation of the public space and of the joining of existent strategies associated with that space with the responses of the local population. Where are the hidden elements? In all the unknown or masked processes and causes of specific processes that generated the current situations, in all the strategies developed in Traian Square, in the uncertainty of its future.

For the future of the square, it is of great importance to create a better collaboration not only between political actors but also between the local population and local authorities. Even though it may be difficult to reach a complete consensus in terms of how the square should be perceived and re-built, a collaboration between the different actors and socio-economical groups may be an important factor for solving the real problems and creating a community. When this main difficulty is solved and people feel that their opinion is taken into account, their interest in participating may be higher, they may become more conscious about the advantages that Traian Square has, the image of the neighbourhood can be improved, and, thus, the segregation, marginalization, and discrimination tend to diminish or even disappear. Then, in a normal situation, actions for re-creating place identity and social cohesion can be done, while people are not focused anymore on finding solutions but on participating in the community's life.

References

Ahmed, A.T., S.A. Mohammed, and D.R. Williams. 2007. Racial discrimination & health: Pathways & evidence. *The Indian Journal of Medical Research* 126: 318–327.

Akkar Ercan, M. 2011. Challenges and conflicts in achieving sustainable communities in historic neighbourhoods of Istanbul. *Habitat International* 35: 295–306.

Arapoglou, V.P. 2006. Immigration, segregation and urban development in Athens: The relevance of the Debate for the Southern European Metropolises. *The Greek Review of Social Research* 121: 11–38.
Au, K.H. 1998. Social constructivism and the school literacy learning of students of diverse backgrounds. *Journal of Literacy Research* 30: 297–319. https://doi.org/10.1080/10862969809548000.
Ballard, R., A. Habib, I. Valodia, and E. Zuern. 2005. Globalization, marginalization, and contemporary social movements in South Africa. *African Affairs* 104: 615–634.
Blanc, M. 2002. Strategies for the social regeneration of disadvantaged neighbourhoods in France (1977–2002). In *Soziale Stadt — Zwischenbilanzen*, ed. U.J. Walther, 211–228. Wiesbaden: VS Verlag für Sozialwissenschaften.
Bolt, G., R. Van Kempen, and M. Van Ham. 2008. Minority ethnic groups in the Dutch housing market: Spatial segregation, relocation dynamics and housing policy. *Urban Studies* 45: 1359–1384.
Broman, C.L., R. Mavaddat, and S.Y. Hsu. 2000. The experience and consequences of perceived racial discrimination: A study of African Americans. *Journal of Black Psychology* 26: 165–180. https://doi.org/10.1177/0095798400026002003.
Buruleanu, D.N., and F. Medeleț. 2004. *Timișoara – povestea orașelor sale*. Timişoara: Mirton.
Camera Deputaților. 1995. Law no.112 of 25 November 1995. www.cdep.ro. Accessed 10 June 2019.
Cândea, M., S. Costachie, C. Braghină, and D. Zamfir. 1999. *Dicționar de geografie umană*. București: Corint.
Cottone, R.R. 2001. A social constructivism model of ethical decision making in counseling. *Journal of Counseling and Development* 79: 39–45.
Cucu, A. 2011. *Cetatea și harta cetății Timișoara, Studiu tehnic și istoric*. Timişoara: Orizonturi Universitare.
Cuțara, A. 1998. *Timișoara – monografie artistică*. Timişoara: Amarcord.
Duncan, O.D., and B. Duncan. 1955. A methodological analysis of segregation indexes. *American Sociological Review* 20: 210–217.
Ergun, N. 2004. Gentrification in Istanbul. *Cities* 21: 391–405.
Fahmy, E., D. Gordon, and S. Cemlyn. 2002. Poverty and neighbourhood renewal in West Cornwall. www.bris.ac.uk/poverty/region.htm. Accessed 06 June 2020.
Forrest, R., and A. Kearns. 2001. Social cohesion, social capital and the neighbourhood. *Urban Studies* 38: 2125–2143.
Gail Jones, M., and L. Brader-Araje. 2002. The impact of constructivism on education: Language, discourse, and meaning. *American Communication Journal* 5: 1–9.
García Pérez, E., and J. Sequera Fernández. 2014. Gentrificación en centros urbanos: Aproximación comparada a las dinámicas de Madrid y Buenos Aires. *Quid* 16 (3): 44–61.
Gyula, D. 2011. *Ghidul Timișoarei*. Timişoara: Tempus.
Hansen, N., and K. Sassenberg. 2006. Does social identification harm or serve as a buffer? The impact of social identification on anger after experiencing social discrimination. *Personality and Social Psychology Bulletin* 32: 983–996. https://doi.org/10.1177/0146167206287639.
Harris, T., and D. Weiner. 1998. Empowerment, marginalization, and "community-integrated" GIS. *Cartography and Geographic Information Systems* 25: 67–76. https://doi.org/10.1559/152304098782594580.
Hațegan, I. 2009. *Turist prin Timișoara*. Timişoara: Orizonturi Universitare.
Hays, R.A., and A.M. Kogl. 2007. Neighborhood attachment, social capital building, and political participation: A case study of low- and moderate-income residents of Waterloo, Iowa. *Journal of Urban Affairs* 29: 181–205.
Heckathorn, D.D. 2011. Comment: Snowball versus respondent-driven sampling. *Sociological Methodology* 41: 355–366. https://doi.org/10.1111/j.1467-9531.2011.01244.x.
Hirschle, J., and T.M. Kleiner. 2014. Regional cultures attracting interregional migrants. *Urban Studies* 51: 3348–3364.

Iglesias Pascual, R. 2017. Subjetividad socioespacial y gestión de la diversidad en las principales ciudades españolas. Lo que los índices de segregación no pueden reflejar. *Boletín De La Asociación De Geógrafos Españoles* 75: 247–270.

Ilieșu, P. 2004. *Timișoara: The history of a European city*. Timişoara: Planetarium.

Lavía Martínez, C. 2009. Inmigración extranjera en Bilbao: una aproximación a la medida de la segregación residencial. *Zerbitzuan* 45: 83–98.

Lewicka, M. 2005. Ways to make people active: The role of place attachment, cultural capital, and neighbourhood ties. *Journal of Environmental Psychology* 25: 381–395.

Lundberg, B., L. Hansson, E. Wentz, and T. Björkman. 2008. Stigma, discrimination, empowerment and social networks: A preliminary investigation of their influence on subjective quality of life in a Swedish sample. *International Journal of Social Psychiatry* 54: 47–55. https://doi.org/10.1177/0020764007082345.

Major, B., W.J. Quinton, and S.K. McCoy. 2002. Antecedents and consequences of attributions to discrimination: Theoretical and empirical advances. *Advances in Experimental Social Psychology* 34: 251–330. https://doi.org/10.1016/S0065-2601(02)80007-7.

Maloutas, T. 2004. Editorial: Urban segregation and the European context. *Επιθεώρηση Κοινωνικών Ερευνών* 113: 3–24.

Massey, D.S., and N.A. Denton. 1988. The dimensions of residential segregation. *Social Forces* 67: 281–315.

Mummendey, A., and M. Wenzel. 1999. Social discrimination and tolerance in intergroup relations: Reactions to intergroup difference. *Personality and Social Psychology Review* 3: 158–174. https://doi.org/10.1207/s15327957pspr0302_4.

Munteanu, I., and R. Munteanu. 2002. *Timișoara – monografie*. Timişoara: Mirton.

Murphy, E. 1997. *Constructivism: From philosophy to practice*. Washington: ERIC Institute of Education Sciences.

Pager, D., and H. Shepherd. 2008. The sociology of discrimination: Racial discrimination in employment, housing, credit, and consumer markets. *Annual Review of Sociology* 34: 181–209. https://doi.org/10.1146/annurev.soc.33.040406.131740.

Pan Ke Shon, J.L. 2007. Residents' perceptions of their neighbourhood: Disentangling dissatisfaction, a French survey. *Urban Studies* 44: 2231–2268.

Pasztor, G. 2013. *Segregare urbană – Abordări cantitative și calitative ale fenomenului la nivelul orașului Cluj*. Cluj-Napoca: Rezumatul tezei de doctorat, Universitatea Babeş-Bolyai.

Popescu, C. 2017. Retrocedările în Timişoara – multe clădiri înstrăinate, puţine recâştigate. www.banatulazi.ro. Accessed 10 June 2019.

Portal Legislativ. 1948. Law no.119 of 11 June 1948. www.legislatie.just.ro. Accessed 10 June 2019.

Raskin, J.D. 2002. Constructivism in psychology: Personal construct psychology, radical constructivism, and social constructionism. *American Communication Journal* 5: 1–25.

Robinson D., and K. Reeve. 2006. *Neighbourhood experiences of new immigration: Reflections from the evidence base*. York: Joseph Rowntree Foundation.

Schwandt, T.A. 1994. Constructivist, interpretivist approaches to human inquiry. In *Handbook of qualitative research*, ed. N.K. Denzin and Y.S. Lincoln, 118–137. Thousand Oaks, CA: Sage Publishing.

Slater, T. 2013. Expulsions from public housing: The hidden context of concentrated affluence. *Cities* 35: 384–390.

Thornicroft, G., D. Rose, A. Kassam, and N. Sartorius. 2007. Stigma: Ignorance, prejudice or discrimination? *The British Journal of Psychiatry* 190: 192–193. https://doi.org/10.1192/bjp.bp.106.025791.

Tobin, K., and D. Tippins. 1993. Constructivism as a referent for teaching and learning. In *The practice of constructivism in science education*, ed. K. Tobin, 3–23. Washington DC: American Association for the Advancement of Science.

Torres-Pérez, F. 2007. *Nous Veïns a la Ciutat: Els Immigrants a València i Russafa*. Valencia: Publicacions de la Universitat de València.

Ültanır, E. 2012. An epistemological glance at the constructivist approach: Constructivist learning in Dewey, Piaget, and Montessori. *International Journal of Instruction* 5: 195–212.
Van Ham, M., and T. Tammaru. 2016. New perspectives on ethnic segregation over time and space. A domains approach. *Urban Geography* 37: 953–962.
Vaske, J.J., and K.C. Kobrin. 2001. Place attachment and environmentally responsible behavior. *The Journal of Environmental Education* 32: 16–21.
Williams, D.R., and J.J. Vaske. 2003. The measurement of place attachment: Validity and generalizability of a psychometric approach. *Forest Science* 49: 830–840.

Foreign Solo Female Travellers' Perceptions of Risk and Safety in Turkey

Bahar Kaba

Abstract Based on the qualitative methodology, the content analysis of 24 foreign solo female travellers' online blog narratives, this paper aims to understand foreign solo female travellers' perceptions of risk and safety in a less conservative Muslim destination, Turkey. The research result revealed that in terms of risk and safety, pre-travel Turkey was perceived as a dangerous or unsafe and uncertain destination by these travellers. However, later, following personal experience, nearly all women travellers agreed that it is a safe country for travelling alone. Additionally, they described their trip as a favourite, great, memorable, comforting, heart-opening experience and recommended it to potential solo travellers. Besides, Turkey's destination image was constructed positively by the travellers' post experiences. For them, the country was mostly represented by hospitable and friendly people, plentiful cultural and historical attractions, beautiful landscapes and natural attractions, good tourism infrastructure and security. On the other hand, some foreign solo female travellers experienced gender risk, including advances from men or unwanted attention and verbal harassment. To negotiate it, they developed various strategies such as dressing modestly, avoiding hanging out late at night, using common sense, ignoring or being rude to men, avoiding eye contact with men, using fake wedding ring.

Keywords Solo female travel · Perception of risk and safety · Destination image · Netnography · Blog narratives · Turkey

1 Introduction

Travel and mobility have become increasingly everyday practices for many (Cohen et al. 2013), and it sometimes seems as if all the world is on the move; on the one hand, asylum seekers, refugees, students, retirees; on the other hand; residential tourists, lifestyle travellers, budget travellers, backpackers, voluntary tourists, leisure travellers, solo travellers, etc. (Urry 2007). As declared by Cohen and Cohen (2012:

B. Kaba (✉)
Faculty of Letters, Department of Geography, Ege University, Izmir, Turkey
e-mail: bahar_1203@windowslive.com

M. Krevs (ed.), *Hidden Geographies*, Key Challenges in Geography,
https://doi.org/10.1007/978-3-030-74590-5_23

2): It is a world of accelerated economic, social and cultural change, driven by the process of globalization, rapid technological progress and the communication and information revolutions, and these forces have led to an increase in the tempo of life, a collapse of time and space, a cultural pluralization, a dedifferentiation of social domains and fragmentation of lifestyles.

As a result of these macro shifts and micro individual factors, most people prefer to travel solo without tours, friends, partners or family, and 'solo travel' has become relatively popular for the past couple of years. Especially, 'women' increasingly prefer to travel solo and 'solo women travel' has become a new trend globally (Jordan and Aitchison 2008; Seow and Brown 2018; Yang et al. 2018). 'Learning about different cultures', 'the ability to be free and flexible', 'social contact or sociality', 'education-learning', 'in search of self-understanding' and 'escape' are the main motivations for women to travel alone (Bond 1997; McArthur 1999; Gibson and Jordan 1998; Wilson and Harris 2006; Chiang and Jogaratnam, 2006; Yang et al. 2015; Ahokas 2017; Seow and Brown 2018).

As a growing phenomenon, there has been a steadily growing interest in the tourism, leisure and geography literature about solo female travel (e.g. Jordan and Gibson 2005; Wilson and Little 2008; Jordan and Aitchison 2008; Yang et al. 2018; Su and Wu 2020). These researches have shed light on *experiences* (McArthur 1999; Wilson and Little 2008; Jordan and Aitchison 2008; Myers 2010) *motivations* (Wilson and Harris 2006; Chiang and Jogaratnams 2006; McNamara and Prideaux 2010), *constraints* (Wilson 2004; Yang and Tung 2018), *safety or risk perceptions* (Ahokas 2017; Toh et al. 2017; Valaja 2018), and *risk-taking behaviour* (Awang and Toh 2018; Yang et al. 2018) of women with respect to solo travel. Also, they generally focus on *western* and recently *Asian* (Zhang and Hitchcock 2017; Seow and Brown 2018; Yang et al. 2018) contexts and samples.

While specific destinations such as India, Malaysia, Egypt are examined by researchers (e.g. Brown and Osman 2017; Thomas and Mura 2018), scholarly work on foreign, solo female travellers' perception and experiences in Turkey, which is a less conservative Muslim destination, is absent. Therefore, the research will contribute to solo female travel literature from a different geographical context, i.e. Turkey, which is both a western and also a middle-eastern country, providing a unique context for the examination of the issue. As Wilson and Little (2008) cited that negotiating movement through unfamiliar spaces and places can be a daunting experience when choosing to travel alone, investigating solo women travellers' experiences in Turkey, as a less conservative Muslim destination, will help to illuminate how gender shapes women's interactions with unfamiliar space and how they negotiate movement in this patriarchal culture.

Despite the growing processes of globalization and technology in every part of life and increasing mobility of people, some geographies remain hidden due to lack of information, uncertain image or it is not experienced individually. Especially, uncertain or negative images of place can restrict some groups from access to these places and cause remain hidden or unknown of these geographies. As it is known, the 'image of place' has a key role in people's spatial behaviour, such as the decision to stay or go, the decision on where to go, which route to take and the decision on how to get there (Kitchin 1994). For instance, images of places like positive or negative and safe or unsafe affect solo women travellers' routes or destination choices. Specifically, regardless of whether real or perceived, the presence of risk has the potential to change the nature of travel decisions; when risk perceptions or safety concerns are introduced into travel decisions, they have the potential to override factors and alter the context of conventional models of decision-making, causing travellers to amend their travel plans (Sönmez and Graefe 1998). For example, prior to travel, some women perceive certain travel destinations (such as Middle East, South America, Africa,) as unsafe and their own access to such destinations as limited (Wilson and Little 2008). Especially, the association of male violence with certain environmental contexts has a profound effect on many women's use of space, and women develop individual images of places where they fear assault resulting from their experience in the space, and secondary information (Valentine 1989).

Besides specific travel destinations such as Mexico, Vietnam, Thailand, Malaysia, Bali, India and Singapore (Yang et al. 2018), Turkey is becoming popular among solo travellers. As a travel destination, the country offers diverse cultural and historical heritage, beautiful landscapes and natural diversity and so on, which attract women travellers. However, after some travellers were raped or murdered, and bump attacks occurred in Turkey in recent years, the country was represented generally in the news or travel-related organizations as an *unsafe* country. For example, according to International Women's Travel Center (IWTC 2018), Turkey is on the list of the eight most dangerous places for women travellers.

In this context, the objectives of the study are to investigate foreign female travellers' perceptions of risk and safety, risk reduction strategies and revisiting intention with respect to solo travel in Turkey, by analysing the content of blog narratives that are fundamental in the construction of tourism experiences (McCabe and Foster 2006) and tourism places. The study also aims to understand Turkey's image as a tourism destination and hidden meanings of country by foreign, solo female travellers. Examining women travellers' narratives can help to recognize the hidden geography of the country and reconstruct the country's image beyond the media news. Also, understanding these perceptions and images can be useful both for the actual or potential solo travellers, and the country as it tries to develop a better tourism environment in terms of safety and quality, etc. Making visitors feel secure and safe before and during a vacation can be critical in terms of a destination's international competitiveness (Huan and Beaman 2004).

2 Perceived Risk and Safety

High perceived risk and safety concerns have appeared to become a central issue of visitors decision-making evaluations (Kozak et al. 2007: 234). They are likely to influence travel decisions and take the foremost importance in travel decisions due to their ability to change the travel intention or destination choice (Sönmez and Graefe 1998). Besides, perceptions of risk and safety can influence *the destination's image.*

For instance, research by Chew and Jahari (2014) revealed that perceived socio-psychological and financial risks influenced both cognitive and affective destination images. Similarly, Khan et al. (2017) tested young women's travel behaviour based on cognitive and affective perceptions of the destination and perceived risks. Their results revealed that the dimensions of perceived travel risks and travel constraints have negative effects on cognitive and affective destination images.

'Perceived risk' is defined as a consumer's perception of the overall negativity of a course of action based upon an assessment of the possible negative outcomes and the likelihood that those outcomes will occur (Mowen and Minor 1998: 176) and it has several dimensions such as financial risk, health risk, physical risk, political instability risk, psychological risk, satisfaction risk, social risk, terrorism risk and time risk (Roehl and Fesenmaier 1992; Sönmez and Graefe 1998; Yang et al. 2018). Especially, after the 9/11 terrorist attack in 2001, followed by a myriad of tragic incidents on both global and regional scales, such as the SARS outbreak in 2003, the Indian ocean tsunami in 2004, the Arab Spring uprisings in 2010 and the recent Paris attacks in 2015 and the Istanbul Atatürk airport attack in 2016 (Yang et al. 2018), the connection between tourism and terrorism has been highlighted and the sense of risk in travel has been aggravated. This has led to ever more stringent security procedures in global tourism, which put increasing constraints on the comfort, ease and freedom of travelling (Cohen and Cohen 2012). After all, tourism scholars have recognized the impact of risk perception on travel and decision behaviour, because tourists are likely to avoid destinations with a greater perceived risk (Kozak et al. 2007). The role of the different dimensions of perceived risk or safety in travel decision-making has been examined in various researches (Sönmez and Graefe 1998; Kozak et al. 2007; Gupta et al. 2010; Fuchs 2011; Yang et al. 2015; Chew and Jahari 2014; Khan et al. 2017).

Furthermore, the effect of gender on risk perception or 'gendered risk perception', which has been ignored by existing tourism risk literature, has begun to attract people's attention recently, with researches related to female travellers (Green and Singleton 2006; Khan et al. 2017, 2019), especially solo female travellers. Previously, the female travellers' risk perception related to different geography contexts was examined by researchers. For instance, Amir et al. (2015) examined the perception of international women travellers on safety and security in Kuala Lumpur. Their research demonstrated that 99% of respondents perceived that walking around in Kuala Lumpur during day time was very safe, but this perception decreased to 15% during night time. Similarly, Thomas and Mura (2018) explored foreign female travellers' perceptions of safety in India and showed that the bloggers portrayed India as a challenging destination, mainly due to its cultural contrasts and vibrant society. Recently, Kour and Gupta (2019) analysed the risk perception and travel intentions of solo women travellers to Kashmir, and they found that physical, financial and performance risks are negatively affecting the travel intentions of solo women travellers.

In another respect, a considerable number of studies focus on women's perception of solo travel and constraints. For instance, Wilson and Little (2008), explored solo women's travel experience under the concept of 'the geography of women's fear', and they revealed that these women perceived travel fears relating to the perceptions of others, i.e. susceptibility to vulnerability, a sense of restricted access and a feeling of conspicuousness. Previously, Toh et al. (2017), Ahokas (2017), Awang and Toh (2018), Valaja (2018), and Yang et al. (2018) also examined solo women travellers' risk perceptions of solo travel and their risk reduction strategies.

According to these studies, women have been subjected to gendered included risk when travelling alone in tourism spaces and places. For example, Valaja's (2018) research results revealed that their gender and their decision to travel solo influence female solo travellers' risk perceptions; especially gendered risks concerning sexual and physical harm are perceived to be heightened when travelling alone as a female. Similarly, a research conducted by Yang et al. (2018) showed female solo travellers perceived gender induced risks to be the highest, including unwanted gazes, sexual harassment and rape.

For dealing with real or perceived risk and safety, women adopt various risk reduction strategies. As the literature indicated, most common risk reduction strategies included 'not visiting destinations perceived as unsafe' (Yang et al. 2018) or 'leaving places where they felt to be in danger' (Wilson and Little 2008), 'avoiding isolated places and not going out at night' (Jordan and Aitchison 2008; Yang et al. 2018), 'dressing modestly' (Wilson and Little 2008; Thomas and Mura 2018; Yang et al. 2018), 'creating relationships' (Ahokas 2017) or 'pretending to be married' (Thomas and Mura 2018), 'behaving according to local female norms' (Wilson and Little 2008), 'remaining aware at all times' (Wilson and Little 2008; Thomas and Mura 2018), and 'avoiding alcohol and drugs' (Thomas and Mura 2018).

2.1 *Perceived Risk and Safety in Turkey*

Turkey is a tourism destination located in between Western Asia and Southeast Europe (Fig. 1). With its 8000 km coastline, its attractive Mediterranean climate and its beautiful beaches, the spectacular geological formations, such as the Pamukkale Travertines and the Fairy Chimneys of Cappadocia, the cultural diversity that reflects the various cultures and lifestyles, Turkey's appeal as a tourism destination is increasing significantly (Alvarez 2010), and the country has become a very competitive tourism destination over the last two decades (Atadil et al. 2015). According to statistics revealed recently by the United Nations World Tourism Organization (UNWTO 2019), Turkey is sixth out of the top 10 destinations that receive 40% of worldwide tourist arrivals. The country was visited by almost 46 million tourists, and gained 26,220-million-dollar international tourism receipts in 2018 (UNWTO 2019). Although statistics are not known, solo women travellers are a significant part of the international tourists who have visited the country in recent years.

Despite these positive developments in the Turkish tourism industry, there are some important issues that can affect international tourism in Turkey. For example, two solo women travellers were raped and killed; Pippa Bacca in 2008, Sarai Sierra 2013. Also, Turkey experienced a series of disasters such as the Ankara Station

Fig. 1 Location of Turkey

attack in 2015, the Ankara Güvenpark attack in 2016, the Istanbul Atatürk airport and the İstiklal street attacks in 2016. As mentioned before, political instability, health, safety and security concerns in the destination, whether real or perceived, have a direct bearing on the formation of negative images which in turn influence destination choice decisions (Sönmez and Graefe 1998).

Although there is no specific study related to the solo women travellers' image of Turkey, previous studies generally stated that Turkey has a *negative* destination image regarding its *safety*, based on various international samples, especially, those who have not visited the country (Baloglu and Mangaloglu 2001; Sönmez and Sirakaya 2002; Tasci et al. 2006). For example, the image of Turkey investigated by Sönmez and Sirakaya (2002) among potential travellers is particularly based on the United States, and founded that the respondents had a negative perception of Turkey's safe and hospitable environment. Similarly, in the research conducted by Tasci et al. (2006), Turkey's image was measured using a student population from the US, and the research indicated the worst rating for the safety and security dimension. Furthermore, the study also showed positive attributes of Turkey's image, respectively, unique culture/customs, amount of cultural/heritage attractions, scenic beauty, variety of outdoor activities, exciting features, peoples' friendliness/hospitality. In addition to these, Alvarez and Korzay (2011) stated that the perceptions of Spanish respondents regarding Turkey as a tourism destination were relatively positive, with the most positively rated items being the perceptions regarding the historical attractions, cultural attractions and natural environment; whereas, there was a more negative perception regarding Turkey's cleanliness and hygiene, infrastructure and personal security.

3 Methodology

Destination image of Turkey generally was measured by quantitative methods. However, to understand solo women travellers' perception deeply in the context of one place, Turkey, and to show unknown aspects of hidden geographies of the country by solo women travellers' narrative, the qualitative methodology, which has no intention to the generalization of the research results, was chosen in the research. Thus, a netnography method was used in the study (Kozinets 2006; Kozinets et al. 2014), a form of ethnographic approach. Netnography enables researchers to access the community members' knowledge online which, in turn, helps to provide in-depth insights into the consumers (Kozinets 2015: 2) and it involves gathering data from various online sources such as social networking sites, chat forums, weblogs and blogs (Zhang and Hitchcock 2017: 320). Blogs, particularly, are about expressing the inner experiences of travellers and sharing them with others who are interested and 'who care enough to read an account of a person's thoughts and feelings' (Conhaim 2003: 27). Furthermore, blog narratives are likely to influence the listener's or reader's future image of a destination (Volo 2010).

The research data was collected between April and July 2019. In the stage of data collection, the Google search engine was used for searching the words 'solo women traveller in Turkey', 'solo female traveller in Turkey'. Through purposive sampling, a total of 24 blogs were selected. When narratives of blogs started to repeat the same theme, the sampling was limited by 24 blogs. Bloggers were included to analye the following two necessary conditions; having travelled to more than one city in Turkey, and giving information about the perception of risk and safety in Turkey.

The bloggers were mostly from western countries; ten from the USA, three from Germany, two from the UK, two from Australia, two from Canada, one from South Africa and four not declared. Istanbul, Nevşehir (Cappadocia), Denizli (Pamukkale) and the Mediterranean and Aegean coastal areas (Muğla, İzmir, Antalya, etc.); these are the locomotives making Turkey one of the leading countries in terms of tourism arrivals and receipts (Alvarez 2010), were most visited a few main destinations (Fig. 2). Also, almost all blog narratives were entered in the last 5 years.

In the data analysis process, the 'content analysis' method was used which involved the processes of familiarization, coding and categorizing the data (Braun and Clarke 2006; Krippendorff 2018). The text consists of a total of 17,679 words that were read and re-read several times to notice initial ideas. After the reading, the data was divided into three main sections, and themes and sub-categories were identified.

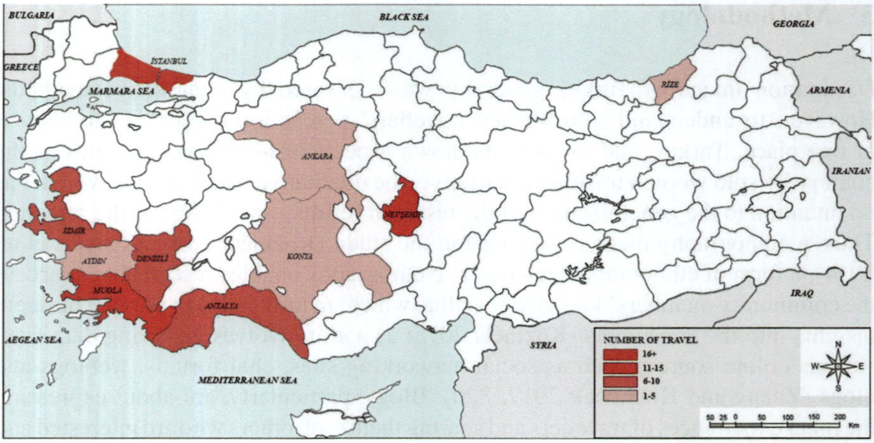

Fig. 2 Visited destinations by solo female travellers

4 Findings and Discussion

In this section, three themes derived from the narratives of solo women travellers are presented: (1) general perception of risks and safety, (2) gendered risk perception and risk reduction strategies and (3) destination image and revisiting intention.

4.1 *General Perception of Risk and Safety*

One of the objectivities of the study was to understand the international solo women travellers' perception of risk and safety in Turkey. The result highlighted that before the solo trip, some bloggers referred to Turkey as *dangerous* or *unsafe* (9) and an *uncertain* (2) destination.

'As a relatively conservative Muslim country with a culture that is vastly different to that which we are accustomed to in the west, Turkey has always been a travel destination that has some element of uncertainty around it when it comes to the ease of traveling around the country as a solo woman' (B 1, The UK).

Furthermore, half of them were warned by their family or friends. As Sönmez and Sirakaya (2002) mentioned, it may be because of the previous wars in the region and terrorist activities that Turkey is generally perceived as being unsafe. Furthermore, from a geographical perspective, Turkey is often associated with other Muslim countries in the region and therefore it is viewed as more similar to Middle Eastern countries, such as Iraq, Iran, Syria, Egypt, Algeria and Morocco, than to other neighbouring Mediterranean countries such as Greece (Alvarez 2010: 129). As seen in examples:

'I remember planning my first ever solo trip to Turkey and my friends' reactions when I told them I was going there on my own for a few months: "Oh my God! Why Turkey?" "Isn't it dangerous?" "You're going to a Muslim country alone? Are you crazy?' (B 6, The USA)

'Prior to my trip: When I have told people of my Turkey vacation plans, they would ask me if" Are you sure you want to go there?" or "Is it safe?" or "You aren't traveling solo in Turkey, right?". Their comments also included "I wouldn't travel in Turkey as a female alone" or "You know that Turkey is at high risk and it isn't safe to travel there' (B, 13 The USA).

Moreover, the analysis of narratives showed that after the trip, most travellers (17) who were warned and were concerned prior to travel, mentioned that they felt safe: 'It's embarrassing to think back on how apprehensive I felt about traveling to Turkey. After the news about the attacks for a few consecutive years, I feared there could be another attack at any moment. Which is silly, because there's a gun attack where I'm from (USA), every week it seems. But I also thought that maybe it would be hard for a solo female traveller. But when I got to Turkey… It immediately felt safer than NYC. Safer than much of Europe. I was shocked. And felt ashamed that I let media generalizations get the best of me the way they do to so many other countries' (B 23, The USA).

'As a woman, I never felt unsafe traveling through Turkey by myself, Traveling to Turkey is the same as traveling to almost any other country, and there's nothing especially harmful about it for women' (B 3, The USA).

Also, some of them (6) stated that media exaggerates the news reported about Turkey: 'As I've mentioned is the case with several travel destinations around the globe, so often the media exaggerates the situation in Turkey and makes it seem like a dangerous, inaccessible warzone. However, the fact of the matter is that life goes on as normal in Turkey, as it does anywhere else in the world 95% of the time' (B 1, The UK); 'All too often the media feeds us the concept that Muslim countries are unquestionably risky destinations, instilling fear in us and creating an idea that these countries are not worth visiting' (B 14, not declared).

4.2 Gendered Risk Perception and Risk Reduction Strategies

Besides the general perception of safety and risk, gendered risk was also observed in the narratives of most participants. For example, in most blog narratives, *advances from men and unwanted attention* (8) were discussed by solo female travellers in relation to gendered risks similar to the work of previous researchers (Brown and Osman 2017; Seow and Brown 2018; Yang et al. 2018; Valaja 2018; Xie 2019). In this respect, blogger 1 expressed her personal experience when she travelled alone in İstanbul:

'The main problem that I had during my time in Turkey-particularly in Istanbul, was with constant unwanted advances from Turkish men. Especially in Istanbul, men would come and sit at my table in coffee shops and restaurants sans invitation and try it on. Alternatively, they would just latch onto me while I was walking and refuse to take no for an answer when I told them that I wasn't interested. I even received dozens of messages from men on my Instagram when I was using the #Istanbul hashtag for my stories (how desperate is that?!)' (B1, The UK).

Also, the result shares a similarity with Thomas and Mura (2018) finding; *verbal harassment* indicated by two bloggers:

'What I have found most difficult about traveling alone in Turkey is the amount of men that approached me. As soon as I arrived in Istanbul, I started constantly getting approached. As I walked the streets, I would hear "Hello lady", "beautiful lady", "darling", "where are you from", "are you lost", "where are you going", "excuse me, hello", "this way princess", "every day I see you, why have you not visited me yet" etc.' (B 13, The USA).

Besides, some bloggers' narratives showed that participants' mobility was limited because of a *gendered space* and *time risk* (Jordan and Aitchison 2008; Wilson and Little 2005, 2008). Participants specifically stated rural places and night time. For instance:

'One of the most important safety tips I can pass along is to be aware of the hour of sunset. Many rural towns which are bustling during the day can turn into ghost towns once darkness sets in. You don't want to be the only person walking down the street, even if it is only 9 pm! The Black Sea coast is currently inundated with Russian prostitutes that are referred to as Natashas. To avoid being mistaken for one of these "ladies of the night", try not to travel to this area alone. And finally, Istanbul can feel safe because there are so many people milling about at all hours of the day. Don't be fooled' (B 8, Canada).

A few studies (e.g. Yang et al. 2018; Seow and Brown 2018) mentioned that being an Asian female traveller has an effect on gendered risk. For example, Seow and Brown (2018) cited that being an Asian solo female traveller in a western environment further intensified the feeling of vulnerability. Different from literature, an important point emerging from the blog narratives is that *being a western female* traveller in a non-western place would cause gendered risks. As were clarified by these examples:

'Unfortunately, many Turkish men are under the false impression that Western women are "easy", as that's what it seems like from what they've seen in movies and on television' (B 3, The USA).

'It's not really commonplace for a woman to walk around unaccompanied so a lone foreign woman is a perplexing concept for them. Unfortunately, like several countries around the world, Turkish men seem to correlate western women with what they have seen in the movies, i.e. that we are easy. Traditional gender roles still apply in Turkey and in some places (certain forms of public transport and places of worship for example), the genders are still segregated' (B 1, The UK).

Besides, the narratives showed that the *possibility of sexually assaulted* is always contemplated by solo women travellers, as mentioned in 'Thomas and Mura' (2018) study. For instance, blogger 13 mentioned fear of harassment in her case (The USA):

'My experience in Turkey can best be described as a love-hate relationship. I loved the beautiful landscape of Cappadocia, the bustling city of Istanbul, and all the Turkish cuisine. But I hated the uneasiness that I felt and concern for my safety.

The problem is that so many men are trying to get your attention it begins to feel overwhelming. I just felt constantly harassed. Where it crossed the line for me was the men that would just start walking alongside you especially at night. It did make me fearful on how to get away from them. I have probably watched too much Dateline and murder shows for my own good so I jump to the extremes in my mind. Turkey's current reputation of being unsafe certainly didn't help ease my mind. No matter the reason, I just couldn't put my guard down enough to enjoy walking the streets of Turkey alone as a female solo traveller'.

Unsurprisingly, every blogger provided a list of tips and warnings to avoid or minimize the probability of becoming involved in cases of gendered risk. For instance, *dressing modestly* (16) was the most common strategy they used. We can assume that solo women travellers were aware of some cultural dress code in Turkey (especially more conservative parts of the country) and tried to respect it.

'Keeping in mind these cultural perceptions, it could help to dress a little more conservatively than usual, especially in the less touristy central and eastern parts of the country. It's not necessary to cover your hair unless you're in a mosque, but wearing long sleeves and not showing cleavage would be a good idea' (B 3, The USA).

'In large cities such as Istanbul and Ankara, modern neighbourhoods (usually the ones where you'll find bars and restaurants) don't have much of a dress code and are comparable to any European city. However, in more residential areas, especially more conservative neighbourhoods, it's wise to keep that short skirt in your luggage and cover up just a bit more in order to avoid awkward gazes. So, if you're traveling around Turkey, make sure to dress modestly, with the exception of the summer cities by the coast (such as Bodrum, Izmir, or Antalya) where the dress code is very relaxed' (B 5, Germany).

Moreover, *avoiding hanging out late at night* (8) and *using common sense* (4) are the other common strategies mentioned by participants:

'As with most destinations around the globe, you should (as I did) exert basic common sense when travelling alone in Turkey, i.e. don't go wandering around by yourself late at night or after dark and be mindful of who you trust' (B 1, The UK); 'When you're travelling, you should behave how you would at home, more cautious, in fact, than you'd be in a situation where you felt comfortable. When I travel solo, I always make sure to buy my own drinks, keep a close eye on my glass, and stop drinking when I feel a little tipsy. I'm wary of anybody who's overly friendly towards me, whether it's day or night' (B 12, The UK).

Also, three bloggers talked about *avoiding eye contact with men* and *ignoring or being rude to men*:

'This is not just for Turkey, but for many places, even in the US, eye contact can mean an invitation or flirting. This was something I never did until this trip, always staring people down and smiling at everyone. I can't help it, I'm a smiler. But I tried it in the markets in Istanbul and it worked phenomenally. Just stare straight ahead. Works like a charm' (B 10, The US); 'Turkish men do flirt a lot, so don't be surprised to hear catcalls or be approached, sometimes with persistence; either ignore them or say some firm words in Turkish from your guidebook and get the message across' (B 14, not declared).

Using fake wedding ring is the other strategy mentioned by one participant:

'As I mentioned above, I spent 10 days sailing around the southwestern coast of Turkey, and it was right at the end of our trip when disaster struck. Our yacht ran out of fuel, thanks to a faulty gauge that told us we had plenty left, and it happened when there was a dearth of wind. Fortunately, I wasn't on board as I was day tripping to a nearby village, but unfortunately, I now needed to find somewhere to stay for the night while they worked on getting back to shore. I elected to stay in a midrange hotel rather than a hostel, and I decided to wear my ring on my wedding finger, something I've never done while traveling before. I felt a little nervous about being the lone woman in the hotel and had read that other travel bloggers often did this to ward off unwanted attention from men. It's probably something I'll never do again' (B 12, The UK).

Generally speaking, the research results are consistent with previous results (Jordan and Aitchison 2008; Wilson and Little 2008; Ahokas 2017; Brown and Osman 2017; Yang et al. 2018; Thomas and Mura 2018). Strategy narratives show that foreign solo female travellers develop mostly global strategies that can be applied almost everywhere. In other words, these strategies are not only valid in Turkey, but also in other countries. However, one that can be specific to a less conservative Muslim destination is dressing more modestly as used in another more conservative İslamic destination, i.e. Egypt (Brown and Osman 2017).

4.3 Destination Image and Re-visiting Intention

As mentioned before, although, there is no specific study related to foreign solo women travellers, according to research into Turkey's image measured by potential international tourists, the country suffers from a negative or uncertain image (Sönmez and Sirakaya 2002; Tasci et al. 2006). However, the perceptions regarding Turkey become significantly more positive depending on the travellers' familiarity level with the destination (Baloglu 2001) or after the tourists visit the destination (Alvarez et al. 2009). In this context, it can be said that the destination image which solo female travellers have of Turkey, is constructed positively post-travel, as can be seen in Table 1. For example, after travelling, several bloggers mentioned that Turkey has *hospitable and friendly people*:

'Turkey is an exciting country to explore with a culture of warm, welcoming people' (B15 not declared); 'Turkish people are very friendly and accustomed to social interaction. They may smile, they may ask where you're from, but that's all I experienced. I would smile back and I found I was never barraged with questions, instead left to be' (B 4, The USA).

The other common images shared by bloggers are *plentiful cultural and historical attractions*, and *beautiful landscapes and natural attractions* in the country, as expressed by blogger 22:

Table 1 Destination images of Turkey

			Images				
	Countries of participants	Writing year	Hospitable and friendly people	Plentiful cultural and historical attractions	Beautiful landscapes and natural attractions	Good tourism infrastructure (accommodation, transport, etc.)	Good security
1	United Kingdom	2018	X	X	X		X
2	United States	2019	X	X	X		
3	United States	2019	X	X	X	X	X
4	United States	2018	X	X			
5	Germany	2017	X				
6	United States	2018	X				
7	Australia	2013					
8	Canada	–		X			
9	United States	2018	X			X	
10	United States	2013	X				X
11	Australia	2018					
12	United Kingdom	2019					
13	United States	2018					
14	Not declared	–				X	
15	Not declared	2015	X	X	X		
16	Not declared	2018					
17	Germany	2013					
18	United States	2017					
19	Germany	2016	X				
20	Not declared	2017	X	X	X	X	
21	Canada	2014	X	X	X	X	

(continued)

Table 1 (continued)

	Countries of participants	Writing year	Images				
			Hospitable and friendly people	Plentiful cultural and historical attractions	Beautiful landscapes and natural attractions	Good tourism infrastructure (accommodation, transport, etc.)	Good security
22	United States	2018		X	X	X	
23	United States	–					
24	South Africa	2016					
Total			13	9	7	6	4

'For a country so rich in culture, heritage, and history, Turkey has unfortunately seen a decline in travel and tourism over recent years. However, Turkey is not a country to be avoided. Brimming with natural beauty, abundant traditions, diversity, and a multitude of aromas, Turkey is a mysterious and intriguing place for solo female travel' (The USA).

Also, Turkey is generally described as *an amazing, beautiful, mysterious-intriguing, great, fascinating, favourite* country by bloggers; 'Based on my experience, Turkey is an amazing country worth exploring' (B 18, The USA); 'Turkey is a great destination for solo female travellers and a perfect introduction to the Middle East' (B 10, The USA).

Tasci et al. (2006) mentioned that among a student population from the US who have never visited the country, Turkey has a 'stereotypical image' such as Middle Eastern, Muslim, Arabic destinations, and thus a desert climate. It can be said that Turkey is also generally identified as a dichotomy, like *conservative-liberal, western-eastern* and *peaceful-chaotic* in most narratives, after the personal experience.

'Turkey is an enigma. It is at once both modern and ancient, Eastern and Western, peaceful and chaotic' (B 3, The USA); 'Remember that although liberal, Turkey is still a Muslim country. In the beach resorts and tourist areas then people are more accustomed to seeing western women in summer clothes but in central and Eastern Turkey, it is much more conservative and so you should dress modestly' (B 1, The UK).

After the trip, most bloggers said it was *a favourite, great, memorable, comforting, heart-opening* experience and suggested the country for other solo travellers to visit. It also seems that the possibility of being exposed to sociocultural practices perceived as contradictory and very different from those of their country of origin allows them to live intense experiences (Thomas and Mura 2018). It can be read from examples:

'So, you see, traveling as an SFT in Turkey is not scary or intimidating at all. In fact, it's quite the opposite—an enriching, fun and heart-opening experience. Whether it's your first or fiftieth solo trip, it is a country that should definitely be on your list!' (B 6 from the USA).

'If you're contemplating whether or not Turkey is safe or for you, stop right now. Turkey is for anyone that wants to have the best and most adventurous 10 days of their life and trust me, when you get there, you'll quickly see how safety is a non-issue as long as you use street smarts like you would in any city around the world' (B 20 not declared).

Also, some bloggers mentioned that they will revisit the country: 'There is no problem whatsoever about travelling alone in Turkey. I plan to return in a not so distant future and visit more of the country, especially the southwest: places like Ephesus, Pamukkale, the turquoise coast around Antalya, and especially Cappadocia' (B 2, Canada); 'Thank You! I love Turkey. Been there twice now and would go back again' (B 17, The US).

Furthermore, one blogger (B 19, Germany) stated that she settled in Turkey: 'I sold up and started to see the world. I also took up residence in different countries until I hit Istanbul. Love at first sight. I was living abroad in Turkey for four years, buying a small condo in Didim on the Aegean Sea'.

5 Conclusion

Based on the content analysis of 24 online blog narratives, this study explored foreign solo women travellers' perception of risk and safety in Turkey. The research result revealed that in terms of risk and safety, pre-travel Turkey was perceived as a dangerous or unsafe and uncertain destination by travellers. However, later, following personal experience, nearly all women travellers agreed that it is a safe country for travelling alone. Besides, Turkey's destination image was constructed positively by travellers' post experiences. The country was represented mainly by 'hospitable and friendly people' and 'plentiful cultural and historical attractions'. In this context, findings support prior studies (Öztürkmen 2005; Tasci et al. 2006; Alvarez and Korzay 2011) that examined Turkey's destination image according to international tourists. Different from existing literature, it is also understood that Turkey is described generally with 'positive connotations' such as amazing, beautiful, mysterious, intriguing, great and fascinating. Generally speaking, the result supports prior studies that mentioned how experience influences the destination's post-visit perceived image (Phelps 1986; Baloglu 2001; Akhoondnejad 2015).

Also, it is understood that some foreign solo women travellers experienced gender risk, including mostly 'advances from men' or 'unwanted male attention'. Furthermore, the 'possibility of being sexually assaulted' is always contemplated by solo women travellers and to avoid or minimize it, they developed various coping strategies. The most common coping strategy used to minimalize gendered risk was 'dressing modestly' that also indicated a previous study conducted in İslamic destination, Egypt, by Brown and Osman (2017).

As mentioned before, the destination image influences the revisiting intention and recommendation intention (Chen and Tsai 2007; Phillips et al. 2013; Akhoondnejad 2015). In this context, it can be said that after travelling, most travellers described their trip as a favourite, great, memorable, comforting, heart-opening experience and suggested the country to potential solo travellers. Furthermore, solo women travellers stated that they intended revisiting the country.

This study contributes specifically to the current body of literature on solo female travel, generally in gender risk and tourism research, with a focus on solo female travellers' experiences in different geographical contexts, and a less conservative Muslim destination. However, there is a limitation to this research that needs to be addressed which is that nearly half the research sample consisted of American women. Therefore, the perception of risk and safety associated with a destination can change according to the research sample, solo female travellers' risk and safety perception should be analysed in different contexts and with different examples.

References

Ahokas, S. 2017. *Safety of female travellers*. Dissertation. Haaga Helia University of Applied Sciences.

Akhoondnejad, A. 2015. Analyzing the pre-travel, on-travel, and post-travel behaviors of Iran's first-time visitors. *Journal of Travel & Tourism Marketing*. https://doi.org/10.1080/10548408.2014.957796.

Alvarez, M.D., K. Inelmen, and Ş. Yarcan. 2009. Do perceptions change? A comparative study of pre- and post-travel perceptions. *Anatolia: An International Journal of Tourism and Hospitality Research* 20: 401–418.

Alvarez, M.D. 2010. Marketing of Turkey as a tourism destination. *Anatolia: An International Journal of Tourism and Hospitality Research* 21: 123–138. https://doi.org/10.1080/13032917.2010.9687094.

Alvarez, M.D., and M. Korzay. 2011. Turkey as a heritage tourism destination: The role of knowledge. *Journal of Hospitality Marketing & Management* 20: 425–440. https://doi.org/10.1080/19368623.2011.562433.

Amir, A.F., M.N.I. Ismail, and T.P. See. 2015. Sustainable tourist environment: perception of international women travelers on safety and security in Kuala Lumpur. *Procedia - Social and Behavioral Sciences* 168: 123–133. https://doi.org/10.1016/j.sbspro.2014.10.218.

Atadil, H.A., E. Sirakaya-Turk, and V. Altintas. 2015. An analysis of destination image for emerging markets of Turkey. *Journal of Vacation Marketing* 23: 37–54. https://doi.org/10.1177/1356766715616858.

Awang, K.V., and J.Y. Toh. 2018. A review on Malaysian women's leisure and perceptions towards solo travelling. *International Journal of Engineering & Technology* 7: 139–142.

Baloglu, S. 2001. Image variations of Turkey by familiarity index: Informational and experiential dimensions. *Tourism Management* 22: 127–133.

Baloglu, S., and M. Mangaloglu. 2001. Tourism destination images of Turkey, Egypt, Greece, and Italy as perceived by US based tour operators and travel agents. *Tourism Management* 22: 1–9.

Bond, M. 1997. *Women travelers: A new growth market*. https://www.hotelonline.com/News/PressReleases/PataWomenTravellers_Nov1997.html. Accessed 15 March 2020.

Braun, V., and V. Clarke. 2006. Using thematic analysis in psychology. *Qualitative Research in Psychology* 3: 77–101.

Brown, L., and H. Osman. 2017. The female tourist experience in Egypt as an Islamic destination. *Annals of Tourism Research* 63: 12–22.

Chen, C.F., and D. Tsai. 2007. How destination image and evaluative factors affect behavioral intentions? *Tourism Management* 28: 1115–1122.

Chew, Y.E.T., and S.A. Jahari. 2014. Destination image as a mediator between perceived risks and revisit intention: A case of post-disaster Japan. *Tourism Management* 40: 382–393.

Chiang, C.Y., and G. Jogaratnam. 2006. Why do women travel solo for purposes of leisure? *Journal of Vacation Marketing* 12: 59–70. https://doi.org/10.1177/1356766706059041.

Cohen, E., and S.A. Cohen. 2012. Current sociological theories and issues in tourism. *Annals of Tourism Research* 39: 2177–2202. https://doi.org/10.1016/j.annals.2012.07.009.

Cohen, S.A., T. Duncan, and M. Thulemark. 2013. Introducing lifestyle mobilities. In *Lifestyle mobilities: Intersections of travel, leisure and migration*, 1–18. Farnham: Ashgate.

Conhaim, W.W. 2003. Personal journals: New uses for an age-old practice. *Information Today* 20: 27.

Fuchs, G. 2011. Low versus high sensation seeking tourists: A study of backpackers'experience risk perception. *International Journal of Tourism Research* 15: 81–92.

Gibson, H., and F. Jordan. 1998. Shirley valentine lives! the experiences of solo women travelers. Paper Presented at The Fifth Congress of The World Leisure and Recreation Association, Sao Paulo, 26–30 October 1998.

Green, E., and C. Singleton. 2006. Risky bodies at leisure: Young women negotiating space and place. *Sociology* 40: 853–871. https://doi.org/10.1177/0038038506067510.

Gupta, A., D.J. Gupta, and N. Arora. 2010. The relationship between perceived travel risk, travel safety, travel anxiety and intentions to travel: A path analysis study of domestic traveller in India. *International Journal of Tourism and Travel* 3: 25–35.

Huan, T.C., and J. Beaman. 2004. Contexts and dynamics of social interaction and information search in decision-making for discretionary travel. *Tourism Analysis* 8: 177–182.

IWTC. 2018. Eight worst countries for women tourists. http://www.internationalwomenstravelcenter.com. Accessed 12 March 2020.

Jordan, F., and C. Aitchison. 2008. Tourism and the sexualisation of the gaze: Solo female tourists' experiences of gendered power, surveillance and embodiment. *Leisure Studies* 27: 329–349. https://doi.org/10.1080/02614360802125080.

Jordan, F., and H. Gibson. 2005. We're not stupid . . . but we'll not stay home either: Experiences of solo women travelers. *Tourism Review International* 9: 195–211.

Khan, M.J., S. Chelliah, and S. Ahmed. 2017. Factors influencing destination image and visit intention among young women travellers: role of travel motivation, perceived risks, and travel constraints. *Asia Pacific Journal of Tourism Research* 22: 1139–1155.

Khan, M.J., S. Chelliah, F. Khan, and S. Amin. 2019. Perceived risks, travel constraints and visit intention of young women travelers: the moderating role of travel motivation. *Tourism Review* 74: 721–738. https://doi.org/10.1108/TR-08-2018-0116.

Kitchin, R.M. 1994. Cognitive maps: What are they and why study them? *Journal of Environmental Psychology* 14: 1–19.

Kour, P., and S. Gupta. 2019. Analysis of travel risk perception and travel intentions among solo female travellers towards Kashmir as a destination. *International Journal of Research in Commerce & Management* 10: 14–16.

Kozak, M., J.C. Crotts, and R. Law. 2007. The impact of the perception of risk on international travellers. *International Journal of Tourism Research* 9: 233–242.

Kozinets, R.V. 2006. Click to connect: Netnography and tribal advertising. *Journal of Advertising. Research New York* 46: 279–288. https://doi.org/10.2501/S0021849906060338.

Kozinets, R.V. 2015. Netnography: Understanding networked communication society. In *The international encyclopedia of digital communication and society*, ed. Q.H. Anabel and S. Luke. https://doi.org/10.1002/9781118767771.wbiedcs067.

Kozinets, R.V., D. Pierre-Yann, and E. Amanda. 2014. Netnographic analysis: Understanding culture through social media data. In *The SAGE handbook of qualitative data analysis*, ed. F. Uwe, 262–275. London: Sage.
Krippendorff, K. 2018. *Content analysis: An introduction to its methodology*. Los Angeles: Sage.
Mcarthur, M. 1999. *Out of place: Gender, identity, and space, and the experiences of contemporary solo women travelers*. Dissertation. Trent University.
McCabe, S., and C. Foster. 2006. The role and function of narrative in tourist interaction. *Journal of Tourism and Cultural Change* 4: 194–215.
Mcnamara, K.E., and B. Prideaux. 2010. A typology of solo independent women travellers. *International Journal of Tourism Research* 12: 253–264.
Mowen, J., and M. Minor. 1998. *Consumer behaviour*. New Jersey: Prentice Hall.
Myers, L.M. 2010. *Women's independent travel experiences in New Zealand*. Dissertation. University of Sunderland.
Öztürkmen, A. 2005. Turkish tourism at the door of Europe: Perceptions of image in historical and contemporary perspectives. *Middle Eastern Studies* 41: 605–621.
Phelps, A. 1986. Holiday destination image: The problem of assessment. *Tourism Management* 7: 168–180.
Phillips, W.M.J., K. Wolfe, N. Hodur, and F.L. Leistritz. 2013. Tourist word of mouth and revisit intentions to rural tourism destinations: A case of North Dakota, USA. *International Journal of Tourism Research* 15: 93–104. https://doi.org/10.1002/jtr.879.
Roehl, W.S., and D.R. Fesenmaier. 1992. Risk perceptions and pleasure travel: An exploratory analysis. *Journal of Travel Research* 30: 17–26.
Seow, D., and L. Brown. 2018. The solo female Asian tourist. *Current Issues in Tourism* 21: 1187–1206. https://doi.org/10.1080/13683500.2017.1423283.
Sönmez, S.V., and A.R. Graefe. 1998. Influence of terrorism risk on foreign tourism decisions. *Annals of Tourism Research* 25: 112–144.
Sönmez, S., and E. Sirakaya. 2002. A distorted destination image? The case of Turkey. *Journal of Travel Research* 41: 185–196.
Su, C.P., and T.C. Wu. 2020. The Dark side of solo female travel: Negative encounters with male strangers. *Leisure Sciences*. https://doi.org/10.1080/01490400.2020.1712277.
Tasci, A.D.A., S.U. Meydan, and S.T. Cavusgil. 2006. Destination image: the case of Turkey. *Tourism Analysis* 11: 79–93.
Thomas, T.K., and P. Mura. 2018. The 'normality of unsafety' foreign solo female travelers in India. *Tourism Recreation Research*. https://doi.org/10.1080/02508281.2018.1494872.
Toh, J.Y., K.W. Awang, and J. Bojei. 2017. Generation y Malaysian women's perception towards solo travel. *Asia Pacific Journal of Innovation in Hospitality and Tourism* 6: 45–54.
UNWTO. 2019. *International tourism highlights*. https://www.e-unwto.org/doi/pdf/10.18111/9789284421152. Accessed 10 April 2020.
Urry, J. 2007. *Mobilities*. Cambridge: Polity Press.
Valaja, E. 2018. *Solo female travellers' risk perceptions and risk reduction strategies as expressed in online travel blog narratives*. Dissertation. Lund University.
Valentine, G. 1989. The geography of women's fear. *Area* 21: 385–390.
Volo, S. 2010. Bloggers' reported tourist experiences: Their utility as a tourism data source and their effect on prospective tourists. *Journal of Vacation Marketing* 16: 297–311. https://doi.org/10.1177/1356766710380884.
Wilson, E., and D.E., Little. 2005. A 'Relative escape'? The Impact of constraints on women who travel solo. *Tourism Review International* 9: 155–174.
Wilson, E.C. 2004. *A journey of her own?: the impact of constraints on women's solo travel*. Dissertation. Griffith University.
Wilson, E., and C. Harris. 2006. Meaningful travel: Women, independent travel and the search for self and meaning. *Tourism* 54: 161–172.
Wilson, E., and D.E. Little. 2008. The solo female travel experience: Exploring the 'geography of women's fear. *Current Issues in Tourism* 11: 167–186.

Xie, Z. 2019. *Asian female solo travellers*. Dissertation. Auckland University of Technology.
Yang, R., and V.W.S. Tung. 2018. How does family influence the travel constraints of solo travelers? Construct specification and scale development. *Journal of Travel & Tourism Marketing* 35: 507–516. https://doi.org/10.1080/10548408.2017.1363685.
Yang, E.C.L., S.P. Sharif, and C. Khoo-Lattimore. 2015. Tourists' risk perception of risky destinations: The case of Sabah's eastern coast. *Tourism and Hospitality Research* 15: 206–221. https://doi.org/10.1177/1467358415576085.
Yang, E.C.L., L.C. Khoo, and C. Arcodia. 2018. Constructing space and self through risk taking: A case of Asian solo female travelers. *Journal of Travel Research* 57: 260–272.
Zhang, Y., and M.J. Hitchcock. 2017. The Chinese female tourist gaze: A netnography of young women's blogs on Macao. *Current Issues in Tourism* 20: 315–330. https://doi.org/10.1080/13683500.2014.904845.

Hidden Geographical Contexts of a Visible Post-war Landscape: A Case Study in Bosnia and Herzegovina

Marko Krevs, Ranko Mirić, and Nusret Drešković

Abstract The aim of this study is to discuss selected hidden geographical contexts of a visible post-war landscape and their influence on explaining what is perceived in the landscape. Hidden geographies refer to the spatial location/distribution of phenomena in the landscape that remain hidden from the observer until they are revealed to him by his cognition. In order to simulate such a perspective in the discussions as well as in the interpretation of examples of revealing hidden geographies, a narrative/mental simulation method is used. The process of constructing explanations for the visually perceived landscape is demonstrated using a model from constructivist psychology, in which prior knowledge, learned rules, and current experiences/learning are confronted with perceived reality. The examples of revealing hidden geographies are observed in a selected area in Bosnia and Herzegovina and shed light on topics such as depopulation, ethnic structure, war destruction, post-war renewal and return of displaced people, and interethnic tensions. Multiple sources (of varying reliability) and methods were used to present these examples, ranging from the collection and interpretation of statistical data, maps, publications, personal online stories, even ethnically biased media news or personal blogs from the internet, to the authors' own field observations and interviews with locals. The discussions and the examples of hidden geographies provide a range of evidence that post-war landscapes are full of hidden contexts which affect the observer's explanation of his visual perception of the landscape, as much as they influence the lives of people inhabiting it, and their natural environment.

M. Krevs (✉)
Department of Geography, Faculty of Arts, University of Ljubljana, Askerceva 2, Ljubljana, Slovenia
e-mail: marko.krevs@ff.uni-lj.si

R. Mirić · N. Drešković
Department of Geography, Faculty of Science, University of Sarajevo, Bosnia and Herzegovina, Zmaja od Bosne 33-35, Sarajevo, Bosnia and Herzegovina
e-mail: rankom@pmf.unsa.ba

N. Drešković
e-mail: nusretd@pmf.unsa.ba

M. Krevs (ed.), *Hidden Geographies*, Key Challenges in Geography,
https://doi.org/10.1007/978-3-030-74590-5_24

Keywords Post-war landscape · Multi-ethnic · Hidden geography · Population processes · Settlements · Constructive psychology · Bosnia and Herzegovina · Bosansko grahovo · Livno

1 Introduction

The post-war landscape as an object of perception, after a civil war in a multi-ethnic region, defies superficial stereotypical explanations, even when we have access to some basic information about historical and current events and processes. Some of the geographic information in the context of what we perceive is inaccessible even through in-depth investigation. Other geographies are potentially discoverable with some effort, and still others are accessible to our explanation based on our sensory information and prior knowledge when visiting such a place.

The aim of this study is to discuss selected hidden geographical contexts of a visible post-war landscape and their influence on explaining what is perceived in the landscape. A selected area in Bosnia and Herzegovina serves as an example. The war to which this study refers took place between 1992 and 1995, a quarter of a century ago. But traces of another war, between 1941 and 1945, also persist in the studied landscape, sometimes in a way that makes it difficult to distinguish between the two. These traces are not only physical but also cultural and personal. It seems that the cultural and personal traces are often deeper and even more persistent than the visual, physical ones. Ethnic identification and related collective mindset, combined with deep personal emotions based either on the war experience or inter-generational transfer, are powerful drivers of processes in society as well as in the physical landscape. These are the main drivers of hidden geographies that form and function as material or non-material components of a landscape. In this study, we look for examples of the visible, measurable, or otherwise detectable manifestations of material and non-material geographies that function as hidden—to (almost) every one or to an occasional visitor.

The study area is located in the western part of the Federation of BIH, in a part of Canton 10 adjacent to the Croatian state border, the canton with the lowest population density in the Federation by far (Fig. 1). The criteria for selecting the area were the multi-ethnic population before and after the war in the 1990s, major war activities that took place there, and intense population processes in the post-war period.

A systematic and comprehensive exploration of the hidden geographies behind the reading of the investigated landscape is beyond the aim of this study. Instead, a specific approach based on the method of *mental/narrative simulation* is used to provide a series of examples of diverse hidden geographies. The narrative with a foreign visitor in the main role makes it easier for the reader to follow the discussion. The visual perception (by the visitor) with almost no prior knowledge and the resulting explanation of the perceived reality are strongly contrasted with provided examples of such geographies revealed by the authors of this chapter. This contrast expands the range of possible hidden geographies, which helps to provide a richer foundation for further discussion.

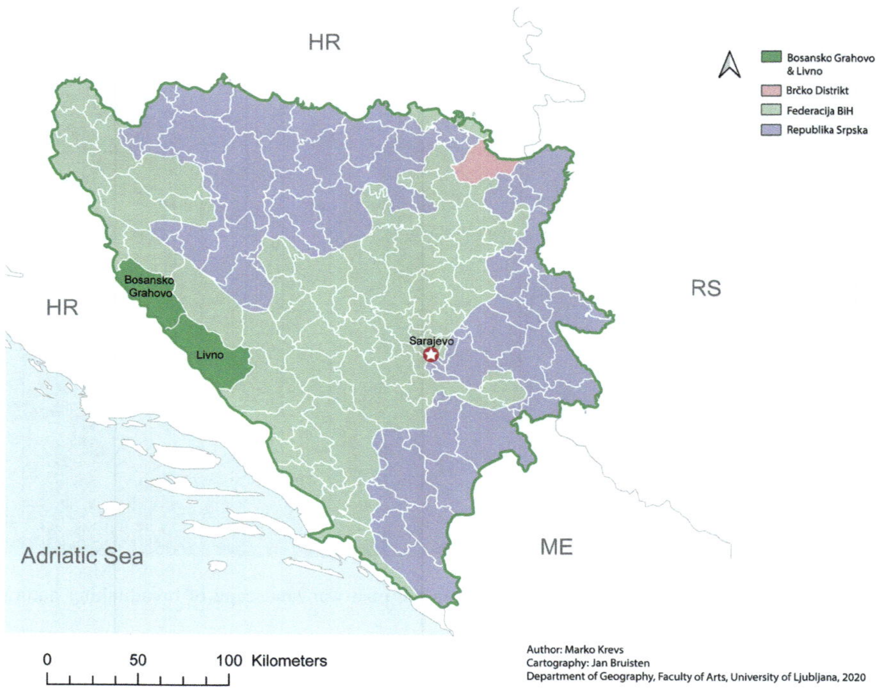

Fig. 1 The area in focus within Bosnia and Herzegovina: municipalities Bosansko Grahovo and Livno

Several visible elements/features of the landscape under study are affected by the war—temporarily or for a long time in the future. It appears that the natural environment has been thriving after the war, due to severe depopulation and decline in economic activity. Nature has reclaimed some formerly populated parts of the area, which was a natural tourist attraction (Fig. 2) already before the war as it is the largest karst *polje* in the world located below the highest mountain in Croatia. Consequently, the focus of the visual perception considered in the discussions is on the man-made landscape, mainly on the population and settlements, assuming that they were among the main immediate targets and suffered the greatest impact of the warfare.

In order to achieve the aim of the study the following goals are pursued:

- establishing the theoretical framework of the study, mainly by explaining the applied understanding of hidden geographies and the model of constructing the explanations of the landscape based on what is (visually) sensed/perceived by the visitor;
- setting the method of narrative simulation used as the backbone of the implementation of the investigation, and explaining multiple methods used to reveal some of the previously hidden geographies;
- suggesting potential, assumed hidden geographies in the area under study;

Fig. 2 Birdwatching and educational tourism in the post-war landscape of breathtaking natural beauty (*Photo*: Marko Krevs 2019)

– recognising, revealing, and discussing selected examples of hidden geographies as a part of the demonstration of possible explanation of the perceived landscape.

2 Theoretical Backgrounds

Two models are considered as a theoretical background for further discussion. The first serves to set a psychological framework for the discussion of the construction of explanations from visual perception and prior knowledge. The second provides a model of the hierarchical layers of hiddenness of geographies as understood in this chapter.

From Senses to Explanations There is a wealth of literature on sensing and perceiving the environment, from the more philosophical (as in Serres 2008) to the psychological (as in De Young 2013), geographical (as in Tuan 1977, 1990), and geoinformatic (e.g., relating to virtual sensing of the environment as in Kang et al. 2020). In this study, we take visual perception as a factor, one of the steps, one of the information sources within a process of constructing explanations. The model presented in Fig. 3, which links sensory perception, prior knowledge, and current experiences/leaning to the construction of explanations is based in part[1] on Gregory's

[1] Gregory uses the model to demonstrate "ins-and-outs: black box of vision", that leads to qualia (sensations, the internal and subjective component of sense perceptions that arise from stimulation of the senses by phenomena) and the "flagging the present", avoiding time-confusion with the past.

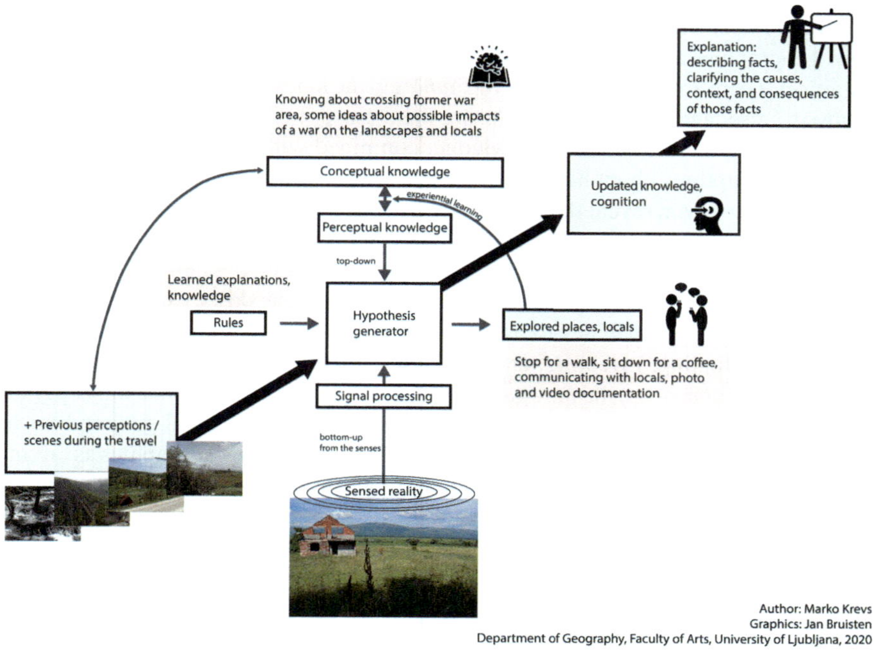

Fig. 3 A model of constructing explanations of the landscape based on sensory information, prior knowledge, and other accessible sources of information during a journey; based on a psychological constructivist approach by Gregory (1997)

(1997) constructivist psychology model of visual perception. An explanation, and before that knowledge/cognition, is therefore based on the interplay of prior knowledge, past and recent experiences including current sensory perceptions to provide information, and knowledge to clarify the causes, context, and consequences of that information.

Hidden Geographies The concept of hidden geographies is discussed in detail in the first chapter of this book (Krevs). In this specific study, geographies refer to the spatial location/distribution of phenomena in the post-war landscape and beyond, in areas indirectly affected by the war, such as places receiving refugees/displaced persons from these areas. Several kinds of hiddenness of geographies are discussed in a later section of this chapter, representing all four layers of the above-mentioned conceptualisation of hidden geographies: undiscovered, uncognised, unpublished, and deliberately hidden geographies.

A particular perspective arises from the combination of the two models, which links phases of learning/constructing explanations with various kinds of hiddenness

"Perceptions are largely based on the past, but recognizing the present is essential for survival in the here and now" (Gregory 1997). In Gregory's constructivist theory of perception, past knowledge and experience are essential in making sense of our environment. In our model, qualia as an outcome of the modelled process are replaced by updated knowledge/cognition and, eventually, explanations if they are carried out.

of geographies. The latter is essentially seen from a *personal cognitive point of view*: geographies are hidden from the observer until they are revealed to him by his cognition. There are *undiscovered geographies* in the landscape, undiscovered by him or anyone else, and his journey will bring them closer to his senses so that he might reveal some of them. His prior knowledge, combined with skills of deduction, may provide assumptions about hidden geographies which he could search for (*assumed geographies*) or even reveal them (*revealed geographies*). His logical thinking skills and learned rules of explanation may include practical ideas, such as where and how to look for (geographic) information and how to recognise whether it is relevant or not. All this, enhanced by testing and improving these skills while travelling to a particular place, contributes to assumed or revealed geographies and to personal strategies for finding, revealing, and explaining them. When the observer is in the explored area (or nearby), he may gather information from locals, which may upgrade his cognition from assumed to partly known geographies or reveal geographies he was not even looking for. A similar impact on revealing hidden geographies as talking to locals can come from accessing online or other information before or during the travel.

3 Methods: Narrative Simulation and Revealing Hidden Geographies

Two methodological approaches used in the rest of the discussion require justification: the method of narrative simulation and the methods used to reveal some of the previously hidden geographies.

Narrative Simulation Mental or narrative simulation[2] (Escalas 2013) imitates the circumstances and process by which the studied landscape is visually observed and explained. Readers develop "mental representations of the described events by combining textual information with their own world knowledge, personal experiences, and causal inferences" and experience "the narrative events as if the events were happening directly" to them (Costabile 2020). In this chapter, narrative simulation provides the framework in which the examples of the hidden geographies are illustrated through descriptions and simple narratives that the authors draw from different sources and observations. The method is intended to help the readers to immerse themselves in the narrated simulation, to take on the role of the main character (a visitor from Western Europe), and to observe the differences between what is seen in the selected landscape and the slices of reality provided by the authors (of this chapter). The personal perspective on the hiddenness of geographies points out *circumstantially hidden geographies*, hidden due to the circumstances encountered by a visitor/observer in the narrative simulation, in addition to the more general, collective understanding of the concept of hidden geographies.

[2] Taylor and Schneider (1989) describe mental simulation as the imitative mental representation of some event or series of events. It enables people to rerun past events, possibly altering them, and to project multiple versions of future events. Consequently, mental simulation is important for preparing for future events and interpreting past events.

The narrative of the mental simulation is as follows:

> *A visitor from Western Europe decided to go to Central Dalmatia and took an alternative route, roughly parallel to the main tourist road, via Bihać, Una National Park, Drvar, Livanjsko Polje, towards Makarska. On the road through the karst landscape of silent natural beauty, as he approaches and then crosses Livanjsko polje, the visitor sees the contrasts in the socio-geographical characteristics of the landscape which can hardly be explained by his experiences from home or deduced from his general geographical knowledge. Abandoned buildings or deserted villages are followed only a short distance away by lively settlements with renovated houses and children playing in the yards, or villages with only a few buildings showing signs of a permanently present population. These settlements, at least at first sight, share similar geographical circumstances and development opportunities. Visitor's basic knowledge of the (near) history of these areas suggests that what he sees could be interpreted as a consequence of the 1990s war. But many questions remain unanswered, such as: Whose buildings were demolished? Where are the former inhabitants now? Did they survive, and return to their homes after the war? Who lives in the rebuilt settlements today? What do the locals do for a living? Is the context of what the visitor sees just a consequence of the hidden geography of the ethnic structure of the 1990s or even from earlier times?*
>
> *In search of answers to such questions, the visitor occasionally stops along his way, observes the landscape, talks with the locals he meets in a restaurant or store, maybe even browses through a guidebook or information accessible on the internet, written in English. And by combining his rather modest and superficial prior knowledge with his observations and other information gathered during his short journey through the visited area, he constructs explanations of what he knows about the visited landscape and its inhabitants.*

Revealing Hidden Geographies Not even a detailed study of the available information could fully answer the visitor's questions in this narrative, and that is not the aim of this investigation. On the other hand, there are many hidden geographies related to the present and the history of the area under study, about which the visitor knows nothing, and with which he might come into contact as he travels through the area. Some of them are pointed out in the next section.

Revealing geographies in this chapter serves mainly to support narrative simulation by illustrating known geographies, as opposed to the visitor's assumed, superficially informed interpretations of a visible landscape. As a side effect, it also reflects the opportunities a visitor has to enhance his understanding and interpretation of the observed landscape—if he was interested enough and the circumstances allowed access to the relevant information as used in the revealing of geographies presented in this text.

To reveal a hidden geography is to provide previously hidden geographic information about it, on a map or in some other form. Some hidden geographies cannot be revealed because they have not yet been discovered or the information is not available or cognisable for other reasons. But to reveal some—those that were selected for this study—multiple sources and methods were used, from the collection and interpretation of statistical data (mainly Popis 2013), maps (ibid.), publications, personal online stories, even ethnically biased media news or personal blogs from the internet, to the authors' own field observations and interviews with the locals. Revealing reaches different levels, from recognising hidden geographies to uncovering facts about them, from representation in the form of narrative or description to field documents (including photographs) or maps based on accessible data.

The reliability of some sources is rather questionable, even data from key institutions can vary in quality considerably. As an illustration, the Municipality Development Strategy of Bosansko Grahovo (Strategija… 2015), prepared in the collaboration with UN Trust Fund for Human Security, provides three different population figures in this municipality: 2.024, 3.091 and "not more than 1.500", delivered from the estimate of Federal Statistical Office (Federalni zavod za statistiku), the preliminary results of the 2013 census (same source) and the estimate of the municipal administration, respectively. Quantitative data are used only in part of the further discussion and only for general illustration. Nevertheless, we assume that errors in data from the same source are approximately evenly distributed and do not affect the geographies we are revealing. It is expected that qualitative data, e.g., from the interviews, personal opinions expressed on social media, or ethnicity-based media news, are biased as they represent subjective, personal, and political views. However, in this study, they are used as such to represent real and strong opinions or emotions that reflect and influence social and landscape reality, even if they are biased or false.

4 Discussing Hidden Geographies of a Post-War Landscape

In addition to the obvious expectations that a post-war landscape should have a concrete, visually perceivable signature of wartime activity, many less visible hidden geographies are *assumed*. They are related, e.g., to unclear connections between perceived physical features in the landscape and past and present processes and to a barely visible wide range of socio-economic and personal geographies. This section takes a divergent approach, emphasising the potential diversity of hidden geographies in post-war landscapes, thus linking the theoretic background to the empirically supported discussion in the next section.

Following the hierarchical four-layer model from the chapter on conceptualisation of hidden geographies (Krevs), hidden geographies can be undiscovered, uncognised, unpublished, and deliberately hidden. Examples of the four variants of *hiddenness from the visitor in our simulation* are given in Table 1. Many of these geographies are hidden because they have not yet been discovered (*undiscovered geographies*), or they are inaccessible to the virtual visitor for a number of possible reasons. Therefore, we can only assume their existence but remain at a rather low level of their characterisation and representation.

The *uncognised geographies* are not accessible to the visitor's knowledge or higher levels of cognition. However, such geographies may be known to the authors of this chapter when they discover them during their field observations or acquire them published (made available to the public, either in visual or other forms accessible to human senses and perception). The examples in the next section are examples of virtual revelation of geographies that are known to the authors of this chapter to the visitor. For the latter, previously hidden, uncognised geographies become known geographies.

Hidden geographies related to *emotions*, such as sense of place, attachment to place, home, family, or ethnicity, are particularly difficult to recognise or reveal, even for the local people. "On the surface, the discipline of geography often represents us with an emotionally barren terrain, a world devoid of passion, spaces ordered solely by rational principles, and demarcated according to political, economic, or technical

Table 1 Examples of geographies assumed to be hidden from the visitor to a post-war landscape

Hidden geographies layers	Examples of geographies assumed to be hidden
Undiscovered	Personal geographies (historical and current), material and non-material (including geographical stereotypes, geographies of values, senses of place[3]) Geographies of acts of war for which there are no witnesses, or the witnesses have died/emigrated/keep information to themselves Assumed geographies of places where displaced people have settled, of current ethnic tensions/tolerance, or even of current real population quantities and structures
Uncognised	Published geographic information, especially if assumed to be unreliable, even if accurate (e.g., automatically assumed to be biased because published in an ethnic group's opinionated media) Online geographic data may remain uncognised if not sought and studied, but also because they are not in a form accessible to the general public (e.g., complex use of online maps, statistical databases for an ICT-illiterate reader)
Unpublished	Hidden by non-publication of geographical information known to some or many (usually the locals), inaccessible to visitors or the non-resident population A commonly known—but unpublished—part of the collective sense of place, e.g., based on ethnic perspectives—positive or negative (topophilia, topophobia), attachments to places, place-based identities Geographical information on migrations that is not published due to lack of official/registered traces, or simply remains out of focus; neglected/ignored related issues reflecting local/regional consequences of such migrations
Deliberately hidden	Places associated with interpersonal, interfamilial, interethnic hostile acts—from the recent or distant past Places related to military or armed actions (by any of armed forces or individuals) kept hidden by agreed/requested secrecy Geography of results of covert policies of ethnic groups against other ethnic groups—affecting, e.g., access to employment or building permits

[3] Personal geographies (Harmon 2003), emotional geographies (Davidson et al. 2007), senses of place (Eyles 1985) , topophilia, topophobia (Tuan 1977, 1990), attachment and alienation, the loss and identification (Žižek 1990, Ruann and Hogben 2007) are phenomena with which even locals might have problems to draw maps about. Among relevant emotions in the studied context are also such as fear of returning, fear of place-related or residents-related memories, fear of current/other residents, or of problems brought by the new circumstances.

logics" (Bondi et al 2007: 1). This is not surprising because emotions are never easy to define or demarcate, and they are "not easily observed or mapped although they inform every aspect of our lives".

The emotional geographies are obviously among those least visible but heavily present in areas to which at least some people feel attached. In a post-war landscape, shared by multiple ethnic groups in the past and also in the present, such emotions are obviously multiplied in intensity, spatially overlapping and often inevitably confronting. As some of the emotional geographies persist and become collective (especially if ethnically-based), they can bring about deeper changes in the seemingly unchanging relationships between the landscape and its inhabitants after the war, or more specifically in an urban setting, "between the built fabric and those who inhabited the city" (Alic 2019). Such changes bring "new modes of thinking and interpreting" the landscape, not only to the researchers but above all to the inhabitants living within the changing physical and human environment. The changed senses of places produce new overlapping and exclusive hidden geographies learned only by locals or those who participate in the community and learn from other people's behaviour and communication. The signals in such communications may be very strong, but hardly perceivable by those external to the mentioned communication, even more so to an occasional visitor.

In a post-war situation many extreme distantiations in the behaviour of ethnic groups—distancing from others who are represented negatively—combine ethnic images and place images into "landscapes of exclusion" (as named by Sibley 1995). Even such strong attachments of "repulsion and desire, fear, and attraction to people and places" (ibid.) may remain hidden to non-locals.

The visitor might only see the results of such *emotions in the local community* and *social tensions* indirectly, for example, through the low return of the displaced population to certain settlements or the messages written on the walls (if he understands them). A logical approach of the visitor to improve his ability to perceive and explain what he encounters in the visited multi-ethnic places inflicted by war is to learn the *signs of the presence of different ethnic groups.* From non-material cultural features, such as languages and dialects, food rules, specific body language, to material culture signs, such as specific ethnic/religious clothing, architectural specificities, religious and graveyard objects, and symbols (e.g., types of religious buildings, forms of crosses, or presence of the star and crescent). *Landscape iconography* (Hoelscher 2020) in particular may seem a promising aid in reading, describing, interpreting multi-ethnic landscapes, e.g., by linking the recognition of ethnic-specific objects to the active demonstration of the ethnic group's presence. But in some places in the study area, the ethnic structure has changed, even to the point of the complete disappearance of a particular ethnic group, leaving behind its landscape iconography. In such cases, applying the interpretations brought by the landscape iconography would lead to wrong conclusions and explanations. In other places, where ethnic/national symbols have gained their presence in frequency and size, the application of landscape iconographic logic could be very helpful for the visitor.

Many of the geographies that the authors of this chapter at least partly reveal below may remain hidden from the visitor. In reality, it would be quite unlikely that a visitor would make the effort to search for information in such various sources, accessible mostly in Croatian, Serbian, or Bosnian. However, the illustrated contrast between the visitor's visual perceptions and the authors' constructed explanations serves the purpose to demonstrate examples of the *circumstantial, functional hiddenness* of various geographies.

5 Hidden Geographies in a Post-War Landscape: Examples in the Area Between Bosansko Grahovo and Livno

The selection of examples of hidden geographies relates to the following topics: depopulation, ethnic structure, war destruction, post-war renewal and return of displaced people, and interethnic tensions. These are the hidden geographies that the visitor might try to reveal, at least the part and in such detail that it might support his explanations of what he encounters in the landscape.

Hidden Geographies of Depopulation The study area, covered by the municipalities Bosansko Grahovo and Livno, is one of the most sparsely populated areas in Bosnia and Herzegovina, with continuous general depopulation since the 1960s, especially in the rural areas since the 1990s, due to its natural characteristics (the altitude of most of the settlements is between 500 and 1000 m), socio-economic isolation and peripheral location in relation to the main national development centres (Pobrić and Avdić 2020) and, of course, due to the impact of the war.

Out of 94 settlements (Fig. 4), 83 have been losing population to a greater or lesser extent since the 1990s, almost all of the settlements being in the municipality Bosansko Grahovo. This is also reflected in the extremely unbalanced age structure of the municipal population, with only 3 pupils in the first grade of primary school in the 2017/2018 school year, and the dramatic lack of working population (ibid.).

Dramatic depopulation[4] (Fig. 6) affected the entire area (Table 2), with only few exceptions. The 11 settlements with a growing population are all rural or suburban. The population of the biggest urban settlements in the studied area has declined, in Livno from just over 10.000 in 1991 to less than 8.000 in 2013 and in Bosansko Grahovo from about 2.100 to 650 (Popis 2013).

Despite the sparsely populated landscape, the visitor can easily notice the effects of the depopulation as soon as he enters almost any of the settlements. Not only from the visual signs of war destruction, which are more evident in the areas with Serbian majority, but from many abandoned houses and gardens with no sign of regular use.

Hidden Geographies of Ethnic Structure The population of the area is composed of all three major ethnic groups: Croats, Serbs, and Bosniaks (Table 2). But

[4] The Institute for Social and Political Research (Milović and Sitarski 2014) classified municipalities into three groups, based on the change in the population between the 1991 and 2013 censuses. The municipality of Livno was placed in the first group, with a change of less than 10%, compared to the pre-war population in 1991. The municipality Bosansko Grahovo, which suffered the second largest decrease in the whole of BIH, was placed in the third group, with the most dramatic, more than 25% decrease in population.

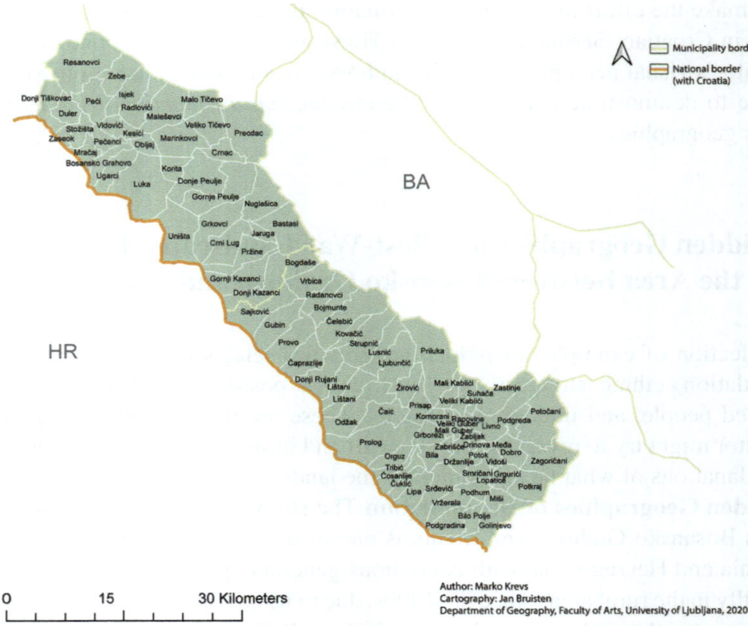

Fig. 4 Settlements in the municipalities Bosansko Grahovo and Livno (*Data source* Popis 2013)

their spatial distribution is not even: in the municipality Bosansko Grahovo and in the western part of the municipality Livno the majority of the population is Serbian, and in the rest of the municipality Livno Croatian at the time of both censuses, 1991 (before the war) and 2013 (the first census after the war) (Fig. 5). Only one—the largest—settlement in the entire area, Livno, had a Bosniak majority in 1991, and none of the settlements in 2013.

Recognising the presence of particular ethnic groups from the visible landscape is possible in principle if one can read the visual communications and symbols of an ethnic group's presence. But the limitations of the method of ethnic iconography have already been pointed out. In a multi-ethnic area such as the one in this study, the ethnic structure may change—and it did after the wars in the 1940s and 1990s. These changes were spatially uneven. The ethnic-specific signs in the landscape may remain; therefore, their interpretation—being the identifiers of an ethnic group's presence—requires some caution. A mosque is only a sign of the presence of the Muslim population, but not necessarily of the Bosniak majority (Fig. 7). Another visible aspect of the presence of an ethnic group is publicly displayed messages such as graffiti, political or advertising posters, and also the road signs indicating the names of settlements. The content of these messages may not be very informative to the visitor unless it contains symbols or phrases he recognises, such as national flags, coats of arms, or words that allude to ethnic identity. In the studied area, the

Table 2 Changes in population and its ethnic structure between 1991 and 2013 in municipalities Bosansko Grahovo and Livno (*Data source* Popis 2013)

Municipality		1991	2013	Index (period 1991–2013)	% Change (period 1991–2013)
Bosansko Grahovo	Population (total)	8311	2449	29.5	−70.5
	Bosniak	12 (0,1%)	6 (0,2%)	50.0	−50.0
	Croatian	226 (2,7%)	393 (16,0%)	173.9	73.9
	Serbian	7.888 (94,9%)	2.028 (82,8%)	25.7	−74.3
	Other	185 (2,2%)	22 (0,9%)	11.9	−88.1
	Area (total)	783,17 km^2			
Livno	Population (total)	40,600	34,133	84.1	−15.9
	Bosniak	5.793 (14,3%)	4.047 (11,9%)	83.2	−16.8
	Croatian	29.324 (72,2%)	29.273 (85,8%)	118.8	18.8
	Serbian	3.913 (9,6%)	438 (1,3%)	13.5	−86.5
	Other	1.570 (3,9%)	375 (1,1%)	28.2	−71.8
	Area (total)	988,65 km^2			

Cyrillic alphabet used in such public communication may help the visitor recognise the presence of the Serb population (Fig. 8).

Hidden Geographies of War Devastation The 1990s war lasted from April 1992 to December 1995. The biggest war activities affecting the area under study took place towards the end of the war, in 1995 (Operation Summer'95, 25–29 July 1995; Operation Storm, 4–7 August 1995; Operation Maestral 2, 8–15 September 1995), and were performed by Croatian Army and Defence Council forces (in case of Operation Storm also by Croatian Special Police) with support of the army and police forces from BiH against Armies of Srpska Krajina and Republika Srpska. The sides that were formed in these war conflicts are important in understanding contemporary interethnic tensions, which do not necessarily coincide with the geography of ethnic majority. Croatian influence or even dominance in the studied area persists, as shown below. The visitor will (expect to) see these impacts of the war quite clearly in the landscape. The areas that will look almost untouched by the war will most probably be the Croat majority areas, as in most of the Livno municipality (with exception of the NW part with the Serb majority).

In addition to the recent war, earlier history, especially during the World War II period, brought many interethnic hostilities as well, mainly from Croatian (by

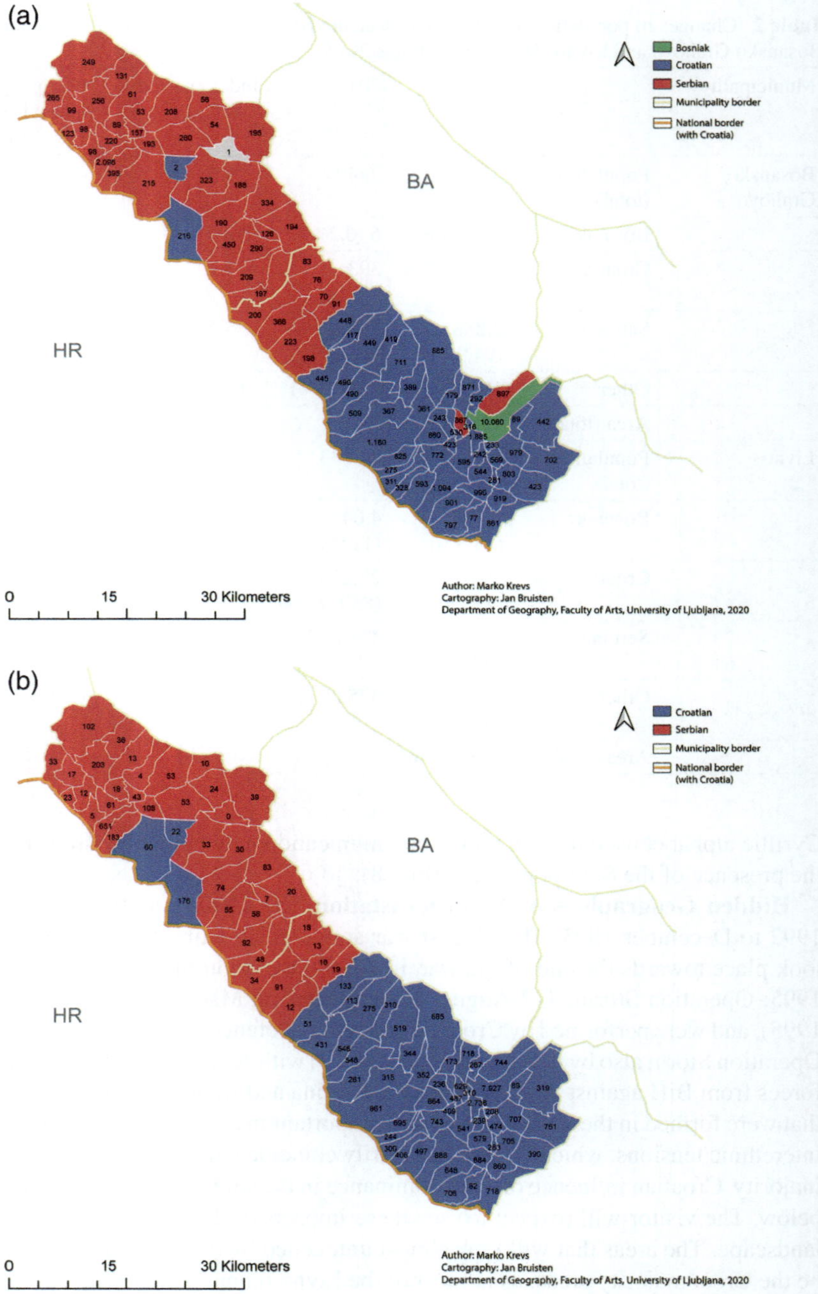

Fig. 5 Population and ethnic majority by settlements in **a** 1991 and **b** 2013 (*Data source* Popis 2013)

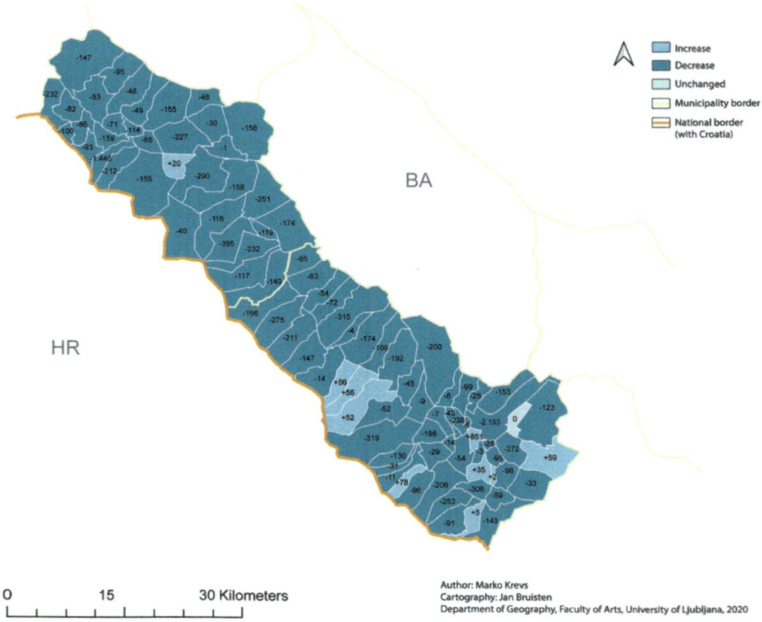

Fig. 6 Changes in population between 1991 and 2013 by settlements (*Data source* Popis 2013)

Fig. 7 The mosque may show the presence of the Muslim population, but not necessarily its majority—Livno case (*Photo* Marko Krevs 2019)

Fig. 8 Road signs with Cyrillic alphabet show a strong presence (usually the majority) of Serb population before the war in the 1990s, but in some settlements this does not necessarily correspond to today's reality (*Photo* Marko Krevs 2019)

Ustashe) or Serbian side (by *Chetniks*), which can hardly be forgotten or ignored by the local population. Such historical knowledge does not justify further hostilities or interethnic pressures but can contribute to a complex understanding of interethnic and socio-economic relations with the (visible) landscape, even today. The visitor can try to identify/guess the period of destruction in the built landscape, for example, by regarding the degree of decay of the ruins or by the size of the vegetation that overgrows the roofless houses. In order to provide a detailed explanation, he would have to combine various sources, as the authors have done in the following examples.

In 1941 the Chetniks slaughtered or banished the majority of Croats and burned down their houses in Bosansko Grahovo, Obljaj, Luka, Korita, Ugarci, and Crni Lug. Torturing the Roman Catholic priest to death was also the end of the parish that had existed since 1863 (Pokolj…; Bosansko…; several details in Nazor 2013), as a part of the systematic ethnic cleansing of the Croat population in Western Bosnia and Lika. In one of the villages mentioned above, Crni Lug (Fig. 13c), the entire population was Serbian after the World War II. During the war in the 1990s, the Serbian population fled and their houses were burned down; the return of the displaced population is slow (Crni Lug 2012; Povratak u općinu Bosansko Grahovo). Another settlement, Korita, is one of the few in the studied area whose population grew after 1991 (Figs. 5, 6, and 9). It is one of the three settlements with Croat majority in the municipality Bosansko Grahovo with otherwise Serb majority.

Fig. 9 Korita, a settlement SE of Bosansko Grahovo, is an example of an island of Croatian majority population within the territory with Serbian majority. After the 1990s war, Croats resettled the area of their annihilation in World War II (*Photo* Ranko Mirić & Nusret Drešković 2019)

At about the same time, in 1941, the Ustashe slaughtered hundreds of Serb residents of Čelebić (Fig. 10; Hronologija…; field research: 396 names on the memorial plaque in the village are referring to these events), which still has a Croat majority but has been suffering severe depopulation after the 1990s.

Hidden Geographies of Post-War Renewal and Return of Displaced The return of former residents (displaced during the war) to their war-torn homes is quite

Fig. 10 Wartime destruction of 1941 and the 1990s **a** and signs of a returning population **b** in Čelebić; a Serb majority settlement before 1941, about one third in the 1990s, and now less than 10% Serb population (*Photo* **a** Ranko Mirić and Nusret Drešković, **b** Marko Krevs 2019)

difficult to register and keep up to date. An obvious reason—the lack of a reliable population register—may be partially substituted by indirect information from the Union for Sustainable Return and Integration in BiH on the use of special funds for the renewal of houses and resettlement (Povratak u Bosni i Hercegovini). Although the Union tries to identify and prevent misuse of the funds, it reports cases when the received funds were used in other places than allocated, even for religious buildings.

Maps of allocated funds and data on returning persons/households are therefore at least partially misleading, and as a consequence, this hidden geography is only partially revealed. On the other hand, such information, even if not entirely accurate, provides a rich insight into a very sensitive process that is difficult to measure. When explaining a local situation of the returning displaced population, the visitor does not actually have the problem of reliability of information from the sources mentioned above. He is in contact with the real landscape, observes the real geography, and easily recognises the renewed buildings, well-kept gardens, and other visible traces of current settlement, especially in areas with major war destruction.

The settlements of Livno municipality with the Croat or Bosniak majority were much less devastated during the war in the 1990s than the settlements with the Serb majority. Apart from some signs of abandoned (but habitable) houses, there are almost no visible traces of the war in the landscape in the Croat and Bosniak majority areas (Fig. 11), while the Serb majority settlements—in both studied municipalities—still show severe damage (Figs. 12, 13, and 14; see also maps in Figs. 5 and 6).

For example, in Radanovci (Figs. 8 and 12), 9 of the 26 destroyed houses were renewed, the average age of the population is over 72 years, and there is no interest in additional returns (Povratak u općinu Livno). In the largest centre in the studied area, Livno, most of the displaced inhabitants are Serbs. Only 10 Serb families returned

Fig. 11 Signs of permanent residents and continuous life in the settlements NW of Livno, the area of Croatian majority (*Photo* Marko Krevs 2019)

Fig. 12 The consequence of which war events is the perceived destruction—from the World War II or the 1990s war? A house in Radanovci, the area with the Serbian majority in the 1990s (with 70 inhabitants), whose population has dropped to less than 20 by 2013 (*Photo* Marko Krevs)

(in the same time period, 280 Bosniak and 100 Croat families returned), while the majority of Serb houses were sold. No Serbs are employed in public administration, and many local roads remain unmaintained (ibid.; Položaj Srba… 2019).

Hidden Geographies of Interethnic Tensions An expected example of hidden geographies refers to interethnic tensions in various dimensions of everyday life. The majority in an area usually prioritises its ethnic group and other ethnic groups may feel or be oppressed. For example, Bosniaks point to the predominant Croat influence on education in Livno (Preživljavanje… 2018) and Serbs to the difficulties in finding jobs in the studied areas, "Serbs have to give up three jobs to Croats to get one" (Položaj… 2019). This also happens in the areas with the Serb majority. One of the reasons, which was also confirmed by the interviews with the local population and the field observation, is the Croats' privatisation of the few surviving economic activities, such as the separation of construction gravel and sand in Pržine and a sawmill in the municipality Bosansko Grahovo (Fig. 15). Serbs also point to the lack of renovation of almost completely devastated municipality centre Bosansko Grahovo, which still looks "like it did in the years after Oluja" (Grahovo… 2018; Fig. 14), the aforementioned military operation Storm.

All geographical explanations of the ethnic tensions and complaints are not equally foreseeable. In 2011 the Croat members of the Livno municipal council called for a referendum on the municipality's secession from the Federation of BiH and annexation to Republika Srpska (Dijanović 2011). Although the idea did not receive much attention except in the media, it clearly shows how far from the geographical realities our interpretations can be, based on what is visible in the landscape or on logical reasoning.

Learning about hidden geographies of interethnic tensions and details of war devastation/events are the most difficult (of those presented here) tasks for the visitor. Explanations of other hidden geographical contexts of the visible post-war landscape, such as depopulation, ethnic structure, even post-war renewal, and return of displaced people, maybe constructed on the basis of accessible information, at least to some extent.

Relevant geographical information about the tensions between ethnic groups and the reflections of these tensions in their every day lives may remain hidden even to a visitor who knows the local culture, including the language(s) necessary to understand often disguised messages in media news, social media discussions, or other

Fig. 13 Donji Kazanci **a**, Donje Peulje **b**, and Crni Lug **c** are the examples of settlements with extreme population decline and war destruction of the built landscape, on the territory of the Serbian majority population SE from Bosansko Grahovo (*Photo* **a** Ranko Mirić and Nusret Drešković, **b** and **c** Marko Krevs 2019)

Fig. 13 (continued)

public communications. Constructing explanations based on the geographic contexts of specific examples of war devastation or its consequences is even more difficult. The more personal information from the people involved in these events that a visitor needs to construct his explanation, the more difficult the task. Relevant information is often scattered in various personal histories that can be highly unreliable and biased. Or worse, those who know do not share the information or are no longer alive. Such undiscovered or deliberately hidden geographies can remain hidden to everyone, not just the visitor in our narrative simulation.

6 Conclusion

This study shows that post-war landscapes are full of hidden contexts that affect both people's lives and their natural environment. The discussion focuses on the identification of hidden geographies and their relation to the visible landscape and providing information to reveal them.

To support this discussion, two theoretical frameworks are used: a model of constructing explanations of the landscape based on what is seen/visually perceived by the visitor, based on a similar model from constructivist psychology, and a model of the hiddenness of geographies. The latter is taken from the chapter on conceptualising hidden geographies (Krevs). The layers of hiddenness of geographies provided structure in a more theoretical part of the discussion suggesting examples of assumed

Fig. 14 25 years later, the signs of war are clearly visible, even the most important institutions such as the high school and the building of the Cultural Centre (only a part is visible on the right side of the photo) **a** and the municipal building **b** are still in ruins (*Photo* Ranko Mirić and Nusret Drešković 2019)

Fig. 15 A rare visible example of revived economic activity in the western part of the area under investigation, a case of sawmill in Bosansko Grahovo (*Photo* Ranko Mirić and Nusret Drešković 2019)

geographies in a post-war landscape. The empirical examples in the later discussion only partially followed this structure, as most of the examples presented were based on or derived from published information. The examples belong to the wide group of uncognised geographies—which are hidden because the observer does not know about them. Only a few examples could be classified as undiscovered geographies, revealed through the results of field observation or interviews with local people.

The model of constructing explanations of what is sensed in the landscape refers mainly to recognition of geographies and their revelation from the point of view of an individual observer. The strength of the model lies in its clear illustration of the interrelationships between several factors and the temporal dimension of the process, such as in the acquisition of prior experience and knowledge and in the iterative construction of knowledge and explanations. The suggestions for the observer that accompany the empirical examples of revealing particular hidden geographies in this chapter refer to different parts of this model, such as "rules" (what to observe in the visible landscape in order to detect a specific phenomenon), field exploration and contacting locals (in order to find direct or indirect information on hidden geographies), or explanation (about possible explanations for what is perceived in the landscape).

An unconventional narrative/mental simulation method provides an experimental observational filter for theorising the assumed hidden geographies as well as for the presented empirical examples. Its main function is to demonstrate circumstantial problems experienced by a virtual visitor while constructing sensible explanations based on rather basic prior knowledge, sensory information, and other relevant information accessible during the travel. It also enables the representation of a wider range of hiddenness of geographies, determined not only by the geographies themselves but also by the knowledge and active exploration/learning of the observer.

Many answers regarding the identification and revealing of the hidden geographies in post-war landscapes remain barely touched or untouched, such as: How relevant is a particular hidden geography and to whom? Would revealing a particular hidden geography trigger positive or undesirable consequences? Could a reliably measured and represented (previously hidden) geography prevent certain undesirable activities in the landscape or society? How can we improve the publication and interpretation of geographic information, especially on sensitive issues, in a way to provide active support in solving problems in a post-war multi-ethnic society?

With so many obvious impacts of war in a landscape, a search for hidden geographies behind the visible ones seems redundant. However, as we hopefully succeeded to show, these hidden contexts can play very powerful roles in the landscape and society. We believe that a very crucial aspect for the future research of hidden geographies is to put it in the service of higher goals such as post-war reconciliation. Post-war landscapes and societies could definitely benefit from such a development.

References

Alic, D. 2019. The landscape of war and the transformations of the public sphere. *Open House International* 44 (2): 62–67.

Bondi, L. et al. 2007. Introduction: Geography's 'Emotional Turn'. In *Emotional geographies*, eds. Davidson, J., Bondi, L., Smith, M. Aldershot and Burlington: Ashgate: 1–16.

Bosansko Grahovo. https://hr.wikipedia.org/wiki/Bosansko_Grahovo. Accessed 15 January 2019.

Costabile, K.A. 2020. Mental simulation in narrative comprehension and construction. *Psychology of Consciousness: Theory, Research, and Practice* 7 (4): 363–375. https://doi.org/10.1037/cns0000216.

Crni Lug. 2012. Udruženje Ognjena Marija Livanjska. Association of Serbs from Livno field, their descendants and admirers; 10 Nov 2012. https://omlivanjska.com/2012/11/10/crnilug/. Accessed 15 January 2019.

Davidson, J., L. Bondi, and M. Smith, eds. 2007. *Emotional geographies*. Aldershot and Burlington: Ashgate.

Dijanović, D. 2011. Hrvatska općina Livno traži pripojenje tzv. Republici Srpskoj! Portal hrvatskog kulturnog vijeća, 18 October 2011. https://www.hkv.hr/izdvojeno/komentari/d-dijanovi/9334-hrvatska-opina-livno-trai-pripojenje-tzv-republici-srpskoj.html. Accessed 14 April 2020.

Escalas J. E. 2013. Imagine yourself in the product: Mental simulation, narrative transportation, and persuasion. *Journal of Advertising* 33, 2: 37–48. https://doi.org/10.1080/00913367.2004.10639163.

Eyles, J. 1985. *Senses of place*. Warrington: Silverbrook Press.

Federalni zavod za statistiku. Sarajevo. www.fzs.ba

Grahovo–mjesto koje se još nije oporavilo od "Oluje". *Frontal*, 6 August 2018. https://www.frontal.ba/novost/93084/grahovo-mjesto-koje-se-jos-nije-oporavilo-od-oluje. Accessed 14 April 2020.

Gregory, R.L. 1997. Knowledge in perception and illusion. *Philosophical Transactions of the Royal Society London B* 352: 1121–1128. https://doi.org/10.1098/rstb.1997.0095.

Harmon, K. 2003. *You are here: personal geographies and other maps of the imagination*. Princeton: Princeton Architectural Press.
Hoelscher, S. 2020. Landscape Iconography. In Kobayashi, A. (ed.) *International encyclopedia of human geography* (2nd edn): 113–119. https://doi.org/10.1016/B978-0-08-102295-5.10198-2.
Hronologija ustaških zločina 1941. https://sh.wikipedia.org/wiki/Hronologija_usta%C5%A1kih_zlo%C4%8Dina_1941. Accessed 15 Jan 2019.
Kang, Y., F. Zhang, S. Gao, H. Lin, and Y. Liu. 2020. A review of urban physical environment sensing using street view imagery in public health studies. *Annals of GIS* 26 (3): 261–275. https://doi.org/10.1080/19475683.2020.1791954.
Milović, T., and M. Sitarski. 2014. Analiza dostupnih rezultata popisa stanovništva u BIH 2013: godine. Idpi–Institute for social and political research, Mostar, BIH. https://www.idpi.ba/analiza-dostupnih-rezultata-popisa-stanovnistva-u-bih-2013-godine/. Accessed 16 April 2020.
Nazor, A. 2013. Protuhrvatski ustanak u srpnju 1941. Dnevno.hr. https://www.dnevno.hr/kolumnisti/protuhrvatski-ustanak-u-srpnju-1941-91851/. Accessed 15 Jan 2019.
Operation Maestral 2. https://en.wikipedia.org/wiki/Operation_Mistral_2. Accessed 16 April 2020.
Operation Summer '95. https://en.wikipedia.org/wiki/Operation_Summer_%2795. Accessed 16 April 2020.
Operation Storm. https://en.wikipedia.org/wiki/Operation_Storm. Accessed 16 April 2020.
Pobrić, A., and A. Avdić. 2020. Depopulacija i starenje stanovništva u Kantonu 10–Bosna i Hercegovina. Zbornik radova 5. Kongresa geografa BIH, Sarajevo. (in press)
Pokolj u Bosanskom Grahovu 27. srpnja 1941. https://hr.wikipedia.org/wiki/Pokolj_u_Bosanskom_Grahovu_27._srpnja_1941. Accessed 15 Jan 2019.
Položaj Srba u Livanjskom kantonu. Runić: Za jedno radno mjesto moramo dati tri. Srpskainfo, 12 July 2019. https://srpskainfo.com/polozaj-srba-u-livanjskom-kantonu-runic-za-jedno-radno-mjesto-moramo-dati-tri/. Accessed 14 April 2020.
Popis 2013 u BIH. www.statistika.ba. Accessed 14 April 2020.
Povratak u općinu Bosansko Grahovo. Povratak u Bosni i Hercegovini. Accessed 14 April 2020.
Povratak u općinu Livno. Povratak u Bosni i Hercegovini. http://uzopibih.com.ba/povratak/livno.html. Accessed 14 April 2020.
Povratak u Bosni i Hercegovini. http://uzopibih.com.ba/povratak/opis-projekta.html. Accessed 14 April 2020.
Preživljavanje Bošnjaka u Hercegovini. Preporod. Islamske informativne novine, 25 Jan 2018. https://www.preporod.com/index.php/sve-vijesti/drustvo/teme/item/8509-prezivljavanje-bosnjaka-u-hercegovini. Accessed 14 April 2020.
Ruan, X., and P. Hogben. 2007. *Topophilia and topophobia: Reflections of twentieth-century human habitat*. London and New York: Routledge.
Serres, M. 2008. *The five senses. a philosophy of mingled bodies*. London: Continuum.
Sibley, D. 1995. *Geographies of exclusion*. Routledge: Society and Difference in the West.
Strategija razvoja opštine Bosansko Grahovo 2016–2020. godine 2015. UNTFHS–UN trust fund for human security. http://www.bosanskograhovo.ba/ba/filedownload.php?did=200. Accessed 15 Sept 2019.
Taylor, S.E., and S.K. Schneider. 1989. Coping and the simulation of events. *Social Cognition* 7 (2): 174–194.
Tuan, Y. 1977 (ninth printing 2002). *Space and place: The perspective of ex-perience*. Minneapolis: University of Minnesota Press.
Tuan, Y. 1990. *Topophilia: A study of environmental perception, attitudes and values*. New York: Columbia University Press.
De Young, R. 2013. Environmental psychology overview. In *Green organizations: Driving change with IO psychology*, ed. Ann H. Huffman and Stephanie Klein, 17–33. New York: Routledge.
Žižek, S. 1990. Eastern Europe's Republics of Gilead. *New Left Review* 183: 50–62.

Index

M. Krevs (ed.), *Hidden Geographies*, Key Challenges in Geography,
https://doi.org/10.1007/978-3-030-74590-5

GPSR Compliance

The European Union's (EU) General Product Safety Regulation (GPSR) is a set of rules that requires consumer products to be safe and our obligations to ensure this.

If you have any concerns about our products, you can contact us on ProductSafety@springernature.com

In case Publisher is established outside the EU, the EU authorized representative is:

Springer Nature Customer Service Center GmbH
Europaplatz 3
69115 Heidelberg, Germany

Batch number: 10732076

Printed by Printforce, the Netherlands